T0100122

A HISTORY OF THEIR OWN

A HISTORY OF THEIR OWN

Women in Europe From Prehistory to the Present

VOLUME I

REVISED EDITION

BONNIE S. ANDERSON

JUDITH P. ZINSSER

NEW YORK OXFORD

OXFORD UNIVERSITY PRESS

2000

Oxford University Press

Oxford New York
Athens Auckland Bangkok Bogotá Buenos Aires Calcutta
Cape Town Chennai Dar es Salaam Delhi Florence Hong Kong Istanbul
Karachi Kuala Lumpur Madrid Melbourne Mexico City Mumbai
Nairobi Paris São Paulo Singapore Taipei Tokyo Toronto Warsaw

and associated companies in

Berlin Ibadan

Illustration credits follow the Index.

Published by Oxford University Press, Inc.,
198 Madison Avenue, New York, New York 10016

Library of Congress Cataloging-in-Publication Data

Anderson, Bonnie S.
A history of their own : women in Europe from prehistory to the present / Bonnie S. Anderson, Judith P. Zinsser. — Rev. ed.
p. cm.
Includes bibliographical references and index.
ISBN-13 978-0-19-512838-3

1. Women—Europe—History. 2. Feminism—Europe—History. I. Zinsser, Judith P. II. Title.
HQ1587.A53 1999
305.4'094—dc21 98-46743
CIP

Printed in the United States of America

Contents

Illustrated sections follow pages 6, 92, 186, 276, and 358.

Acknowledgments

With the publication of this second edition of *A History of Their Own* we are pleased to acknowledge the individuals and institutions who helped to make it possible. First, thank you to all those who, through their letters and e-mail messages, eloquently expressed their support for this project. Second, our thanks to Gioia Stevens, our editor at Oxford University Press, whose cheerfulness, competence, and calm efficiency made our revisions a pleasure to complete. Third, thank you to Laura Kitch, Acting Provost of Brooklyn College, CUNY; John H. Skillings, Associate Dean, and Charlotte Newman Goldy, Chair of the History Department at Miami University, who gave us the time and resources we needed. Fourth, the task of surveying and assessing the vast literature of recent European women's history would have been impossible without the tireless efforts of Jennifer M. Morris and Jeri L. Schaner. Jenny L. Presnell was never too busy to answer a bibliographic question; Elizabeth A. Smith helped us meet last minute deadlines. Finally, we wish to thank our colleagues and friends, who continue to give so generously of their time, affection, and encouragement—in particular, Mary E. Frederickson and Roger J. Millar.

Introduction

For the first edition of *A History of Their Own* in 1988 we stated goals remarkably similar to those of forward-looking women in mid-nineteenth-century Europe and the United States. "It would be really fine if someone would undertake to write a historical work on the position of women in society—how it has been developed from the earliest times up to today," wrote the German feminist Louise Otto in 1844,

> The lack of such a work is always strongly felt, but even more now—to put lessons from the past on a firm historical foundation, to seek prospects for the future, even more to have women begin to feel that they . . . are not just wives and mothers of the people, but half of this people themselves. What has passed for the history of women until now is only biographies of famous women, saints, princesses, heroines, etc.[1]

In the 1980s, we also wanted to legitimize the writing of women's history, to include lives and accomplishments long forgotten, to compensate for the absence of half of humanity from the historical record.

In many ways this project has been successful. During the last decade, scholarship on women has grown dramatically for every region and in every historical era. Because of this growth, the very existence of the history of women no longer needs to be justified as it did in Louise Otto's time. Few would now dare suggest, as they did twenty years ago when we first began the research for *A History of Their Own*, that women "had no history," or that they had achieved little worthy of inclusion in the historical record. Today, the study of women's past has become an accepted field within the discipline, and its practitioners have risen to the top ranks of the profession. Women's history courses are an accepted part of high school, college, and university curricula.[2]

Academic series of books in women's history, rich monographs, specialized encyclopedias, dictionaries, and bibliographic guides, new journals in women's history, internet groups, and web sites have proliferated. Growing networks of scholars, both in this country and throughout Europe, attest to the vitality of the subject. Topics that had to be pieced together in the 1980s are now recognized subfields in their own right: the history of the family, of sexuality, of violence against women, of laws and customs governing their lives. Subsequent research has confirmed analyses we assembled in order to write a work of synthesis: from the patterns of women's religious participation to the gendered dynamics of the witchcraft persecutions, from the shift in male roles when commerce made "providers" out of warriors to the connection between feminist demands and women's participation in political revolutions. Almost all of the principal figures singled out in our narrative history now have their own biographies and often scholarly editions of their writings. Talks that we cited in our notes, ones given at the Berkshire Conference on the History of Women, have become full-length monographs; long-range, general studies have finally been published—for example, an essential survey history of Catholic nuns, and analyses of women's lives in nineteenth-century Russia.[3]

In other ways, success has eluded feminist historians. Most works in women's history are found not in the history sections of libraries and book stores, but instead are catalogued under the impossibly vast category of Women's Studies. Scholarship in the field is read primarily by other women's historians or by feminists. Too many traditional historians make only the most cursory changes to their old narratives. European women now have their own histories, but this knowledge of their past has not significantly altered general accounts of European history. The question with which Louise Otto closed her thoughts on women's history in 1844 still applies today: "Shouldn't the female sex be given more attention in a general history of civilization than they were before?"

Women's historians of all regions of the world, not just Europe, have critiqued the male-centered nature, the subjectivity, the inadequacies of these supposedly gender-neutral prevailing accounts of the past. But however clearly the political nature of historical narratives and the choices and exclusions made in them are demonstrated, the full integration of women's history into all kinds of historical writing remains, in the words of the U.S. historian Anne Firor Scott, an "unfinished business."[4] The burden of proof still lies with advocates for the inclusion of women and the dynamics of gender, not with those who oppose it. Most traditional historians continue to insist that their version of the past, which suppresses women's history and ignores gender, presents an "objective," "apolitical" account of all that is significant.[5]

Nowhere is this resistance more obvious than in European history textbooks. Although much lip service is paid in explanatory introductions to the need for a new, more inclusive history, the canons of the male-focused narrative still govern periodization, the organization of sections and chapters, the choice of heroes and the occasional heroine. At best, women are subsumed under "social history," implying that their contributions to politics, the economy, or intellectual and spiritual life have been negligible. Too often, women then appear only as victims, the subjects of laws and customs that constrained and denigrated them. Some textbooks tell far more about men's views of appropriate female behavior than women's actual lives.[6] At worst, too many European textbook authors have simply added a few paragraphs about women to the ends of pre-existing chapters, pigeonholing the entire female sex as a "minority," an afterthought, an inconsequential and marginalized group whose lives exist outside the main story of Western Civilization.

None of the leading European history texts offers an innovative framework that views women as active participants with men in all areas of human endeavor. None analyzes men as a group or explicitly identifies what men as men have done. None examines the creation of historical definitions of masculinity, none explores how male-centered customs and practices became institutionalized and codified, none demonstrates how concepts of sexual difference have operated as social forces. Instead, the male has been universalized, so that accounts of men's achievements are assumed to be a complete history of the European past. What women accomplished despite constraints, what they made of their lives around and about the traditional men's narratives, has been omitted. There is no history of Europe that adequately describes the past experiences of women and men.[7]

As a result, women still need a history of their own that provides a continuous, female-centered narrative of Europe's past. A number of fine histories of European women now exist, but they either cover only a few centuries or particular groups and nationalities. Anthologies of articles present case studies across time but offer only brief transitional essays to link their disparate parts. Oxford University Press decided to publish a new edition of *A History of Their Own* because it provides a true synthesis, an unbroken, readable account of the European past from prehistory to the present that consistently puts women at the center of the narrative. Freed by its unique organization from the forced exclusions of traditional periodization, this two-volume work tells the familiar stories of the aristocrats and queens, but from a female perspective. In addition, by focusing on women previously ignored, like peasants and domestic servants, *A History of Their Own* provides a truly feminist account of European women's and men's collective history.

Choices that we made in the writing of these two volumes have become models for the field. Many historians have repeatedly called for reorganization of the standard divisions that shape our views of the past, chronological designations like "Middle Ages," "Renaissance," and "Industrial Revolution," but few have actually done this.[8] Instead, the traditional units continue to shape histories of women, however exclusionary or distorting they may be. Freeing European women's history from what the literary critic Jane Marcus calls "the yoke of male periodization" was the most liberating aspect of our collaboration.[9] By forcing ourselves to go against our training, we could re-examine standard historical events and thus reconceptualize the past from women's perspectives. In our reconceptualization, however, we did not abandon chronology. Each chapter is internally chronological, but the same event may appear more than once, as it affects different groups of women. Industrialization, for instance, had an entirely different impact on elite women, peasant women, and women of the modern cities, so it appears in each of the three chapters that focus on these categories. The Renaissance, the Enlightenment, the French Revolution, the World Wars are all dealt with in this way. As traditional historical periods and events receded in significance, others grew in importance. Factors often ignored in histories of men, whether contraception or clothing, diseases or the design of houses, proved crucial in women's lives.

Equally liberating was our decision to change standard historical words and phrases. All historians are trained to examine documents critically, to distinguish between actual reality and what is being said about that reality by a specific group of people. In English, supposedly gender-neutral terms like "peasant" or "revolutionary" appear inclusive but the descriptions that follow never mention women. Other nomenclature based on the unspoken assumption that male experience is everyone's experience also distorts the reality of women's lives. "Working class," for instance, implies both that women in other categories did not labor and that women's lives in this social stratum mirrored those of men. We used descriptive phrases like "women of the people" and "women who earned income outside the home" to convey the reality of poor urban women's lives more accurately. In addition, we reversed traditional modes of expression, writing of "women and men," "queens and kings," "mothers and fathers" in order to make women the focus of our narrative and to counter the weight of a male-oriented past and male-dominant forms of expression.

Rejecting the male-centered biases of our discipline, we turned to other fields for methodological and analytical tools to break what the feminist historian and theorist Joan Wallach Scott calls the "epistemological frame of orthodox history."[10] Anthropology proved the most fruitful. The beginning of each section of this work uses the anthropological technique of "thick description." Women within a particular category in specific geographic localities are described in "cross-cultural montages" generated from all kinds of sources—from folklore, archaeology, and art history to sociology and economics.[11] This technique also allows us to highlight a single woman's life or production—her weaving, her basketwork, her painting, her diary. We then used this "text" to provide insight into the overall social structures and cultural contexts in which she lived. Thus, well-known heroines illumine the lives of unnamed women in similar circumstances. Joan of Arc sheds light on other peasant women's place in society, while Anne Frank's brief life illustrates the tragedies of the Nazi genocide.

In addition to giving us descriptive and narrative techniques, anthropology provided new categories of organization. The concepts of "place" and "function" allowed us to abandon the inadequate and inappropriate periodization of men's histories. Placing women in specific geographic and institutional contexts, identifying them according to their broad functions within European society, revealed the unity of certain groups over time and across the continent. Peasant women, usually rendered invisible in historical narratives, emerged as a separate group whose similarities outweighed geographic, ethnic, or temporal differences. Women within the Christian churches constituted another category unified across time and secular boundaries by place and function. In the modern era, these concepts allowed us to distinguish between different types of experience in the same place and era. In the nineteenth century, the lives of poor urban women differed so markedly from those of peasants that they constituted separate categories, even though the same woman may have lived first in the countryside and then in the city.

These categories also enabled us to assess the long trajectories and patterns of European women's experience over time. We had hoped to find a "Golden Age" for women, a time when European women were not subordinated to and valued less than men. While the possibility of a matriarchal culture in prehistory cannot be ruled out completely, we discovered no era in the historical past in which women dominated.[12] This unequal relationship between women and men, present in the earliest written documents of European culture—the Hebrew Bible, Homer's epics, Roman law—intensified as time went on. The nineteenth century marked the nadir of European women's powers and opportunities. In earlier eras, alternative authorities and customs, as well as regional, governmental, and religious variations, created a range of circumstances that enabled some European women to achieve relative independence and relative dominance. Gradually, however, the growing centralization, rationalization, and uniformity imposed in government, law, the economy, and religion worked to erode these options and further limited women's lives.

The centuries from the Renaissance through the Enlightenment broadened possibilities for most men, giving them greater access to education and more choices of occupation. The opposite happened for women. New national law codes denied them control of their property and earnings, gave primary authority within the family to the husband alone, outlawed any efforts by women to control their fertility, and barred them from higher education and the newly defined professions. During these centuries, increasingly polarized images of the physical and psychological differences, both real and imagined, between women and men, between the "feminine" and the "masculine," justified these growing disparities. Female incapacity and male authority came to seem self-evident and natural.

The popular nineteenth-century ideal of the "angel in the house," a woman happily limited to the care of her household and children, offered a more restricted life to women than even the didactic treatises of previous centuries. The reality was always different from the imagined ideal. The majority of Europe's women continued to earn income; some "angels of the house" created paths out of the parlor and into the world. Even so, the concept of domesticity, with its constraining assumptions and definitions, remained for women of all classes and in all circumstances. The creation of "women's movements" in the nineteenth century occurred in part as a response to this narrowed view of women's capacities and activities.

In the twentieth century, women have changed laws and modified institutions. They, like European men, have benefited from prosperity, universal medical care, and technological progress. Most European women today enjoy full rights of citizenship, have access to education and employment, live longer, and face fewer risks from sexual activity and childbearing than women in earlier ages. While neither complete equality nor a realistic assessment of the value of women's contributions to European life has yet been realized, women's circumstances and opportunities have improved. In large part, change has come because of the effectiveness of the Women's Liberation Movement of the 1970s and 1980s. As the anthropologist Kathleen Gough observes, "It is not necessary to believe myths of a feminist Golden Age in order to plan for parity in the future."[13]

In writing about each category of women, certain questions guided our research. First, how had ordinary women lived? What tasks filled their days? What motivated their actions and determined their attitudes? Second, how to explain the startling contrasts between women's and men's lives in the same eras? Why had laws, economic systems, religion, and politics excluded European women from the most valued activities in life? How had cultural attitudes evolved that defined women, and qualities identified as "feminine," as innately inferior and placed all things female in a subordinate relationship to men and all things male? Why had men created or acquiesced in this inherently unequal system of social relationships? Perhaps more importantly, why had most women accepted or been forced to accept these limitations, which devalued their activities, denigrated their nature, and subordinated them to men?

Third, we looked at the exceptions—those women who achieved prominence and were included in traditional histories: St. Bridget of Sweden, Queen Isabella of Castile, Mme. de Pompadour, Florence Nightingale, Marie Curie. Why had these women gained recognition? Were they exceptions because of their character or historical circumstance? Finally, we studied those women, like Christine de Pizan, who first publicly questioned women's disadvantaged and denigrated status. Why did some women question all women's subordination? How had they come to identify with all women and to work for expanded opportunities for their sex? How and why did feminism begin and where might it lead, as it calls into question the basic values of European culture and society?

The answers to these questions led us to the central thesis of these volumes: that gender has been the most important factor in shaping the lives of European women. However, not all women's experiences are alike. Our narrative recognizes the gulf between a woman in medieval France and a woman in modern England, between a fifteenth-century female merchant and a twelfth-century day laborer, between a German Social Democrat and a Soviet Bolshevik. Our method of organizing women into separate categories graphically indicates the significance we accord these differences. However, underlying these differences are similarities decreed by gender. Throughout the centuries we found an awesome similarity in the effects of gender on European women's lives, in the continued power of the denigrating qualities classified as "feminine." Unlike men, who have been primarily identified by class, ethnic origin, or historical era, European women have traditionally been seen first as female, a separate category of being. As the French socialist Louise Michel wrote in 1885, it has been "painful" for us "to admit that we are a separate caste, made one across the ages," but as we compared our findings from studies of different eras, classes, and ethnic circumstances, no other conclusion was possible.[14]

Part I of *A History of Their Own*, "Traditions Inherited," speculates about women's and men's lives in prehistory and the origins of European culture's largely negative views of women and their subordinate status. It then examines the Greek, Roman, Hebrew, Celtic, Germanic, and Christian traditions about women and their relationships to men, traditions already in place when Europe emerges as a more recognizable entity in the ninth century. Part II, "Women of the Fields: Sustaining the Generations," surveys the lives of European peasant women into the 1980s. Because they make up the vast majority of Europe's women until well into the eighteenth century, we placed their narrative first, thus highlighting their numbers and affirming their significance. Our account gives priority to the constants in their experiences over local differences in geography, custom, patterns of landholding and trade. Part III, "Women of the Churches: The Power of the Faithful," shows how Christianity provided a unique environment for European women. From the early centuries of the religion's growth, through the Reformations of the sixteenth and seventeenth centuries, women could gain authority and relative autonomy not possible in other circumstances.

Part IV, "Women of the Castles and Manors," argues that the lives of Europe's noblewomen from the ninth to the seventeenth centuries are connected because of their elite status and their function as "custodians of land and lineage." While these women sometimes acquired power and acted in place of men, they remained vulnerable because of their gender. Part V, "Women of the Walled Towns: Providers and Partners," distinguishes urban women of the twelfth to seventeenth centuries from their rural counterparts. From the poorest day laborer to the wealthiest merchant's wife, townswomen participated in the significant economic developments of their era: the formation of guilds and the evolution of commercial capitalism. Neither, however, freed them from the constraints of circumstance and attitude that traditionally limited women's lives. These five sections comprise Volume I.

While our first volume focuses on the centuries before 1600 and our second on those after, this division is not rigid. Traditional chronologies are not the organizing principle of this work; the categories of place and function demarcating women's lives are. Thus although Part VI, "Women of the Courts: Rulers, Patrons, and Attendants," appears as the first section of Volume II, it describes court life from the fifteenth through the eighteenth centuries. We argue that the growth of dynastic monarchy created special circumstances in which some women had opportunities to become educated, to write, to exercise political influence, and, in a few instances, to rule. Part VII, "Women of the Salons and Parlors: Ladies, Housewives, and Professionals," examines the lives of economically privileged women from the late seventeenth century to the present. Ideals of domesticity and the realities of better standards of living distinguished these women's lives from those in other classes. Some women turned these conditions to advantage, using their moral and material authority to play active roles outside of their homes.

Part VIII, "Women of the Cities: Mothers, Workers, and Revolutionaries," deals with the lives of everyday urban women in the same centuries, focusing on their participation in economic, social, and political movements. Here we pioneered the thesis that urbanization was more important than industrialization in shaping these women's lives. The chapters grouped under the heading "Women of the Cities" parallel our earlier category, "Women of the Fields." Together, urban and country women comprise the two most numerous groups of women, and so are near the beginning and end of the two volumes.

Part IX, "Traditions Rejected: A History of Feminism in Europe," mirrors the first section of Volume I, "Traditions Inherited" when Europe coalesced. Beginning with the writings of the courtier Christine de Pizan in the fifteenth century, this final section views European feminism as a series of repudiations of the negative traditions that limited women's lives.[15] In this process, a women-centered view of the world, which is still being elucidated and realized today, evolved into feminism. This new edition of *A History of Their Own* concludes with an epilogue on developments in both Western and Eastern Europe since the collapse of communism in the late 1980s. The conservative shift in politics, the shrinkage of the welfare state, the rise of unregulated market economies, the continuance of primary responsibility for child-raising and housework still differentiate women's lives from those of men in the same nation, ethnic group, social stratum, or even family.

Aside from our central thesis about the significance of gender in European women's history, no other aspect of our book has been so controversial as our emphasis on the continuities in European women's lives throughout the centuries.[16] These two convictions are linked. There are variations in European women's lives across time, place, function, and circumstance. They were active participants in political, social, economic, and religious change. But these variations and this participation did not alter women's status relative to men. Despite dramatic transformations of European culture and society from the ninth to the early twentieth century, the meanings given to sexual difference and to "feminine" and "masculine" identities worked to maintain the disadvantaged status of women.[17] The medieval historian Judith Bennett describes a "patriarchal equilibrium" creating an overall pattern of European women's history that resembles a dance, in which the steps and rhythms, partners and groups may change, but the men always lead.[18]

Distinguishing variation from genuine transformation, we argue that a number of important aspects shaping women's lives have remained unchanged over time. Until the last decades of the twentieth century, all women were defined by their relationships to men. Many women—far many more than men—remain in the historical record only as men's women: the daughters of Priam, Lot's wife, and the mother of the Maccabees are but a few of the earliest examples. A woman is first identified as her father's daughter, her husband's wife or widow, her son's mother. No matter what the era in European history, what their class or social rank, what their nationality or ethnic group, most women have lived their lives as members of a male-dominated family. Even those who lived more autonomous lives as part of women's spiritual communities were defined by their rejection of earthly marriage. Nuns, as members of religious orders, were described as the "brides of Christ."

These definitions, as historians of sexuality and of the family have demonstrated, constrained women and men. The "family" protected by law and custom, the union of a woman with a lawful husband for the purpose of procreation, presumed the heterosexuality of both partners and dictated their primary functions and roles. In the male-headed family, child rearing and maintenance of the household have always been gendered, seen throughout Europe's history as women's preordained, biologically appropriate tasks.

Defining women's primary duties as care of the family and the home has not precluded other work. In all historical eras, the vast majority of European women have labored at other chores and assumed other responsibilities.[19] They have worked in the fields. They have earned wages. They have generated additional income for their families. Weeding, reaping, sewing, knitting, cleaning others' homes, raising others' children, working in factories or offices, women's labor has made the continuance of their families possible. This "double burden" of caring for a family and home and earning additional income has characterized the lives of most European women and differentiated them from men. It is women, not men, who have these multiple responsibilities and must find work compatible with these duties or arrange for substitutes to care for their children and their household while they earn income.

In addition, "women's work," whether in the home or outside of it, has traditionally been valued less and considered less important than men's work. Raising children and maintaining the home have been taken for granted and have never been valued as much as labor that men perform, whatever it may be. Paid labor available to women has usually been less prestigious than men's, has traditionally required less formal training, and has been more vulnerable to fluctuations in the economy. As a result, when they have been paid for their work, women have consistently received between one-half and two-thirds of what men earn. Sometimes connected by scholars to different economic systems, this factor has always been present in European history. In reckonings of female and male worth in the Old Testament, in the manor rolls of noble households, in account books of sixteenth-century merchants, in payrolls of nineteenth- and twentieth-century factories, women received less than men. The amount that they are paid may vary: labor shortages or economic regulations may raise women's wages, but, so far, they have rarely equaled those of men. As historians of European women's labor have demonstrated, there was no "Golden Age" of women's household production. Commercial capitalism brought different activities and relationships within the marketplace to women, but not the transformation of the underlying gendered patterns of social interaction and the institutions that protected them. All of these factors shaping women's work limited European women's lives by curtailing their opportunities and resources.[20]

Some women maneuvered around these limits or found ways of setting one institutional constraint against another—the aristocratic woman who sought the Church's support in her choice of a husband against the dictates of her family; the royal women who ruled as queens in their own right; the merchant's wives who managed a husband's fortune after his death; the successful court musicians, poets, and artists. Even they, however, were subject to the most damning aspects of gender: European culture's largely negative views of women. Considered innately flawed, less valuable, and thus inferior to men, all women were supposed to be subordinate to men. This subordination seemed part of the natural order. A woman who did rule over men, who held a dominant role, whether from a throne or within a family, was seen as "unwomanly," a danger to the universe's natural hierarchy, which made man come first.

These cultural views, expressed in the earliest writings of the Hebrews, Greeks, and Romans, changed remarkably little over time. The biblical injunction to Eve that "your desire shall be for your husband, and he shall rule over you" (Genesis 3:16) is repeated in every era and every European nation. The view that "the best woman is she who is silent"—first written down in ancient Greece—reappears often in European men's writings about women. The assumption that only men are truly human—that "a hen is not a bird and a woman is not a person," as the Russian proverb explains—echoes throughout European history. No woman could escape the impact of these views completely. Of all the factors that have limited women's lives, these negative cultural traditions, these negative constructions of what it means to be female, have proved the most powerful and the most resistant to change.

But they have never been all-powerful. Throughout the centuries is also scattered the evidence of European women's agency, of the multiplicity of ways in which they gave value, beauty, and power to their lives. Many took pleasure and pride in their reproductive and nurturing role, in their daily tasks, however mundane. Sadly, much of women's creation has been anonymous and evanescent. Yet it is evidence none the less: the basket of willow branches created to gather food, the weaving in hand-dyed wools which clothed Europeans in the early centuries, the lace tablecloth for a daughter's trousseau, the household objects and children's toys designed to make life easier and more pleasant.

Although most of Europe's women accepted the institution of the male-dominated family for its guarantee of subsistence, an approved partner for life, and a sense of being protected from forces beyond their control, they have not just been victims. Resistance can take many forms. Even when unable to see beyond their culture's attitudes, they mastered the strategies of those in subordinate positions, manipulating, pleasing, enduring, surviving.[21] Some claimed spiritual or moral authority as women, drawing on those religious or ethical traditions that empowered women rather than subordinated them. There is magnificence in the fragments of Sappho's poetry, in Hildegard of Bingen's visions, in Marie de Gournay's defense of women, in Paula Modersohn-Becker's self-portraits, in Mo Mowlam's negotiations in Ireland.

Our belief in women's abilities to create such excellence, to expand the boundaries of human creativity and endeavor, to transcend conditions that seek to limit and control them underlies every section of these volumes. We have never held that women are determined by "their essence" to remain in certain roles or to fulfill certain functions—a charge that our emphasis on continuity has sometimes prompted.[22] Instead we have affirmed women's diversity and applauded their accomplishments. Increasingly, women's varied contributions have affected the lives of all. In 1998, Mary Robinson, as the United Nation's High Commissioner for Human Rights, brought her feminist sensibility to international affairs. While she was president of Ireland from 1990 to 1997, Robinson once explained: "A society that is without the voice and vision of a woman is not less feminine. It is less human."[23]

Throughout our collaboration on *A History of Their Own*, we have taken heart from Virginia Woolf's vision in *A Room of One's Own* of a future in which a woman with the talent of Shakespeare could flourish. "My belief is that if we live another century or so, if we have the habit of freedom and courage to write exactly as we think," she predicted in 1929, a way for this genius can be prepared.[24] With these volumes we make our contribution to this collective enterprise. We look forward to the creation of a world in which women and men will acquire "the habit of freedom and courage," as well as the means and opportunity to succeed as they choose.

Notes

1. Louise Otto cited in Ruth-Ellen Boetcher Joeres, *Die Anfänge der deutschen Frauenbewegung: Louise Otto-Peters* (Frankfurt-am-Main: Fischer Taschenbuch Verlag, 1983), p. 82.
2. For a survey of the changes in the academy and the profession for women's history and women historians, see for example, Judith P. Zinsser, Part III of *History and Feminism: A Glass Half Full* (New York: Twayne Publishers, 1993).
3. Jo Ann McNamara, *Sisters in Arms: Catholic Nuns through Two Millennia* (Cambridge, Mass.: Harvard University Press, 1996); Barbara Alpern Engel, *Between the Fields and the City: Women, Work, and Family in Russia, 1861–1914* (New York: Cambridge University Press, 1994).
4. Anne Firor Scott, "Unfinished Business," *Journal of Women's History*, vol. 8, no. 2 (Summer 1996), pp. 111–20.
5. For discussions of this problem, see the special issue of *History and Theory: Studies in the Philosophy of History*, Ann-Louise Shapiro, ed., vol. 31, no. 4 (December 1992) published in an expanded version as *Feminists Revision History* (New Brunswick, N.J.: Rutgers University Press, 1994).
6. For discussion of this in specialized works as well, see Amanda Vickery, "Golden Age to Separate Spheres? A Review of the Categories and Chronology of Women's History," *Historical Journal*, vol. 36 (1993), pp. 413–14.
7. See for example, current editions of Lynn Hunt, Theodore R. Martin, Barbara H. Rosenwein, R. Po-chia Hsia, and Bonnie G. Smith, *The Challenge of the West: Peoples and Cultures from the Stone Age to the Global Age* (Lexington, Mass.: D. C. Heath and Co.); Donald Kagan, Steven Ozment, and Frank M. Turner, *The Western Heritage* (New York: Macmillan); Mark Kishlansky, Patrick Geary, and Patricia O'Brien, *Civilization in the West* (New York: West Publishing Co.); Jackson J. Spielvogel, *Western Civilization* (New York: West Publishing Co.).
8. Joan Kelly and Gerda Lerner were the first to articulate the need for new periodization in their now famous essays: "Did Women Have a Renaissance?" in *Women, History & Theory: The Essays of Joan Kelly* (Chicago: Chicago University Press, 1984 [1977]); "The Challenge of Women's History," in *The Majority Finds Its Past: Placing Women in History* (New York: Oxford University Press, 1979).
9. Jane Marcus, "The Asylums of Antaeus: Women, War, and Madness—Is There a Feminist Fetishism?" in H. Aram Veeser, ed., *The New Historicism* (New York: Routledge, 1989), p. 140.
10. Joan Scott has discussed this dilemma in many of her essays. See for example, "The Evidence of Experience," in James Chandler, Arnold I. Davidson, and Harry Harootunian, eds., *Questions of Evidence: Proof, Practice, and Persuasion across the Disciplines* (Chicago: University of Chicago Press, 1994), pp. 367–69, 372–73, 376, 378. See also, *Only Paradoxes to Offer: French Feminists*

and the Rights of Man (New York: Cambridge University Press, 1996), p. 124.

11. Judith Lowder-Newton uses this phrase. See her essay "History as Usual? Feminism and the 'New Historicism,'" in Veeser, pp. 153–54.
12. Anthropologists have documented parity between women and men, and societies in which women exercised significant direct power, but not equality as it has traditionally been defined in Western cultures. For essays exploring this question, see for example, Rayna R. Reiter's collection, *Toward an Anthropology of Women* (New York: Monthly Review Press, 1975) and Peggy Reeves Sanday and Ruth Gallagher Goodenough, eds., *Beyond the Second Sex: New Directions in the Anthropology of Gender* (Philadelphia: University of Pennsylvania Press, 1990).
13. Kathleen Gough, "The Origin of the Family," in Reiter, p. 54.
14. Louise Michel, *The Red Virgin: Memoirs of Louise Michel*, Bullitt Lowry, and Elizabeth Ellington Gunter, eds. and trans. (University: University of Alabama Press, 1981), p. 139.
15. Joan Kelly enunciated this idea in "Early Feminist Theory and the *Querelle des Femmes*, 1400–1789," in *Essays*, pp. 65–109. This is also a central thesis of Gerda Lerner's *The Creation of Feminist Consciousness from the Middle Ages to 1870* (New York: Oxford University Press, 1993).
16. On the difficulties of accounting for continuities and changes at the same time, see Sandra E. Greene, "A Perspective from African Women's History: Comment on 'Confronting Continuity,'" *Journal of Women's History*, vol. 9, no. 3 (Autumn 1997), pp. 95–104.
17. For a clear formulation of questions about the meanings assigned to sexual difference, and the gendering of identities to women's disadvantage, see Ava Baron, "Gender and Labor History: Learning from the Past, Looking to the Future," in Ava Baron, ed., *Work Engendered: Toward a New History of American Labor* (Ithaca, N.Y.: Cornell University Press, 1991).
18. Judith M. Bennett, "Theoretical Issues: Confronting Continuity," *Journal of Women's History*, vol. 9, no. 3 (Autumn 1997), p. 86.
19. Even though Europe's prescriptive literature might endorse the idea of a "separate sphere" as an isolated, protected women's world, the concept had no reality in fact. A number of historians have written on this misconception. See for example, Linda K. Kerber, "Separate Spheres, Female Worlds, Woman's Place: The Rhetoric of Women's History," *Journal of American History*, vol. 75, no. 1 (June 1988), pp. 9–39; Dorothy O. Helly and Susan Reverby's introduction in *Gendered Domains: Rethinking Public and Private in Women's History* (Ithaca, N.Y.: Cornell University Press, 1992).
20. See Vickery, pp. 401–4; see also Bennett, "Theoretical Issues," p. 86. Note that Bennett, in her discussion of medieval English brewsters, does not see these constraints as specifically designed to affect women. Rather "[t]hese factors

affected some women differently from others, but they affected all women to some extent. These factors shaped the lives of men as well as women, but they constrained most women more than most men. And these factors grew from fundamental institutions of English life at the time, institutions that were much more than mechanisms for the subordination of women," pp. 87, 86.

21. Anthropologists have made systematic studies of this phenomenon in contemporary cultures. See for example, James C. Scott, *Domination and the Arts of Resistance; Hidden Transcripts* (New Haven: Yale University Press, 1990).
22. For a discussion of the need to avoid the extremes of "essentialism" or "relativism," see Linda J. Nicholson, *Feminism/Postmodernism* (New York: Routledge, 1990), p. 9.
23. Cited in Alida Brill, ed., *A Rising Public Voice: Women in Politics Worldwide* (New York: The Feminist Press at the City University of New York, 1995), p. 155.
24. Virginia Woolf, *A Room of One's Own* (New York: Harcourt, Brace & World, 1957 [1929]), pp. 117–18.

I

TRADITIONS INHERITED

•

ATTITUDES ABOUT WOMEN FROM THE CENTURIES BEFORE 800 A.D.

1

BURIED TRADITIONS: THE QUESTION OF ORIGINS

ONE OF THE OLDEST portraits of a European woman is a carved ivory head found at Dolní Věstonice in the present-day Czech Republic. Dating from about 26,400 B.C., the small carving depicts a female face topped by a bun of hair.[1] The woman's features are delicate and singular; the head is that of an individual. Her brow and mouth twist up to the left, and her nose has a small bump at the end. This head is unique among the many figurines, tools, and bones found so far at Dolní Věstonice, where a group of about 100 to 120 women and men had established a permanent settlement during the severest period of the last Ice Age. They lived in five or six large huts, each about 40 square meters, and hunted the large animals—woolly mammoth, horses, reindeer—that roamed the tundra around the site. They built their houses by permanent springs of water with indoor hearths, where meat was cooked and clay objects were fired.

Thirteen years after the ivory head was found, a female skeleton was unearthed nearby whose skull had a defect on the left side which could have produced the twist of the woman's face up to the left in the ivory carving. The skeleton may well be that of the woman depicted in ivory.[2] The buried woman, estimated to have been 5 feet 3 inches tall and about forty years old, had been carefully placed in a prepared grave on the western edge of the settlement. She was lying on her side, facing west, with her legs drawn up. Her body had been sprinkled with red ochre and two mammoth shoulder blades were placed on top of her, one with lines carved on it in irregular patterns. Her stone tools and an arctic fox's paws and tail were buried with her; the fox's teeth were in her right hand. The grave and the ivory carving are all that have endured of her life, and they provide a fascinating glimpse of a woman's existence tens of thousands of years ago.

Such glimpses into human life in Europe before the invention of writing have tantalized thinkers concerned with human nature and human society.

Was this woman unusual? Probably. While the use of ochre and her burial position were common in this era, the mammoth bones were not. But what does the carving of her signify? Did she rule? Was her status high? Was she subordinate to men in her group? How did these people live, and do their lives illumine our own? Are there patterns of human nature which have never changed? For the historian, the absence of written materials makes a sure answer to such questions virtually impossible.

Yet the issue of the status of women in prehistory is important because the first written documents of the Greeks, Romans, and Hebrews show women subordinated to men. The difficult question of the origins of this subordination has been made all the more complicated by the tendency of thinkers to formulate universal theories on very little evidence. Traditionally, arguments from nature, religions, and prehistory were used to justify women's subordination; conversely, similar arguments came to be used by feminists to assert that such subordination was culturally imposed and should end. The debate took on a new dimension in the second half of the nineteenth century, when the application of evolution to the development of human society produced a series of works which argued that human society had evolved from a matriarchal past to a patriarchal present.

The most influential writers were J. J. Bachofen, Lewis Henry Morgan, and Friedrich Engels.[3] A number of theories were advanced to explain the supposed replacement of female dominance (matriarchy) by male dominance (patriarchy). Bachofen speculated that "mother right," in which women controlled religion, property, and marriage, had fortunately been replaced by "father right," in which men rule, a male god is worshipped, and "the man's spiritual principle dominates."[4] Morgan connected a male-dominant monogamous marriage to the rise of "civilization" and the end of incest. Engels argued that the creation of private property controlled by the male head of the family led to "the overthrow of mother right" and "the *world historical defeat of the female sex.*"[5] Others theorized that men's discovery of their role in conception led to the end of both goddess worship and high female status in prehistory. All these analyses attempted to draw moral precepts from human evolution. Until the last few decades, thinkers differed on whether such evolution had been for good or ill, but they tended to accept the same universalist premise—that female dominance had been replaced by male dominance. Opposition came mostly from those who believed men had always been dominant and that the evidence for matriarchy was deficient.

Recent work in archaeology and anthropology leads away from universal explanations. Both women's roles in early societies and the absence of male dominance in all societies have been explored.[6] Attention has focused on the circumstances that make female subordination more likely rather than on justifying contemporary patterns by appealing to the distant past. The most likely reason for female subordination is the development of intergroup competition and warfare, usually as a response to stressful ecological and social circumstances. This can account for the subordination of women in Greek, Roman, and Hebrew writings, because all these cultures were warrior cultures. But these writings appear relatively late in human development. To search for the origins of female subordination, historians have turned to other disciplines for answers. The most important in casting light on relations between women and men in prehistory have been archaeology, biology, psychology, and anthropology.

Archaeological Evidence

Archaeology examines the durable remains of the human past. Although many fascinating objects depicting or pertaining to women have been found, little can be currently deduced about their cultural significance, or the relations between the women and men who produced them. For many years, for instance, the importance of hunting—usually performed by men—was overemphasized, since weapons and bones endured and figured prominently in archaeological digs. Gathering, by which peoples who hunt procure most of their food, was undervalued, since baskets and carriers and the vegetables and fruits gathered did not last. Recent studies of the teeth of early humans show that the bulk of their diet was vegetable—presumably gathered by the women of the group.[7] But the status of these women gatherers cannot be known. Were they highly valued because they provided most of a group's food, or taken for granted because they procured everyday sustenance rather than the exceptional supplement of animal protein? The same problem in interpretation applies to objects, especially if they are not obviously useful. The function of axes, needles, and blades can be assumed, but the significance of objects like the ivory carving of the woman's head can only be guessed.

Other figurines found at Dolní Věstonice raise many questions but yield few answers. There were bone and ivory pendants shaped like breasts, vulvas, and penises. There were clay figurines of bears, rhinoceroses, lionesses, and humans. Of all these figurines, the one that has received the most attention from scholars is a small, red clay statuette of an abstract female figure. The figure stands naked and has no lower arms, feet, nose, mouth, hair, or genitals. Her arms, eyes, spine, navel, legs, and buttocks are marked out by sharp grooves. Four grooves also line her back, two on either side of the spine. She is fat, with large, pendulous breasts and broad hips.[8]

This figurine is but one of the scores of the so-called "Venuses" found in numerous European sites from Spain to Siberia, all dated within a few thousand years of each other.[9] Many speculations have been advanced about them, most as facile as the view that allowed these figurines to be called "Venuses." They have been described as pregnant women and called fertility objects, although few are discernibly pregnant and gathering/hunting peoples tend to limit their fertility rather than enhance it. They have been called mother goddesses, although there is no evidence of what the people who made them believed. These objects tell the historian very little about women or their relationship to men in these early societies.

The Dolní Věstonice woman and the "Venus" figurines date from Europe's early Stone Age, twenty-odd thousand years ago. Until about 6000 B.C., humans on the continent of Europe lived like their forebears at Dolní Věstonice, by gathering and hunting as the climate warmed and the tundras gave way to forests and woodlands. Other, later settlements yield equally intriguing objects. One culture, at Lepensky Vir on the banks of the Danube, placed carved yellow sandstone boulders in its houses; some were decorated with abstract designs, others carved into fishlike creatures; one had the outline of female genitals incised on it. What these boulders signified remains enigmatic. Other early cultures have left behind multitudes of figurines: birds, fish, women, men, abstract shapes, and unrecognizable objects. Attempts to classify and understand them remain speculative.[10] Other evidence from cave paintings and carvings suggests that at least some of these early peoples on the European continent understood both the female and male roles in reproduction. Animals were grouped seasonally, showing copulation in the springtime and pregnant females in the summer. If women or goddesses were worshipped in early European cultures, this worship probably did not rest on women's supposed total control over reproduction, as was often assumed by earlier scholars.[11]

Women's status and relationship to men also remain mysterious in the archaeological remains of groups who shifted from gathering and hunting to farming and domesticating animals, a change that began about 6000 B.C. in Europe. Early farming communities in the Middle East show some evidence of goddess worship; early communities in Europe reveal neither the veneration of women nor their denigration. At Nea Nikomedia in northern Greece, for instance, people farmed, wove cloth, and made pottery. One large building in the community contained two greenstone axes, three blue-green serpentine frogs, and seven abstract human figures, five of women and two of men. Property may have been privately owned, a factor Engels and other Marxists have connected to the subordination of women. But the status of women at Nea Nikomedia can only be guessed at; there is no evidence of either women's dominance or their subordination.[12]

Just as farming has been used to explain the origin of female subordina-

Venus" figurine from Dolní Věstonice
;,400 B.C.

2. Woman with a water jug leading a little boy. Greek vase, fifth century B.C.

reek women spinning and weaving cloth. Vase attributed to the Amasis painter c. 540 B.C.

4. "Bride," symbolizing women's h
able roles. This appears opposit
"Flute Player." Relief from the Lu
Throne c. 460–450 B.C.

5. Presumed to be the Celtic Queen Medb (Maeve). Gundestrup cauldron, first century A.D.

"Flute Player," symbolizing dis-
norable roles for women. "Flute
ls" often served as prostitutes. Re-
f from Ludovisi Throne c. 460–450
:.

7. Cleopatra of Egypt. Silver tetradrachma, first century B.C.

8. Relief of two Amazons fighting, commemorating the release from service of two female gladiators. Greek, first or second century A.D.

9. Eve, the Tree of Knowledge and the Serpent. Relief from the Cathedral of Saint Lazare, Autun c. 1120 A.D.

10. Mary Magdalen at Jesus' feet, and his follower Martha in the background. Stone relief from Semur-en-Auxois, early thirteenth century A.D.

tion, so has the rise of the state. But the evidence of the first state systems in what is now considered Europe—Mycenae and Crete—is also enigmatic with regard to women's status. These early states built palaces, controlled large surrounding areas, could support people with specialized occupations, and invented writing, much of which remains undeciphered. Some of the translated Mycenaean tablets show that women workers in the palace received the same rations as men and that women were equally represented among the religious staff, which points to women having equal status to men.[13] But few Mycenaean objects portray women, and without more information to provide context, those that do have no clearer a meaning than the carved ivory head from Dolní Věstonice. For example, a unique small ivory statuette from about 1500 B.C. depicts two women and a boy. The women kneel side by side and one has her arm on the other's shoulders. They are dressed alike, in elaborately ruffled skirts and intricate necklaces; one has long hair, the other's is bound or short. The little boy stands between them, and all the figures seem affectionate and serene. But who are these women and what do they signify? Are they human or divine? Do they represent Demeter and Koré (Persephone), her lost daughter, present in later Greek and Roman culture? Do they signal a high status for women in Mycenae? None of these questions can be answered at present.[14]

Though more extensive, the archaeological evidence of Cretan culture is no more enlightening with regard to women's status and relationship to men than that from Mycenae or earlier cultures. This has not stopped some from postulating the existence of a matriarchal Cretan society, which worshipped a peaceful mother goddess whose benevolent rule was overthrown by patriarchal Northern invaders.[15] Such speculation is fascinating but unsubstantiated. From about 3000 B.C. to 1200 B.C., Crete produced writing and sophisticated and refined seal rings, vases, bronzes, figurines, and frescoes, many of which portray women. But women's role and status in Crete remains enigmatic.[16] For example, archaeologists have reconstructed two small (34.2 centimeters, 20 centimeters) faience statuettes of elaborately dressed women holding snakes, but there is no way of telling if these were representations of snake goddesses, or priestesses performing some ritual, or simply of women with snakes. Although most Cretan seal rings depict animals, a number show large women with uncovered breasts, often surrounded by smaller females and males, animals and trees. Their significance is unknown. The famous bull-leaping fresco at Knossos shows one red and two white human figures on either side of a bull. The figures could be female or male.[17] There is some evidence that agriculture was a male activity and weaving a female one: The Harvester Vase shows men carrying a variety of agricultural tools, and traces of looms were found in some quarters identified with women in the palaces.

But separate functions or activities do not determine the status of women or their relationship to men. It is not even known who ruled in Crete. Analogy, tradition, and the male name "Minos" as King of Crete are the only evidence for male rulership.[18]

Cretan religion, or indeed any European religion before deciphered writing, also sheds no light on the question of the status of women and their relationship to men. There is no clear evidence, as there is in contemporaneous cultures in the Middle East and Egypt, of powerful female goddesses or one omnipotent Great Mother. A study of 103 small human figurines found in Crete revealed that 37.3 percent were female, 9.2 percent male, 40.7 percent human with no sexual characteristics, and 12.8 percent so fragmented their sex could not be determined.[19] Even from such extensive urban sites, with so many kinds of archaeological evidence, no one can deduce with certainty the roles of Europe's women in prehistory or the origins of their subordination to men. As the historian Sarah B. Pomeroy writes, "The question is open and may never be answered."[20] To shed more light on this issue, historians have turned to evidence from biology, psychology, and anthropology pertaining to the roles and relationships of women and men.

Biological Evidence

In the past twenty years, biologists have stressed the differences between humans and primates, particularly with regard to the social relations between females and males. The evolution of both primate and human bodies reveals a wider and more complex range of behavior for both sexes than was previously thought. While many primate species show females subordinated to males, not all do. Patterns of dominance are complicated: In a number of species, for instance, young males defer to both older females and older males. Primate evolution seems to have selected for females that are "highly competitive, socially involved, and sexually assertive individualists," as the primatologist Sarah Blaffer Hrdy writes.[21] Earlier, simplistic equations of humans to primate species where the sexes differ markedly (like baboons, whose females are half the size of males) have been superseded by more sophisticated studies of a wider range of species. But little can be applied with certainty from primate behavior to human actions, because humans differ so markedly from primates in key areas of sexual and social behavior.

Of all the primates, only humans have a menstrual rather than an estrus cycle. In estrus, both female receptivity and male arousal are hormonally triggered. Intercourse inevitably occurs at a fixed time in the cycle; the female can become pregnant only then. With menstruation, there is no equivalent compulsion, and humans, unlike other primates, have intercourse when their culture deems it appropriate.[22] Human females also can become pregnant many more times a year than primates, a prosurvival evolution. All human societies are based on families, usually composed of the mother, the father or father-surrogate, and children. Primates show far wider variations. In some species, the husband/father role is nonexistent; in others, monogamy and male aid in child rearing is normal. Then, while primates share with humans "millions of years of natural selection for social intelligence," the "human world is radically different from that of other primates," largely because of the human brain.[23] Language and cultural variation have molded human behavior far beyond primate models. Primatology does not currently explain why women should be subordinate and men dominant in many human societies.

Neither does biology. Increasingly sophisticated studies of human anatomy and physiology reveal no clear reason for either sex dominating the other. Biological evidence does explain differentiation of function, but little more. In contrast to many animal species in which there are dramatic anatomical differences between females and males—in coloration, size, and strength—human females and males differ relatively little from each other. (See Table 1.) The organ which most distinguishes humans from other animals is the brain, and innate intelligence does not differ according to gender.[24] Many of the character traits associated with one sex or the other have been shown to be the product of culture rather than biology. A classic case involved identical male twins, born in the United States in the middle of the twentieth century, one of whom was accidentally castrated as an infant. Differential upbringing alone produced a "girl" twin and a "boy" twin by age five. Since these changes preceded hormonal production, they point to a great deal of behavior being controlled by culture rather than nature.[25]

Women bear children and men do not, but childbearing does not automatically imply physical weakness or inferiority. In many animal species the females are the primary hunters; in many human societies women, whether pregnant or not, perform the major share of heavy labor. Cultures tend to assign tasks to one sex or the other not on the basis of strength, but on compatibility with child care. The female capacity for lactation has led to women being primarily responsible for the care of nursing infants, but the primary role of women in parenting older children seems to be culturally rather than biologically determined. No biological basis for a "maternal instinct" has been found except immediately after birth. Even in animals, mothering is largely learned rather than instinctive behavior.[26] Women's association with childbearing, nursing, and early child care does not automatically lead to either social dominance or subordination.

Biology distinguishes four physiological differences between females and

Table 1. SOME PHYSICAL DIFFERENCES BETWEEN THE SEXES

	Female	Male
Fetus		
	xx chromosome	xy chromosome
	Less susceptible to sex-linked diseases	More susceptible to sex-linked diseases
	No androgen production	Androgen production to change female matrix to male
	By third month, divergence of sexual structures, but organs are analogous	
At Birth		
	Development one month "ahead"	Development one month "behind"
	More resistant to disease	Less resistant to disease: one-third more boys than girls die in 1st year
	Shorter, lighter	Longer, heavier
	Smaller hearts and lungs	Larger hearts and lungs
	Higher percentage of body weight is fat	Higher percentage of body weight is muscle
	Shoulder and hip sockets, pelvic bones shaped differently	
Ages 1–8		
	Similar hormonal production and similar physical development	
Puberty		
	Cyclical hormonal production of estrogen and progestin starts, leading to menstruation	Noncyclical production of testosterone starts, leading to the production of sperm
	Breasts and body hair grow	Voice changes and body and facial hair grow
	Distinctive shoulder-pelvic shape develops: / \ or \ /	
Maturity		
	Capable of gestation, birth, and lactation	Capable of insemination
	36% of body weight is muscle	42% of body weight is muscle
	10% shorter and lighter	10% taller and heavier
	Lower metabolic rate	Higher metabolic rate
	Higher percentage of body weight is fat: greater endurance of extreme temperature	Lower percentage of body weight is fat: less endurance of extreme temperature
	More endurance, but less at some heavy physical exertion	Less general endurance, but more at some heavy physical exertion
	Distinctive shoulder-pelvic shape: / \ or \ /	
Aging		
	Decreased hormonal production leads to lessening of sexual differences	
	Menopause	"Male menopause"
	10% longer-lived	10% shorter-lived

males which may contribute to female subordination, although they do not determine its appearance. If conditions of diet, health, and exercise are equal, women will be, on the average, 10 percent shorter, lighter, and weaker than men.[27] Some women will be stronger than most men; some men will be weaker than most women. But in addition, the hormone testosterone, produced exclusively by males from puberty on, has been associated with "a differential readiness to respond in aggressive ways."[28] Male strength and aggression may lead to male dominance of women but need not do so: Male dominance is not universal in human society, and male aggression is culturally controlled in a variety of ways.

On the other hand, only women have the ability to give birth. In addition, if conditions are equal, women will be longer-lived than men, by a factor of 10 percent.[29] Like male strength and aggression, female childbearing and longevity could lead to dominance. But such dominance by females or males because of their physiology alone is neither inevitable nor universal. In biological terms, either sex could dominate the other, but it is not determined that either will, nor even that one must do so. Given this, how can the fact that in most societies women are subordinated to men, or at best, equal to them, be explained? Why are men more likely to subordinate women than women to subordinate men? While biology offers no conclusive answers, psychology can shed some light on this question.

Psychological Evidence

In contrast to archaeology and biology, psychology offers convincing explanations of men's propensity to dominate over women. From its earliest days in the nineteenth century until the present, psychology has explored the psychic and social implications of sexual differences. While Sigmund Freud and many of his followers argued that women did not develop as completely as men and must "accept the fact of being castrated," later theories have stressed male fear and envy of women. Instead of seeing women as "castrated men," men have been viewed as "psychosexually frailer" than women.[30] This "frailty" is seen especially in a boy's passage to manhood and in a man's sexual vulnerability. It may explain men's need to subordinate women.

One consequence of female mothering—virtually universal in human societies—seems to be that girls have an easier time becoming women than boys have becoming men. A girl can naturally identify with her primary parent, the mother, and her passage to womanhood is made relatively easy by both this identification and a clear physical signal—menstruation. The girl child's need to separate herself from her mother does not become tied up with establishing her own female gender identity, as the sociologist Nancy Chodorow argues.[31] The boy must break his initial identification with his mother and realign himself with the men of his group. Menstruation and her consequent ability to bear children give a woman an obvious function and value to her group; men have no natural equivalent. Women become women and gain value by natural physical processes: menstruation, conception, childbirth, lactation. Men seem to have compensated by inventing analogous social rituals which mark their passage from boyhood to manhood. Frequently painful and mysterious, these rites of passage often involve symbolic equivalents to menstruation or childbearing: bloodletting, scarification, the courageous bearing of pain. In most cultures, the ritual passage to manhood is more difficult than the passage to womanhood. Men usually become men by performing deeds, by doing rather than being.[32] Such difficulties may make male envy and fear of women more likely than female envy and fear of men.[33]

In addition, adult men are genitally vulnerable in ways that women are not. The male genitals are more exposed than the female's, and male fear of genital injury or castration is present in many cultures.[34] Women seem to have no equivalent fears. While rape exists in many societies, it does not seem to carry the negative psychological weight for women that castration does for men. Women can feign or hide sexual arousal; a man's arousal is quickly evident and often beyond his conscious control. In sexual intercourse, a woman can experience multiple orgasms; a man has a more limited orgasmic capacity. Moreover, a man may be impotent, which among other things will prevent him from inseminating. Even if a woman is not sexually aroused, she is still capable of intercourse, conception, and motherhood. Motherhood gives a woman value and a function; fatherhood is far less significant in most cultures. Women know that their children are their own; men must rely on more indirect proof of their paternity. Even when clear on his role in conception, a man cannot be positive of his paternity. As the psychologist Erik Erikson has written, "Behind man's insistence on male superiority there is an age old envy of women who are sure of their motherhood while man can be sure of his fatherhood only by restricting the female."[35]

For all these reasons, many psychologists have argued that men have a greater fear of—and thus, need to dominate and control—women than women have of men.[36] Women's ability to give birth and greater ease in sexual performance may have led to male awe and worship of them in humanity's distant past, but it is far more likely that it led to envy, fear, and resentment. "Boys and men develop psychological and cultural/ideological mechanisms to cope with their fears without giving up women altogether," writes Nancy Chodorow. "They create folk legends, beliefs, and poems that ward off the dread by externalizing and objectifying women."[37] Such psychological formations account easily for the greater prevalence of female subordination in human societies and the devaluation of women found in the earliest writings of the Greeks, Romans, and Hebrews.

These psychological formations may also account in part for warfare. Women's ability to bear children may have led to a male need to achieve and create in an area where men were clearly superior to women. No area of human endeavor provides this so fully as hand-to-hand combat. For in such fighting, men's 10 percent advantage in size and strength becomes crucial. One root of the origin of warfare may be men's need to act in an area in which their superiority to women and necessity to society were paramount.[38] The role of warfare in human society has primarily been studied not by psychology, but by anthropology. There is also additional anthropological data that bears on the origins of women's subordination to men.

Anthropological Evidence

Anthropology studies contemporary peoples and uses cross-cultural analysis to attempt to explore patterns of human behavior. Extrapolating backward in time to deduce the behavior of prehistoric peoples is problematic. All known modern cultures have been influenced by outsiders (even by the presence of the anthropologists collecting data), and modern ways of living may not be the same as those of prehistoric peoples. But anthropologists have become aware of these problems, and in recent years, earlier generalizations about relations between the sexes have given way to more complex and sophisticated studies. There have been attempts to compensate for past biases in the collection and interpretation of data. As a result, female subordination is increasingly seen as a variable rather than a universal human condition. While no cultures have been found where women dominate, there is ample evidence of societies where the sexes are either "integrated and equal" or "separate and equal."[39] Interest has turned from asserting or denying the universality of female subordination to studying those factors that make it more or less likely. Attention has turned especially to the implications of the sexual division of labor.

All human societies specify certain tasks as appropriate for one sex or the other. The division is not totally arbitrary. In all known human societies, females have primary responsibility for early child care; males have primary responsibility for big-game hunting.[40] But the division of labor alone does not automatically mean that the labor of one sex is more highly valued than that of the other. The division of labor leads to female subordination only when societies are subjected to specific kinds of social stress. The most crucial factor seems to be pressure on the environment, which leads to competition within the group or with neighboring groups for diminished resources. Although the connection between diminished resources and female subordination was neither inevitable nor automatic, female subordination appeared only where there was such ecological and social stress.[41]

Basing her conclusions on a survey of 186 cultures, the anthropologist Peggy Reeves Sanday hypothesizes that in the earliest human societies, before population pressure, women and men may have lived in a relatively egalitarian fashion. But as their numbers grew, and hunger, forced migration, or war against other groups became the means for the group's survival, the tendency to subordinate women became more likely. In such societies, female subordination was rationalized and justified. Women came to be portrayed as dangerous and in need of control. Menstrual taboos appeared, aimed primarily at protecting men from "contamination." Childbirth was treated as a handicapping experience, and the group sometimes practiced selective female infanticide. Males were supposed to be active and aggressive, females passive and obedient. These traits were taught to children and became so accepted that they were often seen as innate and natural to the two sexes. Once in place, female subordination seemed both correct and inevitable. In addition, it was powerfully reinforced by the appearance of warfare.[42]

As resources necessary to survival diminished, groups competed both internally and externally for them. It is probably under such conditions that early warfare developed. In early warfare, weaponry was simple and it was the weapons of the hunt that became the weapons of war. Most societies teach weapons skills only to their male children. Male monopoly of these weapons and the skills to use them can easily lead to male dominance of women, either through action or the threat of force.[43] No culture is known in which women are trained to be as warlike and aggressive as men, and in most warlike cultures only males are urged to be aggressive. Females are trained to be submissive and obedient to males.[44]

Not all cultures, even those who war, develop a system of male dominance. "Whether male dominance is part of the solution to stress depends on the previously prevailing configuration of culture," writes Sanday. But many cultures move from participation in warfare to male dominance to the creation of a warrior culture.[45] In such cultures, war was the most important male activity. In war, the male risked his life for his family and group and thus had a valued function.[46] In warrior cultures, women are subordinated and are valued less than men. But women have accepted such subordination and devaluation. Women have rarely sought to be warriors or to fight to the death.[47] Instead, they have depended on the protection of male warriors. Once male warfare is present, a woman needs protection from other male warriors, especially if she is pregnant or caring for young children. The price for such protection has often been her subordination, but it is a price that must be paid for her own and her group's survival.

The evolution of warfare and the development of some societies into warrior cultures explains the subordination of Europe's women by the time writing occurs. War existed among Europe's prehistoric societies: Most Mycenaean objects found, for instance, are weapons or portray warriors. Once a warrior culture developed, it became an almost inviolable system. It ensured the group's survival in what had become a hostile world. Its values were passed on to future generations and came to be seen as both natural and inevitable. A group's beliefs, stories, and religions justified and glorified war and male warriors. The earliest writings of European culture do this. The Greek epics of Homer (written down in the eighth century B.C.), the Twelve Tables of Ancient Rome (c. 450 B.C.), and the Pentateuch of the Hebrews (written down between c. 1150 and c. 250 B.C.; primarily known today as the first five books of the Old Testament of the Bible) all portray warrior cultures in which the subordination of women is well-established.

Written Evidence

These early writings of the Greeks, Romans, and Hebrews form the matrix of later European culture. The *Iliad* and *Odyssey,* the laws of Rome, the first five books of the Bible, shaped the views of later European generations. These writings remained revered and even sacred long after the Greeks, Romans, and Hebrews had ceased to dominate their regions. Through these writings, Greek, Roman, and Jewish views of women and their roles would be transmitted throughout the centuries to later eras, where they had a powerful and far-reaching influence in European history. The images they provided, the morals they drew, the values they embodied became traditions inherited by European women up to the present day. Earlier Mycenaean and Cretan writings had little later influence because their writings became indecipherable; slightly earlier Middle Eastern texts from Egypt, Sumer, and Babylon had virtually no influence on later European culture. But the Homeric epics, Roman laws, and Hebrew Bible did, and they transmitted the views of warrior cultures in which women are valued less than men and are subordinated to them. These were the values later generations of European women and men inherited.

The basic premise of a warrior culture is that men are intrinsically more valuable and more important than women. Thus, women are supposed to be subordinate to men, and this subordination is rationalized and justified in a variety of ways. Women are supposed to be less powerful than men and so, although these writings often portray powerful women (and in the Greek writings, goddesses), even the most powerful goddess is and is supposed to be subordinate to the most powerful god. As in heaven, so on earth: Powerful mortal women of high rank and status are shown, but they are usually subordinate to an even more powerful man. While women were honored if they remained within the relatively few roles allowed them—wife, mother, widow, and in Greece and Rome, priestess—they are overshadowed by males and their concerns. The first writings of each of these ancient cultures have a male content and focus. War is a major topic, and there is little material on women. What there is portrays some powerful women and goddesses, but they are ultimately controlled by male figures. Still, the existence of any powerful female is important, and their existence has led historians of antiquity to argue that women's status overall may have been higher in these early warrior cultures than in the societies that followed them.[48] The actual status of women in early Greece, Rome, and Israel did not endure, but the images and values presented in Homer, Roman law, and the Hebrew Bible did. Although images of some powerful women were transmitted by these writings, their overall message was that if women were not subordinated to men, danger and even chaos would result.

In the writings of Homer, the need for female subordination can be seen even in the portraits of powerful and attractive goddesses. While each goddess is important and while goddesses compose half the Greek pantheon, there is no goddess who is the equal of either Zeus or Apollo.[49] More limited in their attributes, more restricted in their sexuality, the goddesses probably mirror the accepted view of human women as well. Their images embody men's fears that unless female power is controlled by a male principle, women will be dangerous to men. The images of these goddesses: Hera (Juno), Aphrodite (Venus), Athena (Minerva), Artemis (Diana), Hestia (Vesta), and Demeter (Ceres) in their Greek and later Roman forms continued to be significant centuries later in European culture, long after their worship had ended.

The desirability of their ultimate subordination can be clearly seen in the portrait of Hera in the Homeric writings.[50] Figuring prominently in the *Iliad,* Hera joins with Aphrodite in supporting the Greeks against the Trojans, and her vindictiveness leads to the total destruction of Troy. The goddess of fertility and motherhood, Hera is the patron of marriage. Both Zeus's sister and wife, she is portrayed as beautiful but nagging. Zeus complains that he can keep her "but barely in my power, say what I will."[51] She embodies female beauty and sexual power, which she uses when she decides to seduce her husband so as to influence the war. Making an alliance with the gods Sleep (Hypnos) and Love (Aphrodite), she dresses for the occasion, seeking, as so many European women would in later ages, to succeed by her beauty:

> Hêra,
> having anointed all her graceful body,
> and having combed her hair, plaited it shining
> in braids from her immortal head. . . .
> Then she hung
> mulberry-colored pendants in her earlobes,
> and loveliness shone round her. A new headdress
> white as the sun she took to veil her glory,
> and on her smooth feet tied her beautiful sandals.
> Exquisite and adorned from head to foot
> she left her chamber.[52]

Helpless before this beauty and sexual power, Zeus, "the Father," is "subjugated by love and sleep."[53]

But Hera's victory is only momentary. While the father of the gods is tormented by his wife, it is clear that ultimate power is his. "Fine, underhanded work, eternal bitch!" he threatens, reminding her of his superior physical power,

> Do you forget swinging so high that day?
> I weighted both your feet with anvils,
> lashed both arms with golden cord
> you couldn't break, and there you dangled
> Under open heaven and white cloud.
> Some gods resented this,
> but none could reach your side or set you free.[54]

Zeus's affection is reserved not for his wife, but for his daughter, Athena. Most especially Zeus's child, she is born of no mother, but from his own forehead. She is partly male in nature, combining elements considered both masculine and feminine in these early cultures. The goddess of war as well as of weaving, she often wears male garb. In the *Odyssey,* she takes the form of a man (Mentor), and when she goes into battle in the *Iliad,* she casts off her robe and puts on a shirt, breast armor, and a golden helmet. The goddess of wisdom and agriculture, her costume suggests both sexes. She is usually shown wearing a woman's dress with male breast plate, helmet, and spear. Like a male in many ways, she is seen as the goddess most beneficent to men. Like her sister goddesses Artemis and Hestia, she is a maiden, remaining virginal in all the legends told of her.

Her opposite is Aphrodite, sometimes considered the child of Zeus, sometimes born from the sea foam. The goddess of love and beauty, Aphrodite's power is the power of sexual attraction and erotic madness. Even the gods cannot withstand her when she chooses to use it, and its power to disrupt order and hierarchy is graphically portrayed. Aphrodite gives her "pieced, brocaded girdle" to Hera to enable her to seduce Zeus: "Her enchantments came from this allurement of the eyes,/ hunger of longing, and the touch of lips/ that steals all wisdom from the coolest men."[55] Nor are women immune: Helen's flight to Troy was inspired by Aphrodite, and her influence on both mortals and immortals was seen as irresistible for good or ill.

Artemis, Hestia, and Demeter appear hardly at all in the *Iliad* and *Odyssey;* however, they are the subjects of Homeric hymns and figure in early Greek art. Artemis the huntress is Apollo's twin, but she has far fewer powers and attributes than he. Virginal and childless herself, she presides over menstruation and childbearing. Associated with the moon and the bear, she embodies a wild female power. Hestia is the opposite: Represented only as a flame, she is the center and spirit of the home. "She has her place in the center of the house to receive the best in offerings," says the Homeric hymn to her.[56] The *Homeric Hymn to Demeter* shows the power of "fair-crowned Demeter" as both the goddess of grain and a mother. Able to win her daughter back from Hades, the king of the underworld, for two thirds of the year, Demeter embodies the power of fertility and motherhood.

Goddesses needed priestesses to lead their worship, and in both Greece and Rome a few women served as priestesses, although the preferred role for them was that of wife and mother. Those few might possess a great deal of power. The six Vestal Virgins, mentioned in the oldest laws of the city of Rome, were exempted from the male guardianship applied to all other women. The priestesses of Vesta (Hestia), they guarded the city's sacred flame, spending thirty years of their lives as virgins in the service of the goddess and acquiring influence among the city's leaders.[57] Greek women also served as priestesses to goddesses like Athena and Hera. But the Hebrews worshipped one male god, and masculine pronouns are used with "God" in the original texts. He is served exclusively by male priests with one possible exception—the figure of Miriam, Moses' sister. Miriam leads the women of Israel in song after the Red Sea is crossed, but when she later challenges Moses' authority, God punishes her with leprosy.[58]

The preferred and most common role for all women above the rank of slave in these Greek, Roman, and Hebrew writings was that of wife and mother. As a warrior's wife, mother, or daughter, a woman had a valued and honored role, so long as she remained within the family and marriage. In Homer's *Odyssey,* Penelope, the faithful wife, who remains loyal to Odysseus even after twenty years of separation, is extolled as an ideal daughter, wife, and mother. She is

a valiant wife!
True to her husband's honor and her own,

Penélopê, Ikários' faithful daughter!
The very gods themselves will sing her story
for men on earth—mistress of her own heart,
Penélopê![59]

Penelope's steadfastness in resisting her numerous and importunate suitors is rewarded by a happy reunion with her husband. They both weep for joy and spend the night talking and making love. Odysseus himself extols the glories of marriage: "The best thing in the world" is "a strong house held in serenity/ where man and wife agree."[60] Even in the *Iliad,* which concentrates on war rather than home, the marriage of Hector and Andromache is consistently presented as a love match and provides a counterweight to the marital discord of Hera and Zeus.

But like the goddesses, women were supposed to be ultimately subordinated to men. Honored if they remained within the duties allotted to wives and mothers, they were chastised if they ventured beyond them. The most famous scene between Andromache and Hector takes place on the ramparts of Troy, where she urges him not to risk his life and her future by leading an attack beyond the walls. "Lady, these many things beset my mind no less than yours," he replies,

But I should die of shame
before our Trojan men and noblewomen
if like a coward I avoided battle,
nor am I moved to. Long ago I learned
how to be brave, how to go forward always
and to contend for honor, Father's and mine.[61]

Hector concludes by telling Andromache to leave war to him; she should return to her spinning. At first, this passage may seem to imply separate but equal roles for women and men: She spins at home, he fights abroad. However, this is a situation where Andromache's own life is at stake. Regardless of a happy marriage, her opinion is dismissed, and she is told to tend to her woman's work. The situation and even the words are paralleled in the *Odyssey,* when Telemachus dismisses his mother Penelope from the crucial contest with her suitors:

Return to your own hall. Tend to your spindle.
Tend your loom. Direct your maids at work.
This question of the bow will be for men to settle,
most of all for me. I am master here.[62]

Penelope is the most admired woman in Homer's works. Even she is subordinated to her son in a vital issue where her life and interests are directly involved (if a suitor wins, she will be forced to marry him). Honored if they remained within the relatively small territory of home and family, women in warrior cultures were also confined to that territory.

In Hebrew writings as well, women are defined in relation to men and are seen as the objects of male desire. In the book of Genesis in the Bible, Adam calls Eve "bone of my bones, and flesh of my flesh" and it is stated that a man "cleaves to his wife and they become one flesh."[63] Abraham's wife Sarah is "a woman beautiful to behold."[64] Abraham's son, Isaac, "loved" his wife Rebekah, who "alighted from the camel" when she first saw him; Isaac's son Jacob serves fourteen years to win the "beautiful and lovely" Rachel, "whom he served seven years for . . . and they seemed but a few days because of the love he had for her."[65] Each of these women is described in terms of her maternity, but their ability to birth children is dependent on God, who has the power to "open" their wombs and enable them to conceive. Eve is given her name by Adam because it means "mother of all living." Sar'ai (Sarah) laughs when God tells her she will bear a son at ninety, but it happens, and she becomes as God predicted, "a mother of nations." Isaac's prayers that Rebekah conceive are fulfilled with twin boys; Rachel's plea, "Give me children or I shall die," is answered with a son, Joseph.[66] Barrenness was a woman's failure. When Rachel conceives, she says, "God has taken away my reproach." Conversely, mothers were supposed to be respected. One of the Ten Commandments given Moses is to "honor your father and your mother."[67]

Despite such precepts, women in warrior cultures were inevitably valued less than men. "Women" is an insult in the *Iliad* and the Old Testament, applied to cowardly men.[68] Even within the family, women were less significant. The lengthy genealogies in Homer and the Old Testament rarely give the names of mothers and present lineage solely through the male line. In traditional Roman custom, only boys were given individual names; girls bore just the family name and were distinguished from each other by nicknames. A law attributed to Romulus, the legendary founder of the city of Rome, compelled citizens to rear "every male child," but only "the first-born of the females."[69]

Being valued less contributed to and supported a system of female subordination. In Homer's writings, in Roman law, in the Old Testament, women are explicitly subordinated to men, and this relationship is justified in a variety of ways. The fourth of the Twelve Tables of Rome is called Paternal Power and gives the father sole and absolute authority over his children. In the fifth table, it is specified that "because of their levity of mind" all women (except the six Vestal Virgins) shall be under a man's guardianship; both guardianship and inheritance passed through the male line to male relatives. *Paterfamilias,* the father of the family, was an important legal concept in ancient Rome; there was no female equivalent—*materfamilias* was an honorary title given to the father's wife, but carried no power or rights.[70]

In the Old Testament, a woman's word stands only if her husband or father do not disagree with her; otherwise, their views prevail. In a valuation of females and males at various stages of life, males are consistently valued higher: 3 shekels for a girl baby, 5 for a boy; 30 shekels for a woman, 50 for a man. In the story of Lot, Lot's goodness is demonstrated in part by his refusal to give two male guests to an angry mob; he offers his two virgin daughters instead.[71]

Although two versions of the creation of human beings are given in Genesis, the more egalitarian one, in which God created "man in his own image . . . male and female created he them" tended to be ignored in favor of the fuller, older version. There, Eve is created from Adam's rib, and it is she who is "beguiled" by the serpent into disobeying the Lord. The principal basis for female subordination is given in God's punishment of her:

> I will greatly multiply your pain in childbearing;
> in pain you shall bring forth your children,
> yet your desire shall be for your husband,
> and he shall rule over you.[72]

God punishes Adam "because you have listened to the voice of your wife" and so broke God's rule about not eating the fruit of the tree of knowledge of good and evil.[73]

Female subordination is also expressed in the double standard of sexual behavior allowed women and men. The practices of these early warrior cultures catered to male sexuality. Normally, the men of a defeated group were slaughtered and the women enslaved. These slave women were then used by the victorious warriors as concubines and servants, and male sexual use of any slave woman was taken for granted. The precipitating cause of action in the *Iliad* is Agamemnon's decision to take Achilles' slave girl, Briseis, as a replacement for his own Chryseis, who had been ransomed by her father. But slave women were themselves not allowed such freedom. In the *Odyssey,* Telemachus hangs the slave women who had sex with Penelope's suitors; their crime is seen as disloyalty as well as unchastity.[74]

In the Laws of Rome and the Old Testament, women were condemned and punished for behavior allowed men. Laws attributed to Romulus specified that a husband could kill his wife for adultery or wine drinking, and a Roman history of that early period states that when Egnatius Metellus "took a cudgel and beat his wife to death because she had drunk some wine, not only did no one charge him with a crime, but no one even blamed him."[75] Another Law of the Kings stated that wives were not allowed to divorce their husbands, but that husbands could divorce wives "for the use of drugs or magic on account of children or for counterfeiting the keys or for adultery." The Twelve Tables mention only the husband's right to repudiate his wife.[76]

In the Old Testament as well, only the husband has the right to divorce, and a divorced woman is described as "defiled."[77] There is an elaborate test called "the law of the jealousies," which allowed a husband to check if his wife had been faithful. The wife was given the "water of bitterness that brings the curse"—if she drank it and nothing happened, she was considered innocent; if "her body shall swell and her thigh shall fall away" she was considered defiled and "a curse among her people."[78] There was no reciprocal test of a husband's fidelity. While the Hebrews were unusual among warrior cultures in calling for the death of both adulterers, woman and man, female unchastity was more severely punished in some instances. If a woman were not a virgin, she could be stoned to death; if a man raped a virgin, he could make restitution by paying her father and marrying her.[79] In the early books of the Old Testament marriage demands fidelity from the wife, but not the husband; the husband commits adultery when he has sex with another man's wife, not when he has sex with another woman. There are numerous instances of respected patriarchs having two wives or a wife and a concubine. Abraham has intercourse with Hagar because his wife Sarah is barren; Jacob marries Leah and Rachel and takes two of their maids as concubines as well.[80] This double standard is embodied in the language used to describe husband and wife; the husband is the *baal* or master of his wife rather than just her husband.[81]

In Greek and Roman law as well as the Old Testament, prostitutes, but not the men who used them, were condemned as outcasts and inferiors. A Roman law from about 700 B.C. forbade a "concubine" to touch the altar of Juno, and Athenian laws from the fifth century B.C. excluded prostitutes from many women's religious cults.[82] In the Old Testament, there are both slave concubines and prostitutes. "Whore" and "harlot" are used as terms of ultimate abuse, and some biblical verses called for prostitutes to be burned to death, especially if they were priest's daughters.[83] There is no condemnation of the men who bought their services.

In addition, the Old Testament classified all healthy adult women as ritually impure when they menstruated. In a section dealing with disease and bodily discharges, menstruation is described as "her impurity" and makes a woman and everything she touches or sits on unclean for seven days. (Ejaculation, in contrast, rendered a man unclean only until the evening of the day it occurred.) If a man touches her, he is unclean until evening; if he has sex with her, he is unclean for seven days. Her period makes a woman automatically unclean for seven days; if she has other bleeding or discharge from her vagina, then she, like a man with a bodily discharge caused by disease, is unclean until seven days after it ended.[84] If a woman gives birth to a son, she is unclean for seven days and cannot go to the temple for thirty-three days; if she gives birth to a daughter, she is unclean for fourteen days and cannot go to the temple for sixty-six days.[85] This ritual impurity turns both menstruation and childbirth, especially of a daughter, into contaminating events. These laws have the effect of devaluing women as well. Although menstrual taboos were not so strict or widespread in Greece and Rome, they are described in the writings of early scientists like Aristotle and Pliny.[86]

Historians of women have noticed that, although the earliest writings of Greece, Rome, and Israel contain and justify female subordination, they are not misogynist; they do not express a hatred of women in general. That came later: in the Greek poetry of Hesiod and Semonides, writing in the seventh and sixth centuries B.C.; in the satires and poetry of first century A.D. Imperial Rome; in Jewish and Christian interpretations of the Old Testament ranging from the second century B.C. to the third century A.D.[87] Then women would be stigmatized as innately evil. The creation of woman would be seen as a punishment for man, and woman would be identified as the enemy of both men and civilization. She would be seen as the source of disease and trouble, be equated with various despised animals, be described in the most disgusting of terms. Female subordination, with its lesser valuation of women, led easily to such misogyny. Exceptional circumstances might favor other roles and images for women. Imperial Rome also saw the achievement of great female power; some Jewish and Christian authorities stressed women's equality; a century after Hesiod's diatribes against women, the poet Sappho was honored for her verse. But female subordination remained in place. It is the first and most important tradition inherited by women when a truly European culture coalesced in the ninth century A.D.

2

INHERITED TRADITIONS: THE PRINCIPAL INFLUENCES

A DISTINCT European culture emerged with the coronation of Charlemagne in 800 A.D. and his creation of an empire in the early ninth century. Writers and artists, philosophers and theologians used ancient sources and texts for prescriptions about women and perpetuated or revived ancient views of women. This new Europe drew on Greek, Roman, Celtic, Germanic, Hebrew, and Christian sources for its attitudes and institutions. Used ahistorically, uneven in their influence, these ancient sources continued to shape European culture and society for centuries.[1]

Greek images and beliefs about women were both preserved and transformed by the Romans, who carried them to the outer reaches of their vast empire. Roman attitudes toward women, chiefly in the form of laws and customs, remained influential in Europe long after the collapse of the Roman Empire in the course of the fifth century. The new invaders, the Celtic and Germanic peoples, learned writing from the Romans and Christian missionaries and eventually recorded their own laws and epics in which women figured. The Hebrews, scattered throughout the Mediterranean world in the diasporas of the Roman period, continued to influence the new European culture and its views of women through the acceptance of their sacred writings as the Old Testament of the Christian Bible. Christian influence, initially favorable to women but later restrictive, grew unevenly from the first century on. Persecuted by the Roman Empire in the third century A.D., Christianity became the Empire's preferred religion in the fourth century. Between then and the ninth century, it spread by conquest and conversion across the continent and the new Europe of the ninth century identified itself as Christian. Christianity both perpetuated old traditions about women and added new customs.

Before the advent of Christianity, women's experience varied greatly within different cultures. Greeks of the fifth century B.C. contrasted the circumscribed lives of Athenian wives with the freer life of Spartan women. A wealthy Roman matron of the first century A.D., able to transact business without a male guardian, bore little resemblance to her own slave woman. A Jewish woman living in cosmopolitan Alexandria in the second century A.D. had little in common with her biblical counterpart in pastoral Judea a millennium earlier. Celtic carvings and Germanic grave sites reveal wealthy, powerful women absent from the dry paragraphs of the law codes. The lives of women in these early cultures have been pieced together by historians. While much of their experience did not endure and had no influence on later European culture, many traditions concerning women were preserved and perpetuated. Despite differences of culture, law, and circumstance, the Greek, Roman, Celtic, Germanic, and Hebrew traditions about women are more similar than they are different. The bulk of the pre-Christian traditions inherited about women subordinate and limit them, seeing women as inferior to and dependent on men. Some of these pre-Christian traditions, however, consist of images and eras in which women's equality, power, and independence flourished. Never totally erased, they too form part of the legacy of Europe's women.

Christianity, largely because of its initial emphasis on the equality of all believers and its divergent views about sexuality, came to change a few of these earlier traditions about women. Within Christianity itself, women were allowed a far wider range of activities in the early years of the religion than in its later centuries. Although women's roles were severely restricted when the Christian Church became institutionalized in the fourth century, the wider roles for women mentioned in the New Testament remained part of the European cultural heritage. The pattern of freedom followed by restriction, set in the early Christian Church, provided a model for later Church actions regarding women.

It is these sources—Greek and Roman, Celtic and Germanic, Jewish and Christian—which furnish the traditions inherited by European women when Europe coalesced in the ninth century. Many of them, whether restrictive or liberating, remained part of European culture for many centuries, and in some cases, until the present. They are the foundation of attitudes and circumstances that governed most European women's lives; they form the basis of customs and laws; they lay behind the views which both subordinated and empowered women in the succeeding centuries. These traditions comprise the rest of this section. Since the sources for societies before the ninth century are limited and uneven, the focus in the remainder of this section is not on women in each culture, but rather on those traditions pertaining to women that shaped European women's lives in the ninth century and beyond.[2]

3

TRADITIONS SUBORDINATING WOMEN

THE EARLIEST, and in many cases, most sacred writings of the cultures before Christianity—the Greek, Roman, Hebrew, Germanic, and Celtic—enshrined female subordination. As the centuries passed, belief in female subordination endured and acquired the authority of hallowed tradition. Female subordination limited women's roles and function; it defined their essential nature and the proper use of their bodies. It passed intact to the new European culture, which emerged in the ninth century.

"All men's thoughts have been shaped by Homer from the beginning," a Greek wrote in the fifth century B.C., and this remained true in future centuries. Close to half the papyri found in Egypt between the fourth and fifth centuries B.C. were Homeric works or commentaries on them. Homer's epics continued to influence Europe's culture.[1] Roman law was based on early precepts and cited precedents as justifications. Celtic and Germanic peoples did the same. In 643 A.D., when King Rothair of the Lombards wanted laws written down, he investigated "the old laws of the Lombards known either to our self or to the old men of the nation," and this corpus was added to in the following centuries.[2] For the Hebrews especially, military defeat and colonization by Greece and Rome led to increased reliance on "the book," which distinguished them as a people, so that both the Old Testament and the commentaries that developed on it built on the basic ideas about women expressed in the Pentateuch. In each case, the basic premises of female subordination—that women were by nature dependent on and inferior to men—were repeated and elaborated. As time went on, these premises came to have the power of axioms: They seemed natural, inevitable, and in some cases God-given. Each of these cultures argued that woman's physical body—her menstruation, her uterus, her ability to give birth—by definition excluded her from war, law, government, and much of religion. Each argued that woman's body necessitated that she be confined to the protected sphere of the home if possible. Each gave the men of her family (or male guardian) authority and power over her, and each saw her life as almost exclusively connected to the family. Each valued women less than men, sometimes so much so that more boys were raised than girls. Each excluded women from the activities they deemed most important, whether it was warfare, philosophy, or the study of sacred books.

The premise of women's innate inferiority was hardly ever questioned, even by thinkers who questioned other basic premises of life. "The male is by nature superior, and the female inferior; and the one rules and the other is ruled," stated the Greek philosopher Aristotle in the fourth century B.C. "This principle of necessity extends to all mankind." This inequality is "permanent" because the woman's "deliberative faculty" is "without authority," like a child's.[3] Adapted by later Jewish and Christian thinkers, Aristotle remained the scientific authority for Europe into the sixteenth century.[4] Like the Greek philosopher, Roman jurists stood by traditional premises where women were concerned. In the first century B.C., Cicero argued that "because of their weakness of intellect" all women should be under the power of male guardians; in the third century A.D., Ulpian asserted that guardians were necessary for all women "on account of the weakness of their sex and their ignorance of business matters."[5] In Celtic Ireland, law codes ranked women as "senseless," like slaves, prisoners, and drunks.[6]

Jewish thinkers built on the same premises about the female's inferior nature. In the first century A.D., Philo, a Jewish philosopher influential among both Jews and Christians, commented on Genesis:

> The soul has, as it were, a dwelling, partly women's quarters, partly men's quarters. Now for the men there is a place where properly dwell the masculine thoughts [these are] wise, sound, just, prudent, pious, filled with freedom and boldness, and kin to wisdom. . . . And the female sex is irrational and akin to bestial passions, fear, sorrow, pleasure and desire from which ensue incurable weakness and indescribable diseases.[7]

In the same era, the Jewish historian Josephus stated that in the eyes of Jewish law, "The woman . . . is in all things inferior to the man."[8] In Hebrew law, this inferiority implied incapacity. Women, children, and slaves were classed together and women were forbidden to testify "because they have light, i.e. flighty, minds."[9] By the first century A.D., Jewish men in their morning prayers thanked God "who has not made me a woman." The equivalent female prayer was to thank God for having "made me according to thy will."[10]

In the Mishnah, the collection of Jewish legal commentaries, there are two sections which deal specifically with females, one on women and one on menstruation.[11] The classification of menstruation as unclean, based on Leviticus, remained fully in force in later centuries and was the subject of commentaries, which elaborated the rules about contamination from menstruating women. Women's periodic "impurity" was further reason for excluding them from religious duties within the temple. The belief that healthy women become contaminating once a month, because of a natural process that cannot be controlled, inevitably contributes to the view that women are by nature inferior to men.

Other cultures offered their own evidence that women's menstruation demonstrated their basic inferiority to men and their unfitness for male roles. In the Celtic epic, *Táin Bó Cualinge* (*The Cattle Raid of Cúchulainn,* written down in the seventh or eighth century), Queen Medb (Maeve) of Cruchan is presented as a powerful, beautiful, and wealthy woman. A formidable warrior and general, she seems an equal opponent for the hero Cúchulainn, until the climactic end of the epic. Then, while in battle, Medb gets "her gush of blood" and has to leave her war chariot to "relieve herself." Cúchulainn, the hero of the epic, comes up behind her at this point and captures her. The scene of the defeat "is called Fual Medba, Medb's Foul Place, ever since," states the epic.[12]

Greek and Roman scientific and medical texts often described menstruation as a mysterious, dangerous, and contaminating event. The male physicians whose writings formed the influential Hippocratic Corpus (largely from fourth-century B.C. Greece) described the menses as blood which might wander through the body; it might cause consumption if it entered the lungs.[13] The Corpus tended to assume that menstruation was controlled by the moon and that all women menstruated at the same time of the month, a belief perpetuated by Aristotle.[14] All manner of supernatural powers were ascribed to menstrual blood. Aristotle wrote that a menstruating woman could make a clean mirror "bloody-dark, like a cloud" because the menstrual blood passed through her eyes to the surface of a mirror.[15] Aristotle also believed that women had fewer teeth than men, and such glaring errors in the work of a normally careful observer point to Aristotle's prejudices about women overriding his evidence.[16] Pliny the Elder, the first century A.D. Roman authority on natural history, wrote that women menstruated more heavily every third month. He also perpetuated popular beliefs about menstrual fluid, some of which would remain current for centuries in Europe:

> Contact with it turns new wine sour, crops touched by it become barren, grafts die, seeds in gardens are dried up, the fruit of trees falls off, the bright surface of mirrors in which it is merely reflected is dimmed, the edge of steel and the gleam of ivory are dulled, hives of bees die, even bronze and iron are at once seized by rust, and a horrible smell fills the air; to taste it drives dogs mad and infects their bites with an incurable poison. . . .[17]

The persistence of such beliefs in the face of ample evidence to the contrary effectively denigrated those processes and organs unique to women. Beliefs about the uterus and reproduction were even more derogatory. Greek and Roman men who wrote on science and medicine took the male as the standard and saw the female as an inferior variant. "The female is as it were a deformed male," asserted Aristotle in his treatise on reproduction, "and the menstrual discharge is semen, though in an impure condition; i.e., it lacks one constituent, and one only, the principle of Soul."[18] Galen, the eminent physician of the second century A.D., argued that women were men turned outside in. The ovaries were "smaller, less perfect testes" and woman's lack of "perfection" compared to man was explained by the need of the species to reproduce.[19]

Taking the male organs as standard and viewing the female organs largely by analogy, these ancient writers enshrined the belief that the womb "wandered" around the body like an "animal." As the penis becomes "rebellious and masterful, like an animal disobedient to reason, and maddened with the sting of lust," wrote the fourth-century Greek philosopher Plato,

> the same is the case with the so-called womb or matrix of women; the animal within them is desirous of procreating children, and when remaining unfruitful long beyond its proper time, gets discontented and angry, and wandering in every direction through the body, closes up the passages of the breath, and, by obstructing respiration, drives them to extremity, causing all varieties of disease.[20]

This theory of the "wandering womb" proved remarkably persistent. Mentioned a number of times in the Hippocratic Corpus of the fourth century B.C., it was strenuously restated by Aretaeus of Cappodocia in the second century A.D., who wrote that "on the whole, the womb is like an animal within an animal" because it roamed about the body but could be attracted to its rightful place with sweet odors.[21] In both Greece and Rome the symbol of the erect male organ signified good fortune and was often placed before homes and in gardens; the symbol for the female genitals identified brothels.

Seeing the uterus as a revolting "animal within an animal" contributed to the denigration of woman's role in conception, also a common feature of Greek and Roman writings on reproduction. In Aristotle's writings and the Hippocratic Corpus, the woman is seen almost exclusively in her reproductive role and yet her contribution to reproduction is usually considered far less important than that of the man, whose "strong sperm" will produce a boy and "weak sperm" a girl.[22] Barrenness was the woman's fault, however. These views that men were primarily responsible for conception contributed to and were often connected by ancient authors to women's innate inferiority. Plato began his section on the "wandering womb" by explaining that men were created first; women were the offspring of those men "who were cowards or led unrighteous lives" and were thus a symbol of the degeneration of the human race.[23] A female was a female because of her "inability to concoct semen," an inability Aristotle thought came from the "coldness of her nature." Menstrual fluid was deficient semen; woman was a deficient man, able to supply only the "matter" of a fetus, to which the superior male contributed "form" and "soul."[24]

The belief that women were "cold" and "moist" while men were "hot" and "dry" came from Hippocrates; as in Aristotle, "cold" was seen as inferior and was used to prove the female's inferiority to the male. "The female is less perfect than the male for one, principal reason," wrote Galen in the second century A.D., "because she is colder."[25]

Valued less than men, women survived less often to adulthood than men. In Greece and Rome especially, there is evidence that fewer girls were raised than boys. The normal ratio of girls to boys at birth is 100 to 105. In both the city of Athens and the Roman Empire, there are population figures which show the ratio skewed far to the male side. Ancient Greek censuses, lists of citizens and gravestones show a far larger number of males than females, and a recent estimate argued that 10 percent of the female babies of the city of Athens were not raised by their families.[26] Even taking into account underreporting of female births and deaths, censuses from the Roman Empire of the second to fourth centuries A.D. show women greatly outnumbered by men: in Roman Egypt, 100 women to 105 men; in Roman Spain, 100 women to 126 men; in Rome itself 100 women to 131 men; in Roman Italy and Africa, 100 women to 140 men.[27]

Infants were rarely killed outright; rather, they were "exposed"—left outside, on garbage heaps or in public places, in the hope that a passerby might rescue them. Brothel owners collected infant girls and raised them to be prostitutes.[28] In Rome, evidence for the routine exposure of female infants is direct. From as early as the Law of the Twelve Tables, which required a father to raise all his sons, but only one daughter, to a history written in the early third century A.D., which remarks that "there were far more males than females" among the Roman nobility of Augustus' era, the practice of raising more boys than girls was taken for granted.[29] "I beg and beseech you to take care of the little child, and as soon as we receive wages I will send them to you," a Roman husband wrote his wife, Alis, in Egypt in 1 B.C. "If—good luck to you!—you bear offspring, if it is a male let it live; if it is a female, expose it."[30] In both Greece and Rome, girls and women were assumed to need less to eat than boys and men, and so they were usually given less. In the fourth century B.C., Xenophon wrote that outside of Sparta, Greek "young girls who are destined to become mothers and considered well brought up are stinted in their basic diet and eat as little as possible of other foods."[31] In Rome, food assistance was provided far more often and for longer periods of time to boys and men than girls and women. The Roman bread dole was only for men, and the records of a typical fund in northern Italy of the second century A.D. show 246 boys aided, but only thirty-five girls.[32] Even where the effort was made to support equal numbers of girls and boys—as in a fund established in the second century A.D. by the wealthy heiress, Caelia Macrina—girls' grants were between 20 and 40 percent less than boys and ended two years sooner.[33] These practices and the hazards women faced in childbirth shortened women's life expectancy. While women under conditions of relative equality tend to outlive men by a factor of 10 percent, throughout antiquity women's life expectancy was five to ten years shorter than a man's.[34] The system of female subordination deprived some women of life itself.

Outside Greece and Rome, infanticide and exposure were condemned, but girls and women continued to be valued less than boys and men. Both the Hebrews and the Celts set monetary equivalents for their people to serve as a measure of compensation in case of injury or death. Both set values for females lower than those for males. Of the Germanic tribes, the Lombards, Burgundians, and Anglo-Saxons valued women equally with men of the same status. Among women, variations came because of the ability to bear children. For example, among the Salian Franks, the wergild (valuation) of a girl before menarche and a woman after menopause was a third that of a menstruating woman.[35]

Women's Approved Roles

Those women who grew to maturity in these cultures found in place a complete system of values and institutions, which assumed they would function only in their approved roles of daughter, wife, mother, and widow. Women were defined by the family and, within the family, by their relationship to the men of the group. Life within the family protected and supported a woman. Laws rewarded and institutionalized her dependence and subordination. Each of these early cultures explicitly excluded women from activities outside the family, activities these cultures valued most. Powerful cultural messages reinforced this division of roles and activities, excluding women from the important areas assigned to men: warfare, government, philosophy, science, law, and in some cases religion. In classical Greek culture, the male was identified with civilization, reason, and order, the female with nature, emotion, and chaos. Men could be expected to apply reason and logic to life, to control emotion and instinct; women would give in to impulse and selfish needs.[36] Although some Greek playwrights created sympathetic and powerful women in their dramas, classical scholars still debate whether Athenian women were even permitted to watch these plays. Most often, the notion of a woman attempting to assume male roles was considered intrinsically humorous, as in Aristophanes' comedy *The Congresswomen (Ecclesiazousae).* Pericles' famous Funeral Oration, in which the Athenian leader extolled male bravery and pursuit of glory in war, advised women that the greatest glory "will be hers who is least talked of among the men whether for good or for bad."[37]

Roman culture similarly disapproved of women in men's roles and continued to base itself on the power of the father. In Vergil's *Aeneid,* Aeneas becomes a hero and a worthy founder of the city of Rome in part by resisting the temptation posed by Dido, a powerful foreign queen. Instead he chooses the docile and obedient Lavinia as his wife. She is appropriately silent, in the manner urged on Greek and Roman women, and never utters a word in the epic. The expansion of elite Roman women's economic and political power during the early Empire (first century B.C. to first century A.D.) became a major criticism of the Empire itself; only a degenerate government, argued the historians Livy and Tacitus, would have allowed women such power.[38]

Among the Hebrews, her sex excluded a woman from the study of the Torah and the Talmud (the Pentateuch, its commentaries, and the laws), which was a primary religious obligation of Jewish men. Although Rabbi Ben Azzai argued, around 200 A.D., that a father had an obligation to teach his daughter Torah, subsequent practice supported the opinion of Rabbi Eliezar that "whoever teaches his daughter Torah is in effect teaching her lasciviousness."[39] The Palestinian Talmud argued that it was "better the words of Torah were burned than put into a woman's keeping" and the Babylonian Talmud asserted that "There is no wisdom in woman except with the distaff" [used to spin yarn].[40] Women's function was to keep a home which abided by Jewish law and to pass traditional practices on to their children. Mothers educated daughters until they were married and sons until they reached the age of seven and went on to men's schools. But even within the home, a Jewish woman was subordinated to the male's dominance in religion. Able to light the Sabbath candles, a woman was forbidden to say the prayers blessing the bread she had baked or the wine she had poured. The Talmud says "cursed be the man who lets his wife recite the blessing for him on Friday night."[41]

Celtic and Germanic cultures excluded women from warfare, the most honored male activity. One Lombard law of the seventh century specified that if a woman participated in a brawl, no penalty fine need be paid if she were injured, since "she had participated in a struggle in a manner dishonorable to women."[42] In the Celtic epic, the *Táin Bó,* Queen Medb's beauty, leadership, and heroism in battle cannot override the fact of her womanhood, made graphic by her menstruation. When she has been defeated, she speaks with Fergus, up to that point her loyal lieutenant:

> "We have had shame and shambles here today, Fergus."
> "We followed the rump of a misguiding woman," Fergus said. "It is the usual thing for a herd led by a mare to be strayed and destroyed."[43]

Excluded from the public powers reserved to men, women in these early cultures had their own proper roles and functions. Defined in relation to men, women were categorized primarily by their sexual activity with men. From the earliest writings of these cultures, there is a strict division between good women and bad women in this regard. A good daughter was a virginal daughter. When she lost her virginity (and remaining virginal for life was an option only for the tiny number of priestesses whose service required virginity), she became either a wife—a good sexually active woman—or a prostitute—a bad sexually active woman. A wife had sex with one man, which defined her as chaste, a prostitute with many, which defined her as wanton. "A woman's greatest virtue is chastity," stated a Pythagorean text from third to second century B.C. Italy, and this sentiment was repeated in all these cultures.[44] A major inherited tradition of European women was that they would be defined as good or bad, proper or improper, respectable or outcast, because of their sexual connections with men.

Virginity and chastity were seen as inextricably connected to obedience. Training a girl to be obedient, especially to her father, would ensure that she would maintain her proper behavior within the family as a virginal daughter and later as a chaste wife and mother. Laws institutionalizing her subordination to men set the pattern of restrictions for the European culture to come. Within the family, a woman had little more authority than a child. She was always subject to her nearest male relative, who had authority over her person and property. Marriage meant the transfer of this authority from one male to another. "No free woman who lives according to the law of the Lombards . . . is permitted to live under her own legal control, that is, to be legally competent" went the seventh-century Lombard law stating this. "But she ought to remain under the control of some man or her king."[45] Typical of laws in all these early cultures, guardianship of women was assumed to be the best way of maintaining order in the family and society. The ideal woman subordinated her feelings, instinct, and judgment to her father, husband, or male guardian. The ideal woman in all these early cultures was a woman who willingly subordinated herself to the men of her family.

Other laws and legends sought to ensure a daughter's obedience. Greek dramatists saw the conflict between a daughter's obedience and ethical behavior as tragic: Iphigenia must die as a sacrifice to raise a wind for her father's fleet; Antigone must die if she is true to the demands of her religion and buries her brother against her uncle's order. Roman law gave the father power to kill his daughter if he thought it fit. Not used in historic times, this power was the focus of a number of legends. Livy recounted the tale of Horatia, who mourned the death of her brother's enemy. He killed her, and the murder was upheld when their father stated in court that Horatia deserved to die.[46] Hebrew scholars focused on Eve's disobedience in eating from the tree of knowledge as the source of evil. Lombard law stated that if a daughter opposed her father or brother, he could do whatever he wished with her property.

Most of the writings and legends about daughters in Greek, Roman, Hebrew, Celtic, and Germanic cultures focus on the girl's virginity. In Homer's *Odyssey,* Princess Nausicaa, an ideal young woman, tells Odysseus that

> I myself should hold it shame
> for any girl to flout her own dear parents,
> taking up with a man, before her marriage.[47]

Solon, the lawgiver of sixth-century B.C. Athens, forbade all sales of children into slavery except the sale of a girl who had lost her virginity.[48] A daughter's virginity was inextricably connected to her family's honor, and intercourse without her guardian's approval tainted that honor. "Virginity is not entirely yours," went a wedding hymn by Catullus, the Roman poet of the first century B.C. "One third your father owns, one third your mother,/ one third alone is yours: Don't fight with two/ who've sold a son-in-law their rights to you."[49] Roman law called for the death of a daughter who lost her virginity before marriage; Hebrew law allowed the death of a betrothed daughter who lost her virginity to a man other than her fiancé before the marriage; she could be stoned to death for "playing the harlot in her father's house."[50] Both Roman and Hebrew law sought to distinguish between cases where a girl could have avoided rape by crying for help without risking her life, and those where she could not be expected to save herself. In the latter case, she was seen as an innocent victim and not punished; in the former, she could be executed.[51] Both Roman and Hebrew writers used the wanton daughter as a symbol of their culture's decline; in the book of Isaiah in the Old Testament, for instance:

> The LORD said:
> Because the daughters of Zion are haughty
> and walk with outstretched necks,
> glancing wantonly with their eyes,

> mincing along as they go,
> tinkling with their feet;
> the Lord will smite with a scab
> the heads of the daughters of Zion,
> and the LORD will lay bare their secret parts.[52]

Germanic laws fined the man who made sexual advances or gestures to a warrior's daughter. Like Greek, Roman, and Hebrew laws, they viewed the loss of virginity outside marriage as an injury to the guardian, not to the young woman. The compensatory fines should she be so "corrupted" went to him.[53] Willingness on the part of the young woman negated the fine.[54]

The few fragments of ancient writing by women which deal with virginity reveal their acceptance of these attitudes. Temple dedications, especially to Artemis, show that Greek women regularly made an offering just before their marriage. Often this included the toys of her girlhood, which would end when she married and lost her virginity legitimately:

> Timareta before her wedding has dedicated to you, Artemis of the Lake, her tambourine, her pretty ball and the caul that upheld her hair, her dolls too, and their dresses: a virgin's gift, as is right to a Virgin. Artemis, hold your hand above the girl and purely preserve her in her purity.[55]

Part of a wedding hymn by Sappho, the Greek poet of the seventh century B.C., has a bride ask, "Virginity O my virginity/ Where will you go when I lose you?" A second voice answers "I'm off to a place I shall never come back from."[56] Another fragment reads "Do I still long for my virginity?"[57]

The ideal daughter would be beautiful as well as obedient and virginal. Descriptions of youthful feminine beauty are remarkably uniform in Greek, Roman, Celtic, and Germanic writings. The ideal features included fair hair, pale skin, and red lips. Found in Greek novellas, Roman poems, and Germanic epics, such descriptions echoed the following one, which is Celtic:

> The color of her hair was like the flower of the iris in summer or like pure gold after it has been polished. . . . White as the snow of one night were her hands, and her lovely cheeks were soft and even, red as the mountain foxglove. . . . Her eyes were as blue as the hyacinth, her lips as red as Parthian leather. High, smooth, soft and white were her shoulders, clear white her long fingers. . . .[58]

Beauty was honored in itself; in a daughter, it would aid in making a good match.

The expected goal of all women who were not slaves was marriage, and in both ancient Greek and Hebrew, the word for "woman" was also the word for "wife." They went together in all these cultures. Menarche, the start of menstruation, meant the girl was ready for marriage, and in the ancient world, as throughout European history, menarche usually occurred between twelve and fourteen years. Families with property were eager to make the alliance marriage created with other families as soon as possible.[59] Various laws forbade marriage when the bride was under twelve; an eighth-century Lombard law specified that the date should be at the end of her twelfth year "because we know there have been many controversies over this matter, and it appears to us that girls are not mature before they have completed twelve years."[60] Among the elite of these cultures, there was often a disparity in the ages of the couple. Women's deaths in childbirth ensured that men remarried far more often; at his second or third marriage, the groom would almost always be far older than the bride. The marriage of a girl with a man who was her senior by a number of years was often considered desirable. It both mirrored and perpetuated the ideal of wifely subordination within the marriage. Aristotle said the ideal marriage was between a bride of eighteen and a groom of thirty-seven.[61] Jewish wives called their husbands "my master"; husbands called their wives "my daughter."[62] The husband would guide and teach his wife; the wife would bear healthy children for her husband.

The arrangements for the marriage of a daughter in these early cultures set the pattern for the centuries to come. There was nothing casual about the transfer of authority over a young woman from one man to another. The bride's family and the groom or his family negotiated the marriage. There was concern that the young woman not be forced to marry against her will, especially in Hebrew, Celtic, and Germanic cultures.[63]

In addition to the transfer of authority, marriage involved the exchange of gifts, compensation to the bride's family for her loss, and provision for her in the new relationship with her husband. In the wealthier classes of some of these early societies, the bride came to the marriage with a portion of her family's wealth.[64] In Greece and Rome, the bride's family gave her a dowry, which would be returned to her or her male guardian if the marriage ended, through divorce or death.[65] Among the Hebrews, the bride received a payment from her husband on marriage that was written into the marriage document (ketubah) specifying her husband's obligations to her.[66] Celtic and Germanic custom usually also involved a gift from the groom to the bride, to be used for her support when he died. These cultures also provided a gift from the groom's family to the bride as well. Increasingly, this "bride gift" became her property. In Lombard law of the seventh century, the bride's father retained the gift; a century later, it was largely hers.[67] Germanic brides received a third payment, the *morgengabe* or "morning gift," given to her by her husband after the night on which the marriage was consummated. For a queen, the *morgengabe* could be substantial. In the sixth century A.D., for instance, the Merovingian King Chilperic I gave his queen, Galswintha, the cities of Limoges, Bordeaux, Cahors, Béarn, and Bigorre as a *morgengabe.*[68]

All these cultures expected the bride to bring personal possessions to begin her new household: linens, clothing, kitchen utensils—and in royal or monied circles, these could be lavish and substantial.[69]

Both the betrothal (the occasion of the agreement on the marriage contract) and the marriage (the joining of the couple) were considered events worthy of celebration, and feasting and drinking followed the ceremonies. Each culture had rituals and symbols thought to bring good fortune and fertility to the union, and many passed later into European culture. For instance, the custom of placing a ring on the third finger of the bride's left hand came from Rome; it was believed that a nerve ran directly from that finger to the heart. Though marriage was viewed first as a social and economic union, these early societies passed on to European culture the expectation that the wife and husband would find affection and pleasure together.

Love might not be expected, but even in an arranged marriage it was hoped for, and a number of Greek and Roman epitaphs testify to love between wife and husband. "Theodorus, my husband," reads one from an Athenian colony, "I pray that late though it be, you will come and I shall meet you and we shall share our bed, so that we shall forget our misfortune."[70] A plasterer in Roman Lyons raised a tombstone "to the eternal memory of Blandinia Martiola . . . his wife incomparable and most kind to him." She died at eighteen and had married at thirteen. "You who read this," the epitaph ended, "go bathe in the baths of Apollo as I used to do with my wife—I wish I still could."[71]

Epitaphs and funeral tributes give evidence of a wifely ideal, which some deceased women were said to have fulfilled. "Praise for all good women is simple and similar, since their native goodness and the trust they have maintained do not require a diversity of words," stated a Roman son at his mother's funeral in the first century A.D.:

> My dearest mother deserved greater praise than all others, since in modesty, propriety, chastity, obedience, woolworking, industry and honor she was on an equal level with other good women, nor did she take second place to any woman in virtue, work and wisdom in times of danger.[72]

The good wife causes her husband's "property to grow and increase, and she grows old with a husband whom she loves and who loves her, the mother of a handsome and reputable family," declared a poem from Greece of the sixth century B.C.[73] A similar, but more detailed tribute comes from the Hebrew Book of Proverbs in the Old Testament:

> A good wife who can find?
> She is far more precious than jewels.
> The heart of her husband trusts in her,
> and he will have no lack of gain.

> She does him good, and not harm,
> all the days of her life.
> She seeks wool and flax,
> and works with willing hands.

The wife's work is described further: She provides food for her household and gives her servants their orders; she buys a field, plants a vineyard, and makes a profit; she spins and makes clothing, some for her household, some for sale.

> Strength and dignity are her clothing,
> and she laughs at the time to come.
> She opens her mouth with wisdom,
> and the teaching of kindness is on her tongue.
> She looks well to the ways of her household,
> and does not eat the bread of idleness.
> Her children rise up and call her blessed;
> her husband also and he praises her:
> "Many women have done excellently,
> but you surpass them all."[74]

The Celts and Germans emphasized the wife's role as the warrior's supporter. In the epics, mothers of heroes like Ness of the *Táin Bó* or Frigga in Norse legend protected their warrior sons. An eighth-century Anglo-Saxon riddle told how the "loyal kinswoman" cared for the warrior: [She] "wrapped me in clothes and kept and cherished me,/ enfolded me in a protective cloak/ As kindly as she did for her own children. . . ."[75] Warriors' wives were expected to celebrate victories, as the Danish queen in the Germanic saga *Beowulf* does by providing a lavish banquet. A warrior's wife was expected to mourn the death of her spouse; as in earlier centuries, his defeat could mean her enslavement or capture.

In these early cultures, the ideal wife was expected to be sexually faithful to her husband, to be fertile and bear healthy children, preferably boys, to manage the household or perform its duties herself, and to support her husband's military endeavors. Always seen in relation to her husband, her life was expected to be spent ministering to his needs.

All of these early cultures decreed that, just as a daughter should be virginal, so a wife must be chaste. Just as a daughter's loss of virginity dishonored her father, so a wife's infidelity dishonored her husband. Of greatest concern was the chance of her becoming pregnant by another man and her passing off the illegitimate child as her husband's. "She gave me children just like myself," was a common tribute in wives' epitaphs from Roman husbands. All these early cultures used both coercion and praise to ensure the wife's fidelity. Adultery remained primarily a woman's crime; a man could commit it only by having sex with another man's wife, not with another woman. All had harsh laws to punish a woman's sexual infidelity. In Athens, the male seducer was also guilty; the husband could kill him and cast out the offending wife. "The lawgiver seeks to disgrace such a woman and make her life not worth living," explained an orator in the fourth century B.C.[76] Hebrew law gave the husband the right to kill both the woman and the man.[77] Roman law became less harsh as the centuries passed; initially the penalty was death, later it changed to exile, later still to loss of property to compensate the dishonored husband.[78] Celtic and Germanic laws allowed murder or execution of the offending wife or, as in the case of the Anglo-Saxons, allowed the husband to discard her and to collect her monetary value (wergild) from her lover. Thus, the husband could negotiate for a new wife.[79]

Wifely fidelity was also encouraged by cautionary fables, which extolled chaste women. The Roman legend of Lucretia and the Hebrew tale of Susannah and the Elders figured prominently in European art and literature in future centuries. Both stories were remarkably similar. In each, a beautiful young wife is imperiled by an evil man (or men) who threaten to lie and say that she has already had sex with them in order to force her to give in to their sexual demands. Susannah proclaims her innocence and is saved by divine intervention. Lucretia, who yielded after being threatened with a dishonorable death—her royal antagonist declared he would show her family her corpse and that of a male slave in bed together—redeemed herself by suicide, an honorable death in Roman culture. Even though innocent, she dies rather than "provide a precedent for unchaste women to escape what they deserve."[80] The moral was clear: A virtuous wife should die before being unfaithful to her husband.

This fidelity and exclusivity did not apply to the husband. Except among the Hebrews, where a husband's infidelity was disparaged in the centuries after 800 B.C., a double standard prevailed, and husbands were routinely expected to have sex not only with their wives, but with slavewomen and prostitutes.[81] "We [male citizens of Athens] have courtesans for the sake of pleasure, concubines for the daily care of the body, and wives to breed legitimate children and be a trustworthy guard of possessions indoors," stated the orator Demosthenes in the fourth century B.C.[82]

The wife's duty to breed legitimate children remained one of her most important functions. All these cultures assumed that childlessness was the woman's fault, and all allowed a husband to divorce her for her supposed barrenness.[83] A barren woman was a pathetic figure in cultures that valued fertility, and in Greece, Rome, and Israel, barren women sought divine aid: praying and sacrificing to goddesses associated with fertility in Greece; participating in fertility rituals and cults in Rome; beseeching God for children in Israel. Since fertility was associated with divine approval, the barren woman was doubly disfavored and was looked down upon by women as well as men. Hannah's "rival used to provoke her sorely, to irritate her, because the Lord had closed her womb," states Samuel I in the Old Testament, and even her husband's attempt to comfort her, asking, "Am I not more to you than ten sons?" could not console her.[84]

In all these early cultures, divorce was relatively easy for men to obtain, especially on the grounds of the wife's adultery or barrenness. The tradition prevailed even after the Christian Church sought to end all divorce in Europe. In classical Athens, divorce was usually initiated by the husband.[85] Early Roman law gave the right to divorce only to husbands. Women could divorce by the last century B.C., but there is evidence that such divorces were often engineered by the wife's father. Easy divorce primarily benefited men in the family, who sought to establish new political alliances by marrying the women of their allies.[86] Divorce remained unequal throughout Roman history; a fourth-century law of the Emperor Constantine allowed a man to divorce his wife for adultery, procuring, or poisoning; he could also divorce her on "light grounds" if he agreed to give up her dowry and not remarry for two years. The wife could divorce a husband who was a murderer, a poisoner, or a grave robber, but if she acted on any other grounds, she was deported.[87] Among the Hebrews, divorce was the sole prerogative of the man.[88]

In addition to being faithful and fruitful, the wife had a third responsibility to her husband. She must see to the household. These early cultures passed on to European society the traditional standard that a good wife saw to her family's basic needs, whether she was the wealthiest Roman matron supervising her slaves or the poorest member of a Germanic clan working her household garden. Early cultures freed their male elites from work, but even in the most exalted families, women were expected at least to supervise their slaves and perform woman's traditional labor, spinning and "making wool."[89] When Helen is first seen in the *Odyssey,* her servant brings in "the silver basket/ rimmed in hammered gold, with wheels to run on. . . . heaped with fine spun stuff, and cradled on it/ the distaff swathed in dusky violet wool."[90] Her yarn and tools are royal, her distaff is golden, but she is still expected to spin yarn like a poor woman, whom Homer invokes in a simile in the *Iliad:*

> Think of an honest cottage spinner
> balancing weight in one pan of the scales
> and wool yarn on the other, trying to earn
> a pittance for her children: evenly poised
> as that were these great powers making war.[91]

Throughout antiquity, cloth making was always women's work, so much so that in later European cultures, the "distaff side" signified the females of a family. Spinning is "the work they understand . . . [and] the work considered most honorable and suitable for a woman," stated the fourth-century B.C. Greek historian, Xenophon.[92] Roman epitaphs for wives used the statement "She kept the house and worked in wool" so often that it became a routine and expected tribute.[93] The Jewish housewife in the Book of Proverbs "puts her hands to the distaff, and her hands hold the spindle."[94] By the late Roman Empire, women were so often associated with textiles that the word *gynaecea* (women's places) was used in legal contracts to designate weaving, spinning, and dyeing establishments.[95] Of all this labor, what remains is only the tradition of women making cloth. "Of the millions of yards of stuff woven in the thirteen centuries our period covers," write the authors of *Art of the Ancient World,* "only a few square inches survive so that we have practically no direct information of what they were like."[96]

The vast majority of women in these cultures worked the land. One tradition inherited by European women was that they would be paid less than men when they worked for wages, even when they performed the same labor. As early as the eighth century B.C., the Greek poet Hesiod advised a farmer on how to save money. In winter, he should "let go the hired man; hire a childless girl."[97] In addition, women were expected to cater to their husbands. "The poor, not having any slaves, must employ both their women and children as servants," wrote Aristotle in the fourth century B.C.[98]

In the cities, women performed a wide variety of activities to earn income. Women's "double burden"—performing housework and childcare *and* earning extra income—is another tradition which dates back to these early cultures. Most of these means of earning became traditional female occupations as well. Women sold food, clothing, and trinkets in the public markets and streets. They provided food and lodging. They managed brothels and worked as prostitutes. They hired themselves out as wet nurses (wealthy Greek and Roman women often hired poor women to suckle their children for them). They also worked as midwives; throughout the centuries, European women usually gave birth at home, assisted by other women. The presence of a trained midwife was sought after. Male doctors were only called in for dire emergencies or to supervise births in the wealthiest families. By the second century A.D., the Roman doctor Soranus specified the characteristics of a good midwife:

> She will be unperturbed, unafraid in danger, able to state clearly the reasons for her measures, she will bring reassurance to her patients, and be sympathetic. . . . She must love work in order to persevere through all vicissitudes (for a woman who wishes to acquire such vast knowledge needs manly patience).[99]

Soranus described another technique which European midwives used throughout later centuries: measuring the dilation of the cervix with their fingers.

Women worked at such occupations only in the poorer classes; wealthy families sought to ensure their daughters' freedom from such labor by providing her with a dowry. In addition to managing her husband's household, a wife was expected to be a political and dynastic link between families. In times when small numbers of noble families or a particular dynasty ruled, marital alliances to bolster or protect political power became common. Most often in such cases, the man's needs were dominant, and he could take new wives (either divorcing the old or adding to his previous tally) as he deemed necessary to preserve his political interests. The Macedonian, Ptolemaic, and Seleucid dynasties of the fourth to first centuries B.C. used women this way; so did the powerful Roman families of the late Republic and early Empire (the first centuries B.C. and A.D.). Ambitious men of that era made and unmade betrothals and marriages to mirror and confirm their shifting political allegiances. Julius Caesar broke a childhood engagement to Cossutia when he was fifteen, because she did not belong to the politically powerful equestrian class. His first marriage, to Cornelia, allied him with her powerful father, L. Cornelius Cinna, who had been consul in Rome. After Cornelia's death, Caesar married Pompeia, the daughter of another powerful consul, but divorced her when she was suspected of adultery. (Caesar himself was known for his numerous affairs with both women and men.) This left him free to make another political alliance; his third wife, Calpurnia, was the daughter of another important consul. Caesar sealed his political alliance with Pompey by marrying his daughter Julia to him—she became Pompey's fourth wife.[100]

Of the Germanic cultures, the Franks forced alliances and acquired wealth by abducting women and marrying them. Charlemagne, emperor of the Franks in the eighth and ninth centuries, had four wives successively chosen and discarded either for reasons of state or because he disliked them. His grandsons each married a woman to help in controlling the third of the empire he was to inherit. Lothar took an heiress of Alsace, Pepin a wife from the Chartres area, and Louis a kinswoman from East Austrasia.[101]

Occasionally, powerful women of these cultures attempted to manipulate this system of dynastic alliances for their own ends. They parlayed their marriages into bids for power, if not for themselves, then for their husbands, sons, or brothers. But a woman who attempted to do this moved beyond the allowed limits. She was acting contrary to her culture's role for her and was subject to the severest criticism and punishment if her hold on power faltered. The tradition the lives of such women transmitted was the danger of letting a woman seize political power. This legacy can be clearly seen in the life of

Agrippina II (15–59 A.D.), the Roman empress whose career demonstrates both the achievements and limits of this sort of power.

The daughter of a granddaughter of the Roman Emperor Augustus and his adopted grandson, Agrippina II was raised in the highest political circles. Married at thirteen to the son of a powerful consul, she bore him a son—the future Emperor Nero—when she was twenty-seven. That same year, her brother became the Emperor Caligula. He exiled Agrippina II and her sister; a few years later, Agrippina's husband died. In exile for four years, she returned to Rome after Caligula's death and married her second husband, a wealthy senator who left her a fortune when he died. Agrippina was thirty-two then, and she made her boldest bid for power by courting the new emperor, Claudius, her uncle, whose powerful wife Messalina had also died recently. She managed to persuade Claudius to make her his empress and fourth wife in 49 A.D.—a marriage which required a special senatorial decree because of its incestuous nature. Claudius gave her a great deal of power; he allowed her to influence his decisions, ordered that she be called "Augusta" (the first empress to be given this title while still alive), and made her son his heir.

Five years later, Claudius was dead—perhaps poisoned, perhaps by Agrippina. Her seventeen-year-old son became emperor and, for the first years of his reign, allowed his mother power; she advised him on crucial matters, listened to delegations from behind curtains, and was portrayed on coins equally with him. He then turned against her. She went to live with her first husband's mother, and it was there that she wrote her memoirs, which have been lost. In 59 A.D., five years after his accession to the throne, Nero arranged for his powerful mother's murder. She was forty-four, and her dying words were supposed to have been "Smite my womb!" in recognition that her son had ordered her death.[102]

In his history of the era, Tacitus commented on Agrippina's death: "Of all things human the most precarious and transitory is a reputation of power which has no strong support of its own."[103] Women bidding for dynastic power found that this power ultimately lay with men, who could rule without them if necessary. Moreover, women who achieved influence in dynastic and political circles were severely criticized, both in their own day and for centuries afterward.[104] Roman historians linked female influence to violence and decadence, as did Gregory of Tours in his account of the sixth-century Frankish queens, Brunhild and Fredegund. Their bloodthirstiness and treachery are emphasized as they are described deceiving men, murdering, and inciting war to keep themselves or their kin in power. Like Agrippina II, Brunhild, the surviving queen, is shown dying horribly; she was dragged to death at seventy.[105]

By attempting to gain political power, a woman contravened the activities allowed a wife and mother. When she entered the political arena, she became anomalous and exposed. The women who seized power in these early dynasties left their granddaughters at best an equivocal heritage; the memory of female power was always overlaid with strictures against it, and the lives of powerful women were used to illustrate the dangers of permitting women to exceed their proper roles and functions by venturing into areas meant only for men.

Instead, the ancient ideal for wife and mother held firm. From the earliest records, this wifely ideal had been extolled by men. As far as can be told, it was also accepted by women; within the home, as a cherished wife, a woman achieved the safest and most secure existence possible. The role of the chaste wife and mother, who was prolific, hardworking, and devoted to her housekeeping and her family remained one of the most powerful traditions inherited by later generations of European women.

Women's Dishonorable Roles: Slave, Prostitute, and Concubine

As old as the tradition of respect for the good wife was the tradition of contempt, fear—and desire—for the woman who broke the rules and had sex with more than one man. As a slave woman, a prostitute, or a concubine, a woman was cast out from the traditional protections accorded chaste wives and mothers. Women who used their sexuality to augment their power were stigmatized as prostitutes no matter how high their social rank; the Romans called Cleopatra *regina meretrix,* "the prostitute queen," because of her sexual liaisons with Julius Caesar and Mark Antony.[106] Hebrew prophets often made no distinction between prostitution and adultery, seeing in women's independent sexual behavior an ultimate betrayal and subversion of values. "Plead . . . that she put away her harlotry from her face, and her adultery from between her breasts," wrote Hosea. Hosea, Isaiah, and other Hebrew prophets of the period of exile used "harlotry" to condemn Israel's religious apostasy—"the land commits great harlotry by forsaking the LORD."[107]

To men, such women were simultaneously appealing and dangerous. One reason powerful women were criticized was for using their sexual attractiveness to influence men. From the earliest writings of these cultures, men had expressed fear of the power of women's sexual attraction over them. The solution of these early cultures to this problem consisted in trying to divide women into two separate and distinct categories: the wife and the whore. A wife should be obedient to her husband and follow his lead, even in the bed. Independent female sexuality was stigmatized as characteristic of a prostitute. In Plutarch's *Moralia,* for instance (a first to second century A.D. Greek work generally positive about women's capacities and talents), Plutarch states that a wife

> ought not to shrink away or object when her husband starts to make love, but not herself to be the one to start either. In the one case she is being over-eager like a prostitute, in the other case she is being cold and lacking in affection.[108]

Slave women were the most vulnerable; the earliest writings of all these cultures show slave women serving warriors' sexual needs. But as sexual partners to the men of the household, slave women were also in a position to use their sexual attractiveness to increase their status and security. Greek and Roman slave women might eventually gain their freedom; Celtic and Germanic slave women might advance beyond the position of concubine. The sixth-century Frankish Queen Fredegund, for instance, began life as a slave, attracted King Chilperic from his first wife, and married him. Such women naturally incurred suspicion and distrust from the wives and daughters of the household. Functioning as household prostitutes, they undermined the hard-won status of the virginal daughter and the chaste wife.

Slave women were not condemned for having sex with their masters, only for trying to influence them because of it. European men attempted to control women's sexual power over them by confining it. They restricted the wife and regulated the prostitute. (The line between slave and prostitute blurs in these early cultures; slaves could function as household prostitutes; prostitutes were often slave women.) Not only did men in these early cultures classify any woman of independent sexuality as a prostitute, they condemned the prostitute as a despicable and dishonorable woman. A good woman had sex only with her husband or master; a woman who had sex with more men threatened the social balance. This attempt to restrict women's sexuality, however, conflicted with the double standard of sexual behavior, which allowed men in these early cultures sex with women other than their wives. The result was the institution of female prostitution, in which women provided sex to men in exchange for money, but were simultaneously despised and stigmatized for doing so. This tradition endured into the later European centuries, as did many other aspects of prostitutes' lives.

Institutionalized prostitution in cities is present in the earliest records of the Greeks and Romans. Solon, the sixth-century B.C. Athenian lawgiver, was supposed to have established a large brothel so that "young men could thus work off their lust without disordering families."[109] "If there is anyone who thinks that young men should be forbidden to make love, even to prostitutes, he is certainly a man of stern righteousness," argued Cicero on behalf of a client in first-century B.C. Rome.

> But he is out of touch not only with the free life of today, but even with the code and concessions which our fathers accepted. For when was that not customary? When was it blamed? When was it not allowed? When was it not lawful to do what is a lawful privilege?[110]

Celtic and Germanic cultures had no need for prostitutes; men used slave women for sex and practiced polygamy when it suited them. Little stigma attached to the men who bought sex from women.

The women themselves were stigmatized, marginalized, and regulated. This tradition endured into the later European centuries. In Athens, prostitutes were forbidden to participate in public religious ceremonies and sacrifices allowed other women; in Rome, they were excluded from important women's cults. A free-born Roman man was forbidden to marry a prostitute or the freed slave of a prostitute or procurer, but a male procurer could become a full Roman citizen. In Hebrew writings, the word "harlot" was female and was used to signify a low and degraded state: "How the faithful city has become a harlot, she that was full of justice!" wrote the prophet Isaiah about Jerusalem.[111]

In Rome, another tradition about prostitutes was established. They were forbidden to wear a respectable matron's dress and instead were forced to don the short tunic and toga usually worn by men.[112] Throughout subsequent European history, men attempted to make prostitutes instantly identifiable by their clothing or some signal of costume; it was justified on the grounds that this would protect respectable women from being approached by men, and would make it easier for men to find a prostitute for hire. Both Greek and Roman cities also founded the tradition of making money off prostitution; magistrates and states established brothels and then taxed them heavily. License fees were also established and became lucrative sources of revenue.

Liable to physical violence and disease, prostitutes also faced the hazards of unwanted pregnancy. In the first century B.C., the Roman writer Lucretius referred to "movements used by prostitutes" in their attempts to avoid pregnancy. Some ancient contraceptives (especially those which blocked the cervix or acted as spermicides) may have been effective, but most were not.[113] In the second century A.D., the Greek writer Lucian wrote a "Dialogue of the Courtesans" which realistically presented a pregnant prostitute:

> And I so near my time! Yes, that is all I have to thank my lover for; that, and the prospect of having a child to bring up; and you know what that means to us poor girls. I mean to keep the child, especially if it is a boy. . . .[114]

While prostitutes remained generally despised and vulnerable, a talented and quick-witted woman could rise to the better role of courtesan. Famous not only for their sexual services, but also for their social skills as hostesses and companions, courtesans moved in the highest circles. In classical Athens, courtesans were the only women permitted to participate in the dinners and symposia which formed a large part of the elite Greek male's social life. Plato praised the intelligence of Aspasia, Pericles' mistress, and she was considered the "most famous woman in fifth-century [B.C.] Athens."[115] A clever woman in the right circumstances could turn her sexual connection with an important man into a secure and even powerful future. Although this happened infrequently, this dream fascinated future generations of poor European girls. Theodora (c. 497–548), for instance, the daughter of a dancer and bear keeper at the Hippodrome in Constantinople, rose in this way from prostitute to Empress of the Roman Empire in the East.

Theodora's father died when she was four, and she soon joined her older sister in performing as a dancer and mime. (Actresses and actors were lower caste in the Roman Empire and were usually ranked on the same level as prostitutes.) Famous for her wit, she performed as a comedienne at private parties. She became the mistress of an administrative official and lived with him briefly in Egypt before returning on her own to Constantinople.[116] It was during this period of her life that she met men on the highest level of society: various religious thinkers, church officials, and the future heir to the throne, Justinian.

Justinian, himself forty, unmarried, and nephew to the Emperor, fell in love with her. To make his marriage with her possible, he amended the old laws which forbade free men to marry prostitutes. When he succeeded to the throne in 527, two years after the marriage, Theodora became his coruler, Empress of the Eastern Roman Empire. Justinian called her "his sweetest delight" and described her as "partner in my deliberations."[117] She shared in his plans and political maneuverings and participated in councils of state, "always apologizing for taking the liberty to talk, being a woman."[118] She had her own imperial seal, her own official entourage, and her own court. During the Nika Revolts of 532, she persuaded Justinian not to flee, shaming him into staying by her speech in the Council of State:

> If I had no safety left but in flight, I still would not flee. Those who have worn the crown should never survive its loss. I'll never see the day when I am not hailed as Empress. Caesar, if you wish to flee, well and good, you have money, the ships are ready, the sea is clear. But I shall stay. I love the proverb: Purple is the best winding sheet.[119]

Theodora was unique in her sympathy for both women and prostitutes. She was instrumental in passing legislation which gave women more property rights and was responsible for an edict which made pimping a criminal offense and banished all brothel keepers from the city. She founded a convent for former prostitutes and was known for buying girls who had been sold into prostitution, freeing them, and providing for their future. Her mosaic portrait (at the church of San Vitale, in Ravenna, Italy) shows her royally bejeweled, wearing an elaborate crown, collar, earrings, necklace, and bracelets. She is robed in black and stares straight with no expression on her face. At fifty-one, she died, and little effective legislation was passed by Justinian after her death.

The most detailed and factual source for Theodora's life is Procopius' contemporaneous *Historia Arcana.* In it, Theodora is presented as a sexual monster, a woman for whom no deed was too infamous, no sexual act too revolting. She is described as at the mercy of her whims and emotions and as a ruthless manipulator of men.[120] She is called a "demon" and is shown presiding over the castration of a young man. All the early accounts of her life which have been preserved stress the lowliness of her origins and the horror of her slow death from cancer. Despite her obvious abilities and influence, it was impossible for men to write realistically about her; moral lessons had to be drawn to discourage other women from the same path. Condemnation, even of the most highly placed courtesans, remained part of European traditions about women sexually active outside marriage.

In male writings about prostitutes in these early cultures, fantasies superseded reality, and the fantasies also endured in later European culture. There was the fantasy that the prostitute was "honest" and really had "a heart of gold," which made her preferable to a coy girl from a respectable family. There was the fantasy that the prostitute would be so pleased by her client that she would not charge her customary fee.[121] Greek plays of the fourth and third centuries B.C., which long remained popular with Roman audiences as well, often showed the hero falling in love with a slave prostitute, who, it turns out, is a virgin from a higher caste who has been accidentally enslaved. The hero rescues her and gains a beautiful and grateful bride.

These were the benign fantasies. Others more malign also became enduring traditions in European culture. Much Roman literature argued that within every respectable woman there lurked a whore, ready to satisfy her lust when the opportunity presented itself. The story of the wife of Ephesus, who has intercourse with the soldier guarding her newly dead husband's tomb, was told and retold in subsequent centuries. (It first appeared in Petronius' *Satyricon.*) Wanton women and prostitutes were major subjects of Roman satire. More specifically, Roman poets singled out the old prostitute for castigation in the most revolting and obscene terms. "Your teeth are black and furrows scar/ With wrinkles all your forehead's length!" wrote Horace in the first century B.C., in a typical example:

> Your filthy private gapes between
> Shrunk buttocks like a scrawny cow's;

> Your chest and wizened breasts are seen
> Like horses' teats, and flabby shows
> Your belly, and your lank thighs strung
> To swollen calves, provoke my wrath![122]

Invective against women, especially old prostitutes or women who dared to be sexually active, was a major subject in Latin literature. These poems express fear and hatred for all women, especially of women's supposed excessive sexuality. Lust drives women to crimes against men: betraying their husbands, prostituting themselves for sex rather than money, being unfaithful to their lovers. In these poems, no crime is too revolting for women to commit; they urinate on altars, conspire with their mothers-in-law to cuckold their husbands, and even have sex with animals to satiate their lust.[123] With such images, the world of reality is left far behind, and we enter a nightmare realm where women are seen as men's enemies. This set of horrifying images is the final tradition that worked to subordinate European women. These early cultures passed on a substantial body of misogynist literature which comprised an important strand in the European cultural heritage.

Misogyny

"Misogyny" means hatred of women, and it was present not only in Roman literature, but in Greek and Hebrew writings as well. These cultures blamed the first woman for bringing evil into the world. At the end of the eighth century B.C., the Greek poet Hesiod wrote an account of creation that supplied images to later European culture. In it, man is created first and lives happily until, in punishment for Prometheus' stealing fire, Zeus creates the first woman, Pandora. Beautiful to look at, she is "the hopeless trap, deadly to men."

> From her comes all the race of womankind,
> The deadly female race and tribe of wives
> Who live with mortal men and bring them harm,
> No help to them in dreadful poverty
> . . . .
> Women are bad for men, and they conspire
> In wrong, and Zeus the Thunderer made it so.[124]

In a second poem, *Works and Days,* Pandora, "this ruin of mankind," opens her "box" and unleashes "pains and evils" among men.[125] Similar to the biblical Eve in her failure to be obedient, Pandora is blamed for all the evils of the world. Later Hebrew writings, possibly because of Greek influence, tended to make Eve into such a figure as well: a woman who gave in to temptation and did that which God had forbidden—eating and persuading

Adam to eat from the Tree of Knowledge of Good and Evil. "From a woman sin had its beginning," wrote Joshua ben Sirach in the second century B.C., "and because of her we all die."[126]

The tradition of blaming women and considering the best of women to be inferior to men was well-established in the ancient world. "Better is the wickedness of a man than a woman who does good; and it is a woman who brings shame and disgrace," declared ben Sirach. Even earlier, in the third century B.C., "the Preacher" had stated woman is "more bitter than death." "One [righteous] man among a thousand I found, but a woman among all these I have not found."[127] The Hebrews also created the legendary figure of Lilith, formed simultaneously with Adam to be his first wife, before the creation of Eve. Lilith refused to be subservient, left Adam, and took vengeance by menacing children and infants.[128]

Monsters imagined by these early cultures became traditional images in later European literature and art. The Greeks created numerous female monsters: Circe, who turned men into swine; Scylla, the nymph transformed into a menacing rock, with snake's and dog's heads growing from her; Charybdis, the deadly whirlpool; and the Sirens, who lure men to death with their beautiful songs. Hesiod tells of Echidna, half girl and half snake, who gives birth to Cerberus, the dog of hell, of the Chimera, of the Sphinx, and other monsters who menace male heroes like Hercules and Jason.[129] He also mentions the Furies (Erinyes), three inexorable females who relentlessly punish wrongdoers. Later Greek myths supplied the Harpies—ravenous, filthy birds with women's heads—and the Gorgons—hideous women with snakes for hair, whose gaze turns men to stone. Medusa, the Gorgon Perseus killed, is portrayed in Greek sculpture as a grotesque face with her tongue hanging down to her chin. The male hero of Greek legends moved through a landscape thronged with female monsters, whom he must defeat or outwit in order to survive. The same was true of Germanic cultures. The Vikings believed in Sigurd, a giant daughter of the gods associated with the underground world of death; starvation was her knife, famine her table, and care her bed.[130] In the epic, Beowulf proves his heroism in part by defeating Grendel's Mother, the old hag, in an underwater battle.

The literature of these early cultures also presents ordinary men in a similarly embattled stance, usually with regard to their wives. They portray wives as idle and vengeful, shrewish and unscrupulous, as figures to be conquered or derided. Semonides of Amorgos wrote a famous satire in the sixth century B.C. classifying women into ten types. Only one, the bee, is good for man. The others are the sow "who sits by the dungheap and grows fat," the "wicked vixen," who knows nothing, the bitch, who is "always yapping," the sea woman, who "raves," the ass, who "puts up with everything," the ferret, who "makes any man she has with her sick," the horse who "proves a plague," and, worst of all, the monkey who is "hideous" and does "the worst possible harm."[131]

Semonides and Hesiod transmuted the hatred of misogyny into the mockery of humor. The nagging, wasteful, wanton wife, who is always talking and is a torment to her husband, became a staple of European humor in the centuries to come. Hesiod argued that a man "couldn't live with 'em and couldn't live without 'em." If a man did not marry, he faced a "miserable old age" with no children to care for him; if he did marry, he "lives all his life/ With never-ending pain."[132] By the fourth century B.C., making fun of wives had become a standard theme in the comedies performed in Greece and later Rome. "I wish the second man who took a wife would die an awful death," went a fragment from fourth century B.C. Athens. "I don't blame the first man; he had no experience of that evil."[133] Husbands complain that their wives never stop talking, long for the day their wives will die, and try to be rid of them if possible. Plautus' popular comedy *The Twin Menaechmi* concludes:

> Menaechmus sells his property
> Cash and no delay
> All must go—house and lot
> Slaves and furniture
> Wife goes too
> if anyone takes a fancy to her.

Misogyny has no female counterpart in European culture. Women almost certainly made jokes about men and probably had their own stereotypes about men's nature. But these were not written down, and so did not become a tradition inherited by later generations. Male humor about women was not balanced by female humor about men in the written record and literature generally. Repeated over and over, misogynist stereotypes came to appear valid and accurate portrayals of women. The following passage, from the Greek playwright Aristophanes, was written in 392 B.C., but was echoed in writings of later centuries:

> Women roast their barley sitting, just as they always have . . . they bake cakes, just as they always have; they nag their husbands, just as they always have; they sneak their lovers into the house, just as they always have; they buy themselves little tit-bits, just as they always have; they prefer their wine unwatered, just as they always have; and they enjoy a good fuck, just as they always have.[134]

For women, the most powerful traditions in Western culture have been those which try to subordinate and denigrate them.

4

TRADITIONS EMPOWERING WOMEN

TRADITIONS SUBORDINATING women were powerful, but not all-powerful. These early cultures also contained images and beliefs that glorified and empowered the female. In addition, exceptional circumstances occasionally allowed women of unusual ability to rise to prominence in fields usually reserved to men. The lives and accomplishments of Sappho, Deborah, Cleopatra, and Boudica are also part of the traditions inherited by European women. Centuries after the old empires and kingdoms had disappeared, women would take inspiration from these female images and figures.

Worship of Goddesses

All these early cultures except the Hebrews gave supernatural powers to female figures.[1] The chief god in the Greco-Roman and Germanic-Celtic cultures was male, but all these groups also postulated powerful female forces who controlled the lives of heroes and sometimes even the gods.

Fate in these early cultures was female, and the ultimate power of life and death, of destiny and necessity was personified as a woman or a group of three women. Associated with time, the Fates were often perceived as spinning, weaving, and cutting the thread of men's lives, transforming cloth making, the most commonplace and ordinary female activity, into a symbol of ultimate power. To the Greeks, the three Fates were Clotho (the Spinner), Lachesis (the Disposer of the Lots), and Atropos (the Inflexible), who cut the thread of life with her shears of death. In Germanic beliefs, the three Fates were the Norns, who sewed the web of fate and cast lots over the cradle of every baby. They were Urda, Verdandi, and Skuld, and represented the past, the present, and the future. Potent but shadowy figures, the Fates provided the matrix which could determine the lives of the most powerful men and even the gods themselves; to the Greeks, the power of Fate (Moira) was even stronger than the gods; the Norns brought about the end of the Golden Age.[2]

In addition, the Romans worshipped Fortune as a female goddess of good luck, and in later European iconography, Fortune is almost always personified as a woman. Fortune was worshipped by Romans at numerous temples and shrines, where her aid was sought for a particular situation in life. *Fortuna Liberum* controlled the destiny of children, *Fortuna Privata,* family life, *Fortuna Virginalis,* unmarried girls, *Fortuna Muliebris,* women, *Fortuna Virilis,* marriage, *Fortuna Publica,* the state, and *Fortuna Caesaris,* the imperial family.

These cultures also connected the power of the earth and its fertility with the female. Greeks and Romans, Celts and Germans, personified the earth as female, and as Mother Earth she continued to be worshipped and placated in farming and fertility rituals which survived for centuries. Often women were perceived as having power in these ceremonies; they might have to go to the fields or stay away from them, touch the seed or leave it alone. Blood might be poured or sacrifices made and in rituals concerning blood, women's menstruation might be seen as a potent force, connecting her to deep sources of power: the earth, the seasons, time, life, and death.

The Greeks and Romans associated goddesses with natural forces and processes: fertility, the moon, cyclical returns and rhythms. In the Mediterranean world especially, goddesses such as Artemis (Diana), Demeter (Ceres), the Great Mother (Cybele), and Isis were worshipped for centuries. Most of the ceremonies of these cults have been lost, for many of their rites were secret and confined to initiates.[3] Among the Roman religions, the cult of Diana of Ephesus was popular, and the multibreasted figure of this goddess of maternity was found throughout the Mediterranean world. There was also a state cult of Ceres (Demeter), in which women, under the leadership of priestesses, celebrated the Mysteries to ensure fertility.[4] By the third century B.C., the cult of Cybele, the Great Mother of Asia Minor, had become popular in the Roman Empire; by the first century B.C., worship of the Egyptian goddess Isis was even more widespread. Evidence of Isis worship has been found from Spain to Asia Minor, from North Africa to Germany.[5] Assimilating the practices of other cults and the attributes of other goddesses to herself, Isis became a popular and powerful supreme goddess. "You who are one and all," went a Roman inscription to her, and a hymn from the first century B.C. endowed her with attributes more often associated with gods than goddesses in these early cultures:

> I gave and ordained laws for men which no one is able to change.
> I am she that is called goddess by women,
> I divided the earth from the heaven.
> I brought together women and men.
> I ordained that parents be loved by children.
> I revealed mysteries unto men.

I caused women to be loved by men.
I made an end to murders.
I am in the rays of the Sun.
I am the Queen of War.
I am the Lord of Rainstorms.[6]

In the centuries immediately before the rise of Christianity, the worship of Isis was the most widespread religion in the Roman Empire.

The material remains of this early goddess worship are impressive, and the temples and statues built to honor them remained potent symbols in the European culture to come. The temple of Diana at Ephesus was considered one of the seven wonders of the ancient world. Numerous statues of Aphrodite endured, sometimes buried for centuries. But in their Greek, and later Roman versions, they became powerful icons of European culture. The Aphrodite of Melos (Venus de Milo) remained a lasting image of female beauty in its prime. Other female figures and forces survived in statuary form: the great Victory of Samothrace, a heroic forward-striding woman with wings, has stood at the entrance of the Louvre Museum in Paris for many years.[7]

Many goddess cults required female priestesses, and as priestesses, women achieved their greatest legitimate power in Greece and Rome. Freed from the restrictions which confined ordinary women, priestesses shared the supernatural powers accorded goddesses. From the Pythia at Delphi, who acted as the oracle of the god Apollo, to the Vestal Virgins of Rome, a few women achieved spiritual power and religious leadership. By the first century A.D., women often held the position of priestess-magistrate, combining religious and secular power. Superseded by Christianity, in which only men could be priests, memories of powerful priestesses endured, both in the temples and statues they left behind—and in warnings against letting women perform such important functions.[8]

Women Warriors

In addition to goddesses and priestesses, these early cultures recorded the presence of warrior women. From the legendary Greek Amazons to the historical figure of the British queen Boudica, they transmitted empowering images to European women of subsequent generations. For whether women actually wished to fight physically or not, the mere mention of legendary women who had done so provided a tradition of strength and competence in the quintessentially male field of warfare.

One of the most powerful and enduring of these traditions was that of the Amazons. Although no certain historical proof of this female society of women warriors has been found, Greek culture is filled with portrayals of them. Almost always shown in battle with men, the Amazons figure in the friezes atop the Athenian Acropolis and Parthenon. They were supposed to have besieged Athens around 1200 B.C., and the fifth-century B.C. Greek historian Herodotus believed them to have come from Asia. Women's tombs from east of the Don River reveal that some were buried not only with jewelry and mirrors, but also with swords, spears, arrows, and in one case, a full set of contemporary armor. While such evidence does not substantiate the all-female Amazon society of Greek legend, it does imply that some women were warriors.[9]

Amazons were portrayed with one breast bare and, later, were supposed to amputate one breast to draw their bows more easily; to mate with men, but have no other connection with them; to give away their boy babies and raise only the girls. Mentioned twice in the *Iliad,* they figure prominently in a number of Greek legends: Theseus married the Amazon Hippolyta; Achilles fell in love with the Amazon queen, Penthesilea, as she lay dying from his blows; one of Heracles' labors was to steal the girdle of an Amazon queen. Male victory over an Amazon proved male superiority. On the other hand, the very existence of such a powerful female image remained a source of inspiration and strength to European women in later centuries.[10] In 1710, for instance, the English writer Mary de la Rivière Manley published a "Table of Heroines" in her magazine, *The Female Tatler.* She included "all the Amazonian Race," and stated that "The pleasing View of so many dazzling Heroigns gave me extream Delight, and made me sensible of the Advantage I possessed in being a Woman."[11]

The Old Testament of the Hebrews occasionally depicted women functioning heroically in roles traditionally allotted to men. Deborah is called a prophetess and a judge although she is female, a wife, and a mother. She rallies the Hebrews against their enemies and is aided by Jael, another woman, who slays the enemy general by hammering a nail into his head. Deborah saved her people. "The peasantry ceased in Israel, they ceased/ until you arose, Deborah,/ arose as a mother in Israel" went a hymn to her.[12] The Hebrews paid a similar tribute to the pious widow Judith, who also slew an enemy general. Pretending she was going to Holofernes' bed, she instead "took down his sword that hung there . . . and she struck his neck twice with all her might, and severed his head from his body." Bringing the head back in a "food bag" carried by her maid, Judith roused the Hebrew army to victory. Afterward, she led "all the people" in a song and dance of thanksgiving, assuming another role traditionally reserved to men. Her reward was fame and the honor of having saved her people: "No one ever again spread terror among the people of Israel in the days of Judith or for a long time after her death."[13]

Germanic and Celtic epics also depicted warrior women. The Valkyries of Norse legend served Odin and selected male heroes for entry into Valhalla. The Celtic epic, the *Táin Bó,* written down in the seventh or eighth century A.D., described a world in which women ruled, fought, and possessed great power. Despite Queen Medb's (Maeve's) ultimate defeat by Cúchulainn, she was a formidable and honored opponent. A wealthy and powerful queen, she ruled equally with her husband. She fought and had affairs independently of him. She is described as

> A tall, fair, long-faced woman with soft features. . . . She had a head of yellow hair, and two gold birds on her shoulders. She wore a purple cloak folded about her, with five hands' breadth of gold on her back. She carried a light, stinging, sharp-edged lance in her hand, and she held an iron sword with a woman's grip over her head—a massive figure.[14]

In addition to such women warriors of legend, Roman historians like Tacitus left accounts of women who fought against Rome. Zenobia of Palmyra and Boudica of the Iceni remained the most famous. Boudica (d. 62 A.D.) led a revolt against the Roman army in Britain. Her husband on his death had willed half his kingdom to Rome and half to his daughters. When the Roman procurator claimed it all, assaulted the queen, and raped her daughters, she joined with neighboring kings against the Romans. Described as big-boned and harsh-voiced, with red hair to her knees, she led an assault from the north which penetrated as far as London. She took poison after her defeat, rather than be forced to march in a Roman triumphal parade.[15]

Two centuries later, the Roman Emperor Aurelian forced captured Goth women to march in such a procession wearing placards proclaiming them to be Amazons.[16] What remained important and empowering to European women about warrior women was not their final defeat, but the fact that they existed at all. Later female artists and writers would use these early women warriors as inspirational examples; see, for instance, the writer Christine de Pizan's use of the Amazons in her *Book of the City of Ladies* in the fifteenth century, or the artist Artemisia Gentileschi's use of Judith in her paintings in the seventeenth century. By assuming the roles normally allocated men, warrior women kept alive the possibility of women functioning successfully as warriors in a warrior culture and succeeding in realms of war and leadership usually restricted to men.

Women in Power: Queens and Empresses

Women warriors remained exceptional, and in these early cultures, strong women more often gained power by using their intelligence and their dynastic and familial connections to become rulers in their own right. Particularly in periods of political transition and uncertainty, inheritance of the throne even by a female seemed preferable to civil war and disorder. This too was a tradition which would endure in the later European centuries. In times of change, men of enterprise could seize the opportunity to rise to power; women of enterprise were able to do this as well, if circumstances favored them. For unlike men, who could rule alone in their own name, royal women almost always needed a male relative whom they could rule with or for. While not all such regencies resulted in female rulership, acting as a regent or ruling jointly (a tradition in the Ptolemaic dynasty of Egypt, established by Queen Arsinoë II in the second century B.C.) enabled some ambitious women to take the power for themselves. Best known of these women is Cleopatra VII (69–30 B.C.), the last queen of independent Egypt, who seized her opportunities to gain the throne and to forestall Roman annexation of her kingdom. The object of much hostile Latin literature, Cleopatra established a tradition of female power in Europe. While later authors—like Shakespeare—focused on her beauty and seductive wiles, the historical record reveals a woman and a ruler of energy and intelligence. Making and breaking alliances, skillfully taking advantage of circumstances, she ruled Egypt for over twenty turbulent years, and her ultimate defeat should not overshadow her accomplishments.

She succeeded to the throne of Egypt in 51 B.C., when she was seventeen and her brother, Ptolemy XIII, ten. Contemporary coins and portraits show her as a forceful-looking woman with a long nose, not a beauty by past or present standards. She favored an alliance with Rome; her brother and the council regents did not and forced her to flee to Alexandria where she recruited an Arabian army to assert her position. It was there that she met Julius Caesar, himself in search of Egypt's wealth, but officially present to reconcile the warring siblings. She was twenty-one; he was fifty-two. Her sexual liaison with Caesar took no special skill—Caesar was notoriously easy to bed—but she was also able to enlist his support for her rulership. Caesar helped her conquer the Egyptian army, established her on the throne with another younger brother, Ptolemy XIV, and considered both herself and Egypt allied to him.

Two years later, in 46 B.C., Cleopatra came to live with Caesar, now the successful dictator of Rome. She brought her infant boy, whom she called Caesarion and declared was Caesar's son. Caesar gave her royal honors in Rome, treating her like the queen she was, and adding a large gilt statue of her to the Temple of Venus he had built. She lived with him in his villa until his assassination in 44 B.C., and the suspicion of her influence may have contributed to the plot against his life. The same year, her brother was assassinated, and it is a credit to her political skill that she was able to maintain herself as queen of an independent Egypt, now ruling jointly with her three-year-old son. Three centuries later, the Roman author Plutarch paid tribute to her:

> Her actual beauty, it is said, was not in itself so remarkable that none could be compared with her, or that one could see her without being struck by it, but the contact of her presence, if you lived with her, was irresistible; the attraction of her person, joining with the charm of her conversation, and the character that attended all she said or did, was something bewitching.[17]

She used her charms in her meeting with Mark Antony, the successful Roman general, in 41 B.C. He summoned her to meet him in Tarsus to discuss an alliance against the Parthians; according to Plutarch, she made the trip into a dazzling encounter:

> She came sailing up the river Cydnus, in a barge with gilded stern and outspread sails of purple, while oars of silver beat time to the music of flute and fifes and harps. She herself lay alone under a canopy of cloth of gold, dressed as Venus in a picture, and beautiful young boys, like painted Cupids, stood on each side to fan her.[18]

She won Mark Antony's affections and bore him twins. She then ruled Egypt alone for three years, consolidating her power. The couple married in 37 B.C. and thereafter ruled as equals. She supported Antony in his struggles against his rival (and brother-in-law—Antony was married already when he wed Cleopatra) Octavian, Caesar's heir and the future Roman Emperor Augustus. Egyptian hegemony was asserted over much of the Middle East, supported by Antony's generalship and Cleopatra's troops and supplies. Together, they proclaimed themselves gods, appearing as Isis and Osiris-Dionysus. From then on, Cleopatra was addressed as *Thea Neōtera,* the younger Goddess, and she may have been indirectly responsible for suggesting this tactic of reigning as a god to the future ruler of Rome, Octavian.

In 32 B.C., Octavian declared war on Cleopatra alone, judging that he had to break her power in the East to establish his rulership of Rome. Cleopatra and Antony raised ships and men and went to western Greece in early 31. There, at the Battle of Actium, they watched their final defeat. Although both Cleopatra and later Antony succeeded in breaking away from the Roman ships, three quarters of their fleet was sunk or abandoned. Both fled to Alexandria and there separately committed suicide. Whether Cleopatra actually used a poisonous snake is still uncertain, but it is known that she dressed in her full royal regalia before she died. Octavian had forced her sister, Arsinoë, to march in chains in a Roman triumphal parade; by dying at her own hand, Cleopatra avoided that fate. Only thirty-two, she left a lasting image of female royal power.[19]

In the centuries that followed, other royal women seized power when times of uncertainty favored this. In both the Eastern and Western Roman Empires, women ruled with male relatives and in their own right, creating a legacy of female empresses. First with her nephew and then with her son, Galla Placidia (c. 388–450) ruled the Western Roman Empire as Augusta (empress). From the age of fifteen she governed as regent for her brother Honorius and shared power with him until his death in 423. She then governed for a time on behalf of her son, Valentinian III. Galla Placidia's stepniece, Pulcheria (399–453), was declared regent for her brother, the ruler of the Eastern Empire, Theodosius II. Acknowledged by contemporaries to be the diligent, forceful member of the family, she continued to rule behind the scenes even after he had come of age. On his death in 450, she declared herself and her husband Marcian his successors, and they ruled together until her death three years later.[20] This tradition continued in the ruthless hands of the Byzantine Empress Irene (c. 753–803). She overthrew and then blinded her son in order to rule in her own name and managed to hold onto her power for five years.

In the European centuries to come, some royal women would continue to seize power when circumstances favored them, living according to traditions which empowered women instead of subordinating them.

Women of Wealth

Cleopatra's power had rested not only on her lineage and intelligence, but also on the wealth her kingdom represented. In other eras of transition, when changing economic and political circumstances favored the amassing of wealth in female hands, some propertied women were able to use their economic power to lead independent and sometimes influential lives. All these early cultures—Greek, Roman, Hebrew, Celtic, and Germanic—allowed daughters to inherit, especially if there were no sons or male relatives. One important tradition passed on to European women in later centuries was that families would strive to keep their property intact, even if this meant giving it to a woman. Among Athenian Greeks and the Hebrews, daughters were forbidden to inherit if sons were alive, but if there were none, the daughter inherited the entire property. She was, however, supposed to marry a kinsman or at least a member of her father's tribe, to keep the property within the original group. Roman law allowed both daughters and wives to inherit. Celts probably divided the property evenly and gave it all to daughters in the absence of sons; Germanic peoples tended to favor sons, but gave their daughters a portion of the inheritance and all of it if there were no sons.[21]

It was in Rome that women were able to accrue the greatest economic power and rights. One of the few times Roman women intervened in politics was in 195 B.C. to demand the repeal of the Oppian Law, which limited their right to use expensive goods, like gold and carriages. According to Livy, respectable matrons

> blockaded every street in the city and every entrance to the Forum. As the men came down to the Forum, the matrons besought them to let them, too, have back the luxuries they had enjoyed before, giving as their reason that the republic was thriving and that everyone's private wealth was increasing every day.[22]

Twenty-six years later, the Voconian Law was passed, limiting the amount that women could inherit. This type of control proved ineffectual in the long run, for in a time of increasing wealth, families tried to keep their fortunes together by giving them to their daughters, if necessary. In the first century A.D., Augustus exempted a number of women from guardianship and thus male control over their property. The law limiting their inheritance fell into disuse, and by the second century, women could make their own wills freely. Eventually, guardianship became limited to minor children.[23]

There is ample evidence that some wealthy Greek and Roman women in these eras managed their own finances freely. In Pompeii, a moneylender, Faustilla, signed notes, had numerous creditors, and charged 45 percent a year in interest. A woman named Julia Felix owned an apartment building and rented rooms, shops, and baths.[24] Many women, including priestesses, acted as public benefactors, and at the height of the Roman Empire, women like men vied in erecting buildings and statues, presenting games and prizes, giving lavish parties and entertainments.

The enduring effects of women's wealth were twofold. First, wealthy women were able to construct and endow public buildings: temples, baths, meeting halls. Carved into stone, the record of their creations and endowments might be publicly visible for centuries, perhaps empowering other women. "You see here, stranger, the statue of a woman who was pious and very wise, Scholastica," read an inscription from Ephesus. "She provided the great sum of gold for constructing the part of the [two public baths] here that had fallen down."[25] Second, wealth could enable a woman to acquire an education, and an educated woman could make an enduring impact in poetry, painting, or philosophy. While very few—either women or men—were literate, leisured, or artistic in these centuries, the memory of those who were became one of European women's most empowering traditions. Through the centuries, examples from the past, especially Greece and Rome, would be used by those who advocated education and wider intellectual and creative roles for women.

Educated and Artistic Women

Literacy was a privilege for women in these early cultures. Eurydice (d. 390 B.C.), the mother of Philip II of Macedon and the grandmother of Alexander the Great, commemorated her gratitude for learning to read on a stone tablet:

> Eurydice, daughter of Irrhas,
> offers this shrine to the Muses,
> Glad for the wish of her heart
> granted by them to her prayer,
> Since by their aid she has learned,
> when mother of sons grown to manhood,
> Letters, recorders of words,
> learned how to read and to write.[26]

Eurydice lived at the start of one of the periods when women were able to accrue wealth, and it was in exactly those periods—the Hellenistic era in Greece (c. fourth to first centuries B.C.) and the late Republican and early Imperial eras in Rome (first century B.C. to the second century A.D.)—that some women were able to become educated. They left records of their presence in science, art, philosophy, and literature.

In the ancient world and for many centuries thereafter, the most likely way for a woman to enter the sciences or the arts was to be born into a family which specialized in those fields. There are some epitaphs to women doctors in Greek and Roman cities from the first century A.D. on; a number of them associate the woman with a medical family. "You guided straight the rudder of life in our home and raised high our common fame in healing—though you were a woman you were not behind me in skill," went a tribute to Panthia from her husband Glycon in the second century A.D.[27] Of the eight women artists mentioned by ancient authors, a number are described as the "daughter and student" of a father who painted.[28] The fullest account is of Iaia of Cyzicus, who worked in Rome about 100 B.C.:

> She used both the painter's brush and, on ivory, the graving tool. She painted women most frequently, including a panel picture of an old woman in Naples, and even a self-portrait for which she used a mirror. No one's hand was quicker to paint a picture than hers; so great was her talent that her prices far exceeded those of the most celebrated painters of her day. . . .[29]

While none of these women's works of art from this era have survived, the tradition they established of women working in the arts endured in later European centuries. On the other hand, the powerful representations and images of women which have endured in the art of these early cultures were made by men.

To some degree, the same is true in philosophy. While the names and some brief accounts of female philosophers are all that remain of their writings, the most extensive argument for empowering women comes from a male philosopher. Despite Plato's misogynistic passages in his other writings, in the *Republic* and the *Laws,* he argued that women and men are the same in nature and therefore should be given the same education. In the *Republic,* there is a sustained argument that in the ideal state women should be equal to men and share in administering the government: "Men and women alike possess the qualities which make a guardian; they differ only in their comparative strength or weakness." These guardian-women are to be spared the "hard work" of child care and will "share in the toils of war and the defense of their country."[30] Written in the fourth century B.C., these utopian ideas were preserved and ensured that in the centuries to come the possibility of female equality would be encountered wherever Plato's philosophy was studied.[31]

Six hundred years later, in the third century A.D., Diogenes Laertes wrote that Plato had two female students at his Academy who dressed as men. Knowledge of other female philosophers comes largely from such late sources and is fragmentary, only hinting at the influence and intellectual prestige of these early educated women. Three women have reputations and are mentioned by name as philosophers in the ancient world: Hipparchia of third-century B.C. Athens; Beruriah of second-century A.D. Jerusalem; and Hypatia of fourth- to fifth-century A.D. Alexandria.[32] All were the subject of pithy anecdotes illustrating their quick wit and command of argument.

"I am much stronger than Atalanta from Maenalus [the famous runner], because my wisdom is better than racing over the mountain," Hipparchia is supposed to have declared. "You don't think that I have arranged my life so badly," she replied to a critic, "if I have used the time I would have wasted on weaving for my education."[33] Beruriah was one of the very few women cited in the Talmud for her scholarship. In one passage, she corrects her husband, Rabbi Meir, for praying for the death of some sinners. Citing verses that said "let the sins (not the sinners) cease," she convinced him to pray for the sinners' repentance, and "he did pray for them and they repented." A number of passages cited her as an authority and start, "Rightly did Beruriah say. . . ."[34]

Hypatia (c. 370–415) left a reputation as a great scholar, although none of her writings have survived. The daughter of a mathematician, she taught both mathematics and philosophy at the university in Alexandria. She was a popular teacher and was known as The Nurse or The Philosopher. Hypatia invented a number of scientific instruments and wrote works on astronomy and mathematics. In 415 the Christian Patriarch of Alexandria incited a mob to attack her; she was pulled from her chariot and killed by the crowds. Both her body and all her books were burned.

Some women's writings—letters and poetry from the Greek and Roman cultures—have survived. The poetry rivals the best written by men of the day, and one of the most important traditions which has empowered women in European history is this heritage of women writing, articulating a woman's view of life. The quality of this writing could not be denied, and the fact that women were able to create great literature, even within cultures that subordinated them, put the lie to assertions of women's innate inferiority. Throughout the centuries, some women were able to create literature which endured. Although much of this writing has been lost, enough was preserved to establish a tradition of women writers, beginning with these early poets. Honored by their own societies, they expressed women's concerns and dreams. In the third century B.C., for instance, the Greek poet Anyte of Tegea wrote of a little girl's sorrow:

> The child Myro made this tomb
> for her grasshopper, a field-nightingale,
> and her cicada that lived in the trees,
> and she cried because pitiless death
> had taken both of her friends.[35]

Writing on their own themes, women chose to focus on love and to say very little about war, the subject of much poetry written by men. In their own day and thereafter, their championing of love and its claims was remarked upon. While male poets wrote about love as well, women created a tradition of female love poetry in which love was the greatest good of all:

> Nothing is sweeter than love, all other blessings
> Come second to it. I have spat even honey
> From my mouth—I, Nossis,
> Say this is so. But one whom Kypris [Aphrodite]
> Has not loved, will never know
> What roses her flowers are.[36]

The Greek poet Nossis of Locri wrote this verse in the third century B.C.; four centuries later, the Roman poet Sulpicia echoed her sentiments:

> This day has brought a love
> it would shame me to conceal
>
> . . . .
>
> If I sin, I glory in sinning:
> I will not wear virtue's mask—
> the world shall know we have met
> and are worthy, one of the other.[37]

Of all the female poets, Sappho of Lesbos—the earliest in time—was the most brilliant. Acknowledged as a poetic genius in her own day, she garnered praise throughout the centuries, becoming an inspiration to women of other eras and other cultures. Plato called her the "tenth Muse," and her writings were cited by other poets and writers on literature as examples to be studied and imitated.[38]

Little is known for certain about her life. Probably born about 630 B.C., she wrote in Greek about two hundred years after the works of Homer were first recorded. She married, although the name of her husband is in doubt, and she had a daughter:

> I have a beautiful child
> who looks like golden flowers
> my darling Cleis,
> for whom I would not [take]
> all Lydia. . . .[39]

Like other poets of her day, she almost certainly performed while playing a lyre (hence the name lyric poetry for her work). "Come, divine lyre," she sang, "speak to me and find yourself a voice. . . ."[40] She considered herself not only a poet, but a good one: "I do not imagine that any girl who has looked on the light of the sun will have such [poetic] skill at any time in the future."[41]

Sappho wrote nine books of lyric poems; only one poem survives intact. Much of her writing was preserved because her phrases were considered exemplary and were cited as the finest illustrations of poetic techniques. Even in fragments, her vivid metaphors and turns of phrase convey her talent. She described Aphrodite's chariot as drawn by "beautiful swift sparrows whirring fast-beating wings [who] brought you above the dark earth down from heaven through the mid-air. . . ."[42] She articulated moods with simple declarations:

> The moon has set and the Pleiades;
> it is midnight, and time goes by,
> and I lie alone.[43]

She used intricate metaphors—"like hyacinth which shepherds tread underfoot in the mountains, and the ground, the purple flower. . . ." went one fragment.[44] And she knew the power of simple repetition:

> Hesperus [the Evening Star], bringing everything
> that shining Dawn scattered,
> you bring the sheep,
> you bring the goat,
> you bring back the child to its mother.[45]

Of all her poetic gifts, it was her skill at love poetry that made her most renowned. Love was her major theme and her only complete remaining poem is a plea to Aphrodite, the goddess of love, to give Sappho her heart's desire, the love of a woman she adores:

> Come to me now again and
> deliver me from oppressive anxieties;
> fulfill all that my heart longs to fulfill,
> and you yourself be my fellow-fighter.[46]

The objects of Sappho's love in this hymn and other poems are women as well as men, and from antiquity on, Sappho was criticized for being "irregular" and a "woman-lover."[47] (The formation of words to define female homosexuality from Sappho's name and birthplace—"Sapphic" and "lesbian"—are from recent centuries.) The criticism did not override the admiration, however, and it is Sappho's descriptions of the sensations and feelings love evokes which have been remembered through the centuries. "Love shook my heart like a wind falling on oaks on a mountain," she wrote, "Love has obtained for me the brightness and beauty of the sun."[48] She described jealousy as "a subtle fire [which] has stolen beneath my flesh" and the sight of her lover with another made her feel "little short of dying."[49]

She honored women, valuing their love and their perceptions. She described women as complete human beings, fully worthy of love and admiration. In her verses, the theories of female subordination and the justifications of female inferiority do not exist and with her creation of a world in which women were cherished equals, she denied those who disparaged women. Even more, she valued the claims of love over all else in life—even war in a warrior culture. In this she helped to create the long tradition of European love poetry, although few poets after her would be women expressing their love for a woman:

> Some say cavalry and others claim
> infantry or a fleet of long oars
> is the supreme sight on the black earth.
> I say it is the one you love.
>
> . . . .
>
> these things remind me now of Anaktoria
> who is far, and I for one
> would rather see her warm supple step
> and the sparkle of her face
> than all the dazzling chariots
> and armored infantry of Lydia.[50]

In one of her verse fragments, Sappho looked to posterity: "Someone, I say, will remember us in the future."[51] And so it was to be: Never entirely forgotten, her name survived through the ages. And as long as Sappho was mentioned, the tradition that a woman could write and write as well as any man survived. Other women would follow her example, writing on their own subjects in their own voices and doing it so well that their writings also endured. In the first century B.C., the Greek poet Meleager praised some of these women and included them in his *Garland,* an anthology of epigrams. He described choosing the poems as "weaving into the garland many lilies of Anyte, many white lilies of Moero, and of Sappho, few flowers, but these few, roses. . . ."[52] The tradition of women writing endured. It was one of the traditions inherited which most empowered European women in the centuries to come.

5

THE EFFECTS OF CHRISTIANITY

Beliefs and Practices Empowering Women

The Greek, Roman, Hebrew, Celtic, and Germanic cultures bequeathed a mixed legacy of traditions to European women of the ninth century. Overall, women's subordination was justified and perpetuated, but also inherited were images and memories of women which empowered them. The life and teachings of Jesus of Nazareth, later institutionalized as the beliefs and practices of the Christian Church, added both to traditions that empowered women and those that subordinated them. Initially, Jesus' words and actions included women in ways that were new and surprising in the Roman-dominated Palestine of the first century A.D. He made little distinction in his teachings between women and men, despite the occasional consternation this may have caused his male followers; when he talked at length with a Samaritan woman, his disciples "marveled that he was talking with a woman."[1] To some of his contemporaries, Jesus appeared to reject traditional ideas of status, of free and slave, subordinate and inferior. He saw no special flaws in woman's nature. He included women in his preaching and allowed them a life and a role outside the family and their relationship to a man. He used his authority to call for a reversal of contemporary and traditional values and attitudes. In his Sermon on the Mount, he favored "the meek, for they shall inherit the earth," and those "who are persecuted for righteousness' sake, for theirs is the kingdom of heaven." He encouraged rich and poor—both women and men—to leave their families and follow him, promising they would find a new Christian family. He promised the "kingdom of heaven"—a life after death—to all who embraced his teachings, regardless of status or gender.[2]

Jesus thus rejected much that earlier cultures had taken for granted. Most significant for future generations of European women, he preached the equality of all believers in his doctrine. By his actions and his words, he rejected the traditional descriptions of females as inferior and undermined the ancient justifications for their subordination. He consistently saw women as created in God's image just like men. He never referred to the secondary creation of Eve from Adam's rib, nor did he ever ascribe any special sin to Eve rather than Adam for the disobedience in the Garden of Eden. The act of baptism cleansed women equally with men from the taint of sin. Women's piety, just like men's, would determine their life after death and enable their bodies to rise into heaven.

Jesus gave equal value to women in other ways. In his parables, he used women as well as men to illustrate the values of faith, humility, and charity that he favored. "Harlots" who believed in him would know salvation before priests who denied him.[3] The poor widow's gift was worth more than that of the rich, for "they all contributed out of their abundance; but she out of her poverty has put in everything she had, her whole living."[4] When he compared faith to "a grain of mustard seed which a man took and sowed in his field," he immediately followed it with an example which would mean more to women: "The kingdom of heaven is like leaven which a woman took and hid in three measures of flour, till it was all leavened."[5] "Five wise maidens" who filled their lamps with oil and were prepared to "meet the bridegroom" exemplified the true believers ready to receive God's blessing.[6] He used the metaphor of birth to describe his mission and promised that his followers would feel the joy of a woman who has given birth.[7] When a bleeding woman sought his help and touched his gown, he not only did not condemn her, but comforted and healed her.[8] He saved the life of one of the most despised figures in Jewish society, the adulteress, also an outcast in many contemporary cultures. Hebrew law called for her to be stoned to death; Jesus said, "Let him who is without sin among you be the first to throw a stone at her," and when no one did, he told her, "Neither do I condemn you; go, and do not sin again."[9]

Most unusual of all his actions with regard to women, Jesus spoke directly to them of his doctrines and accepted women as his special followers along with men. In Roman-dominated Hebrew Palestine, women hardly ever figured as subjects of Jewish rabbinical or Talmudic teachings, let alone as students. In contrast, Jesus first declared his divinity to a Samaritan woman, in a culture in which women were subordinate and Samaritans outcasts. He stated his mission to Martha of Bethany: "I am the resurrection and the life; he who believes in me, though he shall die, yet shall he live, and whoever lives and believes in me shall never die."[10] He raised her brother Lazarus from the dead as proof. Mary of Bethany, Martha's sister, even "sat at the Lord's feet and listened to his teaching." When Martha complained that Mary was not helping her serve the meal, Jesus replied that "Mary has chosen the good portion, which shall not be taken from her."[11] Christian women in the European centuries to come repeatedly cited this affirmation of women's right to learn and preach to justify their studies. Later generations also took his recommendation to John to care for his mother, Mary, as a reminder of the special significance of women as mothers. Early Christians portrayed Mary in their wall paintings and statues as a queenly mother, the Virgin enthroned with the holy infant on her lap, his hand held up in blessing.

Jesus' favoring of women and his encouragement of their untraditional behavior continued throughout his life. Women rather than men played the key roles in the events surrounding his death and resurrection. In describing the crucifixion, the four Gospels report that all the apostles had fled except John, but that the women remained to pray at the foot of the cross and witness his death.[12] The women made plans to prepare his body for burial. Mary Magdalene first returned to the tomb and discovered his body gone. She, "from whom he had cast out seven demons," not one of his male followers, first witnessed his resurrection, the ultimate proof to the faithful of Jesus' divinity. Mark and Luke reported that the male disciples did not believe Mary Magdalene and that Jesus later rebuked them for their lack of faith.[13]

Women's presence in the Gospels, whether as pious example, as mother, or as follower, remained a powerful tradition for women in the European centuries to come. Those books of the New Testament which deal with the years immediately following Jesus' death, the Acts and the Epistles, offered many models of women acting as equals within the new faith. Women of later generations could find in these writings of the first century A.D. justification for roles outside the family and marriage.[14] Accounts of Jesus' life consistently mention five women companions: Mary Magdalene, Mary of Bethany, Joanna, Susanna, and Salome.[15] The verses of the Acts of the Apostles are sprinkled with names of women who helped the new faith after his death. Tabitha, a seamstress living in the Jewish community of Joppa, was called a "disciple" and was considered so essential to her followers that they prevailed upon Peter to bring her back to life.[16]

Despite his other teachings, which pointed to the desirability of women's subordination, the Apostle Paul declared to the Galatian Christian community that "There is neither Jew nor Greek, there is neither slave nor free, there is neither male nor female; for you are all one in Christ Jesus."[17] Paul's Epistle to the Romans mentions thirty-six colleagues, sixteen of them women.[18] Women labored "side by side" with him, and he met and corresponded with them as leaders of their own congregations.[19] In Corinth, Paul taught with a woman, Prisca (Priscilla). She and her husband Aquila had once saved Paul's life, and had probably jointly led their congregation in Rome before fleeing to Greece to avoid persecution.[20] Paul's follower, Thecla of Iconium, left her family to teach with him and became a model in those centuries for devout Christian women.

In the centuries of the persecutions, Christian women other than those associated with the apostles and evangelists played untraditional roles as priests and prophets, roles that would be embraced as powerful traditions by religious Christian women in later centuries. The Gnostics, a scattered group of sects later denounced by the mainstream of Christian leadership, are presumed to have allowed women believers in North Africa to act as priests, leading prayers and baptizing. By the second century, the Gnostics had produced a number of books and teachings which stressed both the active participation of women and the bisexual nature of God. One Gnostic hymn listed the attributes of God in terms reminiscent of contemporary hymns to Isis:

> For I am the first and the last.
> I am the honored one and the scorned one.
> I am the whore and the holy one.
> I am the wife and the virgin.
> I am [the mother] and the daughter. . . .[21]

The Montanists of Phrygia included two women prophets among their leaders in the beginning of the third century. Prisca preached the imminence of the second coming of the Savior and recorded a vision in which she saw Jesus in female form.[22] The other prophet, Maximilla, assumed leadership of the group on the death of Montanus. Under attack, she justified her role by saying, "I am pursued like a wolf out of the sheep fold; I am no wolf: I am word and spirit and power."[23]

Most powerful of the images to be preserved from these early centuries of the Christian faith were the portraits and legends of the women who became martyrs in the persecutions of the Roman Empire. They, more than any other women, became the heroines for future generations, models of independence, of bravery, and of the power faith could give a devout Christian female willing to die for her beliefs.

Until the beginning of the fourth century when their doctrines became the "preferred" religion of the Empire, over 100,000 Christians suffered this fate, from Carthage in North Africa to Lyons in France to Vienna in Austria, as emperors and their prefects tried to extirpate a cult that refused to worship the gods of the Roman state and preached pacifism.[24] The lines between fact and imagination blurred in the glorification of the martyrs as those who died became "saints"—literally, "holy ones"—objects of prayer, the subject of exemplary tales and Christian iconography. In the centuries to come many female saints became readily identified by the symbols of their persecution in a Europe where few could read: Saint Catherine and her wheel, one of the instruments of her torture; Saint Margaret and a dragon, one of the beasts that attacked her; Saint Barbara holding a tower, a miniature of the one in which she was imprisoned by her parents. Many became archetypal models of the way to achieve Christian salvation. The slave woman, Saint Blandina, was whipped, burned, and finally died trussed in a net gored by a bull, but she had exhausted her torturers and won their admiration for her staunch adherence to the Christian faith. Other women left comfortable lives and high positions to become Christians. Saint Agnes, a well-born Roman girl, chose death rather than marriage to a non-Christian suitor. Saint Davia left her priesthood to Minerva; Saint Anastasia left her life as a Yugoslavian noblewoman. Saint Pelagia of Antioch, an exemplary virgin, threw herself off a roof rather than suffer rape. Saint Afra of Augsberg, originally a prostitute, converted and also died for her faith.

Amid the legends and the bits of historical evidence one woman stands out, for she left her own account of the days and hours leading to her martyrdom. Known in the liturgical calendar as Saint Perpetua of Carthage, she died in the persecutions of the first decade of the third century. Perpetua was a model of all that pious Christian women in the subsequent centuries would aspire to: self-sacrifice, courage, and steadfastness. Only twenty-two at the time of her imprisonment and still nursing her infant son, she initially "was terrified, for I had never before been in such a dark hole" as the prison.[25] She refused to listen to her father's pleadings that she recant, and she was baptized in prison, the final act of faith for an early Christian.[26] She nursed her baby in prison and began to have significant dreams, including one in which she saw herself as a man fighting as a gladiator in the arena:

> My clothes were stripped off, and suddenly I was a man. My seconds began to rub me down with oil, as they are wont to do before a contest. . . . We drew close to one another and began to let our fists fly. My opponent tried to get hold of my feet, but I kept striking him in the face with the heels of my feet. . . . He fell flat on his face and I stepped on his head. . . . I began to walk in triumph towards the Gate of Life. Then I awoke.[27]

An unknown observer recorded her death. She refused to dress in the robes of a priestess before entering the arena. Knocked off her feet by the charges of a wild cow, she straightened her clothes, repinned her hair, and walked across the arena to rejoin her companions. At the end, she guided the sword hand of the inexperienced soldier who had been unable to kill her with the first blow.[28]

The opportunities for martyrdom and the creation of Christian heroines continued as women came under attack when the frontiers of the Roman Empire collapsed in the invasions of the fifth century on. Saint Marcella, a Roman matron who studied with the Church Father, Jerome, was killed by the Visigoths when they sacked Rome in 410. Saint Ursula of Cologne (and her legendary 11,000 virgins) probably died at the hands of the Huns about 500. Saint Gudula of Brussels died in a Viking raid about 680. The Vikings came by sea and sailed up Europe's rivers to seize the treasuries of the churches and religious communities. Terrified of being raped, Christian nuns created another tradition of female piety and martyrdom, mutilating their faces in hopes of repulsing the attacker. Saint Usebia did the same when threatened by an Arab (Saracen) invasion of Spain in the eighth century.

With the end of the persecutions, invasions, and the occasions for martyrdom, Christian women found other ways to serve and advance their faith. As in the first decades of the religion, they assumed functions and roles outside those usually assigned to women. They acted like men in converting others to Christianity and so created another tradition for Europe's Christian women. Legends told of Mary Magdalene being shipwrecked and then winning converts in the south of France. Ireland reputedly became Christian through the efforts of two missionaries: Saint Brigid and Saint Patrick. The daughter of a Celtic chieftain and his Christian bondswoman, Brigid (c. 450–523) supposedly refused to marry so she might dedicate her life to the faith. "I never crossed seven furrows without turning my mind to God," she reputedly declared.[29]

Like men, Christian women converted their families as well as strangers. Mothers converted children; wives converted husbands. When these women were members of royal or imperial families, such actions could have far-reaching consequences; a woman could shape the faith of an entire people. Constantine, the Roman Emperor who made Christianity the preferred religion of the empire, may have been influenced by the Christian zeal of his mother, Helena (c. 248–328?). During his reign she worked to convert others, preserved sites associated with Jesus' life, built churches in Palestine and Asia Minor, and intervened in religious controversies. At age seventy, she traveled to Jerusalem in search of the true cross.[30] Female rulers of the sixth- to ninth-century Eastern and Western Empires, like Eudoxia, Pulcheria, Theodora, and Irene followed her example and actively intervened in the religious controversies of their day. Each used her power to favor one doctrine over another, or one Pope over another.[31]

Among the Germanic peoples, royal wives converted their husbands and thus whole kingdoms. In sixth-century France, Clothilde told her husband Clovis, king of the Franks, "The Gods you worship are no good. . . . They haven't even been able to help themselves let alone others." When he blamed the death of their son on her choice of gods, she thanked "the Creator of all things" for welcoming "to His kingdom a child conceived in my womb."[32] Victory in battle after he prayed to his wife's Christian god brought Clovis' conversion. Similarly King Ethelbert of Kent and King Edwin of Northumbria in England converted because of the efforts of their Christian wives Queen Bertha and her daughter Ethelberga.[33]

In the first centuries of the religion, the most holy life for a Christian came to be one of asceticism. These Christians attempted to live not for this world, but for the next. Their goals were spiritual growth and moral improvement, cultivated by prayer, charitable acts, and immersion in the Scriptures through study and discussion. Spiritual growth also necessitated the denial of bodily appetites, which were thought to chain the soul to earth and prevent it from approaching God. Asceticism meant denying the body's needs for food, water, sex, cleanliness, and sleep as much as possible. Some Greek, Roman, and Hebrew men had extolled asceticism and followed this path themselves, but Christian asceticism went further; it extolled not only sexual continence, but sexual virginity. Earlier ascetic women usually married and had children; Christian ascetic women could reject marriage and childbirth. Ascetic Christian women might be widows who had been married and borne children, or they might be women who wished to remain virgins and thus rejected marriage and childbearing altogether. Like men, these women wished to abandon their traditional roles and dedicated themselves to a life of celibacy and pious devotion.

In the early legends about Thecla, she preached with Paul; in later versions, she also became an anchorite, a religious hermit, living alone, praying and fasting. In the fourth century, Bishop Palladius estimated that twice as many women as men had chosen to live as solitary ascetics.[34] But to live alone in the desert posed particular dangers for women, and even male anchorites tended to cluster in groups. Instead, women who wished to pursue such a life began to gather together and to create religious communities, endowed establishments under the supervision of a male churchman. This became the pattern in future European centuries. By 800, monasteries, convents, and abbeys founded by women and men for the female devout dotted the European landscape.[35]

Precedents existed from the first century of Christianity for women leaving their families to join groups dedicated to religion. In the days of Roman persecutions, as "deaconesses" women had taken vows of chastity and had assisted with the baptism of female believers. By the third century, groups of widows had gained official recognition as religious orders designated to pray and care for the sick. The Church Councils of Nicaea and Chalcedon authorized women over forty to pour the wine at communion and to teach newly baptized mothers and their children.[36] It was in the course of the fourth through sixth centuries (called the Patristic Age because of the leadership of the men known as the Church Fathers) that such groups evolved into the kinds of avowed communities that became women's religious orders. In these centuries, pious women and their male mentors established the essential attributes of Christian spiritual life. Like men, women of these religious communities denied their bodies by fasting, celibacy, and the suppression of physical needs and desires. Like men, these women spent their days in study and prayer. For Jerome, the fourth-century Church father, and for European women of future generations, the Roman matron Paula (347–404) and her daughter Eustochium exemplified all that women could achieve in such a life. They became models of behavior for those well-born European Christian women who wished to devote themselves fully to their religion.

In 380 Paula (later called Saint Paula the Elder) abandoned her life of a typical wealthy Roman matron and joined a circle of Christian ascetic women who had gathered around the widow Marcella. Paula had been widowed that year; the wife of a Roman Senator, she had five children and was thirty-three years old. She experienced a religious crisis and walked the streets of Rome giving away her money. When she joined Marcella's group, it had been meeting regularly for thirty years. These high-born ascetic women lived a life almost the opposite of that expected of wealthy Roman women. They did not go to social functions or parties and stayed secluded indoors except for visits to Christian shrines or churches. They ate as little as possible and remained celibate. They did not wash and wore the coarse clothing of the poor. They allowed themselves little sleep and took that on hard mats on the floor. Their only reading was Scripture, and they studied with Jerome while he was secretary to the Pope in Rome.

Jerome praised Paula as an ideal Christian woman:

> She mourned and she fasted. She was squalid with dirt; her eyes were dim with weeping. . . . The Psalms were her only songs; the gospel her whole speech; continence her one indulgence; fasting the staple of her life.[37]

When the Roman synod exiled Jerome in 385, Paula left her three younger children to travel to Egypt, Bethlehem, and Jerusalem with him.

Only Eustochium, her daughter who had also pledged herself to a pious life, accompanied her. Jerome extolled Paula's choice of the spiritual Christian life, a choice made by some women in succeeding generations. "She overcame her love for her children by her love for God," he wrote approvingly to Eustochium:

> Yet her heart was rent within her, and she wrestled with her grief, as though she were being forcibly separated from parts of herself. . . . Though it was against the laws of nature, she endured this trial with unabated faith; nay more, she sought it with a joyful heart. . . .[38]

They settled in Bethlehem and, admiring the monastery established by another wealthy Roman widow, Melania the Elder, Paula supported the founding of four similar communities, three female and one male. She and her women studied Scripture and prayed at intervals throughout the day. On her death, Eustochium succeeded Paula as head of the female monasteries. Jerome praised Paula's studies; he reported she had learned Hebrew so well that she could sing the Psalms without a Latin accent. Her daughter Eustochium edited Jerome's translation of the Bible, the future Latin Vulgate.

Other Church Fathers and leaders helped in the foundation of female religious communities. Augustine, bishop of Hippo, encouraged women to establish groups in Carthage, and his sister, Perpetua, founded an order for women in his diocese. Under the auspices of Bishop Caesarius, all of the major cities of central France in the sixth century created communities for women.[39] In the eighth century, the missionary Boniface founded all the dual monasteries (establishments for both women and men) in Germany. Royal and propertied women and men gave their lands and their wealth to endow these religious communities. Women also founded monasteries for women. Saint Brigid established the religious community for women at Kildare in Ireland in the fifth century and became its first abbess. Saint Gertrude's mother founded Nivelles for her, the seventh-century dual monastery which she ruled near Brussels. Osric, the eighth-century King of Northumbria, founded Gloucester for his sister, Kineburga. By the eighth century, there were over twenty monasteries for women in England and over twenty in the Frankish lands of France and Belgium; by the ninth century, there were eleven for women (out of a total of twenty) in German Saxony.[40]

In France, Queen Radegund (518–587) became the model pious foundress for Europe's women. A princess of Thuringia who had been taken as booty on the death of her parents, she became the fifth-ranked wife of the Frankish king, Clothar I. Dedicated to a religious life since childhood, she quarreled with him over her pious vocation and her desire to spend her time caring for the sick. When she feared he would have her killed, she fled, and with her property and the support of Germain, her bishop, she founded a community for women at Poitiers. As abbess she provided an example to her community, and her chaplain wrote after her death: "Human eloquence is struck almost dumb by the piety, self-denial, charity, sweetness, humility, uprightness, faith, and fervor in which she lived."[41]

From the fourth century on, the support and encouragement of a prominent male Christian leader came to be increasingly significant, as the developing Church created a male hierarchy and tried to enforce uniformity of practice and doctrine. Paula and Jerome had followed the ideas of Melania the Elder in formulating the rules for their religious communities; increasingly, the ideas of male leaders came to be preferred. All insisted on modesty, chastity, and obedience for religious Christian women. All envisaged a life devoid of luxury and devoted to prayer, with the women separated from regular activities and cloistered away from daily contact with others, especially men. In the fourth century, the Church Father Gregory of Nyssa, wrote of his sister, Saint Macrina the Younger (c. 327–379), describing the ideal life she and her women lived which other women should aspire to:

> As if their spirits had been delivered from their bodies by death . . . so their way of life was regulated to imitate the life of the angels. . . . Their existence bordered on the incorporeal nature; a nature freed from human cares.[42]

Between 512 and 534 Caesarius, bishop of Arles, wrote the rule that would spread across the Western kingdoms and become the basis of Christian women's religious orders in the European centuries to come. The devout gave up all their property and took vows to remain cloistered and celibate for life. Their community was to be self-sufficient, with the women performing household tasks themselves. Adaptions of Benedict's rule (formulated for his monastery at Monte Cassino in Italy) in the seventh century to women's groups gave their lives carefully demarcated occasions for prayer and the singing of psalms, with study and duties filling the interstices. These rules later regulated the monasteries, convents, and abbeys of Europe.

By the ninth century, Christian women, their families, and their ecclesiastical mentors had created an alternative to the conventional roles allowed women in earlier societies. In Christian Europe, devout women could leave their families, eschew marriage, and forgo childbearing. They could become "brides of Christ" and reap the rewards of that spiritual union which Jerome had described for Eustochium:

> Let the seclusion of your own chamber ever guard you; ever let the Bridegroom sport with you within. If you pray, you are speaking to your spouse; if you read, He is speaking to you. When sleep falls on you, He will come behind the wall and will put His hand through the hole in the door and will touch your flesh. And you will awake and rise up and cry: "I am sick with love."[43]

They could devote themselves to spiritual concerns. They could fulfill a new role within the protection of their church. They sang of their joy in an early Christian hymn:

> Fleeing from the sorrowful happiness of mortals,
> having despised the luxurious delights of life and love,
> We leave marriage and the bed of mortals and a golden house for Thee. . . .
> Singing a new song, the company of virgins ascending to Thee towards the high heaven. . . .[44]

Beliefs and Practices Subordinating Women

The Christian belief that even the most religious women must live separate, cloistered, and celibate lives, under the watchful guardianship of a male ecclesiastic, became a tradition inherited by European abbesses and nuns of the centuries after 800. Even while they allowed these women to abandon their traditional functions and roles, men like Caesarius and Benedict saw their female followers as different—and inferior—in nature from devout men. They modified their rules to accommodate what they called women's "weaker" nature. For despite the acceptance of some pious Christian women in the untraditional roles of teacher, martyr, missionary, and foundress, other Christian teachings and practices justified both women's traditional roles and their subordination to men. There were three chief sources for this. First, there were the passages subordinating women in the Old Testament, which became as sacred to the Christians as it was to the Jews. Second, traditions subordinating women were present in every culture where Christianity spread—among the Greeks, Romans, Hebrews, Celts, and Germans. Third, there was scriptural justification for women's subordination in the writings of Paul, other apostles and evangelists, and early Christian writers. All these men described women as inferior by nature and thus justified their subordination to "superior" men. Christian traditions subordinating women became an important part of the legacy inherited by later generations of European women.

In contrast to Jesus, the writings of Peter, Paul, and Timothy, and of the Church Fathers like Jerome, Tertullian, Augustine, and John Chrysostom emphasized women's inferiority and declared that women should be subordinate to men. As Christianity became accepted and institutionalized, the equality granted women in its first centuries was disavowed. Paul and other apostles and evangelists separated the women who aided them from the vast majority of women, whom they preferred to remain in traditional female roles. Drawing on Hebrew law and custom, Paul asserted that women should be veiled and

> keep silence in the churches. For they are not permitted to speak, but should be subordinate, even as the law says. If there is anything that they desire to know, let them ask their husbands at home. For it is shameful for a woman to speak in church.[45]

Invoking the second account of Creation, Paul stated that "man is the image and glory of God; but the woman is the glory of man." Referring to Eve's supposed creation from Adam's rib, he added, "For man was not made from woman, but woman from man. Neither was man created for woman, but woman for man."[46] The letter to Timothy repeated this argument in stronger terms:

> Let a woman learn in silence with all submissiveness. I permit not woman to teach or to have authority over men; she is to keep silent. For Adam was formed first, then Eve; and Adam was not deceived, but the woman was deceived and became a transgressor.[47]

Peter was supposed to have told women to be "submissive" so that they could "win over" more converts who would be attracted by their "reverent and chaste behavior."[48] Like earlier Hebrew prophets, he admonished women for adorning themselves and urged them to be modest, invoking the figure of Sarah, who called her husband "lord."[49] Both Timothy and Titus argued that only men could be bishops in the new Christian churches.[50] As part of the New Testament, these prescriptions were perpetuated when and wherever Christian scripture was accepted.

By the fifth century, the male leaders and theologians of Christianity had accepted and restated the most denigrating traditional views of women. All that was inferior or evil they associated with the female, all that was good and superior with the male. In the second century, the Church Father Origen saw male and female qualities in the soul—the male qualities were the superior ones. In the same century, Clement of Alexandria recommended that women wear veils and men grow beards to emphasize the difference between the sexes:

> His beard, then, is the badge of a man and shows him unmistakably to be a man. It is older than Eve and is the symbol of the stronger nature. . . . His characteristic is action; hers passivity.[51]

Following Paul and the other apostles, the Church Fathers often referred to the creation story of Eve fashioned from Adam's rib. Eve's act of disobedience in the Garden of Eden became evidence of all women's inherent weakness and evil, and the principal justification for her eternal subordination to her "natural" superior, the more spiritual and rational male. Contemporary Jewish scholars and rabbis also stressed Adam's relative innocence and Eve's greater guilt, and some of these writings were accepted as orthodox by the Church Fathers as well.[52] The Christian Church Fathers also drew upon the Greek legend of Pandora, which was familiar to them. All stressed the resemblance between Pandora and Eve as bringers of evil to men. From the second century on, Eve was seen by the Christian Church as the source of sin, the temptress of man, and the embodiment of all women. "And do you know that you are [each] an Eve?" wrote Tertullian around 200,

> The sentence of God on this sex of yours lives in this age: the guilt must of necessity live too. *You* are the Devil's gateway. *You* are the unsealer of that forbidden tree. *You* are the first deserter of the divine Law. *You* are she who persuaded him whom the Devil was not valiant enough to attack. *You* destroyed so easily God's image, man. On account of *your* desert, that is death, even the Son of God had to die.[53]

Two hundred years later, in the fifth century, John Chrysostom repeated this argument. "What then about other women, if this was Eve's doing?" he asked, after a passage blaming Eve for corrupting Adam.

> Yes, indeed: they are all weak and frivolous. . . . For we are told here, not that Eve alone suffered from deception, but that "Woman" was deceived. The word "Woman" is not to be applied to one, but to every woman. All feminine nature has thus fallen into error. . . .[54]

This misogynist fear and hatred echoed older images of women's seductive power. It took on new dimensions in the theological treatises of the Church Fathers and passed into both secular and religious European culture as a tradition inherited by women.

In the writings of the Church Fathers, the act of disobedience to God became the "original sin." In describing this transgression, they shifted attention from the act of disobedience in eating the fruit of the Tree of Knowledge of Good and Evil to the first sex act. Increasingly, Adam was seen as relatively innocent and as the embodiment of mind, who had been corrupted by Eve, the temptress and the embodiment of flesh. Augustine enunciated the doctrine, which became Christian dogma. Adam saw Eve naked, and in "just retribution" the man—the personification of mind and spirit—lost control of his body. In the uncontrolled erection that the sight of her inspired, "the flesh began to lust against the spirit."[55] Fear of the male's sexual response became fear of sexuality in general and led to denunciations of the female. The Church Fathers portrayed Eve as object, as the cause of lust and the personification of all that was uncontrolled.[56]

From condemnation of Eve and the lust she inspired in Adam, the

Church Fathers came to condemn all women and the lust they inspired in men. The Church Fathers praised celibacy and wrote with disgust of anything sexual. There was ample Christian precedent for this. "You have heard that it was said, 'You shall not commit adultery,' " Jesus stated in his Sermon on the Mount, "but I say to you that every one who looks at a woman lustfully has already committed adultery in his heart."[57] He declared there would be no marriage in heaven. Jesus himself never married and stated that there would be some men "who have made themselves eunuchs for the sake of the kingdom of heaven. He who is able to receive this let him receive it."[58] Paul acknowledged the danger of "passion" and wrote to the congregation at Corinth: "It is well for a man not to touch a woman."[59] Building on this, the fourth-century Church Father Jerome reasoned that "if it is good not to touch a woman, then it is bad to touch a woman."[60] The Church Fathers ranked the completely celibate life higher than even the chaste married life. In their eyes, women became the means by which men sinned. "All the Devil's strength against man is in his loins," as Jerome explained to Eustochium.[61]

Seeing women as flesh, as potentially dangerous to men, made easy the incorporation into Christianity of all the older beliefs and practices surrounding women's bodies and reproduction. Drawing especially on the Hebrew traditions of the Old Testament, Christian writers gradually asserted that woman's body had the power to pollute. "Nothing is so unclean as a woman in her periods," wrote Jerome. "What she touches she causes to become unclean."[62] Some early Christian congregations followed the Hebrew practice of separating female and male believers. By the third century, menstruating women, even deaconesses, could not approach the altar.[63] By the seventh century, all of the myths about the destructive power of menstrual blood had been revived and reasserted. Bishop Isidore of Seville insisted that the touch of a menstruating woman could prevent fruit from ripening and cause plants to die. Pope Gregory the Great "commended" women who stayed away from church when they were menstruating, but did not insist that they do so.[64]

Childbirth was once again seen as a contaminating experience. By the end of the sixth century, the Hebrew tradition that a woman remained unclean for thirty-three days after the birth of a son and sixty-six days after the birth of a daughter had become Christian practice as well.[65] The Christian ceremony of "churching" evolved. A priest ritually purified a woman after the specified time from the contamination of childbearing and the greater contamination of having birthed a daughter. Only then could she reenter the church and participate again as a member of the congregation.

These denigrating views of women played their part as Christians began to formalize the organization of their Church. Women's inherently "weaker" nature, their role in the fall from divine grace, and their periodic "uncleanness" were cited to exclude women from all the positions of responsibility and leadership they had initially enjoyed. By the third century many congregations, such as the Gnostics, having different beliefs and practices, and with different views on women (they praised Eve, considered Mary Magdalene an apostle, and allowed women priestly functions) found themselves condemned as "heretics" and successfully expelled from the mainstream of Christian orthodoxy.[66] In particular, Christian women were excluded from the priesthood. Paul and Timothy's injunctions against women speaking were cited. Third- and fourth-century ecclesiastical treatises attributed to the apostles and drawing on their authority spoke with a uniformity that became increasingly characteristic of Christianity. They specifically limited the priesthood to men because, as they argued, "the weak [woman] shall be saved by the strong [men]."[67] These documents justified the exclusion of women, even deaconesses, by referring to Jesus' life: "If it had been lawful to be baptized by a woman, our Lord and Master would have been baptized by Mary, his mother. Had Jesus intended women to perform these functions, all of his apostles would not have been men."[68] Both the injunction and the justifications became traditions reverently referred to in subsequent centuries whenever devout European women again claimed religious authority equal to that of men.

Barred from these new functions, women were supposed to be traditional wives and mothers. The appropriate role for Christian women according to male Church leaders was the same as it had been for the Greeks, Romans, Hebrews, Celts, and Germans: She should be an obedient wife and a prolific mother. The Christian apostles and Church Fathers found more justifications and even narrowed the traditional limitations and subordination. "Yet woman will be saved through bearing children, if she continues in faith and love and holiness, with modesty," explained the first-century author of the Epistles to Timothy.[69] The Church Father Ambrose explained that procreation was "the precise function of their [the women's] sex."[70] By the fourth century the Church Fathers commonly referred to God's curse as described in Genesis to justify women's role and subordination. "Man commands, woman obeys, as God said to her at the beginning," wrote John Chrysostom. "Your desire shall be to your husband and he will lord it over you." As wife, the woman would deal with "all that is within the house" and willingly accept her situation for "it is better for you to be under him and to have him as your lord, than that, living freely and on your own, you fall in the pits."[71] The "Constitutions" attributed to the apostles described the ideal Christian wife no differently from her predecessors in previous cultures. She should be "meek, quiet, gentle, sincere, free from anger, not talkative, not clamorous, not hasty of speech, not given to evil speaking, not captious, not double-tongued, not a busybody."[72] Subsequent European generations idealized the same qualities, used the same reasoning, and looked to the same biblical sources to justify the subordination of their wives and the designation of procreation and care of the household as women's appropriate functions.

The Christian leaders extolled the life of wife and mother for women. While insisting on a wife's submission to her husband, they also advocated changes in practice that would grant a woman more equal status and more protection within marriage. Jesus had emphasized the indissoluble nature of the union: "The two shall become one flesh. . . . What therefore God has joined together, let no man put asunder." Christianity opposed divorce and remarriage even if the couple separated.[73] The Church Fathers Chrysostom, Tertullian, and Augustine spoke of the couple as companions. Augustine defined the first duty of marriage to be *fides,* loyalty to each other.[74]

To achieve such companionship and have the marriage endure for a lifetime, Christian churchmen insisted on the consent of both partners. To ensure this, Church councils made provisions against incest and against women or men being betrothed against their will. Christianity advocated a single standard of sexual behavior. Adultery was a sin for both wife and husband. Although the older traditions proved strong and the Church's prohibitions could not be uniformly enforced, the secular law codes of the Roman Empire and the Germanic kingdoms increasingly included these Christian views of marriage.[75]

Augustine ranked loyalty above procreation as the first purpose of marriage; most Christian leaders stressed procreation. Of all the Christian traditions inherited by European women, this one most defined and limited their lives. With sexuality made a potential source of sin, Christian writers beginning with Paul insisted that intercourse should take place only within marriage. It could have only one legitimate purpose: the impregnation of the woman. Feelings of pleasure and sexual arousal became "concupiscence." Intercourse when impregnation was not possible became the sin of "fornication."[76]

Although the Greeks, Romans, Hebrews, Celts, and Germans had valued women for their reproductive ability, all had allowed women some means of control over their own fertility. None condemned contraception or abortion with the vehemence of the Christian Church Fathers. They denounced any form of contraception, even withdrawal. Jerome called contraception "murder" of "a man not yet born."[77] Following this Christian reasoning, Justinian's sixth-century law code for the Roman Empire made abortion "homicide." Churchmen even set penances for miscarriage.[78] The church had taken control of women's fertility. Whatever improvements a wife or mother might have gained from the Christian view of marriage, she lost to the Church's view of her body.[79]

Given these views, women might wish to follow the path of virginity to Christian salvation—the path promised to men "who have not defiled themselves with women."[80] Like men, women could deny their sexuality, remain "innocent," and so achieve redemption. By denying herself intercourse with a man, a woman negated the reproductive function which made her female, and so, in the eyes of the Church Fathers, rose above her "inferior" nature. In this way, according to Christian teaching, a woman could become a man. Jerome, the greatest advocate of virginity for women, explained this transformation:

> As long as woman is for birth and children, she is different from man as body is from soul. But if she wishes to serve Christ more than the world, then she will cease to be a woman and will be called man.[81]

Virginity freed woman from "her weak sex . . . [and] her body, which, by natural law, should have been subservient to a man," as Leander of Seville stated in the seventh century.[82] The respect for a woman who chose a virgin life became a tradition inherited by Christian Europeans. But respect and value came only because the woman had denied her gender. Women's nature, function, and roles were denigrated anew by the glorification of those few women who had rejected them.[83] The ancient definitions of women as honorable wives and mothers or dishonorable prostitutes and concubines remained intact in the Christian era. As virgin, wife, or prostitute, a woman remained defined by her sexual relationship with man.

By the ninth century, Christianity had both empowered and subordinated European women. Gains included the new role of honored virgin, able to live outside the family, and the addition of a new source of authority to which women might appeal. A girl betrothed against her will could cite the customary law and ask for the intervention of the local churchman. A wife about to be divorced by her husband could appeal to the Church to maintain the marriage. In addition, the Church periodically passed through waves of religious enthusiasm and reform. In these eras, as in the first decades of Christianity, women joined new movements and performed a wide range of roles within them. They preached, prophesied, and taught, and thus were allowed a sense of functioning as the equals of male believers.

But these eras of ferment, experimentation, and change passed rapidly. Each was succeeded by a far longer period of consolidation and conservatism. As in the third and fourth centuries, these times of consolidation reinstated—often in even stronger form—the age-old traditions of male dominance and female subordination. Female subordination easily became associated with tradition, order, and orthodoxy. Allowing women more seemed dangerous, unnatural, and counter to standard views. The Church, like the family and the state, fundamentally believed and taught that women were inferior to men and should be their subordinates.

In the end, European women, like the women before them, would live in a culture whose values, laws, images, and institutions decreed their inferiority and enforced their subordination to men. Female subordination was the most powerful and enduring tradition inherited by European women.

II

WOMEN OF THE FIELDS

•

SUSTAINING THE GENERATIONS

1

THE CONSTANTS OF THE PEASANT WOMAN'S WORLD: THE NINTH TO THE TWENTIETH CENTURIES

The Life

Until the last decades of the eighteenth century 90 percent of Europe's women lived in the countryside, dependent upon the land and what it could produce.[1] Monographs about specific communities, about different aspects of their evolution describe regional and temporal variations in these lives. For example, they tell how climate and geography caused differences in seasonal activities and crops produced, in the use of building materials, how ancient customs led to varied patterns of landholding, how combinations of economic, social, and political factors brought the adoption of new attitudes and new methods of cultivation to one region and not another, in one century and not another. Yet for hundreds of years, in thousands of villages across Europe, the constants of peasant women's lives far outweighed the differences of place and time.

Generation after generation their lives bear an awesome similarity to each other. The margins of a fifteenth-century Book of Hours show a peasant woman in the fields in her broad-brimmed straw hat. She has tucked the edge of her overtunic into her waistband so that her legs can move easily. They show a woman churning butter, another catching fruit in her hat in the orchard, another plucking a bird for a holiday meal. In a nineteenth-century painting by the Belgian artist Henry van de Velde, a peasant woman rakes the grain dressed in a pale gray blouse and skirt and a clean, white apron; she wears her reddish hair in a bun high off the nape of her neck. She holds the wooden rake just as the woman of the fifteenth-century Burgundian Book of Hours did. It was long handled and light, so that the women could easily pull the long grasses into piles and lift them onto the wooden carts.

In the 1950s Teresa, a peasant woman from the village of Torregreca in southern Italy, assisted her husband in the harvest in the same way as the woman pictured in an English psalter from the fourteenth century. Both walked behind the man, bent over, catching the armfuls of wheat as they were cut with the sickle. Both skillfully made the stalks into bundles. Teresa lifted the cut wheat from his left arm. She divided the bunches of grain by eye, turned the halves heads to bottoms and with a quick twist of a few strands bound them together and let them fall to the ground.

Others made these images. Peasant women themselves left few records of their lives. Well into the twentieth century the vast majority could not read and could write no more than their name for a village notary.[2] Yet the historian can re-create rural women's world and their concerns. Household accounts, manor rolls, court and parish records, note these women's obligations to the landlord, the fines they owed, the births and deaths of their children. Fourteenth- and fifteenth-century illuminators, artists from the sixteenth to the nineteenth century, pictured them at their seasonal tasks and celebrations. In the nineteenth and twentieth century, folklorists and anthropologists recorded their words, their songs, their stories, and photographed them in their traditional clothing at their traditional tasks.[3]

All testify to the sources of peasant women's strength: their veneration of the land; their acceptance of responsibility for the survival of the family; their willingness to work; the comfort they took from their beliefs. These dictated the rhythms of rural women's lives and the choices they made through the centuries.

First there was the land; still precious in the 1950s when Teresa of Torregreca in southern Italy saw her priorities as no different from those of country women hundreds of years before:

> To me my job is to see this land gets farmed, to raise up the children, make them go to school and teach them what's right and get work, as much work as I can get while I'm still young enough to work in the fields, because that's all I know. . . .[4]

Cettina, also of Torregreca, explained how different the perspective on life was for women. "We have to care," she explained. "We're the ones who have to make do, however it is."[5] Women concerned themselves with the practical aspects of life and accepted responsibility for the survival of the family. "If the father is dead, the family suffers," went a nineteenth-century Sicilian proverb. "If the mother dies, the family cannot exist."[6]

To ensure survival, the peasant woman's life was one of continual labor, as described by the Irish proverb still current in the twentieth century: "One woman in the house be always working."[7] From the time they were small children through maturity to old age, rural women expected to work. They knew no division of labor, no separate spheres for women and men. They worked everywhere. They performed all but the heaviest tasks having to do with the preparation of the fields and the harvest. They helped to plow, to spread manure, to weed, to reap, to thresh. They did all the work of the household. They gathered kindling, hauled water from the well, tended the fire. They gardened, they tended the animals, nursed the children, cooked the food, swept the clay floors. Well into the twentieth century, in addition to the household tasks, the fields, and the animals, they also did extra jobs to make money needed for rent, for taxes, for necessaries. So they hired themselves out as day laborers or laundresses, they sold cheese and butter, they worked their spindles, knitted, made lace. And always there were children to care for, pregnancies, infants to breast-feed, all around and about the other tasks.

The multiplicity and the never-ending quality of their work sets peasant women apart from their contemporaries and endured as a constant reality generation after generation. In a fourteenth-century English poem, Piers the Ploughman calls his wife Dame *Work-while-you've got a chance.*[8] Little differed between his description of the woman waking in the middle of the winter night to "rock the cradle cramped in a corner," rising again "before dawn to card and comb the wool, to wash and scrub and mend . . . and peel rushes for their rushlights—" and the memory of a Breton son of his mother's day at the end of the nineteenth century. She dressed and arranged the folds of her coif (the headcovering commonly worn by Europe's peasant women). Then she tended to the animals—a pig to prepare slops for, a cow to milk. She fed the children and sent them off to school, worked a small piece of lace as she walked back the half-league from where she had pastured the cow. Some housework, some clothes to wash, and the tasks of the day continued:

> get the midday meal ready, crochet as she returned to the field, work the land with as much strength as she could muster, come back pulling the cow by its rope and with a load of hay on her back or a heavy basket on her hand, find her children at home, make them behave and do their homework, mend the worn clothing, rage and fume or laugh heartily, depending on the circumstances, cram some more food down the pig, milk the cow a second time, cook the gruel or potatoes, do the dishes, put the whole brood to bed, tidy up, return to her crocheting or her sewing . . . wait for father and not get to bed until he had.[9]

The Breton woman's children went to school—a change brought about by state-supported public education in the nineteenth century—otherwise her life was little different from that of a peasant mother in fourteenth-century Montaillou in France, in a sixteenth-century English village, in twentieth-century Yugoslavia.

The days, months, and years of seemingly endless and unchanging tasks could lead to resignation. "I have rocked my babe, and willed to it that things be different," went a lullaby a Russian peasant mother sang to her child.

> There'll be some changes,
> There'll be some bread.
> But for me there'll be no change,
> I'll see nothing different,
> I'll eat no bread and salt.[10]

More discouraging, living as she and her family did on the edge of subsistence, the fruits of her labors could be suddenly wiped away. War, disease, and extremes of weather could bring years of catastrophe that tested peasant women's strength and endurance. "When necessity is upon us, we must suffer," went a fourteenth-century English proverb.[11] In the 1950s Teresa received two hectares (about five acres) from the land reforms in southern Italy, which she tended while her husband Paolino worked in Germany. She lived in the fields in a lean-to with no window and the tools stacked in the corners. Two years she plowed and prepared for the harvest and then the hailstorm came:

> My heart pounds and I keep thinking I'm going to be sick. That time I found it all lying on the ground and chewed looking. I came in here [to the storeroom] and sat and cried for hours. Every time I felt a little better and I started to do something—I don't know, mix the feed for the chickens—I'd look out that door and see the fields again and I couldn't help it—I'd cry.[12]

Countrywomen might give way to resignation, to bitterness, to rage, but rarely to defeat. They found ways to explain, to justify, and to give value to their lives. Peasant women took comfort from their faith. They believed that they could influence the uncontrollable, even make sense of and bring order to the often brutal and capricious life of the countryside by spells and prayers, by ancient explanations and rituals. Countrywomen and men made no clear distinctions between material and spiritual reality, between the natural and the supernatural, the living and the dead, the real and the imaginary. This was a world of "wishing" as countrywomen described it in their folktales. This was a world in which ghosts might appear at the fireside for wine once a year, a dog could turn into a raven, the croaking of a frog foretold the future, and a blue bead would keep away "the evil eye." Well into the twentieth century, peasant women and men continued to value and attempted to influence this world of the spirits. Thus they perpetuated age-old beliefs and customs in the modern countryside and in the towns and cities to which they migrated.

This world of older beliefs and customs did not conflict with the comforts of more formal religion. The shrines of Christian saints were built over older places of worship, and the Christian holy days came to coincide with those honoring other goddesses and gods and with days honoring the natural phenomena of the agricultural year, like the solstices. Formal religion gave countrywomen comfort in other ways. It explained and justified the harshness of life, honored their activities, and offered rewards in the future. Divine displeasure could explain a plague of caterpillars or an earthquake. A divine plan could make their place in the village world seem essential and honorable. In a German folktale God told Eve:

> It is proper and necessary for me through your children to provide for the whole earth. Were they all princes and lords, who would cultivate the grain, thresh, grind, and bake? Who would forge, weave, hew, build, dig, cut, and sew? Everybody must have his place, so that one may support the other and all, like the parts of a body, be nourished.[13]

Christianity gave explanations of the creation of the world, the source of death in Eve and Adam's fall from Grace, and promised the redemption of the faithful through the crucifixion of Jesus. Christianity explained that by God's mercy, by prayers, by dedication to the rituals of the religion, all believers could achieve bliss in the Christian heaven.

Whatever happened, however little they had, Europe's peasant women took pride and pleasure whenever they could, balancing the reality of today with hope for the future. "May colorful clothes bedeck your shoulders!/ And good thoughts fill your mind!" went another Russian lullaby.[14] They valued what they had, what they could do, and what they could teach their children. The tales they told stressed ingenuity—it was the "smart and artful" who survived in the German stories collected by folklorists in the nineteenth century.[15] Peasant women emphasized the practical: "Don't promise us a crane in the sky, give us a titmouse in the hand," went a traditional Russian proverb.[16] They had the companionship of other women at the times of greatest trial: in childbirth, in sickness, and in death. And they rarely allowed themselves to question. "People say they can't make out how we lived that life," remembered Mary Coe, born in an English village in 1889:

> But I say, remember, we'd never had anything else. Our parents before had had the same life. You see, when you've never had anything, you never miss it.[17]

"You never miss it," explained Mary Coe, and so it was until the beginnings of industrialization eased some aspects of the countrywoman's tasks. Throughout the centuries peasant women gave their energies to the basic needs of their families, to survival. Even so, they could sometimes make special gestures for their daughters. Mary Coe remembered that when her mother provided the Christmas present of an apple or orange it was "ever so grand."[18]

The Landscape

The landscape the peasant woman inhabited changed little over the centuries. More families were able to live from the land, the uses of the land changed, but for much of European history, it looked as it was portrayed in the fifteenth-century Books of Hours: a village, the fields with no signs of ownership, just ditches, thorned hedges, or woven twig fences protecting the plantings from the animals. Peasant women saw some furrows lying fallow, others sown with grains, others plowed and waiting for the next harvest cycle. They saw vineyards, orchards, a common pasture, the parish church, the fortified manor or castle of the local lord, who retained some of the lands for himself as his domain (demesne), and gave over others as tenant strips that families worked for themselves.

The forest, called "the waste" by rural women and men, filled the rest of the landscape. Well into the seventeenth century, the forest held great significance in their lives. All were menaced by its dangers. In peasant women's folktales the boars came out of its depths and trampled crops or wounded livestock; the hares ate the cabbages of the household garden; countrywomen ground wolfsbane and monkshood into an aconite powder to poison wolves, foxes, and rats. Yet, all could use the forest. From the forest floor peasant women gathered wood for their fires. Along its paths lived men who helped their families: the woodcutter (like the one who saved Red Riding Hood) and the charcoal burners who gave them fuel, the sabot maker who carved their wooden shoes.

Throughout the centuries countrywomen and their families used the forest to build their houses and the lean-tos where they stored food and sheltered animals. Unless unusually prosperous, when they would use stone (like Joan of Arc's fifteenth-century French family and the yeomen farmers of sixteenth-century England) peasant women and their families built in wood and other easily accessible materials like birchbark in Norway, thatch from the meadows and thickets for the roofs of southern England, or the peat "soddies" of western Ireland. All across Europe use of the forest was part of peasants' rights, whether they were serfs or tenants.[19]

The houses the peasants built for themselves were small and simple in design. The lengths of the available timber determined the size of the beams and thus the width of the house and the height of the ceiling. Local stone placed under the key supporting posts kept the walls from sagging. Woven twigs stuffed with moss and overlaid with plaster made from lime, water and sand or clay made the walls—the "wattle and daub" of English folktales. Dirt, clay, or flagstones made the floor. Given the limitations of their building materials, peasants did not change the design of their houses over the centu-

1. Peasant couple harvesting grain. From an Italian manuscript, fifteenth century.

2. Peasant woman making ricotta cheese. From an Italian manuscript, fourteenth century.

3. Danish milkmaid, late nineteenth century.

4. Mother and her child. Hungary, twentieth century.

5. French peasant family, 1912. Photographed by the composer Ernest Bloch.

6. Portrait of a French peasant by Françoise Duparc, eighteenth century.

7. The Witches of Mora. Engraving from Sweden, 1670.

8. Soviet peasant woman operating a cotton picker, 1970s.

9. Women talking in a village square. Southern Italy in the 1950s.

ries. Into the twentieth century, wherever they lived in Europe, peasant women had one large central room, anywhere from 5 to 25 meters long, about 4 meters wide, with one section for people and one for animals.[20] The building would be one story, sometimes with a sleeping loft, sometimes with rooms added at the end of the house when an aging mother and father had given over the land to daughters and their husbands or to their sons. They covered whatever windows they had with translucent substances like waxed cloth, horn, or cut talc. In the warmer climates, there was no covering, and they simply closed the wooden shutters when colder weather came.

Though the centuries would bring changes in cultivation and increases in the population, neither altered the basic environment in which peasant women lived. In the ninth, tenth, and eleventh centuries, they lived with their kin in the least forested areas, tilling the easiest, loosest soils, using natural clearings for their small household gardens where they raised the legumes, roots, and cabbages on which the families subsisted. Sporadic periods of peace and order, warmer weather in the eleventh, twelfth, and thirteenth centuries allowed families to cut and burn into the edges of the forest, clearing new land, strip by strip. They drained inland and coastal marshes with dykes. Gradually the strips joined to become new fields, new fields met old settlements, and families moved across Europe to new valleys. Thus more could live in the old areas. From the fourteenth to the seventeenth centuries, despite less favorable conditions—new kinds of wars, colder, wetter winters—the consolidation of settlements and cleared fields followed much the same patterns: thirty to forty households clustered together in the England of the mid-fourteenth century; sixty-eight households in a southern French village of the mid-seventeenth century; still about fifty families, their houses together surrounded by the fields for cash and food crops, in a Greek village in the 1960s. Only in northwestern Germany and in Scandinavia did peasant families live in sparsely populated groups, and only from the Elbe to the Urals would they see vistas of cleared plains. For the majority of Europe's peasant women, the essential elements of the landscape, like the essential elements of their lives, remained the same through the centuries: the clearly demarcated settled areas, the houses clustered in the village with the fields all around; the forest, and the spires of the market town in the distance.[21]

The Year's Activities

Europe's peasant women and their families marked the cycle of the seasons with the feast days of the Church and the labors that filled out the months of the agricultural year. Feast days and seasonal changes fell into a pattern for Christian Europe: Lammas (1 August) to Michaelmas (29 September), the time of harvest; Michaelmas (29 September) to Christmas (25 December) for preparation of food for the winter; Christmas to Easter, the hardest time of the year waiting for the first grains, corresponding to the Lenten days when fasting became a virtue; Easter or Hocktide (the Monday or Tuesday after Easter) to Lammas (1 August), the summer growing time, with the first of May a traditional date for the celebration of spring and St. John's Day, June 24, midsummer day. For peasant women the sense of beginning, of a new year of possibilities, came not in January, but at harvest time. The illustrations for the fifteenth-century Books of Hours show the barley, wheat, and rye ready to be cut, the hops for brewing, the purple and green grapes clustered on the vines ready for picking, the honey waiting to be collected from the hives. Throughout the continent all celebrated the abundance of the harvest.

Harvest Time

Until the early part of the eighteenth century, a Russian peasant woman killed the goose she had raised for the eighteenth of September to celebrate the coming of the New Year. Into the nineteenth century in England, the goose was fattened on the stubble of the harvested fields for the Michaelmas celebration on the twenty-ninth of September. Others celebrated on All Hallows' Eve, All Saints' Day, and All Souls' Day (31 October, 1 November, and 2 November), Christian days replacing the harvest festivals of earlier faiths. In the sixteenth century near Nuremberg, the peasants danced in the meadows. Into the nineteenth century English families made it a time for revelry, for tricks, for feasting on spiced cakes and frumenty made of boiled wheat, milk, raisins, and currants that the women had dried in the late fall. For English and French villagers harvest was also a time for propitiating the dead and for honoring the local saint with special masses. In Brittany the whole region participated in the procession to the saint's church or shrine and the feasting afterward.

Celebration came after the hard labor in the fields. The peasant woman and her husband harvested their own strips. In addition, as a servile tenant (tied to the landholder by obligations of service and goods owed in kind) she would be summoned to a "bid reap," as it was called in thirteenth-century England. A woman might cut with the hand sickle, but more often reaping was men's work with the heavy two-handed scythe. Instead, the peasant wife, like the woman at the Abbey of Werden in the Rhineland of the ninth century, would "bind the sheaves, collect them and pile them up . . ."[22] These continued to be the countrywoman's tasks, whether in the fourteenth-century French village of Montaillou, on the lands of the fifteenth-century Dukes of

Burgundy, in southern Italy in the 1950s. This was also the time for harvesting flax and hemp—the hemp for making rope and sacks, the flax for linen. Then, with Michaelmas at the end of September, the cycle of plowing for the year began again. The peasant woman and her children herded the cattle into the cut field to eat the stubble. Other fields left fallow would be prepared for planting. According to English tradition, the winter grains had to be in the ground by All Hallows Eve. If not helping with the plowing, the peasant women might bring a midday meal of bread, cheese, and peppers for the men.

Cutting, binding, and hauling the sheaves did not mean the end of the fall labors. Much had to be done before peasant women had flour, the staple of their families' diet. In thirteenth-century England, a country wife and her husband used the oxen to thresh the grain from the stalks; her job was to winnow the grain from the chaff. In nineteenth-century Brittany the men flailed the grain, the women raked and swept it into bags. On a windy day peasant women brought the bags to the field, throwing the grain into the air, catching the chaff on a cloth to use as new stuffing for pillows and mattresses.[23] Once the wheat was separated from the chaff, a peasant wife might store it in baskets she had woven from marsh grasses. From the eleventh century on, she could take it to the grinding mill instead of doing it by hand.[24]

The fifteenth-century Books of Hours show mousetraps, or the family cat sitting by the stone fireplace, all to protect these valuable grains that ensured the family's survival. Into the eighteenth century in France, for instance, grains made up 95 percent of the diet.[25] In the earlier centuries, peasant women usually baked their bread in the lord's ovens. In Vasilika in twentieth-century Greece, they used their own. During the harvest season Greek peasant women rose at two or three in the morning to bake for the whole week and then went to their work in the fields. Aside from coarse black bread, peasant women made a variety of other foods from the grains: oatcakes, pancakes like blinis and crêpes, porridges and gruels. In the fall and winter months rural women added to the family's meals with the vegetables they produced from their household gardens. Vegetables were the other staple of the peasant's diet into the twentieth century. From as early as the ninth century, there is evidence that peasant women grew peas, vetches (wild peas), broad beans, and roots like turnips, onions, radishes, parsnips, leeks, carrots, and beets. From the sixteenth century on, they also had potatoes. In nineteenth-century France, the Breton son remembered that in his childhood "potatoes had kept us alive lots of times."[26]

The peasant mother and her daughters prepared one main meal a day for the family. The hour shifted, depending on local custom. A seventeenth-century Yorkshirewoman had it ready at noon, her eighteenth-century counterpart in France prepared it for five in the afternoon. Throughout Europe peasant women served the same basic fare: bread and a thick broth, much like a stew, made from peas and beans they had dried, from roots they had stored. The big cauldron simmered on the hearth fire for hours; either water was boiled to cook vegetables in cloth bags, or the fire was set in the cauldron itself, while an earthenware pot containing the broth and stewing ingredients was placed inside. Leaving this to simmer, a peasant wife went about her other tasks. In the earlier centuries families drank wine, ale, or cider, all harvested and prepared in the fall months. In the *Trés Riches Heures,* the fifteenth-century Book of Hours of the Duke de Berry, peasant women and men pick the grapes, the women carry the heavy baskets on their heads toward the manor castle. One peasant woman stands adjusting her cloth headdress, her white apron bulges over her blue and red dress as if she were pregnant; she leans favoring one side perhaps to relieve the ache in her back. From as early as the ninth century in France, peasant women and their families helped in the wine making in October and November. Seventeenth-century English brewsters made beer in the fall to sell to other villagers.

What meat countrywomen and their families had was prepared in these months of late summer and fall. The livestock and the fowl were women's responsibility. The peasant wife took charge of the litter of five to six piglets born in the early spring. She weaned them at six months and then in the fall months fattened them on nuts in the forest. In eleventh-century England on the manor of Ramsey Abbey, peasant women collected the acorns, but were allowed only enough to fill "hose of reasonable size."[27] In seventeenth-century Beauvaisis peasants purchased *glandée,* the right to collect nuts. Best of all would be to have the right of pannage, like the women of the Books of Hours who let the pigs run loose and threw sticks into the oak trees to make the acorns fall to the ground.

The sow would be saved for the next year, the litter slaughtered. Thirteenth-century English women called November "blood month." The fifteenth-century Book of Hours of the Duke de Rohan showed a man wielding the long-handled ax to slaughter the pig. A peasant woman used every part of the animal. Into the nineteenth and twentieth centuries, she caught the blood as the throat was slit, adding it to her grains as blood pudding, putting it in casings as *Blutwurst.* She boiled the carcass in water so that she could scrape off the hair for brushes or to add to plaster for her walls. The bladder could be used to store the lard, the trotters boiled for their gelatin. In many villages the butcher went from house to house cutting the meat, which peasant women then stored covered with salt in barrels, or hung from the rafters over the hearth or up the newly cleaned chimney to smoke. The salted and smoked pork would be the family's principal source of protein in the long winter months to come.

Peasant women and their families might have access to seafood. In the fourteenth century references are first made to salting and preserving herring. Salted whitefish and cod might be bought at the local fair, or the men might catch freshwater fish—eels and flukes—with nets and weirs in the mill pond. The men would also be responsible for special meals of small game, rabbits, or wild birds netted on the fall migration.

In most instances, the livestock and fowl proved too valuable to eat, and peasant women had other uses for them. They fattened the cows on the stubble of the harvested fields, sold them at the fall market and purchased new ones in the spring. Thus they did not have to feed the animal over the winter when marsh grass might be the only fodder available. The wife in a German folktale chose to slaughter the cow and sell the hide, first tanning it to pliability by soaking it in water and oak bark. Only well-to-do families ate mutton; for the rest, the sheep were more valuable for their wool or for the price they would bring at the market.

The same was true of the chickens and the geese. Well into the eighteenth century, the chickens and their eggs were kept to pay the dues owed the landlord or for the market. Local fairs came, as they did in fourteenth-century Montaillou in southern France, on All Saints' Day (November 1) when peasant women walked fifteen to twenty miles to sell their surplus. One goose would be for the harvest feast day, in Germany and Denmark for St. Martin's Day on November 11. Peasant women wasted nothing; they plucked the feathers before the carcass cooled. The wings became feather dusters. A Danish wife needed twenty-four geese to make a feather bed filled with down.[28] An English peasant woman dried the bigger feathers overnight in the oven and used them for lamb and baby bottle nipples and for spigots. As the goose roasted over the fire, the peasant woman collected the grease, a lubricant, a waterproof coating, something to protect the pig's skin from the summer sun. Some of the grease would go for a holiday pastry or pie, sweetened with honey from the manor's hives.

The Winter

Christmas and Easter, the two most important festivals of the Christian calendar, framed the hardest part of the peasant woman's year: the months before the spring grains; the months on a Breton farm of the nineteenth century described as "truly black, not only because break of day was late and the night came too quickly, but because of the black cold, the black mud, the black rain, the black wind."[29] European peasants celebrated the twelve days of Christmas, merging the holy days from Jesus' birth on December 25 to January 6, the Epiphany commemorating the visit of the three kings, with older celebrations of the winter and its solstice (December 21–22). Long before Christianity, Europeans burned the "yule" logs and collected the sacred mistletoe.

To this tradition churchmen added ceremonies celebrating the nativity; peasant families added rituals and festivities to brighten and cheer the dark winter days. In fourteenth-century Montaillou in France, Christmas was a market day. On the Dean of Wells's manor in North Curry, Somerset, it was a communal feast, with the wife of John de Cnappe having produced loaves of bread, ale, bacon, cheese, a hen, and two candles for light.[30] Into the nineteenth century in Russia, peasant women broke the seven-month winter with their Christmas celebrations, decorating fir trees, making dough crosses to mark the day. In Caltagirone, Sicily, into the beginning of the twentieth century, peasant mothers and fathers offered their two-year-olds to play the child in the manger as their alms to the church for the Christmas mass.

There would be little for rural men to do in the months after Christmas beyond preparing the fields for spring planting. In England, Candlemas, February 2, commemorating the churching of the Virgin after the birth of her son, signaled the start of spring plowing; the wife and husband plowed the fields together—one day to do one acre.[31] He guided the plow while she led the team of oxen or horses along the strips. By Lady Day, March 25—the Feast of the Annunciation celebrating Jesus' conception—the men had sowed the spring grains. Most of these months, however, peasant families spent indoors, around the central hearth of their main room, the ceiling blackened from its smoke. The hearth, or in a more well-to-do home, the wall fireplace, was their source of heat and light. A woman of central Europe might have a tile stove or a clay oven instead. Peasant women used dung or peat that they had dried in the fall sun for fuel. In France they used wood and claimed the right of estover to collect it from the landlord's forest. Countrywomen had resin from evergreens for torches, the herb mullein dipped in beeswax, or sheep fat, for lighted wicks. In the poorest houses peasant women kept their cows or goats inside for warmth and to make it easier to collect the manure. They used the plants savin and rue to sweeten the air and to keep away the fleas and lice that plagued the family during the long months when they could not bathe. Peasant women knew how to mix the juice of the houseleek and sage with water to ease the itching and pain of the insect bites.

The illustration for February in the Duke of Burgundy's Book of Hours shows the countrywoman's winter world. She sits on a bench in front of the fire with her skirt pulled up, her feet extended to warm them. Her wash hangs from rods suspended along the edges of the ceiling. In Russia the poorest peasants had earthen ledges along the walls to sit on. Most peasant women acquired other furniture and household possessions. A husband might have planed pieces of wood into the family's trestle table, which the wife polished clean with sand or cinders. He might have carved bowls that she stored with earthenware pots and plates in a cupboard. Old barrels could be cut down for stools. Into the nineteenth century the family bed was her major possession. In poor households it might be nothing more than a raised plank set by the chimney for warmth, but in most households the bed was a more elaborate wooden structure, complete with doors and hangings to keep out the cold.

Anything with metal peasant women prized particularly. The thirteenth-century English widow, Christina of Long Bennington in Lincolnshire, noted her ironbound chest in her will. In the sixteenth century, Thomasina Whitehead of Farnsfield in Nottinghamshire listed her three brass pots, her kettle, and her candlestick as part of her wealth; other women like Joan Atkinson and Katherine Francis left their metal and pewter to their daughters as their legacy.[32] In nineteenth-century Brittany, a son remembered that his mother polished the brass nails in the furniture every Saturday morning.

Most of the peasant woman's time during the winter months went to providing cloth for her family. From as early as the ninth century peasant women were responsible for all aspects of fulfilling the family's needs, from collecting the raw wool and flax to finishing the shirt or coverlet. The simple standing loom rested against the wall. In addition to providing for their families, as tenants into the twelfth century in Germany and France, peasant women owed days of service in the lord's sewing rooms or finished pieces of linen or woolen cloth.[33]

The primacy of this female responsibility, especially the making of thread for the cloth, goes beyond written memory. The image of the peasant woman spinning—a long stick to hold the raw wool, her distaff, in one hand, the spindle used as a spool in the other, with the thin thread of wool held taut between them—lasted into the twentieth century. (The spinning wheel was not commonly used by European countrywomen until the sixteenth century.) This simplest method of spinning gave names to the female: the "distaff side," the "spindle side" of the family. In 1378 the rebels of the English Peasant's Rebellion made the image part of their rallying cry: "When Adam delved and Eve span/ Who was then a gentleman?" Nineteenth-century Frenchwomen of Baugeois spun hemp as an offering to the Virgin on their wedding day.[34] Grandmothers told stories and sang ballads of magical ways to spin, cut, and sew. The dwarf Rumpelstiltskin spun gold out of straw for the queen, the maid in "The Elfin Knight" had to sew a shirt "without any cut or hem," to "shape it knife-and-sheerless," and to sew it "needle-threadlesse."[35]

In reality, there was no magic, just the painstaking labor of turning raw wool or flax into a length of cloth that could be used as a covering, or cut and sewn into a shirt or a tunic. The finer strands of wool became worsted, the coarse hairs of the sheep woven and beaten became the matted and teased heavier cloth called felt. The wool peasant women worked by the winter hearth came from the shearing done in the previous spring or summer. One fleece could weigh as much as 18 pounds, but almost half of that was grease.

So the peasant woman's first task was to clean, beat, and then reoil the fleece to keep the fibers elastic enough to spin and weave. Then they picked the fleece clean, carded it, separating the strands with a teazle (thistle) from the fields. After the thirteenth century, they might have carding boards: two wooden paddles with metal teeth that passed one over the other combing the wool in between. With these they made light fluffy rolls that could be spun into thread. A peasant woman might walk about her tasks with the distaff and spindle, or sit on a stool like the woman in the Book of Hours of Catherine of Cleves. As the spinning of one roll into thread was completed, she would twist the end together with the next and make enough to fill a skein—forty rolls. The tighter she spun, the stronger the thread. A good spinner could complete five skeins in a day. Five skeins would be enough to make a sweater. In the earlier centuries, peasant women wove the threads, but from the fifteenth century on, they might choose to knit them.

Peasant women made linen from flax, an annual that they and their families harvested in the fall. The stem became the fiber that could be spun and woven into cloth. A peasant wife began by removing the seeds, soaking the stalks in water, and then drying them to make the outer layer rot and crack, a ten-day to two-week process called retting. Having pulled off the outer covering, she softened the stems by beating them with a mallet. From the fourteenth century, the job was easier, as peasant women used a "flax breaker," two grooved rectangular pieces of wood hinged together, between which the stems were crushed by repeated blows, to break the outer husk. Then peasant women combed and recombed the stems, splitting them into single strands. As with wool, they wrapped the strands around the distaff, drawing out two or three at a time to twist onto the spindle. The resulting thread was stronger than wool, but brittle, making the weaving more difficult. In Greece and southern Italy rural women had cotton as well; they collected and cleaned the bolls, spinning the rolls into thread.[36]

Leaving the wool or the linen thread to their natural grays and browns was the easiest and most common practice. To add color, peasant women collected roots, leaves, and lichens from the forest to use as dyes. They knew which dyes needed to be fixed with mordants to hold the color, and which did not. (Alum and cream of tartar are the modern-day fixatives.) The lichens of northern Europe gave rich brown tones, paler oranges, reds, and yellows. Birch bark made a brown. Madder gave a pale pink, goldenrod and barberry a yellow. Boiled greenwood leaves with a mordant made a green dye. Blackberry would give a lavender or blue, woad a richer hue. For black the women used walnuts or flag iris.

With the skeins of natural or colored thread, the peasant woman was then ready to go to her loom. Into the nineteenth century in many parts of Europe countrywomen worked looms similar to those used in the ancient world, a simple rectangular frame made of wood. They unwound the spindle and measured off the vertical threads in roughly equal lengths. This was the "warp" of the future piece of cloth, one end of each thread attached to the frame, the other to a stone or small clay weight. The peasant woman set the loom upright, tilted against the wall, and let the warp threads go taut, hanging down vertically across the wooden frame. A horizontal loom like the ones in the Books of Hours might simply mean the frame had been placed on legs, like a table with the warp threads still running across it and still held taut with stone weights. To weave the women took more thread and wound it on to a shuttle, which they passed back and forth through the warp threads, thus creating the "weft." This technique turned the spun thread into "plain weave," a rectangular piece of cloth whose size was determined by the size of the frame, perhaps one meter by two meters. Even this simple weaving took time. In the 1960s in a south Italian mountain town, a mother and daughter allowed fifteen to twenty days to make the three pieces of cloth necessary for one sheet, part of the daughter's trousseau.[37] To whiten the linen, women boiled the cloth in lye made from the fire's ashes, washed it over and over again in the local stream, and then left it to bleach in the sun for six to nine weeks.[38]

Anything with more than one color, with any kind of pattern, increased the complexity of the weaving, necessitating more than one shuttle and requiring that the weft threads be changed. At the beginning of the twentieth century in Sicily, the widow Gna Tidda made fabric of this kind. She worked a more complicated horizontal loom with foot pedals. One pedal would lower half the warp threads, she would pass the shuttle through, then press the other pedal to reverse the process. By pulling the beater toward her—a wooden bar suspended from above, with comblike "teeth" that separated the warp threads—she could tighten the threads into closely woven cloth. Gna Tidda could even widen the cloth, keeping track of the changes she made by putting a kernel of corn into her lap for every twenty-five threads added.[39]

In the early centuries peasant women finished the woolen cloth themselves, treading on it in water to fill in the weave, fluffing the wool with teazles to "raise" the fiber, clipping it to make it smooth, then drying it stretched on poles to block it. By the thirteenth century the landlord would have a fulling mill powered by water or wind to do this. The linen would be beaten and trampled but it did not need to be raised, clipped, or fulled at a mill.

Finally, peasant women would be ready to make the articles their families needed. They took the rectangular pieces of cloth and sewed them together to make sheets, coverlets, tablecloths; they cut old pieces as swaddling for a baby. With a minimum of cutting, with ties rather than buttons, they made simple undershifts, or tunics to use as overgarments. A cape or a cloak with a hood would be just another version of strips of cloth joined together. No one had many clothes. In fourteenth-century Montaillou, the wives and husbands took them off when they went to bed, but otherwise wore the same shirts and tunics over and over again. An agreement between a father and his children in thirteenth-century Weedon Beck promised him one garment and one set of linen drawers for the year.[40] The peasants on the Datini estate in northern Italy in the late fourteenth century had one gray tunic each. In fifteenth-century France Joan of Arc spoke of her red woolen dress. Into the 1500s only the length of the garments and the woman's headcloth clearly distinguished her clothing from a man's. In nineteenth-century Russia, both women and men still wore the long shift and the tunic overgarment typical of the earlier eras.

Throughout the year peasant women devoted time and attention to some articles like the special dress for their weddings, or the bedclothes and table linens that would be part of a dowry. As early as the fourteenth and fifteenth centuries in Hungary and other regions of eastern Europe, women embroidered the cloth for their wedding dress, making it look like a fancy woven fabric. In nineteenth-century Sicily young women still embroidered linens; in the 1960s in Vasilika, Greece, they still wove woolen rugs for their trousseaus. Throughout the centuries, the more prosperous the peasant woman's family, the more she would be able to make, the more she would possess. The few English peasant women wealthy enough to make wills, like Christina of Long Bennington in Lincolnshire who died in 1283 and Joan Atkinson of Halam who died in 1561, listed linen sheets, towels, pillows, blankets, bolsters, coverlets, and a tablecloth as their valuables.[41] In eighteenth-century Sologne, France, the bolsters, covers, and featherbeds constituted 40 percent of the family's assets.[42]

The Spring and Summer Months

In the countryside spring created a special excitement after the long gray months of cold, wet days and nights. The margins of the Books of Hours blossom with wildflowers: mallows, primroses, cornflowers, daffodils, iris, bluebells, and daisies. In their tales the peasant women described fields of blue flax in flower, the supple new willow twigs which they wove into baskets. The forest became a welcoming place of dappled sunlight and trees coming into leaf. In their stories the Virgin found wild strawberries there, and hid behind a hazel bush when frightened by a snake. Russian peasant women into the nineteenth century sang to "mother spring" on March 9 to welcome the returning birds.[43]

Beginning with the Annunciation (March 25), the feast day of the Virgin that coincided roughly with the vernal equinox (21/22 March), peasant women and men marked this change of seasons as they had for centuries with a series of festivals celebrating the rebirth of the sun and the fertility of the soil. The Christian church absorbed many of the ancient rituals. In Joan of Arc's village of Domrémy in northeastern France, the dancing and singing by the young people at the "fairies tree" came on the fourth Sunday of Lent. Spring saint's days, like Saint Berlinda's of Belgium, became occasions for drinking and dancing. Even the holy days of the Passion, Palm Sunday, Maundy Thursday, Good Friday, and Easter followed older traditions. All across Europe, from England to Russia, from the thirteenth to the nineteenth century, peasant women dyed eggs, or brought them to the landlord as goods owed for Easter—a word similar in sound to an Anglo-Saxon goddess of spring. In England, Hocktide, the Monday and Tuesday after Easter, became days of playful courtship, the women leading the men around on Monday, the men leading on Tuesday, releasing each other only on the payment of a "fine." May Day survived despite efforts by rulers like Charlemagne and Catholic and Protestant churchmen to ban it and continued to be a time when young people danced around Maypoles, stayed out all night lighting bonfires. Whitsunday (13 June) occasioned a week's holiday in thirteenth-century England. The summer solstice (21/22 June) became transformed into St. John's Day (24 June), and often coincided with the first cutting of the hay.

In this season from Easter to Lammas (1 August) peasant women returned to work in the fields after the men had seeded the spring grains. Rarely did popular beliefs anywhere in Europe allow women to sow the fields for fear of endangering the harvest. Instead, beginning as early as April, women had the more arduous tasks of hoeing, weeding, and mulching the newly seeded crops. They had to do these in addition to their many other chores. The "housewife" of Thomas Tusser's sixteenth-century guide to husbandry had planted her own garden, the flax, and the hemp by the beginning of March. Into the nineteenth century in Ireland women cut the turf for the family's fires in May. In thirteenth-century England peasant women had sheared the sheep by St. John's Day at the end of June.

Spring was the time for the laundry. Lye, the whitener, came from the ashes of the fire. In nineteenth-century Brittany the women did a big wash twice a year, in April and in October.[44] On the first day the clothes went into buckets, covered with ashes and then boiling water. The second day they were beaten with paddles, the third day spread in the sun to dry and to bleach. In the twentieth century the long water trough in the piazza of a Sicilian village was a meeting place, the washing the occasion for the countrywomen to talk, laugh, and gossip together.

Spring and the early summer months brought ways for peasant women to supplement the family's diet and income, in the last lean weeks before the harvest. They cared for the livestock: the newly purchased cow about to calf, the goats and sheep giving birth, the litters of piglets, chicks, and goslings to raise. A Russian peasant woman blessed the animals on St. John's Day. All across Europe into the twentieth century peasant women weaned the nursing young of the cows and goats early and used the milk for making cheese and butter. Some milk, some cheese provided protein for her own family, but most of it had to be sold. As early as the ninth century, Anglo-Saxon women produced enough cheese to make it one of the early exports of England. In Montaillou in southern France in the early fourteenth century, the wives sold cheese at the market fair on Whitsunday. The method for making this valuable foodstuff remained the same throughout the centuries. Peasant women separated out the cream, soured it, thickened it, pressed and aged it. The excess liquid from the pressing—the whey—was a treat for their own children. In much the same manner they made butter, churning the milk until it became heavy and no longer stuck to the paddle. In 1654 a peasant woman sold her butter in 5 pound lots at 6 pence a pound to the Countess of Bedford's cook.[45] The cook also bought her cream at 4–5 pence a pint.

The rhythm of the seasons structured peasant women's lives as it had those of women before them and remained a constant throughout the centuries. Another constant was their reproductive life. As mothers, they bore and sustained the generations that worked the land from the ninth to the twentieth century.

Children and Nurturing

Childbearing and mothering: peasant women took these responsibilities as seriously and as matter-of-factly as they did the success or failure of the crops, the selling of cream, and the making of cheese. With them lay the possibility and the responsibility for the generations to come; with their bodies and the care they gave the infants they bore, they created and sustained the hope of new life. Childbirth and child rearing were constants in the countrywoman's world.

Demographers assume that women's potential for childbearing remained the same throughout the centuries, roughly twenty years. Women, regardless of social position, began to menstruate between twelve and fourteen. (Changes in the age of menarche, the onset of menstruation and ovulation, come with extreme variations in diet. In the second half of the twentieth century, better diet—in particular an increased percentage of body fat—has occasioned earlier hormonal changes and thus earlier puberty. Studies for Germany in 1500 and for Scandinavia in the nineteenth century indicating menarche at sixteen or eighteen appear to be aberrations in the common pattern, caused perhaps by malnutrition.)[46] The end of women's fertility, the onset of the hormonal changes leading to menopause, appears to have been a constant as well. As early as the seventh century, Visigothic laws assumed a woman infertile after forty. The twelfth-century learned abbess Hildegarde of Bingen, in one of her medical treatises, estimated a woman's childbearing years to be over by fifty.[47] The number of children women conceived and carried to term depended on their own susceptibility to disease and the relative ease or difficulty of childbearing for each individual. Demographers estimate that a woman might have five to seven successful pregnancies at two and a half year intervals if she lived a normal life span.[48] From the woman's perspective, she could assume that she would be pregnant or nursing a child for most of her adult life.

Everything in the peasant woman's world favored and valued the wife who could bear healthy children. The Church extolled sexual intercourse for the purpose of procreation and offered heroines who avoided "the shame of barrenness," like Saint Elizabeth, mother of John the Baptist, and Saint Anne, mother of the Virgin Mary, both elderly women miraculously able to conceive.[49] Countrywomen made herbs like feverfew into tonics that they believed would make them fertile. Once pregnant, a peasant woman might be given special privileges: One lord of a German manor, eager for more potential laborers, allowed pregnant women to eat fruit from the orchard and grapes from the vineyard, but only "a small branch with two clusters."[50] They might have fish and game from the lord's preserves and pay less to use his bake oven. They might be allowed to keep the hen due the lord at Shrovetide. Into the late 1930s in Lough in County Clare, Ireland, families expecting an immediate pregnancy excused the new peasant bride from heavy work for the first year of her marriage. Conversely, if her "belly" had not begun to swell by six months, the young woman could expect anger and abuse.

Pregnancy and the birth of the baby were times of special female companionship. The women of the village performed many tasks together throughout the year, working in the fields weeding and raking, walking to the market town, washing clothes at a stream, but the time of birthing was their province, the source of a special bond that only they as women shared. "In her waiting" as contemporaries called it in the thirteenth century, female friends, relatives, and the wisewoman of the village would give advice and then help with the delivery.[51] The men of the family waited once the labor had begun, sometimes adding to the tension. A peasant woman in southern Italy remembered her labor in the 1950s; her mother and sisters had brought the midwife and they would

> stand by the bed and wait to see how far apart the pains were. I'd hear the shuffling in the room and know the men were arriving, one by one. My father, my brothers, my husband's brothers, they all sat there by the fire and drank wine and waited. If you make a sound, if a pain catches you by surprise, or the baby won't come out and you can't stand it and you moan, you've disgraced yourself. You keep a towel shoved in your mouth, and everytime it hurts so bad, you bite down on it and pray to God no noise comes out. I always tied a knot in one end so I could bite real hard, and my sister had a way of crooning and stroking me that made it better.[52]

The ease of childbirth depends on three factors: the physical condition of the mother, the size of her pelvis relative to the size of the baby's head, and the position of the baby in the uterus. A female relative like the Italian sister, or the "wisewoman" (the midwife), was active throughout the birth with words of encouragement, with practical aids. The midwife made herbal teas to hasten the labor: wallflower acted as a diuretic; lady's-mantle and ergot (a fungus found on ripe plants) strengthened the contractions. Herbal ointments might help to soften the cervix. Henbane produced a twilight sleep for a woman in extreme pain. The midwife could determine the position of the fetus from feeling the abdomen and could listen to its heartbeat. With her fingers she measured the dilation of the cervix and thus the progress of the labor. Pushing on the uterus, she helped the final stage of the delivery.

In English villages mothers sat up in bed for the first stages of labor, sometimes on straw to absorb the discharges. They went to moon-shaped "birthing stools" for the last phase of labor. In nineteenth-century Russia, peasant women gave birth kneeling in the warmth of the family bathhouse. Women left to themselves favored these upright positions, which gave them the advantage of gravity and the use of more of their muscles for the expulsion of the head, the final act of delivery.

Though governments like those of France and Russia in the eighteenth and nineteenth centuries tried to decree examinations and licenses for midwives throughout the countryside, women of the villages did not seek training and found the fees high. Instead, midwives continued to gain their special status not because of any formal education but because of their own experiences of childbirth, their knowledge of herbal medicine, and their intuitive sense of measures that eased the process. The midwife might, as in France in the seventeenth century, be paid by the village along with the shepherd and the schoolmaster. In most instances she would be an effective guide and aid to the pregnant woman. (Despite numerous stories, especially from the nineteenth century, about "dirty, ignorant hags" who acted as midwife, doctors rarely, if ever, came to the aid of pregnant women in the villages, so that well into the twentieth century peasant women had to rely on midwives and their female relatives.) Only in the case of a prolonged labor did midwives need extra knowledge and skills. A long labor meant that it would be difficult, even impossible, for the mother to deliver without intervention. Inaction would mean death for the woman and probably for the infant as well.

Prolonged labor occurs because of the size of the baby's head in relation to the mother's pelvis, or because of the fetus' position in the uterus. The first cause was less common. Malnutrition, a common condition among peasant women, worked to the advantage of the mother by lowering the birthweight and size of the child. But transverse and breech presentations were not unusual. In later centuries both necessitated the use of forceps or even surgery. A breech presentation (the most common of the two), with the buttocks or feet emerging first, could be delivered naturally, especially if the child were small. The danger to the mother would come from lacerations to the vagina, unable to stretch gradually to accommodate the head. For the baby, moving backward down the birth canal, there would be the danger of strangulation or oxygen deprivation should the umbilical cord entangle itself around its neck. A transverse presentation with the baby lying horizontally across the cervix would be virtually impossible for a mother to deliver unassisted. She could not by her contractions alone force it into the birth canal. Though the midwife chanced infecting the mother, once she discovered the awkward lie of the fetus, she would probably try to reposition it, either from outside or even within the uterus once the cervix had dilated sufficiently. Medical historians have consistently asserted that knowledge of this maneuvering, commonly called "podalic version," was known to Greek and Roman male obstetricians and then forgotten until the sixteenth century when it was rediscovered by a man, the French surgeon Ambroise Paré. This does women a disservice. A peasant woman, a midwife, could judge the irregular position, and if nothing else, her knowledge of birthing cows and horses would tell her to intervene and give her ideas of how to help the mother's travail.

Prolonged labor, lasting over twenty-four hours, could cause death. After forty-eight hours the mother suffered from dehydration. That and exhaustion could lead to heart failure. Internal tears could cause bleeding. After so many hours, the uterus, a muscle, might rupture and hemorrhage. Then the mother would bleed to death. To save a woman in such circumstances, before so much damage had been done, the infant was sacrificed. The midwife might send for the barber surgeon, if one were available in the village. Otherwise she acted herself, using a variety of hooks and hammers to crush the fetus' skull and thus extract it and the rest of the body from the vagina. Some women survived this procedure. If the child could not be extracted or the mother died before the procedure could be performed, both the Church and custom dictated that her abdomen be cut open and the child taken out. (This was the original cesarean section.) Popular belief decreed that her mouth be kept open to allow the infant's soul to escape should it too be dead.[53]

An old English proverb told a woman she would "lose a tooth for every child you bear."[54] There might be worse consequences, like danger of infection, or from tearing. A Brittany peasant remembered that his grandmother had left her bed too soon—to do the laundry—and died of a fever, the principal symptom of such an infection. In Torregreca in southern Italy, Maria went to the fields to heave the wheat sheaves for the threshing machine just a week after the birth of her sixth child. She hemorrhaged and bled to death just as other peasant women had for centuries.

Infections might have lesser consequences. Gonorrhea and pelvic inflammatory diseases could close off the fallopian tubes with scar tissue and cause intermittent pain, fever, and exhaustion. Malnutrition, a chronic possibility in the peasant woman's world, caused irregular ovulation and cessation of menstruation altogether. Malnutrition, especially protein deficiency, could lead to toxemia, to kidney damage, to high blood pressure in pregnant women and thus to premature birth or the death of the fetus because of abnormalities of the placenta or premature separation of the placenta from the wall of the uterus. Even the most successful childbearers had miscarriages. In 1681, Alice George of St. Giles parish in Oxford was 108 years old, still tall for her time and known to have been able to reap as much as a man in her younger years. She had married at thirty, borne fifteen children, and had three miscarriages.

In the 1930s Lucia of the north littoral of Croatia told the schoolteacher Darinka Host about her experiences: two miscarriages in the first year of her marriage, then five years of barrenness, another miscarriage, and then a son. Eighteen months later she bore another child. Her mother had successfully brought pregnancies to term, but after each birth she "was bedridden for five months." Demographers estimate that even in good health, with easy conception, the harsh conditions of peasant women's lives caused spontaneous abortions—miscarriages—a third of the time.[55] The records of births for seventeenth- and eighteenth-century English and French villages show April,

May, and June the best months for conception for a successful pregnancy. Then the mother would have less arduous labor in the first months, and more than enough food in the later months of carrying the fetus.[56]

Care of the newborn infant followed patterns and beliefs as old as ancient Greek, Hebrew, and Germanic customs. Into the eighteenth, nineteenth, and even the twentieth centuries in some parts of Europe, the midwife bathed the infant, then swaddled it, carefully binding each limb separately, then wrapping a cloth all around. Usually the baby was swaddled with the wrappings changed two or three times during the day for the first four months of life.[57] As late as the 1960s the elementary schoolteacher in a southern Italian village advised that the baby be confined in this way for a minimum of six months. All believed that such binding protected the child from deformity and illness, preserving the straightness of the limbs and the proper placement of each organ. Of all the Europeans, only the Celts were described in the twelfth century by the historian Giraldus Cambrensis as not swaddling their infants and leaving "nature alone" to determine the baby's shape.[58]

Women breast-fed their infants if they could. Only rich women of the castles or the towns could afford a surrogate, a "wet" nurse, to feed their infants. The colostrum and the breast milk, easily digestible and with valuable antibodies, improved the baby's chances of survival. Mothers gradually weaned babies between one and three years of age. Early in the infant's life, the mother gave it solid food, pap (a thin flour gruel) or bread soaked in milk or water. Peasant women might choose to nurse because they believed the practice prevented conception. In many communities the Church approved if a mother wished to remain sexually abstinent during this period.

Peasant mothers took all responsibility for children until about age five. If they had more than one child, they would have the elder watch the younger. Until the child could walk, care would be relatively easy. Mary Collier, an eighteenth-century peasant woman, wrote in verse of taking the babies to the fields:

> [We] wrap them in our clothes to keep them warm
> While round about we gather up the corn [grain];
> And often unto them our course do bend,
> To keep them safe, that nothing them offend.[59]

In Brittany in the nineteenth century mothers left their babies in their cribs, or in the family's enclosed box bed, returning periodically in the course of the day to nurse them. Once the child was able to walk, dangers were ever-present. Even in the 1980s in Denmark, mothers complained that they had to watch their children carefully because of potentially harmful farm machinery. Women of earlier times feared hearth fires, deep wells, and domestic and wild animals.

Although "A soft word will open an iron door" remained a popular proverb in Bosnia into the 1930s, peasant mothers probably ruled their children strictly.[60] In fourteenth-century Montaillou in southern France, mothers put them to bed before serving the husbands their evening meal. It is likely that mothers beat the younger children, fathers the elder. In the 1930s in Croatia, Lucia remembered how hard it had been to care for the other children while her mother was ill: "When first I made the bread, I was still not strong enough to knead it firm, and my mother would rap me over the knuckles. I did the work for all the younger ones and also got beaten for them all."[61] Grandmothers told stories of little girls and boys who suffered hardship and even death because they disobeyed their mothers. Red Riding Hood wandered off the path in the woods despite her mother's warnings and was eaten by a wolf. "The Wayward Child" of another tale "wouldn't do what its mother wanted. For that reason the dear Lord did not look with favor upon it and let it fall ill." Even in death it would not be still, often reappearing to the mother. In the end she beat the apparition's arm with a switch and, "Now for the first time the child had peace under the earth."[62]

Care of children and the family extended beyond feeding, clothing, and discipline. To peasant women fell the responsibility for comforting and curing a sick child, husband, parent, or grandparent. Peasant women had no scientific names for the afflictions that might come to their families. They knew only symptoms: fever, cough, cloudy eyes, headache, abdominal pain, swelling, bleeding, discharges, rashes. Like their mothers and grandmothers they relied on intuition and on the knowledge passed from generation to generation. They could, like the women in thirteenth- and fourteenth-century southern France, call on the services of the village wisewoman, an elder acknowledged to have special skill with herbs and who knew rituals and prayers that could cure. The churchwardens of a seventeenth-century English village paid such women a regular salary just as others paid the midwife.[63]

Healing was a special knowledge. Every symptom from toothache to eye inflammation to convulsions, palsy, and loss of hearing had its appropriate remedies, a whole range of practices that the wisewoman and her contemporaries believed had curative powers. There might be a specific prayer or a ritual like wearing red flannel to draw out the "red" heat of a fever. Or the wisewoman could transfer the "spirit" of the affliction; for example, one woman believed that she could cure consumption by burying an egg in an ant hill—the disease would disappear as the ant hill washed away in the rain.[64] Much of the peasant woman's pharmacopeia remains a mystery. Some sounds like magic, like a spider covered in dough taken as a pill against ague.

Other substances like digitalis (foxglove), hyoscyamine (belladonna), and aspirin (willow bark) are used in modern medicine. In particular, twentieth-century scientists have come to respect the peasant woman's knowledge of when to pick, when to cut, and how to preserve the many herbs, roots, and blossoms that she used. Changes in temperature and light from day to night, from season to season, alter the chemical properties of many plants; for example, the yield of poppies is four times greater at nine in the morning than at nine at night.[65]

Each symptom had its herbal remedy. Children drank calamint tea or inhaled the smoke of burning coltsfoot leaves to soothe a cough or to relieve congestion. The root of marshmallows, oil of thyme, peppermint, and anise seed are still ingredients in medicinal lozenges. Every headache and fever had its specific cure. A sore throat called for periwinkle tea, or a salt water gargle. Lavender stopped dizziness. Herbs like pimpernel and dill helped digestion and eased abdominal pain. Tansy leaf tea slowed the cramps of dysentery. Celandine and wormwood cleared worms from the intestines. Aloe and marigolds are still used in lotions to soothe injured or burned skin.

Centuries of experience, flashes of intuition, acts of faith, the accumulated lore of this herbal medicine gave peasant women a way to deal with the illnesses all around them. They believed that they could ameliorate pain, cure, and thus take some control of the unknown hazards of their world.

Despite all of peasant women's efforts and expertise, death menaced their families. Though rural women bore an average of six children in a lifetime, half of them died before they were twenty. The tenth- and eleventh-century Hungarian cemeteries show one child in five dead before the first birthday, two in five dead before the fourteenth birthday.[66] Studies of French and English village families suggest that until the beginning of the eighteenth century a woman averaging five births would see three of her children die. Twenty-five percent of children died in their first year, in bad times as many as 25 percent more would be dead before their twentieth birthday.[67]

Death was women's special province; just as they oversaw the coming of life, so they waited for its leaving. Peasant women had cowslip wine, herbs like poppy seed, belladonna, mandrake root, and hemlock to ease pain and give sleep. Women prepared the body, whether it was an infant, a child, husband, parent, or friend. They washed the limbs, wrapped the body in white cloth—a winding-sheet, or shroud, that they would probably have woven and hemmed themselves. In 1895 Nell Chapman, new wife to an English village shepherd, made the linen shrouds for herself and her husband as her first act of their new life together. Women had always been the principal mourners, praying over the body, in some villages for nights and days in succession, despite the disapproval of the Church, which decried such "excesses" of mourning from as early as the seventh century. Women sang of death in their ballads and told of it in their tales, especially the death of children. "The Wife of Usher's Well" lost three of her sons "about Martinmas, when nights are lang and mirk."[68] The mother of the German folktale "The Little Shroud" knew quiet acceptance only when her seven-year-old son appeared to reassure her, his shroud soaked with her tears. Then the storyteller imagined him sleeping as he had in life but now "in his little underground bed."[69] Peasant women made offerings at the parish churches, "the mortuary gift," just as their Germanic and Celtic predecessors had buried articles with the bodies of their dead. Into the twentieth century women tended the gravesites.

Countrywomen took responsibility for the health of their families. Aware of the precious quality of life, they honored birth and death with special care and rituals that survived into the modern era, long after the worst dangers of their world had disappeared.

Threats to Survival

The women of the countryside accepted the vulnerability of life. For wherever the preindustrial patterns of cultivation prevailed, peasant women and men lived on the edge of subsistence. Bad fortune periodically endangered all that women valued. The natural forces of bad weather, bad harvest and epidemic diseases inevitably broke the rhythm of their seasonal activities, and challenged their ability to sustain themselves and their families. In those times, peasant women battled for life.

Peasant women knew that the key to survival was the length of the winter and the success of the summer growing season. Into the eighteenth century, natural disaster and its companion, famine, usually came at least once in a generation. In a good year, a seventeenth-century French family would need a minimum of 2 hectares of land to produce enough grain for food, for rents, taxes, tithes, and enough seed for the next spring's sowing. In a bad year the peasant family needed double the number of hectares to yield the same harvest.[70] And there was no more land in a bad year than a good one.

All of the sources describing the lives of countrywomen and men speak with special respect of the winter—of the cold, the wind, the wet. Peasant families counted the passage of years by the number of winters.[71] The gradual cooling of the climate, a drop in the mean temperature of 1 degree centigrade by the late fourteenth century, made the winters worse and signaled the "little ice age."[72] Into the seventeenth century, peasant women in northern Europe saw increased rainfall that caused the rivers and canals to flood and the Baltic to freeze over. In 1690 in Scandinavia, the weather never warmed long enough for the grain to ripen.[73] In Greystoke, England, 1623 was a bad year for peasant women and their families. In September, the parish register noted deaths: a son Leonard on the eleventh, the mother on the twelfth—"Jaine, wife of Anthony Cowlman . . . which woman died in Edward Dawson's barn of Greystoke for want of maintenance."[74]

Warm, dry weather could mean olive trees in the south of England, figs and vines in the north of France. Warming of the climate in the eighteenth century brought a longer growing season and bigger harvests. But heat could also bring drought and wildfires. In sixteenth-century Castile, Spanish villagers had to move as the soil hardened and cracked with lack of rain. In French Beauvaisis, from 1647 to 1651 recurrent droughts and bad harvests forced tenant families into debt to buy food; they defaulted, lost land and tools, and starved.[75]

Worst of all for peasant women and their families, even into the twentieth century, were the early spring months before the first grains ripened. "Many people die in the spring when food is scarcest," explained a Bosnian woman in the 1930s.[76] The seventeenth-century French village records of St. Étienne tell of the weaver Jean Cocu, his wife and four daughters. Together they made 108 sols a week. To survive they needed 70 pounds of bread a week. With bread at ½ sol a pound, they had no difficulty. In 1694 the price was 3.3 sols per pound. By December of 1693 the family had registered at the office of the poor; by March of 1694 the youngest daughter had died; in May the father and then the eldest daughter followed.[77]

For countrywomen changes in the weather had repercussions other than the success or failure of the harvest. Each season had its illnesses. As early as the thirteenth century in Denmark, the long cold months indoors passed smallpox from family member to family member. The lice gave typhus to the adults. Peasant women's children died of croup, diphtheria, whooping cough, from bronchitis, asthma, and rheumatic fever. As women and men passed from fifteen to thirty-five, they died of tuberculosis. In nineteenth-century Brittany, Émile Guillaumin described his wife's death after she had helped with the harvest:

> I got Victoire, who didn't mind at all, to come with me to help load the few stocks we had made the day before on to the cart. She got very hot, then began to shiver under the downpour which fell unexpectedly early: that night she coughed blood, and two days later she was dead.[78]

Pneumonia, pleurisy, and acute tonsillitis, all took lives in the winter. In the sixteenth century influenza claimed lives in France, Italy, Germany, and the Netherlands. Dysentery with its rapid dehydration killed in every century, especially when the population was weakened by food shortages. In nineteenth-century Brittany women and men still dreaded "the cold hell" of the winter months.[79] Warm weather did not necessarily mean health for a family. The hot moist air bred the mosquitoes of malaria. In the nineteenth century summer droughts left bad water and brought cholera and typhoid fever.

From the thirteenth century to the seventeenth century, one disease above all others caused terror. Peasant women and men called it the Black Death and suffered it in epidemic waves, first in the fourteenth century and then periodically into the seventeenth century when it more or less died out. The first European outbreak of the bubonic plague in 1348 coincided with a long period of bad weather and food shortages. This made peasant families particularly susceptible to the new disease strain. In such circumstances the Black Death flourished.[80]

In 1348 the weather was warm and humid in Europe. An English peasant woman remembered that it started to rain on the 24th of June, St. John's Day, and just never stopped.[81] The grain rotted in the fields. The disease came off the merchants' boats with the rats and their fleas to the ports of Italy and southern France, then across the mountains and rivers and into the fishing towns and villages as people went about their regular tasks, not knowing that they carried the infection with them. First the rats died by the hundreds of thousands. In Dorsetshire, England, by the middle of July, the women and men began to sicken and die. The mortality in the first wave, from 1348 to 1350, was 20–25 percent. A peasant woman lost many of her friends, perhaps her parents, her husband, for adults were the first to contract the disease. In the second wave, which came sporadically from 1351 to 1385, the children proved the most vulnerable. The loss of the young and of another 20 percent of the population caused a downturn in the population curve. By the 1380s Europe had lost about 40 percent of its population.[82]

Peasant women first noticed the swelling of the lymph glands under the arms or in the groin. Then spots, sometimes described as blue or black, sometimes as red sores, appeared all over the body. Fever and bleeding from the nose were considered particularly common symptoms of the disease. Death could come in three days. This first wave of the plague developed into the pneumonic form very rapidly, becoming even more deadly. Then the disease was transmitted directly from person to person without the intermediary rodent and fleas needed by the bubonic form of the disease. There would be little or almost no swelling, just a sudden onset of a very high fever, inflammation of the lungs, coughing, spitting blood, and death. Without antibiotics, even in a modern medical facility, the mortality rate for pneumonic plague is 60 to 70 percent.[83]

Manorial and parish officials recorded the deaths of peasant women and their families. In 1348 on a Suffolk manor of fifty holdings, twenty-nine families had been wiped out.[84] By 1364 around Nice in southern France only twelve of the twenty-eight villages had sufficient families to list for the hearth taxes.[85] By 1401 two thirds of the rural communes around Pistoia in northern Italy had disappeared from the records, the lands abandoned.[86] Well into the seventeenth century, a cyclical pattern of the disease was established. Then at the end of the seventeenth century the population's immunity stabilized, and the plague ceased to kill.

Death and harm might come to peasant women and their families not only from natural disaster and disease, but also from the hands of men. Violence was a part of the life of the countryside, both in individual assaults against women in particular and in the indiscriminate horror of warfare. The customary laws of the Germanic and Celtic peoples, the fourteenth- to the seventeenth-century records of the secular and ecclesiastical courts, show that women's bodies and their lives were supposed to be protected. Frankish law set wergilds for killing women. Fourteenth-century ecclesiastical courts of southern France and manor courts in England fined men for rape and put them in the stocks. In 1400 in an Apennines village, a group of men took the priest's sister Bartola, "like a lamb among a pack of wolves" and raped her for two days. Four of them were executed.[87]

Along with the ancient traditions that valued a woman's safety and called for her protection, however, went the equally ancient tradition of protecting men from false accusations. This left the burden of proof of sexual assault on the woman. Burgundian and Lombard customs awarded fines only if the man had not been incited. In seventeenth-century Somerset in England, a woman had to cry out, to make her accusation right away, could never have had intercourse with the man before, and must not become pregnant. All across Europe men believed that conception meant enjoyment on the woman's part.

If the guilty man was a peasant woman's social superior, she had little recourse. Twelfth-century writings of the elite classes suggested a nobleman "use a little compulsion as a convenient cure" for a village maiden's "shyness."[88] In fourteenth-century Montaillou in southern France, the priest notoriously used his position and his ability to bring charges of heresy against parishioners to exact sex from the women of the village. Most vulnerable of all were young women living as servants, either in a household or in a public inn. In the eighteenth century near Nantes, servants assumed their masters wanted them to accept the advances of their friends: Marie Toutescotes allowed M. de Kernoel, a visitor, to have intercourse for "she felt that she should not refuse him being afraid that her master would be angry with her."[89]

While the religious and secular customs of the community sometimes helped women who had been assaulted or harassed, nothing could lessen the impact of war on the countryside and its inhabitants. Wars remained a constant in the world of peasant women and men, changing over the centuries only in the methods and justifications, never in their effects on peasants' lives.

Until the sixth century, the peasant wives of the Germanic and Celtic peoples remained behind to deal with the work of subsistence while the men raided the neighboring settlements each spring. From the eighth to the eleventh centuries peasant women and their families suffered the attacks of distant invaders, the Arabs from North Africa and Spain, the Magyars from the east, the Vikings from the north. The ninth- and tenth-century invasions led peasant families to seek protectors, to tie themselves to a mounted warrior and his followers. As his serfs with obligations and payments in kind, they put themselves under the shelter of his military strength and gained the authority of his "ban" and of his manorial court. Thus they could hope to plant and harvest in relative peace. When the warriors fought among themselves in the eleventh and twelfth centuries, peasants looked to the Truce of God, the promise of bishops that knights would not fight on certain days of the week, at certain times of the year. As early as the thirteenth century, peasants turned to the new royal dynasties—the kings, emperors, and czars—who claimed the lands and the wealth as their own, taking taxes but giving the law and order of royal justice in the royal courts. But then from the fourteenth to the eighteenth centuries, the dynasties fought among themselves.

So the wars continued, as predictable as the coming of spring when warriors and their commanders made their plans for new campaigns. The peasant woman and her husband, her sisters and brothers, her mother and father, her daughters and sons watched the methods of fighting change: The elite professional warriors on horseback gave way to paid foot soldiers and bowmen marching in ranks, and then to artillery and massed armies. Occasionally, peasants armed themselves to try to keep the warring from their villages, but most often they fled, or tried to negotiate to avoid the coming of the soldiers.

From the fourteenth to the seventeenth centuries, free companies of soldiers, like bands of outlaws, menaced the people of the countryside. In the fifteenth century Joan of Arc's family herded their livestock to a fort on an island in the River Meuse to wait out the raiding soldiers of the Hundred Years War. In 1423 Joan's father was the villager who signed the agreement with the leader of a band of soldiers, a yearly fee levied on each household paid so that there would be no pillaging. New wars gave new names to these blackmail payments—*appati, pati, raencons du pays, souffrances de guerre.* Even such arrangements did not always mean safety. A 1439 inquiry into excesses of the Dauphin's soldiers in two French villages reported "women violated, people crucified, roasted and hanged."[90] In their *cahier* (list of grievances) to the Estates General of 1484 the peasants reported: "No region has been free from the continual going and coming of armed men, living off the poor people, now the standing companies, now the feudal levies of nobles, now the free archers, sometimes the halberdiers and at other times the Swiss and pikemen."[91]

Before the twentieth century, nothing matched the horrors of the Thirty Years War of the seventeenth century—the troops of many dynasties marching across the fields: from Austria, France, Hanover and Saxony in Germany, from England, Hungary, Spain, Sweden. The armies came in the spring and the summer, not just bands of soldiers, but small states with thousands of followers to service the fighting men's needs. The commanders Tilly and Wallenstein between them had armies of 48,000—48,000 additional people to live off the land that peasant women and their families tended.[92] Armies raped and murdered. They stole the crops and burned the fields. They killed the livestock. In their wake, armies left typhus and dysentery. In 1618 the population in the areas of heaviest fighting was 21 million; in 1648 when the leaders signed the peace 13.5 million remained.[93] The English ambassador described Neustadt, Germany, in the winter of 1635–1636 after the troops had gone through. The children just sat at the doors of their houses. The dead were found with grass in their mouths. In parts of the fighting areas dysentery hit epidemic proportions and then came the plague. Some 10,000 women, men, and children died in four months.[94] Even in the well-managed professional wars of the eighteenth century, the future Joseph II wrote to his mother, the Empress of Austria, in July of 1778: "The war has begun and the peasants have been pillaged, . . . our armies are still intact."[95]

In the twentieth century, technology and mechanization renewed attacks on civilian populations; the scale of the revolutions and wars meant consequences never imagined before. Peasant women died with their children and their husbands from starvation, disease, and exposure.[96] In World War II peasant women worked the land and cared for the livestock. Without fuel or draft animals, women pulled the plows themselves. When the armies came, just as in the centuries before, Italian peasant women hid food, buried their valuables, and let their cows loose in the woods.[97]

As in past centuries peasant women managed to survive and to provide for their children. Anna Cecchi, wife of the shepherd in a north Italian village, knew when the bombs were coming because the sheepdog started to howl. She and her grandmother took her daughters to a makeshift shelter. "What did I think during the war?" she was asked. "To continue," she answered. "Just to continue."[98] Russian women in the 1950s still knew the song that described war for the women of the countryside:

> And so now the war has ended,
> I alone remain alive.
> I'm the horse, the ox, the housewife,
> And the man and the farm.[99]

Into the twentieth century peasant women sustained their families and protected each new generation, though tested many times over with natural disasters, disease, and war.

2

SUSTAINING THE GENERATIONS

The Family and Marriage

Throughout the centuries a woman and a man survived in the countryside and could hope for the future by living within a family. Only with others could peasant women and men complete the ever-present tasks that ensured subsistence, and make order of the challenges and the periodic chaos of their lives. For women, in addition, the protection of a father, a husband, a son, in theory, made them somewhat less vulnerable to the violence of other men. Following the traditions inherited from earlier centuries, a young woman would be expected to marry, thus creating a partnership with a young man. She did not, however, give up her ties to her own family and might even keep her mother's or her father's name. Thirteenth-century English records list Ciciley Wilkinsdoughter and Emma Androsmayden. In fifteenth-century France Joan of Arc explained to the churchmen who questioned her at her trial that: "I am called sometimes Jeanne d'Arc and sometimes Jeanne Romée," thus acknowledging ties to both her parents' kin.[1] Into the eighteenth and nineteenth centuries in Denmark, Ireland, and France women continued to be known by mothers' and fathers' surnames even though marriage had legally changed them.

Both young peasant women and men saw the advantages of the partnership. In earlier centuries, marriage might mean freedom from the bonds of serfdom. Even as serfs, by their union the couple gained a measure of independence from their parents, adult status within the village community, and the opportunity to create their own future. A French proverb, "Mariage, ménage," meant that with marriage the couple had created their own household.[2] (The evolution of certain words reflects the acquisition of adult status in the languages of both Eastern and Western Europe. Among the thirteenth-century Slavs of Kiev in Russia the words for "woman" and "man" were the same as those for "wife" and "husband." In English the word "woman" combines the word *wif,* meaning female spouse, and *mann,* meaning human being.)

Both wanted to marry. In the 1930s a Bosnian woman remembered her marriage at fifteen: "I had a very nice chance come my way, and knowing I was poor, wasted no time thinking about it. . . ."[3] For the man, joining with the woman—a "housewife," as she would come to be called—had the same practical value and significance. The Sicilian proverb explained, "The man is the soul of the house. A good wife is the primary wealth of the house."[4] In the 1920s a farmer from County Clare, Ireland, insisted:

> Here is something I want to tell you and you can put it in your head and take it back with you. The small farmer has to have an intelligent wife or he won't last long. He may do for a few years but after that he can't manage. . . . There are a thousand ways an intelligent woman makes money . . . and when he is getting a wife for one of his sons, he should look to a house where there has been an industrious and intelligent woman, because she has taught her daughters how to work and that is what is needed.[5]

The landlord—the knight or noble who gave access to the fields—knew the importance of the marriage of his female and male serfs or tenants. He could expect more of a couple. On the twelfth-century English manor of Ashcroft in Somerset, the unmarried holder of five acres owed ½ pence and two and a half hens, his married counterpart 1 pence and five hens. In thirteenth-century Wantage in Berkshire, the married man owed two more hens and an extra tax.[6] Manorial agreements forbade marriage outside the property and required the lord's permission from as early as the eleventh century in England to as late as the 1730s in Poland. To the landlord a marriage off the manor meant the loss of laborers and progeny, in Poland the loss of a cow, as well, for this was the traditional dowry (the gift of the bride's family to the groom). In some parts of Europe, in addition to his permission, the lord used the occasion of the marriage to collect another fee. The English called it *merchet,* and in the eleventh and twelfth centuries, a better off peasant father might owe the equivalent of the value of an ox, in the fourteenth century, three months' income from the fields.[7] Historian Jean Scammel suggests that failure to pay this tax may have given rise to the idea of *jus primae noctis*—that the landlord could claim the first act of intercourse with the new wife. The Celtic epic *Táin Bó* acknowledges that the king always had "the first forcing of girls in Ulster."[8] While no concrete evidence exists, manorial records of the fifteenth and sixteenth centuries suggest the memory, if not the reality, of the custom. A document of 1419 from Bourdet, Normandy, in northern France, lists the *merchet* as 10 sols, a joint of pork and a gallon of wedding drink, and continues, "I may and ought, if it pleases me, go and lie with the bride, in case her husband or some person on his account fail to pay to me . . . one of the things above rehearsed."[9] A 1486 legal decision of Ferdinand V of Catalonia in northern Spain prohibited lords from taking their peasants' wives on the wedding night. A list of sixteenth-century customs from Béarn, Switzerland, states that peasant men, "before they know their wives . . . are bound to present them the first night to the lord . . . to do with them according to his pleasure, or else to pay him a certain tribute."[10] In seventeenth-century France the lord exercised the *droit du seigneur* symbolically by putting his leg in the couples' bed after the feast on the wedding night. The cahiers presented by representatives of the peasants to the French Estates General of 1789 complained of the right still being claimed.[11]

In the German folktales the hero usually fell in love first, but before marrying he would often seek the advice and approval of his parents. The stories mirror the European reality, for throughout the centuries families negotiated the marriage, both the joining of their daughters and their sons and the arrangements for their provision, the dowry from the bride, the dower from the groom. So significant were these arrangements that a thirteenth-century English father registered the agreement at the hallmote. In nineteenth-century Russia the negotiations took place before an official of the village.

Long before churchmen played any part in the ceremonies, the rituals of the betrothal and marriage literally and symbolically signified the obligations of the groom and his family. From the ninth century, the tradition of the Lombards and other Germanic peoples continued throughout Europe: The daughter stood holding her father's hand; he gave it into the young man's. By this simple gesture, her father transferred his guardianship of her and gave to another man the responsibility of protecting and providing for her, for protection and provision were the ways in which peasant families defined their responsibility to their daughters. Other parts of the ceremony symbolized the gift of the groom's family, the dower, the promise of a portion of his goods, and access to his fields, should he die and leave her a widow. The English service kept the Anglo-Saxon words and phrases. "Plighting the troth," meant the pledge, but, in its early sense, "plight" also meant danger and thus implied punishment should the terms of the agreement not be met. In the thirteenth century an English groom would "endow" his bride on the church steps and give the ring as a symbol of the dower. Even the phrase "with this ring I thee wed" emphasized the traditional promise to the bride, for in Anglo-Saxon "wed" also meant "pledge."

The promise of the bride's family, though not represented in the rituals, was no less significant. The mother in a German folktale looked to character and chose as wife for her son the young woman who cut the cheese without wasting any. In reality there was the promise of the young woman's labor and her reproductive capacity, but more would be expected. Throughout the centuries, all across Europe, the tradition of the Celtic, Germanic, and other ancient cultures continued, and the bride brought goods and later money, perhaps a house or rights to land, as her contribution to the marriage. This, her dowry, formed the other part of the negotiations between the two families. Failure to complete the pledge also carried penalties. When in thirteenth-century England, Richard Maud did not provide his sister Avis' dowry, a house worth 40 shillings, a dress, and a cow, he was brought before the manor court.[12]

A daughter was always entitled to a portion of the family's wealth on the death of her father or mother. Gradually, access to the land tended to go to the sons, and the custom arose of giving daughters their inheritance at the time of their marriage, and not as land but as goods or money. Fourteenth-century brides from one village in southern France brought a dowry of new clothing, in one village, a robe and a tunic; in another village, a richer family also gave a chest, two sets of bedding, and 75 livres.[13] Fathers in sixteenth- and seventeenth-century Essex in England gave their daughters money. A poor farmer set aside 26 shillings 8 pence in his will; Ralph Josselin, the yeoman vicar, planned £200 for one girl, Jane, and £500 for the other, Rebecka.[14] In eighteenth-century France, a young girl might not look to her family at all. Some left at twelve or fourteen to work in the towns and then returned in their midtwenties to marry, having earned the dowry themselves.

By the nineteenth and twentieth centuries, the accumulation of the dowry had become a significant burden on the peasant family, yet one willingly undertaken to ensure a proper and adequate future for their daughter as the wife of an appropriately placed young man. There was still the contribution in goods, now sometimes called the bride's trousseau. By the 1930s in Sicily, the numbers of pieces had multiplied to thirty-five or forty. A young woman brought two sheets, four pillowcases, a bedspread, tablecloths, towels, doilies and her lingerie as her *corredo*.[15] Her father and brother had worked to buy the material as a first obligation, even before providing for the son's marriage, for the *corredo* alone assured the daughter's future. From the 1880s in Ireland, raising a dowry became harder and harder, taking as much as ten to twelve years' rent to gain a match with a man of good status. Eventually, a pattern evolved: The family provided for one daughter; the others emigrated.[16] In the 1930s in Croatia, goods and money might not be enough. A young woman from a prosperous family, was known to be a good worker, yet the money her mother could offer counted for nothing. She had brothers, and they, not she, would inherit the rights to the land of the *zadruga* (the family-held property), so her mother could not arrange a good match. The mother lamented her daughter's plight:

> Janja, my child, go ahead, wait no longer. Go ahead little lamb, lower your sights as the good God wants it, . . . So she got married, she is all right but—12 acres are not 25. So you see, if Janja had been my first born, and if I had had no other children. She would have been . . . our heiress. Then, well, then—God knows whom she would have gotten for a husband, but why talk about it, you know how it is.[17]

In the 1950s in Vasilika, Greece, peasant families had to make the same choices for their daughters and sons. They sold the land to raise money for both a son's education and for a dowry of as much as $3,000 to $4,500 to provide a daughter with a suitable husband.[18]

Throughout the centuries the wedding ceremony celebrated the change in status for the couple and the promise of their sexuality. The bride had a special dress. In eighteenth-century Sicily she wore a multicolored one, with a garland of orange blossoms as a symbol of fertility. In Russia and Yugoslavia the fabric had been intricately embroidered. Before the ceremony in Russia, the young woman was ritually cleansed, her hair rebraided from one plait into two. Before the altar both she and the groom wore special crowns kept for the occasion. Their friends teased and taunted them. Young people in sixteenth-century Champagne in France created distractions during the exchanging of vows, interrupted the couple on their wedding night and would not leave until they had been given money, drink, and flowers. In the Côtes-du-Nord in nineteenth-century France, the bride's friends surrounded her hoping to claim the bouquet; the groom and his friends took chase and literally tore her away from her friends. The more disheveled and upset she was, the better sign of her virtue and indication of her future faithfulness.[19]

From as early as the mid-twelfth century, the priest was present as a witness, to lend permanence to the vows, but the ceremony itself usually took place outside the church door. Only in the sixteenth century did the churches come to play a significant part in the ritual joining. The decrees of the Council of Trent in the 1560s made marriage a sacrament for Catholics and gradually all of the ceremony came to take place in the church, not just the celebratory mass. Protestant churchmen added the requirement of "posting the banns," publicly announcing the plan to marry three months before the betrothal. All were meant to emphasize the factors important to the ecclesiastical authorities: the consent of the couple and the indissolubility of the union.

The Church's intentions did not conflict with those of peasant women and men. Marriage was a willing partnership, a union essential for survival. Together couples worked the land on which their lives and the lives of their children depended.

Access to the Land

The new couple could not survive, could not sustain themselves or their future children, without access to fields, to pasture, and to forest. When by their marriage daughters and sons showed themselves old enough to establish a household and thus ready to have their inheritance, rights to the land would be the most significant possession their families could give them. For example, into the eighteenth century in the Waldviertel of Austria, when the young man gained permission to marry, he gained the right to his father's strips. (The fact of transferring strips to a son on the occasion of marriage led to the coincidence of meaning between "husband" and the old term for farmer, "husbandman.") The same custom came to be followed with daughters; at her marriage, a daughter received her inheritance, usually goods and money, but in some parts of Europe, she too received the rights to land.

Twelfth- and thirteenth-century records for Lincolnshire and Somerset attest to female landholders in England, with sons known by their mothers' names and wives acting in the manorial court in land disputes on their own, their sisters', or their husbands' behalf.[20] In some shires when there were no sons, daughters rather than males of another line held the land. Customs varied all across Europe. Sometimes the land went to the daughter already married, sometimes it was divided equally among them, sometimes held in common. Thirteenth-century peasant mothers like Sarah Howe from Cashio, Hertfordshire, and Catherine Leman of Hendringham in Norfolk granted or willed land to their daughters on the occasion of their betrothal and marriage. Women also leased land. In 1272 Lyna did so with her brother William in Methwold, Norfolk. Julian Joye of Burley, Rutland, bought land in 1299 but by local custom could only hold it during her lifetime; she could not will it; rather, on her death, it reverted to the landlord. Though these women may have acted for themselves, families commonly assumed that the husband would make the principal decisions, consulting his wife only when he wished to "alienate" or give up her rights. Should he mismanage the holding, however, she or her family could appeal to the manor court. In 1280 in Hertfordshire, Margaret asked that her land be given back to her. While he was alive she had, as a dutiful wife, consented to her husband's request to alienate the property, now the jury and presiding reeve decided in her favor.[21]

Other parts of Western Europe, often Celtic or formerly Roman regions, followed an even wider range of customs that sometimes gave women access to lands along with the men of the family. From the sixteenth century into the eighteenth century in Franche-Comté, land went to those still in the household, both females and males. In sixteenth-century Romans in southern France, a father could divide his land as he chose, giving it to a daughter if he wished. In the thirteenth- and sixteenth-century custumals (collections of local customs) for Normandy, Anjou, Brittany, Touraine, and Maine, the lands were to be divided equally among all the children. In parts of sixteenth-century Scotland, women kept their own names and were listed in the court records as paying rents and having livestock. The same was true in Santander in northern Spain into the twentieth century; daughters and sons inherited equally from the family holdings.[22] In the areas where servile status (serfdom) prevailed into the nineteenth century like Poland, the Czech Republic, and Slovakia, custom prevented the division of the land and favored sons over daughters. In those regions, as in Croatia and Serbia, it was assumed that a daughter had rights to the land, but lost them on her marriage when she left her family to become part of another.

From the earliest European centuries a peasant woman had access to the land in her own right, or as a wife. Over the centuries, however, she experienced changes in the ways in which access to land was defined and in the ways in which her labors guaranteed subsistence to her and her family. Into the 800s, women and men might share communal access as kin of a Germanic chieftain. They would labor with others, free and equal in status, raising enough to feed themselves and their kin. By the tenth and eleventh centuries, they might have access as serfs, no longer equal but subject to the warriors who gave them protection. The right to use of their own plowed strips came in return for their goods and services, now owed to the warrior, the manor lord, who in turn held the land as vassal to a duke or a king. By the thirteenth century, the couple probably rented their land from that same manor lord, or they might have come to own the fields themselves. Then as tenants or freeholders, their goods and services would be exchanged for money in the newly specialized, commercial economy that had come to characterize most of Europe. Worst of all they might have little or no direct ties to the land and thus no control over the value of their labor or the goods they produced. In the earliest centuries, these were the slaves of the Germanic kingdoms. In the later eras, these were the landless day laborers, the poorest villagers, with the right to little more than the use of a hut, possessing only a few tools and bedclothes for themselves and their children.

Of all the possible ways a woman might live in the ninth century, enslavement gave her the least; as a "bonds-woman" she was in effect removed from her family, with no protection, no liberties, only obligations and responsibilities. The Greeks, the Hebrews, and the Romans had depended on their slaves for domestic and field labor. Germanic and Celtic peoples did the same. The Celtic epic, *Táin Bó,* showed Irish queens and their kings offering women (never men) and cattle as tribute to a conqueror; a bondmaid was considered the equivalent of a stallion.[23] Seventh- and eighth-century inventories of estates in Germany, France, and northern Italy list women and men along with the livestock. The Lombards in the eighth century gave slave women as offerings to the Church. Among the Franks they worked as field laborers, as domestic servants, as sexual partners for their owners. Some women and men of the eighth and ninth centuries put themselves into bondage in exchange for protection. More often among the Franks and with other Germanic groups, women were captured in the regular seasonal raids, bought from Viking traders, bred, or enslaved as punishment for crimes. Among the Danes of the eleventh century, slavery was a traditional sentence for an unchaste woman.

Only churchmen opposed slavery and tried to mitigate some of its worst aspects, especially the sexual vulnerability of the enslaved woman. In the decrees and canon laws of their church councils and in their books of penance, they spoke against enslaving Christians. They advanced marriage when sexual intercourse had taken place and manumission when children had resulted from the union. Gradually the laws of the Germanic peoples came to reflect their conversion to Christianity and thus made some concessions to enslaved women. Lombard law of the seventh century described the ceremony making a woman "folkfree": Taken by the hand to a crossroads as symbolic of her new status, she would be told, "From these four roads you are free to choose where you wish to go."[24]

In the long term, however, altered circumstances, not changes in religion, ended enslavement. As long as slavery suited the economic and political needs and realities of the Roman, Germanic, and Celtic worlds, it flourished. In the uncertain conditions of the ninth and tenth centuries, when all found themselves at the mercy of trained and mounted marauder bands of Arabs, Magyars, and Vikings, a household of slaves became a burden, rather than a sign of wealth and status. Landowners freed women and men, gave them use of the fields in return for seasonal work and a portion of their produce. At the same time free women and men gave up their rights to their fields to successful local warriors, allowing themselves "to be delivered into and consigned to [his] protection."[25] They too owed occasional labor and goods. Both groups—former slaves and former free women and men—came to be known as serfs.[26] Only in Scandinavia, northern Germany, Celtic Ireland and Scotland did the ties of family and circumstances maintain the old freedoms and avoid the institution of serfdom.

Wherever it occurred, the relationship between serf and lord lay somewhere between free and slave. A woman and a man had a claim to the land as if they were free, but they did not own the land and could not leave it. They had the warrior's protection (the lord's "*ban*") and the right to judgment in his court (held in his yard or great hall) but were listed along with the cattle in the inventories of his goods. In the tenth century the Holy Roman Emperor Otto II listed serving women and men in his wedding gift, his *morgengabe,* to his new wife, Theophano. King Louis VII of France in the twelfth century exchanged his serf like a brood mare in his agreement with the abbey at Chartres:

> We make known to all, present and future, that we have granted to the church of the Holy Father at Chartres, Havissa daughter of Renaud de Dambron, wife of Gilon Lemaire, mayor of Germignonville, who was ours in servile status, and we have given her to be owned in person and perpetually, with all the fruit of her womb. The abbot and the monks of the said church have given us in exchange another woman of their *familia.*[27]

Gradually, as the relationship became more clearly defined, women and men were considered servile but nonetheless had increasing control over the terms of their serfdom. The family was tied to the land, but over the centuries, their strips and their rights of access became heritable for a daughter or son. The landholder merely required a payment on accession, the chevage of French manorialism in the twelfth century. As early as the tenth and eleventh centuries, when families commended themselves to the warrior for protection, those with oxen, or a plow, were able to specify what goods and services they would give in return. Such specifications gradually became the custom on manors throughout Europe. By the twelfth century in central Germany, peasants heard such a document, the *Weistum,* read each year. In England in the fourteenth century, the harvest bylaws recorded the duties and goods the serfs owed and the monetary value assigned each one. When governments came to legislate on the subject, they also allowed for the changed circumstances. State law of seventeenth-century Denmark tied the woman and man to the land between the ages of fourteen and thirty-six, but after that they were free to leave.[28]

In Eastern Europe the same kinds of circumstances, the same needs and fears, occasioned the evolution of servile status but centuries later than in the West. In the 1400s in Brandenburg in eastern Germany, princes competing for power favored the efforts of the nobility to control the peasantry. By the end of the fifteenth century, women and men had to find others to replace them should they leave the manor. As a result of the Thirty Years War, many others found themselves forced to accept servile status to survive. In the sixteenth century, statutes in Poland and Romania defined serfdom as having ties to the land and gave protection to the landlord against complaints from his serfs. In Russia and Yugoslavia, serfdom developed around the special village institutions of the *mir* and the *zadruga*. Both bound the woman and the man, not as individuals, but as part of the family belonging to the community.[29]

Serfdom, like slavery, died out in the countryside of Western Europe not because of changes in attitude, but because of new economic circumstances that rendered it obsolete. From the twelfth century on, women's and men's relationship to the land and the landholder changed first to that of tenant, then to freeholder and its opposite, the landless laborer. These changes came with the transformation of their world from relatively isolated agricultural communities to a diverse, interdependent mercantile economy spanning communities, regions, and kingdoms. Money paid and earned took the place of goods and services as the means of exchange and the reward of production. Specialization, not self-sufficiency, became the best use of labor and the land.[30]

In these new circumstances women and men became tenants with leases, owing rents, not goods and services. From the twelfth century on, they no longer gave days for certain tasks, but paid the landlord to be excused from the days of work in the fields. Old money payments like the chevage of Burgundy and the *taille* of Mâconnais became fixed taxes owed at set times of the year, not payments at the will of the landlord with their connotations of dependency or servility. As early as the thirteenth century in France and England landholders eager for money freed whole villages of obligations in return for what English peasants called "liberty pence." By the beginning of the fourteenth century, percentages of the harvest owed at Michaelmas had become money payments, rents, all across Western Europe.[31] Other circumstances accelerated the changes. Shortages of population—from the Black Death of the fourteenth century, from the French religious wars of the sixteenth century, from the Thirty Years War of the seventeenth century—caused the landlords to agree to leases and rents to attract peasants to their untilled lands.

The changes in production and the new circumstances also favored consolidation of holdings. The crosspatch of family strips evolved into fields held by the lord or by one of his enterprising tenants. Common lands turned from pasturage for all families into enclosed fields for one. The women of the more enterprising couples gained from these changes. The couples adept at marketing their produce or increasing the yield of their strips ceased to be tenants and came to own land, as the *fermiers* of France, as the yeomen of England and the *Bauern* of the German countryside. These, however, were the minority. The others, the vast majority, the cottars of England, the *metayers* and *manouvriers* of France, the *Höldner* of Germany and the *mezzadri* of northern Italy lost rights and status. Though they still had access to land and were released from the most onerous services and rents, they owned too little and owed too much to be able to support a family from their fields. They ceased to be self-supporting tenants and became sharecroppers and day laborers, dependent once again on those who owned the land and controlled the distribution of produce.

Gradually the same economic changes came to Eastern Europe, but with different consequences for peasant women and men. There, increased specialization and production had the effect of perpetuating not ending serfdom. For example in fifteenth- and sixteenth-century Hungary, the lords, not the peasants, first took advantage of the improvements in cultivation techniques and technology to increase yields. The landlords not the tenants marketed the resulting surpluses. Peasant women and men found their customary payments in kind and days of labor owed enforced with new vigor. As serfs they lost their rights to use the common lands and forest as their lords consolidated holdings and cleared new fields. All was done to increase production for the local and export markets that had evolved. All made survival for peasant women and their families more difficult. Serfdom and servile status came to an end in Eastern Europe only by royal decree. In the eighteenth century, for instance, the Austrian Empress Maria Theresa on her own authority created tenants, set maximum dues, and limited services.[32]

This polarization—division of the rural population—was already evident by the thirteenth century in Western Europe. In one English village the yeoman family had thirty acres, the cottar wife and her husband, five.[33] Economists estimate that by the beginning of the fourteenth century, 40 percent of Europe's peasants had been reduced to such small holdings that they could not support their families.[34] In New Castile in Spain in the last quarter of the sixteenth century, 60 percent of the population had fallen to the status of laborers, some with no claim to land at all.[35]

The system of *mezzadria* had evolved by the fifteenth century in Tuscany in northern Italy. Couples held so little land that they owed half their harvest just to pay for the use of the oxen and to buy new seed. Families could not survive in such circumstances. In eighteenth-century Altopascio, just in one decade (the 1740s) 20 percent of the families left the land altogether and emigrated to the nearby towns.[36] For those who stayed, from the renters of small holdings to the poorest and most vulnerable with no access to land at all, the traditional ways of life did not provide enough. Peasant families all across Europe needed additional income to survive. The responsibility for earning this money fell to the women. Peasant wives added this responsibility to their other tasks.

Additional Income

In the seventeenth century, the French aristocrat, Mme. de Sévigné, was charmed and excited by the sight of the peasants on her country estate coming in with bags of small coins to pay their rent, 30 francs all in sous.[37] What she found so picturesque meant laborious accumulation of the money sou by sou by her tenants. To raise such a sum took all of a peasant couple's energy and enterprise. In the 1670s in the French villages of Goincourt, Esparbourg, and Coudry St.-Germer, three quarters of the families held less than two hectares of land (about five acres). The rent that had to be paid to landholders like Mme. de Sévigné took the proceeds of 1.5 hectares.[38] To meet their obligations and to survive, peasants had to find other sources of income. A contemporary's description of an eighteenth-century French couple, Marguerite and Covin, explains how they managed. They kept the wheat from their two hectares and the wine from their small vineyard. The money for rent, and all other cash payments, came almost completely from Marguerite's efforts. She sold eggs, milk, butter, cheese, wool, and vegetables—all produced from the livestock she cared for and the garden she tended. She spun, and both she and her husband wove, "piece work" probably, arranged by an itinerant merchant from a nearby town.

This was the solution shown from the earliest manorial records: Peasant women more often than men earned the extra money to ensure the survival of the family as the new commercial economy evolved. Women worked in the fields, performed services for the wealthier members of the village, sold what they could raise or make—tasks peasant women knew from the work of their own households, from the old manorial obligations, now done for others as paid labor. Most commonly, extra money came from women's wages working in the fields, from the sale of the farm animals they raised and the produce they grew. Katherine Rolf, employed by the nuns of St. Radegund's in Cambridge, England, in the year 1449–1450, earned 4.5 pence for four days weeding and 18 pence for twelve days of thatching with the male laborers. The next year she sold chickens to the nuns, helped the candlemakers, combed and cleaned wool, in addition to the field work of threshing and winnowing the grain.[39] As a day laborer she was probably given a midday meal along with her pence.

By the eighteenth century, women and young girls moved with the needs of the planting and harvesting, traveling to pick fruit, to glean in the mown fields. At the beginning of the twentieth century in northern Italy around Pavia almost 40,000 women came from the surrounding areas to weed the rice paddies from late May to the beginning of July.[40] In 1961 in Vasilika, Greece, girls and women from the neighboring villages came to help with the cotton crop, hoeing and mulching in the spring, picking in the late summer. In the old days they had lived with the family for the time they worked, but by the 1960s they could be brought and taken home in the trucks and tractors. That had changed, but nothing else had. Teresa in the south Italian village of Torregreca was proud of the ways she coped:

> Nobody can ever say Paolino didn't get on because he had a lazy wife! Some days I wish I could add. I bet if you count what I've got off the land and the chickens and those years I've had a pig and all my hired work, why, I bet I've brought in more in the last ten years than Paolino has, that is, money we could spend or food we could eat.[41]

Extra sous for a peasant woman's family could always come from hiring out for tasks other than field work. Peasant women performed household duties for the landlord, for the richer tenants and freeholders. The Countess of Leicester's household accounts for 1265 list the five shillings paid the alewife who came in and brewed ale for the noble family and its retainers.[42] In the fourteenth century Jehanneton, a farmer's wife, worked as milkmaid on the country lands of the Ménagier of Paris. She would have supervised everything from the care of the cows to the making of butter and cheese. Four women worked for the sixteenth-century Bedford landowner, John Gostwick, as laundress, waitress, and dairymaids. In seventeenth-century Sussex, for 6 pence the Widow Lewis gathered herbs and attended the sick wife of the manor of Herstmonceux.[43]

Daughters commonly went to work in the households of relatives or of the richer families. Even Ralph Josselin, the seventeenth-century Essex yeoman vicar, sent his daughters out to service, Jane at ten and Anne at fourteen. In eighteenth-century France anywhere from 2 to 12 percent of the rural population (depending on the region) worked as servants.[44] As early as the sixteenth century in England, as a result of the Statute of Apprentices, the constable (the local government official) organized a Mop Fair, where the young gathered and waited to be chosen for work. In the nineteenth century, sisters and brothers stood in the marketplace on the Saturdays of Whitsuntide and Martinmas hoping to be hired with the wages set for six months.

Whatever the task, wherever the place, whatever the century, when it came to her wages, a woman earned one third to one half less than a boy or man. The Englishman, Walter of Henley, in his twelfth-century manual of husbandry, gave the traditional justification used into the modern era. He called women "half men" because he reasoned that they worked more slowly at jobs considered less valuable. He paid 1 pence for three acres of weeding, a female's job. The early fourteenth-century southern French monastery paid the weeder one eighth the wage of the male who mowed.[45] Only a catastrophe as extreme as the Black Death occasioned more equal wages for English and French peasant women. In the plague years the female grape pickers of Languedoc made four fifths the wage of the men, when usually they received only half as much.[46] The intervention of parliaments and kings in England affirmed these inequities; the 1563 Statute of Artificers set the women's wages at one third to one half of the men's, though they worked the fields the same number of hours, at jobs just as arduous.[47]

As household servants, women fared no better. Petronella, the laundress for the Countess of Leicester in the thirteenth century, earned 1 pence a day, and as female labor, was the worst paid of all the household staff.[48] The male servant who worked for Margaret Fell of Swarthmore in the seventeenth century made up to £4 a year while Ann Standish, his female counterpart, made only £1.[49] Worst off of all throughout the centuries was the maid of all work, such as the Fells' Peggy Dodgson, who, like her counterpart in the thirteenth century, still made only 1 to 2 pence a day for a whole range of tasks: laundering, beating flax, spinning, knitting, and work outside at carting and spreading manure, weeding, collecting peat.[50]

When peasant women were not working their own fields or someone else's, tending their own or someone else's household, they found yet other sources of income, especially in the cold, dark times of winter when there was little outdoor work to be done. Usually countrywomen did "piece work," especially that associated with the manufacture of cloth, the earliest of Europe's specialized, commercial enterprises. In the ninth century, peasant women produced everything from the thread to the finished cloth and article of clothing or bedding for their families or for the landlord in his workrooms. With specialization and the expansion of fairs and markets, the landholder ceased to require the female tenants' skills, and even for peasant women's own needs many tasks passed out of their hands. From the middle of the eleventh century, male dyers and fullers finished the cloth. By the early fourteenth century in the southern French village of Montaillou, the weaver was a man and the tailor an itinerant craftsman.

In these changed circumstances, weaving ceased to be a household task except for the family's simplest needs and became, instead, a way to earn income. Women heads of households on the Medici lands of eighteenth-century Altopascio in northern Italy spun and wove to supplement their income. At the beginning of the twentieth century in Sicily, a mother employed Gna Tidda to make her daughter's trousseau. The old widow supported herself this way and worked the three pairs of bedspreads on her fifty-year-old loom.[51]

Into the eighteenth century, spinning, not weaving, would be the favored task for most women; with a spindle and distaff they could work at any time, sitting by the fire or walking the cows to pasture. From the thirteenth century on, they might have a spinning wheel, and with its capacity to spin the thread and wind it on the spindle at the same time, women could work much faster, and fill more spindles in many fewer hours. A single woman might do nothing else. At the end of the 1600s a Picardy spinner could earn 5 sous a day.[52] In the eighteenth century, Swiss peasant families with little land valued their "spinster" daughters over their sons, for they could earn more.[53]

When spinning became mechanized and work for townswomen, peasant wives, mothers, and widows already had other ways of making cloth that they adapted to supplement the family's needs. Sometime in the fourteenth century, a peasant woman invented knitting, weaving the yarn on wooden needles instead of on a loom. A late fourteenth-century German altar panel shows the Virgin knitting a sweater for Jesus. She has worked the wool on four needles and is just finishing the neck. Into the nineteenth century southern Italian and Sicilian women knitted as they had once spun, while going to and from other tasks, and they used the needles to make holes for the macaroni. Knitting became piecework and another source of income, especially with the rise of the professional armies. In the mid-eighteenth century a Danish retired soldier's wife supported them both with her spinning and knitting. By the sixteenth century in northern Italy, Belgium, and parts of France, lace making and straw plaiting (for hats) developed as regional specialties. Women in Normandy, Valenciennes, and Malines in France sold white cotton lace—an intricate combination of knotting, weaving and knitting—that, like spinning, could be done in the midst of other chores. In seventeenth-century England, an estimated 100,000 women and girls earned extra money for their families with this craft.[54] Into the twentieth century, Sicilian peasant women still made lace in the traditional manner and also *sfilato*—removing the threads from a piece of cloth to produce intricate patterns. Into the twentieth century by such efforts peasant women used their energy, their ingenuity, and their skills to ensure the family's survival.

Impossible Choices

Life on the subsistence level meant a life always vulnerable to the forces of famine, war, and disease, with disaster coming at least once in a generation. A peasant couple made order out of this potential uncertainty and chaos not only by altering their legal status, but also by working however and whenever they could. Together they worked for survival. Together wife and husband, mother and father, they planned for the continuation of the family and provision for their daughters and sons when they reached maturity. To ensure survival of this and a new generation, peasant women and men kept both the definition and the size of their family relatively constant throughout the centuries. A woman lived first as a child with her parents, then by her marriage, as a wife and mother in a new household with her own daughters and sons. This pattern of two generations living together, the "nuclear family," existed throughout Europe. It has been documented in ninth-century Frankish lands, the manorial settlements of tenth-century Provence and Castile, thirteenth-century Picardy, the north Italian countryside around fifteenth-century Arezzo, seventeenth-century English villages like Cogenhoe, Clayworth, and Chilvers, in eighteenth-century Germany and Poland. Exceptional circumstances might enlarge the family for a time. For example, in fifteenth-century Tuscany, fathers-in-law remained on as the *capo di famiglia,* retaining authority over the fields even after sons had married and brought their wives home.[55] A brother-in-law might stay and work the land. In eighteenth-century Austria and other parts of central Europe, a woman's household might pass through several phases, first with infants, parents, and grandparents together, then deaths of the old people, then births to her own children.

But overall, the pattern of two generations of the nuclear family remained unchanged. The family that peasant women clothed, fed, and cared for averaged four to six members, parents and their offspring. Historians long held the idea that peasant women and men had lived many generations together, as an "extended family" of aunts, uncles, grandparents, parents, children and so on. In fact, studies of communities have shown that multigenerational families in parts of Russia and the *zadruga* of Serbia, having up to eighty people living together as one household, are the only significant variations from Europe's nuclear family pattern. In the early centuries such large households could not be accommodated in the West; it was only in the relatively prosperous twentieth century that rural families, like those studied in the Netherlands and England, came to live with more than two generations together (20–35 percent of the households).[56]

A family of four to six people meant that the peasant couple had done more than establish a separate household to guarantee the quality of life for themselves and their children. They had also made conscious choices about when to marry, when and how often the wife would conceive, and about how many infants would survive. Always at risk, always practical, they wanted to control the numbers their land and their labors would have to support. Natural factors—times of shortage and disease, of famine and extremes of weather—affected a woman's reproductive capacity. Demographers estimate that a peasant woman would lose one in three of her pregnancies from malnutrition or disease, in most instances.[57] Yet even with such natural factors, Europe's preindustrial population should have grown, and it did not.

Until the industrial age the statistics show a population in balance, with a relatively equal ratio of births to deaths, approximately 3 percent.[58] Only twice—from the eleventh to the thirteenth centuries and again in the eighteenth and nineteenth centuries—is there sustained and, in the second instance, dramatic growth. Evidence from as early as the ninth century shows that peasant women and men—by not marrying, by marrying late, by preventing conception—determined the size of their families and thus the lack of population growth all over Europe.

From the 800s to the 1900s peasants allowed their families to grow only when conditions improved. When conditions worsened family size, either from choice or natural disaster, contracted. On the ninth-century manses (peasant holdings) of St.-Germain-des-Prés, for instance, where there was more land, or status improved, there were more in the household. Similarly, from the eleventh to the thirteenth century with the end of slavery and the relative equality of status in English and French villages, better food and more land, families had more children. In Pistoia in Tuscany in northern Italy the *catasto* of 1427 (a census for tax purposes) shows that the number of children born increased with the richness of the land the family held and the amount the land was taxed. A poor woman and man living in the middle hills typically had one child. The householder and his wife in the more prosperous mountains and on the plains would have three or four.[59]

Unusually high mortality rates encouraged the birth of more children as well. After the Black Death had decimated these same villages in the fourteenth century, the birth rate rose as if in response to the excess mortality.[60] The same happened in seventeenth-century Germany when the Thirty Years War took almost half the population. In less than a century the numbers had been restored to the countryside.[61] More income also meant larger families. Statistics for 1693–94 in three French villages of the Beauvaisis region show correlations between high wheat prices and the number of marriages and conceptions. In the seventeenth century, areas with more opportunities for home industries and thus more income showed greater population growth than the towns supplying the work.[62] Then in the eighteenth and nineteenth centuries with all of the circumstances favoring survival—relative peace, higher yields from improvements in cultivation practices, better distribution of food, extra wages from specialized trade and industrial manufacture—couples allowed the size of their families to increase.[63]

The characteristic pattern that normally results in continuous population growth has women marrying in their early twenties and bearing five to seven children at two and a half year intervals over a period of twenty years of fertility.[64] This did not happen in Europe before the eighteenth century. Couples intervened to prevent the normal increase, a phenomenon known as frustrated fertility. The account in a 1901 newspaper from Lomellina, Italy, of the interchange between a striking female rice weeder and the commander of police suggests how couples intervened. She justified the strike for extra wages because they were not for herself but to pay debts and to feed her children. His response to her and all the women was "that they should not have married and should not have had children."[65] In fact what the police commander suggested so patronizingly had been the stratagem used by peasant women and men from before recorded history. In hard times they did not marry, or they married later, thus curtailing the period of the woman's fertility. Once married, they limited the number of their offspring.

As early as the ninth century, German records of manses show only 28.6 percent of the population married; in northern France at St.-Germain-des-Prés, 43.9 percent; at St. Remi of Reims, 33.4 percent.[66] Families often allowed brothers and sisters to remain within the household, as they did in fifteenth-century Languedoc, but on the condition that they did not marry, thus controlling the number the land had to support and preventing division of the strips to establish new families.[67] When in the late sixteenth century more exact statistics of peasant couples' lives are available, they clearly show the effect of delaying marriage. In the English village of Colyton, from the sixteenth to the nineteenth century, the shift from twenty-seven as a woman's age at her first marriage in 1560 to almost thirty between 1647 and 1719, and then back down to twenty-five by 1837, exactly coincides with periods of population decline and growth.[68] The same happened in a Tuscan village from 1650 to 1750 when the marriage age rose from 21.5 to 26.1.[69] Irish families at the end of the nineteenth century continued to make such choices. Only one girl and one boy could marry and then the girl had become a woman of thirty, her husband thirty-eight.[70]

Once married, with land, having established the new household, the couple did not leave the size of the family, the number of offspring, to chance. As best they could, they controlled the intervals between and the frequency of the woman's pregnancies. Records of French and English villages in the sixteenth and seventeenth century indicate a correlation between good times and the frequency of births, bad times and a decline. In the seventeenth and eighteenth century, the villages of the Basse-Meuse region show net population losses and the key factor appears to be longer intervals between births.[71] In the same centuries in Tourovre-au-Perche the relative wealth of the family affected the couples' decisions. The yeomen, craftsmen, and merchants' wives gave birth every two years, the laborers' wives waited another four months.[72]

Women and men prevented conception through sexual abstinence and, when they had intercourse, by withdrawal (*coitus interruptus,* removing the penis and ejaculating outside the vagina). Like Giles Barnes of Cricket St. Thomas in Somerset in the seventeenth century, the man might be the one to restrict fertility. He married the woman he made pregnant, but after that as he explained to the Church court, "he spent about the outward part of her body." He wanted no more children and threatened to "slit the gut out of her belly" if she conceived.[73] Peasant women had their own ways of avoiding conception and, once pregnant, of aborting the fetus. They believed in douches and purges, spermicides like salt, honey, oil, tar, lead, mint juice, cabbage seed; some abortificants like lead and ergot were effective but dangerous. With enough lead ingested, a woman became permanently sterile. Other substances might purge her system but would not directly affect the pregnancy, such as douches or teas of rosemary, myrtle, coriander, willow leaves, balsam, myrrh, clover seeds, parsley, and animal urine. More effective would be the cervical caps and vaginal blocks mentioned in German and Hungarian sources, like beeswax or a linen rag. Peasant women believed actions prevented conception: drinking cold liquids, remaining passive during intercourse, holding one's breath, jumping up and down afterward.[74]

Individual women's experiences testify to the choices they made. Twentieth-century peasant women speak matter of factly of limiting their pregnancies. In Serbia in the 1930s Vuka explained that she had always hated her husband. He was "like a stone when he gets on me." After the death of her daughter just five days old, she vowed not to have any more children:

> I get rid of them. Twice a year these seven years now . . . How? Oh by myself. I massage myself. I get it away. I boil herbs and steam myself and it comes away.[75]

In a Bosnian village the women spoke of drinking vinegar, of opening the cervix with a spindle.[76] Peppina, a twentieth-century peasant woman from southern Italy, became pregnant twelve times. Then she made other decisions. When her husband was away, she left food for the children, laundered their clothes, and went to town to abort the fetus. She came home, spent three days in bed, and then went back to work.

The Catholic Church approved sexual abstinence, even setting aside holy days as inappropriate for intercourse. To do anything else, Peppina and the women before her, whether Catholic or Protestant, went against official teachings. As early as the ninth century, Christian theologians in their penitentials, their decretals, their commentaries, their papal decrees and encyclicals taught that by intervening in the process of insemination and pregnancy a woman acted as a murderess. Abortion was the first act to receive such harsh condemnation. Using Aristotle's distinction of the animate and inanimate, Gratian in his *Decretals* of 1140 made it infanticide to abort a fetus after sixty-six days for a female and forty days for a male—the periods of time

Aristotle believed it took to create a fully formed infant in the womb. In the enthusiasm for uniformity and purity of the Counter Reformation, Pope Sixtus V in 1588 declared all abortions murder, a definition reaffirmed in 1679 by Innocent XI and in 1869 by Pius IX.

Christian theologians had always insisted that intercourse without hope of procreation was sinful, illicit, an act of lust, not an act for God's purpose. First, Saint Augustine, then Gratian, then Thomas Aquinas condemned withdrawal. As early as the sixteenth century some Catholic theologians acknowledged the need of the poor to limit the number of their offspring, but nothing changed in the Church's policies. In the twentieth century, Pope Pius XI in 1930 was but one of the leaders to condemn any form of contraception. Until the twentieth century, the views of the Protestant Churches were the same as the official Catholic doctrine (now most sects of Christianity leave limitation of births to individuals to decide).

When all else failed, couples had yet another way to limit the number of people they and the land would have to support. Studies of ninth-century manorial rolls at St. Germain-des-Prés, of fifteenth-century Canterbury Church courts, of seventeenth-century Somerset parish records, and interviews with women of twentieth-century Bosnian *zadrugas* all show the same choice. Let the child be born and then let it die. In particular, let it die if it is female.[77]

The evidence for this selective infanticide comes from the astonishingly disproportionate ratios of females to males living in the countryside all over Europe from the ninth to the seventeenth century. The average ratio of females to males at birth is 100:105 with the numbers evening out in adolescence and early adulthood, and the women achieving the higher numbers in old age. On the rich fertile lands of St. Germain-des Prés in the ninth century, an area of intense demographic pressure, the ratio of females to males was 100:115.7–117.1. The poorer the family the more exaggerated the ratio.[78] Fourteenth-century English manorial records show ratios of 100 females to 133 males.[79] For the year 1391–92 a list of the serfs on the southern English lands of John of Hastings shows forty-six females to seventy-eight males.[80] Studies of infant mortality in Somerset in the seventeenth century even suggest the consequences to the next generation: Only one of the three females born would live to bear her own children, yet another effective way of curtailing population growth.[81]

In contrast, during the relative prosperity of the thirteenth century, peasant women and men could provide for more children, and the population grew accordingly. Significantly, the ratio of females to males then became roughly equal. Expansion came because more of the little girls survived, more could grow to adulthood and bear their own children.[82] In fact survival of more infant females probably explains much of the population expansion of the eighteenth and nineteenth centuries, for it was not the number of births that increased, but the number of children who survived. In most of Europe in the late nineteenth century, couples commonly had four to five surviving children.[83]

Secular and religious records tell of other ways peasant women and men controlled the ratio of girls to boys and the size of their family. Accepted custom might simply mean less care. For example, the little girls on the ninth-century manse of St. Victor of Marseille would die because they were weaned after one year, making them more susceptible to disease, while their brothers nursed until they were two.[84] Fifteenth- and sixteenth-century court cases tell of mothers and fathers prosecuted for murdering their infants. At the Essex Quarter Sessions of 1570 three servant women (probably unmarried and desperate about how to deal with an illegitimate child) killed their babies. One slit the infant's throat, another abandoned it in a field, the third left it in a ditch. Violence like this is rare in the records, however. More commonly the Church records tell of "overlaying," a silent and perhaps unintentional death. Joan and Stephen Tiler of Rochester fell asleep and woke to find their daughter "lying [dead] between them in bed."[85]

For the unmarried servant woman with no prospect of a betrothal, the birth of an illegitimate child, whether female or male, meant disgrace in most village communities. Its death meant that she could continue to work and that the shame of her pregnancy might be forgotten. For a married peasant woman, the death of her infant might also have indirect benefits. If she could maintain her breast milk, she had another source of income; she could become a wet nurse. Into the seventeenth century, the wealthier families of the manors and the towns employed countrywomen to breast-feed their babies. In the fourteenth century the merchant's wife, Margherita Datini of Prato, periodically found wet nurses for the wives of her husband's friends. She suggested 12 florin a year for the wages, about the same as for any female servant.[86]

She listed the qualifications in her letters: The wet nurse must not favor her own child, no matter how much it cried, should have moderate-sized breasts so that she would not flatten the child's nose, must not become pregnant easily, and ideally should look like the biological mother. In August of 1398, Margherita Datini wrote: "I have found one in Pizza della Pieve, whose milk is two months old; and she has vowed that if her babe, which is on the point of death, dies tomorrow, she will come, as soon as it is buried."[87] Still in the 1960s when Peppina from a south Italian village had breast milk, she went to Rome to be a wet nurse.

Instead of the wet nurse joining the household, babies might be sent to the country. At the beginning of the fifteenth century, the Florentine merchant Antonio Rustichi noted in his diary that he had sent successive children out to the wife of a farmer, then to the wife of a baker. Increasingly this became the practice, especially in the eighteenth and nineteenth centuries when town and Church authorities took over the care of abandoned children. The older children stayed in the foundling homes in the town, infants went to foster homes in the countryside. No one questioned the high mortality of children at nurse—in the summer months of the nineteenth century as high as 75 percent.[88] The wet nurse could not breast-feed all the abandoned children in her care; the money paid would not feed them, and so they died. Even the infants whose parents were known might be expected to die. French artisans sent their babies to wet nurses and then simply stopped paying them—not necessarily out of callousness but simply a harsh choice: the death of one to preserve the lives of the older siblings. In such circumstances the foundling home and the foster home (also called the "baby-farm") made the impossible easier, the infant's death more distant and thus less painful.[89]

Survival Outside the Family

The circumstances that particularly tried the ingenuity and strength of rural women came when they outlived their husbands. The peasant woman then became a widow—perhaps a young woman alone with small children or an old woman with few, if any, family members to help her. Europe's countryside had many widows, for if a woman survived her child-bearing years, she tended to live longer than the men in her village. In fifteenth-century Tuscany in the villages around Arezzo a disproportionate 6 percent of the population were widows. From 1574 to 1821, 22 percent were widows in a group of English parishes studied by demographers.[90]

The folktales portray her vulnerability. Red Riding Hood's grandmother lived alone outside what English peasants called "the hue and cry" of the village, too far for her screams to be heard when the wolf attacked her. John Langland in his poem of peasant life *Piers the Ploughman* compared the widow to Jesus, forsaken by God.

In theory the dower negotiated for the woman at the time of her betrothal was meant to sustain her and any young children should she be widowed early. The various local practices followed the Germanic customs of the earlier centuries closely. A widow had access to a portion of her husband's lands during her lifetime or until her remarriage—when in theory the new husband would take on responsibility for her provision—and a portion of the movables, the furniture, bedding, whatever articles of the household the couple had. In England in Kent, in Normandy, France, in the Austrian Heidenreich she inherited one third to one half of the land and the household movables. A 1919 Russian statute guaranteed her one seventh of the movables.[91] Customarily, her holding of the land was meant to be temporary, until another male could assume the responsibility, usually her eldest son. In Warbleton, Sussex, in southern England, in 1322 the custom was for widows to "hold tenements in bondage as their bench [right] until the younger son is fifteen years old, and then widows ought to surrender to the younger son, as heir, a moiety [half] of the inheritance. And they will hold the other moiety as their bench if they remain widows. . . ."[92] Into the twentieth century in County Clare, Ireland, a mother held the land until the heir came of age. Restrictions on what a widow might do with the land reflected concern for the male heir, in particular the desire to assure that, should she remarry, the children of the dead husband would not lose their rights. For example, a Russian charter of Pskov took the land from her on remarriage unless it was specifically willed to her.[93] In seventeenth-century Berkshire a woman free tenant might hold land on her remarriage if named in the lease, but the preferred heir would also be named, giving her no right to alienate the property—that is, sell or bequeath it to another.

A death duty, called the heriot in England, would be just the first of the widow's obligations as heir, owed in return for access to the land that she and her husband had worked together. Customarily the landlord took the family's best beast. When Thomas of Merdens in Halton, Buckinghamshire, "entered the way of all flesh" in 1303, the lord claimed his two oxen.[94] If the couple had been too poor to have livestock, in thirteenth-century England or in fourteenth-century Flanders, the widow owed one third to one half of their movable goods: pots, chests, even tunics. The Church also claimed a "mortuary gift" in theory to make up for tithes not paid in the husband's lifetime. Priests in sixteenth-century France refused burial until they had received the "best blanket."[95]

More significant, however, would be the regular tasks and payments owed throughout the year. For the year 1266–67 Matilda on the manor of Clifford in Gloucester had the right to hold the family's twenty-four acres, but as a widow she alone owed all the services, goods, and payments. She probably hired a man to help her with the days of plowing—half an acre a week through the fall, winter, and spring—with the mowing of the lord's fields and the carting of the lord's produce. She herself had done the two days of weeding and the days of lifting hay onto the wooden carts, and perhaps even presented herself for the days—three each week between June and August—of unspecified labor. She owed almost 11 shillings at different times of the year—122 pence when a day of "manual service" was valued at only 3 half pence. This money she would have to raise from her fields, from her labors.

Of her produce, the lord claimed "eggs at Easter at will."[96] To earn extra money Matilda brewed ale. The widows of fourteenth-century Montaillou in France sold cheese, ran the inn, and sold at the fairs. The necessity continued into the twentieth century.

The lords of the manors were skeptical of the widow's ability to fulfill the obligations that went with the strips of field she held in trust for a male. In Ashton, Wiltshire, in 1262 she had to pay sureties of 18 and 20 shillings to guarantee that she would uphold her duties.[97] Landlords preferred a widow to marry, to acquire in its literal sense "a husband who can maintain the land."[98] Sixteenth-century Danish bailiffs in Funen and Jutland tried to force widows to remarry, excusing them of the inheritance payment, the "relief" owed to renew the lease, on the theory that the new husband could pay it. (Though landlords wanted widows to marry, in England they punished those who did so without their permission. In Mapledurham, Hampshire, in 1281 Lucy, widow to Walter le Hurt, lost the claim to her land for "fornication" and marriage "without licence of lord."[99] Others paid fines into the mid-seventeenth century).

As soon as there are records in England, in France, in Spain, from the sixteenth to the twentieth centuries, they show that peasant women remarried much less often than their male counterparts. They might remain unwed, but that did not mean that they always chose to cope with the responsibilities of the land, especially if their children and grandchildren had come of age. "Agnes ate Touresheade" at Hindringham in Norfolk in 1310 "came into full court and said that she was powerless to hold one messuage and one yardland of servile conditions . . . after the death of said Stephen [her husband]." She assigned the land to a granddaughter and her new husband on the understanding that they would provide for her during her lifetime.[100] In 1320 in the court rolls of Dunmow, Essex, Petronilla's son John pledged that he would provide during her lifetime

> reasonable victuals, in food and drink, as befit such a woman, and more over said Petronilla will have a room, with a wardrode, at the Eastern end of said messuage, to dwell therein during . . . [her] lifetime, and one cow, four sheep, and a pig, going and feeding on said half-yardland as well in winter as summer . . . for her clothing and footwear.[101]

Local custom might guarantee her care in return for giving over her rights to the family fields. In Haute Provence, traditionally the land went to the children but the house to the mother, thus ensuring inheritance for the offspring, service for the landlord, and subsistence and shelter for the widow.[102]

A peasant women's husband before his death might have made arrangements with the children. In sixteenth-century England William Brasier left the house to his son but on the condition that his widow have "free bench" (literally the right to a bench before the fire) and food, or £3 a year so that she could live separately.[103] To be in the care of one's children and grandchildren might also mean hardship, the neglect and humiliations of old age. Widows in northern Tuscany in the fifteenth century usually lived with their sons, but became subject to the heir and his wife. The tax records for the region reflect their reduced status, the assessor listed them after the children.[104] In nineteenth- and twentieth-century Croatia and Serbia, even though the communal living of the *zagruda* guaranteed a widow a place, as she became older and older she lost her authority as female head of the household.

Peasant women without family, without access to enough land, perhaps too old for extra labor, had few alternatives to choose from in order to survive. A woman alone might be a widow, a wife abandoned by her husband in bad times, a female member of the wandering poor. In seventeenth-century Denmark, the majority of the indigent in villages were widows with children. In eighteenth-century France a regular member of the village community was the beggar woman, her eyesight gone at fifty, her fingers too stiff to make lace. Until the late sixteenth century landlords and the Church provided some care for such people. In 1315 and 1316 the Cistercians in Brunswick and Bremen in northern Germany are said to have fed 4,000 a day from Lent to the harvest.[105] A harvest code of 1329 for England gave old women the right to "glean," the right to gather whatever they could in the mown fields before the animals were released to graze on the stubble. A landlord's wife appointed an almoner, a member of her household whose job it was to distribute alms and the scraps from the table each day. The Countess of Bedford in the seventeenth century had an almoner who gave shillings to a "poor mad woman," and a "poor distracted woman."[106] On the whole, by the end of the 1500s local and royal governments had taken over what charity there was from individuals and religious orders. The English Poor Laws required the poor to petition the quarter sessions. In the seventeenth century this might be more than a woman could manage. The parish records list "a poor walking woman" and "a poor woman name unknown, who had crept into Mr. Miller's barn." Both died of starvation.[107] The eighteenth-century records for the *bureaux de charités* of France give sums to feed a child and an adult woman, but there was not enough to provide for them all. In the village of Mende the records show 1,000 poor but sufficient funds to give only 100 of them one meal twice a week.[108]

What other recourse did women have? They could become a prostitute or a thief. They could turn to violence against those better off than themselves. In Somerset in the seventeenth century a peasant woman could sell the use of her body for 50 faggots of wood, 6 pence, a 9 shilling coverlet and half a bushel of wheat, but she then risked being brought before the local court for "fornication."[109] Peasant women could infringe on the rights of the landlord and steal goods from their neighbors. These crimes brought them to the courtyard of the lord's château or castle because they had broken the rules of the manor. The accounts of peasant women's violations increase in bad times, and increase as the lords expanded their privileges into the common lands and the forest. In Tooting, Surrey, in 1246 two widows were fined 6 pence each for encroaching on the lord's land. The next year Lucy Rede let her cattle stray into the lord's pasture. At Ruislip in the same year Isabella, a widow, had to pay the 18 pence fine when her son went into the lord's woods—probably to collect fuel.[110] In fifteenth-century England, Agnes of Weldon and her three children stole eight sheaves of grain at the harvest. On the manor of Wakefield during the famine of 1315–1317, a woman was prosecuted for stealing from her neighbor's toft, or household garden.[111] One accused woman in a seventeenth-century Somerset court milked a cow, another took a loaf of bread.

The royal courts of these early centuries heard accusations not for manorial crimes but for crimes against the King's peace, when peasant women acted not alone but with a man—a husband, a father, a lover or sons. The women watched the roadside while the men robbed; the women stopped travelers on a bridge; they sold the stolen goods. In the eighteenth century women led bands of their own. In Brittany, Marion de Faouet and her children robbed travelers on the highway. Marie Jeanne Bonnichon worked the region near Bourges, luring merchants into ambush. Consistently, however, rural women chose crime less often than their male counterparts.

Sometimes peasant women became desperate and enraged. They joined in the rioting and mob violence of the countryside. Women marched into London with John Ball in the rising of 1381. Women marched in the German Peasants Revolt of 1525. In the Heilbronn area in the village of Bökingen, the Black Farm Women *(schwarze Hofmännin),* as they were called by their contemporaries, were thought to have magical powers to inspire others. They led the countryside against the local town council and were with the 8,000 peasants who took Weinsberg.[112] In England women joined the men to protest enclosures; in June of 1641 in Lincolnshire mothers and their sons broke down the fences around the commons to let the cattle in. Near Maldon in Essex, Captain Alice Clark led the male and female weavers in a grain riot. In France in the seventeenth century the same kinds of anger inspired the women led by Branlaire in Montpellier in 1645 to rebel against the royal tax collectors. She insisted that the tax money would take bread from their children's mouths.[113] The same cries against unfair privileges and payments could be heard at the end of the eighteenth century when women joined men in the attacks on the châteaux of the nobility and the burning of manorial records in the early months of the French Revolution. In the nineteenth century countrywomen joined the protests against work lost to machines.

In the late nineteenth and the early twentieth centuries, collective protest turned to strikes, a different tactic occasioned by the same desperation. In 1896 in Tuscany in the Brozzi region of northern Italy, the straw workers watched as the price for their different types of labor dropped from 10 to 7 centesimi, and from 20 to 15. Of the almost 85,000 peasants involved in this craft, ninety-five percent were women.[114] Fifty initiated a strike and, led by Darsena Conti, they went into the countryside.[115] From May 15 to 30 they mobilized crowds, stopped wagons and trains, burned materials, and harassed anyone who would not join them. By the seventeenth of May the men had joined them—over 38,000 people.[116]

By the thirtieth, the government in Rome had called for a commission of inquiry. Accommodations involving the establishment of cooperatives did not help to regularize the prices at an acceptable level, and in September of 1897 the women struck once more to demonstrate their dissatisfaction. With the support of the Socialist Party, another group, the rice weeders of Tuscany, struck in 1901 for more pay as their first demand, then for a nine-hour day. Because the majority were migratory workers, they were at the mercy of the local elite, who could in the last resort have them rounded up and sent home by the police. In 1907 after a third set of strikes and work stoppages, a national bill was passed about their situation. Though of limited value in the long term, it signified acknowledgment of their plight.

However difficult the lives of married peasant women, survival alone presented even greater challenges. Throughout the centuries peasant women in this situation looked to village customs and to the secular and religious institutions for assistance. In the last resort they relied on their own strengths. These less tangible qualities enriched the emotional and spiritual life of others as well.

Giving Value

In addition to providing for their family's physical survival, peasant women also provided moral, emotional, and spiritual sustenance. In the tales they told their children, in the ideals and rituals of their faith, countrywomen gave value to the choices they made and a sense of purpose and accomplishment to the life they lived within the narrow confines of their circumstances and the roles allowed. There would be much in the harshness of life to justify and explain. As a daughter, as the female, the peasant woman's life would never be without responsibility. She would be the first child picked to work and the one always expected to assist in the maintenance of the family. In Limousin at the beginning of the nineteenth century, a French woman described herself as having no children even though she had three daughters "for a daughter is merely a daughter, whereas only a boy has the privilege of being a child."[117] As a wife and mother the peasant woman's responsibilities to her husband and her children consumed her energies.

There would be much in the attitudes of her world that a rural woman would have to ignore or endure. Both the customs of the culture and the Church insisted on the female's innate inferiority, her naturally subordinate status, and the need for her obedience to male authority. Peasant women rarely questioned these assumptions and expectations, or the male fears and ambivalences that underlay them. Rarely did they allow them to break their spirit. Instead, rural women found ways to value what they could aspire to and what they did. They learned to make virtues of the dependent qualities demanded of them.

Peasant daughters heard the denigrating proverbs and tales, the ones selected and refined through the centuries of Europe's oral tradition, from their mother or grandmother. They learned the assumptions of their inferior nature, their failings of character. Peasant women believed that favorable signs in a pregnancy—carrying the fetus high, no morning sickness, and rosy cheeks—meant a boy. Proverbs and tales derided women's intelligence, mocked them as careless, criticized them for speaking too much, and promised punishments for disobedience. But each traditional denigration could be countered in the peasant girl's imagination with other stories. Folktales told of barren couples rewarded with a daughter like Snow White, so beautiful and good that she was as flawless as the white snow that inspired her name. For the dutiful little girl, the patient, devout, and industrious, or the quick-witted and enterprising heroine, the rewards were without measure. Snow White's marvelous innocence gained her the love of a prince. Cinderella's goodness and beauty won the king's son. Beauty's loyalty turned the Beast into a handsome young lord. The obedient daughter who sacrificed her hands for love of her father earned a royal husband.[118] Clever, courageous maidens who performed incredible tasks also won this prize, not just marriage, the expected goal of a young girl, but a marriage way above her station and the promise of a life of riches and love.[119]

As well as folktales, Christian peasant daughters learned the basic premises and tenets of the faith from their mothers. In fifteenth-century France, Joan of Arc recited the Lord's Prayer *(Pater Noster),* the Creed, and the Prayer to the Virgin *(Ave Maria)* to her mother. Her family attended the parish church in Domrémy, listened to a sermon, and received communion, perhaps once a year at Easter, confessing to the priest as had been required by the Church Council of 1215. (The formal outlines of the service existed by the end of the seventh century, though there was still variation in ritual and belief.) Before the sixteenth century and the advent of Protestantism, much was in Latin, but the homilies, the exemplary stories, were told in the vernacular. From these a peasant woman learned of her descent from Eve, the first sinner, of the uncleanness of her reproductive organs, and of her divinely ordained subordination to men. Women found ways to transform or to mitigate each premise, each tenet.

Eve, so often the symbol of evil, became the image of the dutiful mother in peasant women's folktales, questioning the Lord for rewarding her beloved children unequally. The cathedral window at Erfurt from the second half of the fourteenth century showed an Eve spinning with her distaff, the child swaddled in a cradle beside her. Other mothers became ideals for all Christians: Saint Felicitas who watched her seven sons tortured before her eyes, and the first Mother of Heaven, the Virgin Mary. Tales and ballads portrayed Mary as the model of motherly virtues, compassion, and forgiveness. Into the 1970s in a central Italian hill town a pregnant woman felt honored with the affectionate title "Madonina."[120]

Christian peasant women into the twentieth century participated and believed in the rituals meant to cleanse them of sin, of their inherent impurity. Mémé Santerre from the Nord province of France, born in 1891 to a poor family of country linen weavers, wrote of her first communion, "Oh, it was a day like any other, but so beautiful that I still remember it." She was given a new apron from her godfather, material for a new dress from her godmother, a white veil to wear and a gilt candle to carry in the procession to the church. After the service

> we returned to the cellar to weave until time for vespers. When we returned my mother let me wear my new apron. That was the greatest pleasure of the day. I was eleven years old.[121]

With the onset of menstruation, peasant women accepted the tradition of their physical uncleanness. From the earliest days of Christianity, they did not approach the altar or receive communion on those days of the month. Into the 1920s in County Clare, Ireland, girls sat apart from the boys. Into the twentieth century in some parts of Europe, peasant women after giving birth to a child followed a tradition as old as Leviticus. They kept apart from the community and did not attend church until blessed by the priest and ritually cleansed, in a ceremony called churching. In Russia a woman had to keep apart from her family altogether. In Brittany in the nineteenth century, a woman went silently and alone to the church, her friends pretending not to see her as she made her way in her funeral cloak. Blessed by the priest, she removed her cloak and went outside to be welcomed by the people of the village, who acted as if she had just been too busy to see them all this time.[122] Danish Protestant women waited for the pastor to come to their door to bless them. Spanish women in a twentieth-century village knelt in the church with a candle and allowed the priest to bless them and then lead them to the altar.

With equal willingness peasant women embraced other beliefs and rituals that made some sense of the hardships of life and gave comfort and hope. From the earliest books of penance, Christian women accepted the idea of sin, of human transgression and the means offered by the Churches to restore them to grace, to God's favor. By the sixteenth century, a Catholic woman could confess privately to her priest and expiate her sin through a penance he set. She might pay "indulgences" as a penance. As early as the fourteenth century, peasant women said the Virgin's special prayer with a rosary, a string (or chaplet) of beads that could be carried everywhere. Women took responsibility for trips to the shrines of local saints, and for the prayers and rituals that had once honored former deities and spirits of their locality.

Women of the countryside remained loyal to their faith long after peasant men and long after both women and men of the towns. Well into the twentieth century, countrywomen went to church and took part in the ritual because Christianity seemed to give a spiritual value and significance to the hardships and conditions of their lives. In the sermons heard in the parish church, those who suffered and mourned were blessed, their miseries transformed into sources of salvation and eternal life. Thus, a woman of the Bocage region of Normandy in the 1970s could speak of misfortune as "good suffering."[123] Deprivation, self-sacrifice, humility, all were virtues that guaranteed honor and reward to a Christian. Above all, the Catholic faith asked obedience of its most devout followers, and the life of a peasant woman was often synonymous with obedience, learned as a child with her parents, then practiced as a wife to her husband. The rituals of the marriage ceremony established the relationships. In France the bride knelt before the groom; in Russia she took off her new husband's boots at the wedding feast. The exchange of vows, the passing of her hand from that of one male to another, all signified the male's guardianship and authority, her acceptance of the protected and subordinate position. A seventeenth-century wife of a serf would remember "As the tsar answers to God, so a boyar answers to the tsar, a peasant to his Lord, a woman to her husband."[124] A prayer book from 1946 used in a Santander village in northern Spain reminded a wife of her circumstances. When making her confession, she was to consider how well she had obeyed the dictates of the Church and society:

1. Esteem your husband
2. Respect him as your leader
3. Obey him as your superior
4. Reply to him with humility
5. Assist him with diligence.[125]

Little in the peasant woman's society supported deviation from this ideal of behavior. Everything condoned her husband's right to punish her if she did. Irish and French proverbs warned men of the consequences should they be lenient with their wives: "Do not give your wife authority over you, for if you let her stamp on your foot to-night she will stamp on your head tomorrow."[126] Any man who allowed his wife to rule over him in the eyes of his village became the butt of the charivaris (local festivals), forced to ride backward on an ass.[127] Both religious and secular authorities condoned correction and "chastisement," as it was called in Gratian's *Decretals* of the twelfth century. Customs and laws all across Europe accepted and pardoned men who bloodied or even killed their wives. A sixteenth-century French proverb expressed the attitude that continued to be voiced into the twentieth century: "A good horse and a bad horse need the spur. A good woman and a bad woman need the stick."[128] Lean Lizzie, the scold in a German folktale, found herself thrown to the bed, her arms held together, her head pressed into her pillow by her husband, "until she fell asleep from extreme exhaustion."[129] In the 1930s in rural Serbia husbands beat their wives because of complaints from parents, for disobedience, or just to show their power over them. In the 1950s in southern Italy a husband could threaten murder if his wife was unfaithful. Pietro's parting words to Ninetta as he left to work in the north were "I warn you if you put horns on me I'll come back and kill you. Do you understand? I'll kill you."[130]

The majority of peasant women throughout the centuries accepted the circumstances, the attitudes, and the necessities of survival even though they were left vulnerable and subordinate. Occasionally a woman could not. Thirteenth-, fourteenth-, and fifteenth-century religious and secular court records from England tell of peasant women who lost control and turned on their world. They killed their children and husbands. They killed themselves. In 1226 the coroner inquired whether or not Alice de la Lade had "from madness or maliciously and intentionally" killed her child.[131] Margery, the wife of William Calbot, knifed her two-year-old daughter to death and then forced her four-year-old son on to the hearth flames.[132] A Northamptonshire woman whipped her ten-year-old to death. Agnes, the wife of Roger Moyses, killed her son Adam during what the court called "one of her frequent bouts of insanity." In 1316 Emma le Bere, though ill and restricted to her bed, in a "frenesye" killed her children by cutting their throats with an ax. Then she hanged herself.

Bitter, despairing, enraged, peasant women turned on their husbands. The thirteenth-, fourteenth-, and fifteenth-century English records tell of the wife who waited until her husband slept and then cut his throat with a small scythe. Another woman, while sick with fever, killed her husband and then claimed she could not remember anything when she recovered two weeks later.[133] The English judges showed some compassion in the murder of children, but none with violence toward a husband. This threatened the most basic relationships of the society. The fourteenth-century court records called murder of a husband "petty treason," similar to a servant killing a master. In 1726 Catherine Hayes was the last Englishwoman convicted of the crime. She was executed in the traditional way, burned at the stake. Into the eighteenth century Europe's men chose such a dramatic death because burning at the stake seemed the fit punishment for women who had defied the basic premises, patterns, and authority of the culture.

Such openly defiant women were the exceptions, however. The vast majority of European countrywomen found ways to adapt and accommodate, ways that gave them a sense of value and purpose despite their subordinate relationship to men.

3

THE EXTRAORDINARY

Joan of Arc

Joan of Arc (1412–1431), the extraordinary sixteen-year-old daughter of a French peasant family, defied almost every tradition of the peasant woman's world. She disobeyed her parents, importuned those above her station for help, and insisted that she must act outside accepted female roles. Joan told everyone that she had been sent by God to join the army of the King of France and to raise the English siege of the town of Orléans. Everything about her manner, her demands, and her actions was unorthodox. In normal circumstances, the very personification of the insubordinate, disobedient female, she would have been left to the reprimands of her mother and father. But in the midst of the Hundred Years War, the secular and religious leaders of fifteenth-century France listened to the unorthodox. They came to agree with Joan's vision of herself as the young virgin granted access to God through her voices. They came to perceive her as a heroine: the holy maiden warrior, zealous and strong, sent for the salvation of the kingdom. So perceived, Joan's passion, energy, persistence, and ingenuity gained her power and success in roles traditionally reserved to men and to men of higher caste as well.

Joan is remarkable for other historical reasons. Unlike every other woman or man of the early fifteenth century, documents abound to describe and explain her life. By her unorthodox actions and assertions, Joan defied the acknowledged authority of her day, and that authority, literate and careful, recorded the words of her childhood friends, the memories of her family, and of those she fought with. Most unusual, there are also her own words, pages and pages of questions and answers, cross questions and explanations in French notes, in Latin prose. These are the records of her trial, and then of her rehabilitation, a "retrial" twenty years later, when the Church reconsidered her case and reversed the condemnation of her as a schismatic and heretic (as one who would not accept the paramount authority of the Church).[1]

These sources reveal a life initially much like that of other young girls in a French village of the early 1400s. Her godmother Beatrice, wife of the mayor, remembered her spinning, and working at harvest time, even though her father was the most prosperous of the farmers in Domrémy. Her friend, Jean Waterin, said they drove the plow together. The priest, Jean Colin, did not think her especially pious. Then at thirteen she believed that Saint Michael and the angels appeared to her, "towards noon, in summer, in her father's garden."[2] No, she had not been fasting, she explained to the questioner at her trial. She heard them in the forest, when the bells of the parish church tolled. From that moment on, as she explained in March of 1430, "they often come without my calling," and "she had never needed them without having them."[3] The church at Domrémy and the one across the river in Maxey had statues of Saints Margaret and Catherine, and soon she saw their faces crowned with light and heard their voices. With the first appearance of her voices, Joan made a vow of virginity, increased her devotions, made frequent pilgrimages to the next villages, and to the hermitage dedicated to the Virgin Mary, Our Lady of Bermont.

Her voices never spoke of a life of contemplation, the usual response for one so religious, rather of a life of actions inconceivable for a female in the context of fifteenth-century France. Yet their message was clear; Joan repeated it over and over. She was to relieve the siege at Orléans, to see the Dauphin (the Valois heir) crowned at Rheims, to free the Duke of Orléans, to regain the loyalty of Paris. To attempt even one of the voices' tasks—to do anything other than work in the fields or in the household and to marry—set Joan apart and made her a disobedient rebel at risk, a female outside the protective confines and patterns of fifteenth-century village life. Though she kept her visions to herself, by sixteen she had focused all of her energy, her quick intelligence, her strength, and her perseverance on accepting the challenge of their commands.

In the spring of 1428 she lied to her parents in order to try to see Robert de Baudricourt, the local lord her voices had said would take her to the Dauphin. Joan persuaded Durand Laxalt, her mother's cousin, to accompany her to the castle, but De Baudricourt dismissed her request, telling Laxalt "to give her a good slapping and take her back to her father."[4] Joan refused to stay away. Catherine le Royer, the family friend with whom she stayed, remembered the tension, Joan's sense of urgency, "how the time lagged for her as for a woman pregnant."[5] Joan chafed at the delay, insisting she had only one year to fulfill the tasks set by her voices. In January of 1429 she finally convinced her listeners with the simplicity and directness of her assertions:

> before we are in mid-Lent, I must be at the King's side, though I wear my feet to the knees. For indeed there is nobody in all the world . . . who can recover the kingdom for France. And there will be no help . . . if not from me.[6]

De Baudricourt changed his mind, and with his safe conduct and six of his knights at arms to accompany her, riding a horse bought for her by Laxalt, Joan traveled at night across the English-held territory, covering the 500 kilometers to Chinon, the Dauphin's castle, in just eleven days.

Joan had badgered help from De Baudricourt, but others already believed in her and her visions. Jean de Metz, who rode with her to Chinon, gave her the men's clothing she wore when she burst into the court and picked the Dauphin from a crowd of women and men. There, with one gesture after another, with one statement after another, this seventeen-year-old peasant woman convinced the religious and secular elite of France that she was the instrument of their salvation. She reminded them of Merlin's prophecy, revived by the peasant visionary Maxine Robine at the end of the fourteenth century, "that France would be ruined through a woman and afterward restored by a virgin."[7] A gold crown and a buried sword she described appeared as if miraculously. Churchmen of the University in Poitiers questioned her on her faith and became convinced of her sincerity. The dean of the faculty, a Dominican, remembered the interview. When asked if she believed in God she replied, "Yes, more than you do."[8] The Dauphin's mother-in-law, Yolande of Aragon, Queen of Sicily, and her ladies, examined Joan's claim of virginity and established that she was in fact as she described herself, "the maid." Queen Yolande offered money for the expedition. In the end the learned churchmen of Poitiers and the King's Council unanimously endorsed her and her mission.

The war between the feudal monarchies of France and England had come to a stalemate with the Duke of Burgundy allied to the powerful English Earl of Warwick. The French Dauphin Charles's commanders gave him conflicting advice. The English held the capital, Paris. Plans to raise the siege at Orléans had foundered. Joan stood before the Dauphin, a stocky seventeen-year-old peasant, her hair "cut round above her ears" in her rough wool tunic, hose and leggings, assured, brusque, determined, clear on what should be done. She claimed she was a virgin sent by God to save France. To any who still doubted she answered impatiently, "But lead me to Orléans and I will show you the signs I was sent to make."[9] The Dauphin let Joan go to Orléans.

As she had convinced the learned and the courtly, so Joan convinced the soldiers of the army, then the citizens of the towns. One of her earliest supporters, the twenty-five-year-old commander, the Duke of Alençon, described himself as "bedazzled." (He had given her a horse when he had seen her, the evening of the second day at Chinon, wield a lance and run at a tilt.)[10] He remembered the march to Orléans, with about 4,000 men for the campaign, at last enough troops. A later chronicler marveled "We had plenty . . . because everyone was following her."[11] At Orléans the crowds pressed around her, and "felt themselves already comforted and as if no longer besieged, by the divine virtue which they were told was in the simple maid."[12] Winds changed as if by a miracle, and supplies came into the besieged town.

Perhaps most significant of all, Joan convinced the English soldiers, and the citizens in the other towns they held, of her powers. Alençon remembered the evening she had taken the standard and shown herself on the parapet at Orléans "and the moment she was there the English trembled and were terrified."[13] Dunois, another of the commanders, insisted that after Joan's arrival "four or five hundred soldiers and men-at-arms could fight against what seemed to be the whole power of England."[14] In June of 1429 and July the towns of Meung and Troyes surrendered just because she arrived.

As a peasant daughter, Joan knew nothing of fighting and warfare except what she had experienced as a villager with soldiers passing back and forth across her valley, the Meuse. When she was thirteen, the soldiers had burned the church. Twice she had helped to herd the livestock to fortified places of safety. "For fear of the soldiers" her parents would no longer let her tend the livestock alone.[15] In May of 1428 she, her family, and the others of the village had fled Burgundian troops, who burned what they could. It was Joan's vitality, her confidence, her faith in her visions, and the beliefs of those around her that transformed her into an able warrior and a leader among men.

The young peasant woman participated in at least seven military engagements. The King's Councillor wrote of six days and nights with her in the field and commented, "She bears the weight and burden of armour incredibly well. . . ."[16] At Orléans she fought on horseback in an attack on a stronghold. "The first to place a scaling ladder on the bastion of the bridge," she remembered of one encounter in May of 1429.[17] She appeared fearless in battle; Dunois told of one of the assaults at Orléans when she had taken up a fallen standard, rushed to the top of a trench, and then been followed by the men. At St. Pierre le Moutier in October of 1429, her squire, Aulon, described rushing to assist her, when she was fighting almost by herself, and

> taking her helmet from her head, she answered that she was not alone, that she still had fifty thousand men in her company and that she would not leave that spot until she had taken the town. At that moment, whatever she might say, she had no more than four or five men with her. . . .[18]

Joan took punishment like the other soldiers. In the assault on Jargeau in June a blow on the head cracked her helmet. Twice she was hit by crossbolts—once at Orléans in the neck and once in the thigh during an attack on one of the gates of Paris in September of 1429. None of this stopped her.

She had a natural gift for the new methods of warfare. At Troyes in July of 1429 she placed the artillery; she advised with "prudence and clearsightedness" and drew the praise of experienced commanders like Dunois and Alençon.[19] Yet, contrary to subsequent popular memory, Joan never actually took command, rather she assisted, advised, and, perhaps most important, chided those in charge when they delayed or appeared to lose their zeal for the fighting. At Orléans she told Alençon: "Do not have doubts, when God pleases, the hour is ripe. We must act when God wills it. Act, and God will act."[20]

Joan's authority also came from the simple contradictions of her situation: a young peasant woman, a pious virgin, moving confidently and unmolested in the midst of an army. Joan was never shy with her companions at arms. When she met Gobert Thibault, one of the King's squires, he remembered, "she clapped me on the shoulder, saying that she would like well to have men of my sort [fighting with her]."[21] But at the same time, she impressed all with her piety and her virtue. She prohibited pillage, organized confessions and masses for the troops, and dismissed the women who accompanied and serviced them. She visited and encouraged the wounded. She fasted and received communion twice a week. Most miraculous to her contemporaries, she inspired no lust in the men she served with. In camp she tried to have a woman sleep with her whenever possible, but this is not why they let her be. From her first days alone in the field with soldiers, the men believed that something else which they could not explain deterred them. The two knights who originally escorted her to the King assumed they would have intercourse with her, "but when the moment came to speak to her . . . they were so much ashamed that they dared not. . . ." Even though she lay between them in the fields at night, they explained that they had felt no desire.[22] Both her squire Aulon and Alençon, her commander, dressed her wounds, helped to arm her and had, therefore, seen her breasts and bare legs. Aulon described himself as "strong, young and vigorous," yet never "was my body moved to any carnal desire for her. . . ."[23] Dunois remembered the effect of her presence: "As for myself and the rest, when we were in her company we had no wish or desire to approach or have intercourse with women. That seems to me to be almost a miracle."[24]

How then did she lose her power? How did she come to be captured and abandoned by her king? The very circumstances and qualities of character that had given Joan power led to her capture by the English. At first the Dauphin drew on her strength and rewarded her success. Two days after the army lifted the siege at Orléans, his treasurer entered on the accounts for May 10, 1429, a suit of armor for her: "complete harness for the Maid, 100 livres tournois."[25] But his crowning in July of 1429 signaled a new period of military inaction. Joan could not accept this. Of her original tasks, only the recapture of Paris remained.

Joan did not go home. She acted in spite of the King and kept fighting on her own authority. At Compiègne in May of 1430 her remarkable powers seemed to leave her. She and three to four hundred of her men found themselves drawn out and cut off from the gates of the town walls. She was forced back to the bridge near the walls and stranded outside when the town fathers decided to close the gates. A Burgundian account said that Joan was pulled from her horse. Unable to remount, she surrendered to a nobleman. The commander, Jean, Prince of Luxembourg, set her ransom at 10,000 gold crowns, a king's fortune in those times.[26] The Dauphin, now King Charles VII, and his councillors never tried to free her.

Instead the English gained custody of the Maid. They perceived her as a fearsome opponent. In 1429 the Duke of Gloucester had acknowledged Joan's special power and ordered that the captains and soldiers, deserters terrified "by the Maid's enchantments," be captured and punished.[27] Joan's unorthodox behavior, her success in opposing them, became in the eyes of the English sources of terror, "enchantments," and Joan was perceived as the agent of the Devil, not of God. With her capture, the English believed that God had favored them. Now they must discredit Joan and the supernatural power she seemed to hold.

Joan's voices, the source of her extraordinary authority, gave the English the means to discredit her. She claimed the voices came from God, therefore her enemies could allow the Church to examine her. The English controlled Paris and thus the learned ecclesiastics of the university. They had proved allies among Church officials, like Pierre Cauchon, the bishop of Beauvais, who became the principal judge at Joan's trial.[28] Thus inquisitors could be chosen and procedures arranged so as to ensure the desired outcome to the inquiry and the trial.

Cauchon and the English stretched or ignored every rule of ecclesiastical procedure to weaken Joan, to force her to denounce her own power and to deny her voices. The conditions of her imprisonment at the castle of Bouvreuil left her no privacy and vulnerable to sexual assault. Instead of the women usually assigned for a female prisoner, or at the least men associated with the Church, Joan had five English soldiers, three in the room with her even when she slept. At night they fettered her to "a great block of wood."[29] Cauchon hoped to see her intimidated, confused, and confounded. From the ninth of January to the 30th of May 1430, he and the inquisitors questioned her, sometimes twice a day, usually in the morning, beginning at eight, in a room of the castle or in her cell. She had no counsel. When two of the participating friars offered advice, Cauchon reprimanded them, and they fled back to their monastery. Sometimes it was two or three questioners, sometimes the whole panoply of two judges, a prosecutor, eight examiners, three notaries, the bailiff, and forty-eight assessors, all the priors, bishops, and canons who would give the final verdict.[30] In the questioning Cauchon and his prosecutor, Jean d'Estivet, a canon of Bayeux, skipped from topic to topic, tried to trick Joan by altering the context of her remarks, and threatened torture.

Joan withstood all this. From the first days of her capture, in the best tradition of her saints, Margaret and Catherine, she had defied the English and their allies. She had even tried to escape twice before the trial had begun. Perhaps the lives of her two heroines inspired her. Both of these early Christian martyrs captured the popular imagination of the thirteenth, fourteenth, and fifteenth centuries, with the tales of their purity, faith, and bravery. The powerful threatened their virginity and tortured them. Saint Margaret fought off dragons and demons. Saint Catherine converted the learned with the cogency of her arguments.

As her trial progressed, Joan could easily have made parallels with her own situation: her virginity imperiled, torture threatened, the learned sent to confound her, and the prospect of a "glorious martyrdom" for the faith. The churchmen examined her on more than sixty charges. To explain how she could have had such impact, Cauchon and his friends hoped to prove her a witch, a heretic who had made an unholy pact with the Devil. They could not. Nothing she said gave evidence of such an alliance. This did not mean, however, that other equally damning charges could not be found.

Throughout Joan answered as she always had, in straightforward, simple words, speaking as an equal. She questioned the validity of the tribunal, asking at first for an equal number of churchmen loyal to France. She recognized Cauchon as her enemy and like the legendary Catherine and Margaret warned him early in the questioning: ". . . you say that you are my judge. Consider well what you are about, for in truth I am sent from God, and you are putting yourself in great danger."[31] She admitted on May 9 that she might give in when shown the instruments of torture, but, she bravely added, "I should always thereafter say that you had made me speak so by force," thus negating the validity of the confession as evidence.[32] Twice, on April 18 in her cell, on May 2 in the courtroom set up in the castle, in the presence of the whole group of sixty-three churchmen, she refused to submit to the final list of twelve charges.[33] She continued to insist that she was right, and they were wrong.

To do this in the context of 1430 meant, as Master Jean de Bouesque,

Doctor of Theology, explained, "She is schismatic of the unity, authority and power of the Church."[34] The simple stratagem of allowing Joan to speak and react in the formal world of an ecclesiastical inquisition and trial had worked. By the facts of her birth, her words, her actions, she had condemned herself. To the hostile, learned churchmen she appeared a young, illiterate peasant woman who had dressed and fought like a man, all the while insisting that she had received divine revelation. The burden of proof, of justification of such unorthodox behavior and assertions lay with her. Nothing she could say or do would convince them.

The bishop of Lisieux thought "the vile conditions of her person" made everything she claimed and had accomplished suspect.[35] Her voices became "lies," "delusions," errors of faith that caused an endless display of disobedience, first toward her parents and now toward the wisdom and authority of the Church. Her manner of speaking showed Philipert, the bishop of Coutances, that she had "a subtle spirit, inclined to evil, excited by a devilish instinct."[36] Her refusal to wear female clothing throughout her imprisonment became symbolic of her blasphemous attitude and behavior. Even when offered the chance to hear mass and to receive the sacrament at Easter (a request she had made continually since her capture), she refused when told she would have to dress appropriately. She explained that "it was not in her to do it . . . adding that this attire did not burden her soul and that the wearing of it was not against the Church."[37] As with everything that passed between Joan and the churchmen, the issue was simply one of authority: Would she or would she not obey? Again on the 23rd of May she refused to recant when presented with the charges. The next day she was taken to the cemetery of the abbey of St. Ouen for the public reading of her sentence. The pyre for her execution had been prepared in the market square.

The source of her remarkable strength lay in her faith. All those days, standing before the churchmen in her gray tunic, the knee-length cloak, the long black hose and boots, she had been confident of God's presence, his guidance in all she did and said through the mystery of her voices. Perhaps she expected God to save her. When she had asked her voices "whether I shall be burnt" they "answered me that I should trust in Our Lord and that he would help me."[38] But there in the cemetery, on the high scaffolding before the resplendent assemblage of the representatives of the Church, of the English King, of the Prince of Luxembourg, her assessors, the crowd, she doubted. She heard the condemnatory sermon, heard herself declared cast out. Suddenly she spoke up and asked if she might submit to the judgment of the Pope. Refused, she told them that "she wanted to hold all that the judges of the Church wanted her to say and to maintain and to obey their every command and desire."[39] She made her mark on a shortened list of charges confessing her most grievous sin: "claiming lyingly that I had revelations from God and his angels St. Catherine and St. Margaret, and all those my words and acts which are against the Church I do repudiate, wishing to remain in union with the Church, never leaving it."[40] Cauchon announced her sentence, life imprisonment, a "salutary penance . . . that you may weep for your faults and never thenceforth commit anything to occasion weeping."[41] Manchon, the notary who had kept the minutes of the trial, remembered that she seemed almost gay remarking, "Come now, among you men of the Church, take me to your prisons and let me be no longer in the hands of these English." But Cauchon ordered the guards, "Take her to where you found her."[42] That afternoon they shaved her head, she put on the dress they gave her to wear, and then she found herself chained with the same guards in the same cell.

Historians can only speculate on her thoughts in the next days. She had saved her own life, but to what end? The life of solace, confession, and penance she had been promised had proved a fraud. She remained in chains, in the hands of the English soldiers. Three days later on Sunday, May 27, Cauchon learned that she had put on men's clothing again. Twenty years later at the rehabilitation inquiries, no one could tell the churchmen how she had come to find them. When Joan stood again before the churchmen in a man's tunic, she declared herself indisputably an unrepentent relapsed heretic. The only appropriate punishment for such behavior in 1430 was excommunication and death.

For Joan, this action meant reclaiming her faith and her special power. Like Saints Margaret and Catherine, she had chosen the "glorious martyrdom" of the virgin. She must have assumed that imprisonment as Cauchon had arranged it meant the risk of rape at the hands of the soldiers. Rape meant the loss of her virginity. Without her virginity, Joan's religion told her, she would lose her special tie to God: the Maid become "slut," a fallen woman, imprisoned for life without hope of heaven. She explained to the two assessors who came to question her that day that "when she put on women's clothes the English had done her great wrongs and violence in prison."[43] She explained to her questioners that she had given in "for fear of the fire." She told them that now her saints spoke disapprovingly of her actions. "Since Thursday, my voices have been telling me that I have done and am doing a great injury to God by making myself say that what I did was not well done." She spoke with more humility but the message of disobedience was the same: "If I were to say that God sent me, I shall be condemned, but God really did send me."[44]

Cauchon rushed the arrangements and ignored the objections and reservations of the thirty-nine assessors who wanted to give Joan more time to understand the gravity of her words. At nine the next morning, Wednesday, the thirtieth of May, she was taken, dressed in women's clothes, to the market square of Rouen. Cauchon read the indictment:

> You are for the second time a relapsed heretic, like a dog which has the habit of going back to its vomit! . . . You have fallen back into your former sins. . . . We discard you as a rotten member. . . .[45]

The executioner took her to stand on the pyre tied to a stake with wood piled all around her. He set the fire underneath. Guillaume de la Chambre remembered it took almost half an hour before she stopped crying out to the saints and to Jesus and "was choked by the fire."[46] The executioner carefully disposed of the ashes; no trace of her was to remain for the living to venerate.

But people do not always need relics to venerate a special individual. The power once given proved too strong to take away. As the historian Keith Thomas wrote, the legend of Joan's life became "one of the most resonant and flexible symbols in the whole of human history."[47] The secular world embraced her first; the fifteenth-century French chroniclers saw her as God's gift for the salvation of their kingdom. Then the Church bowed to the pressure, first in the 1450s at her rehabilitation, and much later in 1920 with her canonization, not for her visions or her exploits in the name of the faith but for her exemplary life. The Church chose to sanctify her because she had been "a simple, honest girl and a good Christian."[48]

Honoring her in this way, the ecclesiastics focused on her piety and her faith, but they refused to acknowledge a divine source for her voices. The image the Church projected was of a simple shepherd girl, the brave maiden unjustly burned at the stake. All in between vanished. They excised her energy, her military exploits, her male dress, the whole range of her unorthodox, "unwomanly" words and actions. Not so in the popular imagination. The images that had originally empowered her continued. Sixteenth-century France named her Jeanne d'Arc and made her a national heroine. The men of subsequent centuries took her story for their plays and poems, her image for their statues. She became the spirit of France, the maiden, the holy warrior, the Republican and Napoleonic symbol for opposition to the English and for those who would protect France from foreign domination. In the Second World War Charles de Gaulle used her standard, the Cross of Lorraine, as the symbol of Free France.

And some women saw what her extraordinary actions meant for them. Perhaps the first was her contemporary, the courtier and writer Christine de Pizan (1364–1430). An old woman in July of 1429, she heard of The Maid in her retreat at the convent of Poissy. The siege had been lifted at Orléans.

Joan became the subject of her last known poem: Joan in the tradition of Esther, Judith, Deborah, the brave, defiant, powerful woman:

> By whom God restored
> His people when they were hard-pressed. . . .
>
> Ah, what honor to the feminine sex!
> Which God so loved that he showed
> A way to this great people
> By which the kingdom, once lost,
> Was recovered by a woman,
> A thing that men could not do.[49]

The Witchcraft Persecutions

Joan's short life demonstrated the quixotic, potent effects of men's fears of women who confidently claimed authority and who acted independently. Those perceived as supporting men's institutions found themselves honored and valued, those thought of as threatening or opposing dishonored and condemned to death. Joan was one young woman perceived as challenging men's authority. A century and a half later, hundreds of thousands of peasant women came to be seen in a similar way in villages all across Europe, from Finland to Italy, from Scotland to Russia. They became victims of the witchcraft persecutions of the sixteenth and seventeenth centuries—hundreds of thousands of women accused, tortured, executed. How did this happen? The answers lie in a horrible coincidence of new circumstances and old attitudes.

All historians of the witchcraft persecutions have searched for patterns true from one region to another that will explain the outbreaks of hysterical accusations and executions. Why then? Why there? Why did those people accuse? Why were particular women singled out? All regions have not yet been studied. For the areas analyzed the records are by no means complete. Rarely is there the whole story of the women's ordeals from initial accusation, through questioning, torture, judgment, sentencing, and execution. Historians have only begun to piece together the records of the accusations, the confessions, accounts of judges, reports of inquisitors for all parts of Europe. Between 1300 and 1500, there is sketchy evidence of about 500 trials with original depositions for only twenty-one in Germany, France, and Switzerland.[50] Then there is an exponential jump in the numbers for the late sixteenth and early seventeenth century for Western Europe, continuing into the beginning of the eighteenth century in Eastern Europe.[51]

In the 1500s much seemed to conspire to create uncertainty and upheaval. Learned men sensed impending chaos. All of the traditional means of establishing order seemed discredited and useless. Religious reformers and prosletyzers questioned the Catholic faith and its rituals. Protestant sects condemned everything from the authority of the Pope to everyday practices like the saying of the rosary. Printing presses carried the doubts and attacks all over Europe. Religious and civil wars in Germany, France, and Scotland with one prince replaced by another, one faith by another, called all princes and all beliefs into question. Locally controlled markets and trade gradually had given way to great trading centers and monopolies protected by the regulations of dynastic rulers, leaving town elders—the guildsmen and the wealthy entrepreneurial families—without the familiar economic system that had guaranteed them their livelihood and hegemony.

The secular and religious elite searched for ways to reinstitute a sense of certainty, to end the questioning and the changing. Townsmen, royal families, churchmen worked to bring order through definition and conformity. The guilds closed their membership to all but a select few and ruled their towns by fiat. Queens and kings codified regional laws and customs and adopted uniform Roman procedures as they expanded their system of royal courts. Leaders of different sects vied in their demonstrations of religious zeal and in their efforts to establish themselves as the founders of the one "true" faith.

Though by the 1550s all Europeans had long ago accepted the leadership of these institutions—the authority of the Church and the secular rulers—Europeans also had vivid memories of other powers and other beliefs. A prayer to the Virgin or an appeal to the king's court would help, but the old remedies, a spell or a traditional chant, would do no harm. Countrywomen and men believed that this other kind of power was the special province of only a few in the village—the "cunning folk" as they were called in England. Always women were seen as having ties to this extraordinary world of charms, incantations, and spirits. Europeans believed that the magic was there for those who knew how to call upon it and continued to appeal to those who claimed this unorthodox knowledge of and access to the supernatural. For example, in the 1580s the English churchwardens of Thatcham Berkshire hired the local cunning woman to help them find the thief who had stolen the communion cloth.[52]

Women with these powers appeared in the traditional stories of the ancient cultures and in the tales told into the nineteenth century. Their kisses gave their lovers magical powers, their warnings saved the heroes of the sagas. The wisewoman helped sisters find brothers and young maidens win back forgetful princes. "The old woman in the forest" gave her heroine a lover by her powers to transform; a tree became a handsome young man.[53] Just as the legends and sagas tell of women using their special knowledge and skill for good, they tell of those who tapped this power to do ill. The Norsemen of *Njal's Saga* wanted the old woman to use her second sight but feared that her ability to predict the future meant that she could control it. In the tales, the white magic of rescues and transformations had its counterpart in the potions and curses of black magic: witches who would eat Hansel and Gretel, who could turn their stepchildren into a lamb or a fish, who tried to kill their husbands' offspring, who could strip a young man of his strength and potency. Many believed that women could harm the livestock and bought and sold charms against their powers. When a young woman married, German villagers feared the oldest woman in the community would be jealous and bring harm with the evil eye. A piece of bride's cake, in which a silver coin was baked, was given to her to appease her.[54]

Another aspect of the old beliefs and traditions survived. In villages all over Europe this supernatural force for evil was assumed to be available to all women, whether they called upon it or not. Somehow the gift of their reproductive ability made them potentially dangerous. A man could not marry in May, the woman's month, because then he would fall prey to lust and give her power over him. Women's menstrual blood had an ancient reputation as a harmful substance. Germans did not allow a pregnant woman in the stable. Anything having to do with birth seemed to have magical properties, and the midwife privy to its mysteries was thought to possess special powers. When peasant women and men moved to the towns, they took these ancient beliefs and traditions with them. In Paris and Florence in the fourteenth century, men accused women of bringing on lust or impotence with their potions. Beginning in the fifteenth century in French towns and in the sixteenth century in English towns, midwives came under suspicion. They had to take oaths against using magic or sorcery (as the Church called it in France). The English townsmen persecuted the women for "superstitious practices."

From the earliest days of Christianity, churchmen had denied these ancient beliefs and insisted that women could not call up such extraordinary powers. Religious leaders spoke in their books of penance, canon laws, papal and conciliar decrees of "delusions" and "superstitious practices," of women misguided and confused. As the source of misfortune, they offered their believers Satan, the Devil, the personification of evil, the ever-vengeful fallen angel seeking souls for his kingdom, Hell. He, not the wisewomen of the villages, had the power to harm, and he need not be feared for with Jesus' death on the cross, his power had been broken; everyone knew that the crucifix or holy water would frighten him away.[55]

In the uncertain atmosphere and conditions of the sixteenth century, the attitude of Europe's churches changed. Protestant and Catholic leaders came to doubt Christianity's victory over the Devil. They imagined that he had found an opportunity to try for hegemony again. And out of their fear, the religious and secular elite transformed the actions of the wisewomen. The cunning woman's magic, once officially discredited, became in the imagination of Europe's leaders more real and powerful than it had ever been before. Embellished with all of the oldest misogynist mythology, the wisewomen's ways, words, and gestures became evidence of the beliefs and rituals of a heresy that threatened to destroy mankind. The learned of the Churches wrote and preached that the Devil was at large in the world. These women, the cunning folk, had made a pact with him, worshipped at his altar, and had become his agents.

The villagers and the citizens of the towns could accept this picture of the Devil and his followers, and this explanation for the misfortunes of the time, for the everyday hardships and mishaps of their lives. In the past they had sought traditional spells and rituals to protect themselves. But in the sixteenth and seventeenth centuries the elite of the Church and state offered more than the propitiation and counterrituals that had been the familiar remedies against such magic. They claimed to offer an end to the misfortunes forever. If peasant women and men, townswomen and men, would identify the cunning folk, officials of Church and state with all the paraphernalia of legal procedures and scientific rules of evidence would try and execute them. Thus rich and poor, powerful and powerless, Protestant and Catholic tacitly and actively participated in the persecution and murder of thousands of illiterate peasant women.

Jean Calvin, the Protestant leader of Geneva, described Satan and his treachery in his *Institutes:* "We have been forewarned that an enemy relentlessly threatens us, an enemy who is the very embodiment of rash boldness, of military prowess, of crafty wiles, of untiring zeal and haste, of every conceivable weapon and of skill in the science of warfare."[56] Martin Luther, the German Protestant leader, warned, "Let every man think that he himself might have been, and yet may be bewitched by him. There is none of us so strong that he is able to resist him."[57] Catholic leaders like Pope Innocent VIII in his Bull of 1484 had already found evidence of the Devil and of his worship in the north German countryside, followers of what thirteenth- and fourteenth-century theologians had defined as the heresy of "demonology."[58] Innocent saw them in every village: evil people "who have given themselves up to devils" practicing "spells" and "infamous acts," harming animals and crops, giving pain, causing sterility and impotence.[59]

The Pope had been alerted to these lapses in faith by two Dominicans, Henry Kramer (Instituto) and Jacob Sprenger, who had been sent to inquire into beliefs in the German villages. Now with the approval of their findings in the papal bull, they wrote a treatise for their brother clerics describing the newest manifestation of this heresy and its believers. The result, published in 1486, was the *Malleus Malificarum (Hammer of the Witches)*, which combined the traditional attitudes about heresies and the procedures to deal with heretics with ancient beliefs in a Devil, in the power of the supernatural, the magical, and women's special powers.

The traditional condemnations of women fill the pages of the churchmen's treatise and tie all women to witchcraft and sorcery. Witches exist as a matter of faith. Not to believe in them is in itself another heresy. It is self-evident that witches are female. As the Churchmen explained: "where there are many women there are many witches."[60] Men, like Jesus, are protected from the lures of the Devil, but women because they are "feebler both in mind and body," are easy prey just like their predecessor Eve.[61] Sprenger and Kramer, like the learned men before them, assumed that women hated their own weakness, hated not to rule, and became "a wheedling and secret enemy," a creature who "always deceives," bitter, and vengeful who found "an easy and secret manner of vindicating themselves by witchcraft."[62] In the churchmen's imaginations witchcraft came to be defined in the broadest terms. Both white and black magic were potentially evil. Drawing on writers like Horace and Ovid, on the Church Fathers like Augustine and Thomas Aquinas, they filled out the list of what Innocent had called "infamous acts."

The idea of women armed with magical powers and using them against men had a long tradition that lasted through writings, through the popular imagination, and went all across Europe. A Russian chronicle from as early as the ninth century notes, "Particularly through the agency of women are infernal enchantments brought to pass."[63] The sermon of Berthold of Regensburg in 1250 warned:

> Many of the village folk would come to heaven were it not for their witchcrafts. . . . The woman has spells for getting a husband, spells for marriage; . . . spells before the child is born, before the christening, after the christening. . . . Ye men, it is much marvel that ye lose not your wits for the monstrous witchcrafts that women practice on you![64]

In the early days of the Church such women with their "superstitious practices" had been declared needful of penances and admonitions from the parish priest, nothing more. To Sprenger and Kramer and their successors, however, the village wisewomen became the chosen ones of the Devil, his worshippers and his agents.

The Devil won these women as his worshippers, according to the fifteenth-century *Malleus Malificarum,* by playing on yet another supposed defect in the female. Sprenger and Kramer repeated the traditional belief that woman "is more carnal than a man," that she is in fact "insatiable."[65] They imagined women's sexual intercourse with Satan. So joined, these women became the Devil's followers. By calling up his demons, they learned the potions and formulas that gave them their supernatural powers: to fly, to make men impotent and women miscarry, to drive horses mad. These late fifteenth-century inquisitors and those who wrote of this heresy in the sixteenth and seventeenth centuries described its rituals. They drew on all of the most pornographic of the Church's traditional vilifications of heresies and heretics. The churchmen described mockery of the mass, desecration of the host, orgies on a Witches' Sabbath, cannibalism of newborns, gluttony, drunkenness, lewd dancing, intercourse with every variety of creature in every possible position.

Since the *Malleus Malificarum* appeared so early in the age of printing, historians can only speculate about its direct influence. In 1487 the Theological Faculty of the University of Cologne endorsed the treatise, sixteen editions had been published by 1520, a total of thirty-two by 1660 with translations into German and French.[66] There is no question that what they wrote echoed attitudes that had been part of European culture since its beginnings. By the mid-1500s the learned and powerful universally accepted their reasoning. By the mid-1500s no one with secular or religious authority doubted the existence of a heresy of devil worship, no one doubted its practices or denied witchcraft, and all saw its principal believers and practitioners as the illiterate, older peasant women of the countryside. Throughout the sixteenth and seventeenth centuries numerous Catholics and Protestants wrote their own treatises and condemnations, based both on theory and on their practical experience in working to destroy the heresy.[67] All agreed that the only way to be rid of the witches and their master, the Devil, was to kill them. As Georg Pictorius of southwest Germany explained: "If the witches are not burned, the number of these furies swells up in such an immense sea that no one could live safe from their spells and charms."[68] The Frenchman Bodin in his 1580 demonology warned of the consequences to those who did not root out the villainous menace:

> Those too who let the witches escape, or who do not punish them with the utmost rigor, may rest assured that they will be abandoned by God to the mercy of the witches. And the country which shall tolerate this will be scourged with pestilences, famines and wars.[69]

In response, the secular rulers produced laws and ordinances against witches. The first appeared in France in 1490, then in Emperor Charles V's early sixteenth-century codification of Imperial law, the Carolina, then in English laws of 1542, 1563, 1736, Scottish law, a Russian decree of 1653, the Danish King Christian V's ordinance of 1683. All condemned the women to imprisonment or death, as the means to spare their countries the Devil and his agents.

How many women were persecuted? How many accused and how many executed? The historian E. William Monter has called it a time of "lethal misogyny." Men fell victim, but at least two thirds of those who died were women. In the waves of panic and fear, sometimes all were women, sometimes 80 percent or more, always the vast majority.[70] The most conservative estimate of the numbers of European women strangled, drowned, burned, and beheaded is 100,000. Most historians believe there were many more. Southern Germany went from extreme to extreme; between 1561 and 1670 Catholics and Protestants executed more than 3,200 people, usually in groups of twenty or more, 82 percent of them women.[71] In the decade 1581–1591 in French Lorraine, accusations ran in the thousands. Of the 2,000 accused in Lorraine, 90 percent died.[72] The terror in the Jura region of Switzerland at the end of the sixteenth century produced records of almost 900 cases.[73] Persecution came in two waves in England, in the 1580s and 1590s, then again from 1645 to 1647 when 490 were executed in Essex, one third of those accused.[74] Women suffered in clusters of persecution in Scotland: 200 and 300 identified at a time from the 1590s to the 1660s—over 3,000 trials, and 1,000 executed.[75] The numbers go into the thousands in Spain and Poland. In Italy, Denmark, Sweden, Finland, and Russia the persecution lasted into the late seventeenth century, long after the fear had passed elsewhere.[76]

The writings of churchmen produced the explanation for the sense of disorder and confusion, and the description of the heresy; princes of both the Church and state, the men in positions of power and authority, were the ones who encouraged the first accusations and were the most zealous in hunting down the evil ones. Thus they showed their concern for their subjects, their loyalty to the new or the old faith. In the years of hysteria in southwest Germany, 1562–1666, for instance, the prince bishop of Trier first accused the Jews, then Protestants, identifying both groups as witches. Other bishops, counts, judges, and margraves worked to clear out the heresies. In the villages of the Jura region, Switzerland, and the Cambrésis region of Namur, it was the better off who came forward to accuse. In 1609 the inquisitors in Navarre occasioned the spread of panic across the Pyrenees to Spain. In the 1640s in Essex the judge Matthew Hopkins initiated and continued the persecutions. A Special Commission sent to hunt out witches caused the craze in the 1590s in Scotland and gained the help of the local landowners eager to assert their authority and prove their piety. In Navarre and Besançon local royal officials and the seigneur petitioned the king and the Church for help. In the 1680s in Denmark, the local squire of Djursland made the first accusations. In Russia, the waves of persecution began only with decrees from the tsar in 1648 and again in 1653.

Once prompted, the villagers responded. The first witch was always easy to name. Neighbors and relatives accused each other. Once accused, the victim could think of others. Almost always the village named women, often those outcast by choice or by poverty, women living outside the traditional patterns of their village's expectations. Sometimes they were married, sometimes widowed, usually they were the women ready to speak back, to quarrel, to curse those who angered or frightened them. It was the wife of a soldier like Marguerite Tattey or the prostitute Nicola de Galice, a herdswoman, the priest's mistress. It was a diviner, Magdalena Nessler, or Anna Demut, a woman villagers believed could change the weather. It was the midwife of Dichtline. (The mythology surrounding midwives comes from their involvement with the birth process, which in the witch persecutions was transformed into their means of access to infants for the Black Mass. The idea of evil women harming newborns is ancient; for example, Lilith, Adam's first wife, was believed to have taken revenge on the angels who took her children by strangling the infants of others.) Throughout Europe the wisewomen known for their cures and their curses were accused.

The circles of accusations went ever wider. In Neuchâtel in the winter of 1582–83 the woman thief the villagers caught admitted she had attended the Witches' Sabbath and named others, who in turn accused others. Four were named in Ortenau in 1627, who accused others in the neighboring town of Offenberg. By 1631 in Sasbach 150 witches had been identified all the way to the *Stadthalter,* his wife and daughter. The Scottish commissioners in 1697 found Margaret Atkin, who confessed and then offered her special powers to detect other witches.

No one stepped forward to defend the women. Rather the opposite. Peasant women and men sensed the opportunity to be rid of those who had frightened them. Given the opportunity, they came forward eagerly to present compromising memories of the women's activities and words. For example, in March of 1621 Marie Lanechin, the widow of Jean de Baulx, a sixty-year-old woman, had thirteen accusers: a journeyman, a rich tenant, villeins and their wives, and a laborer. They told tales of illness, deaths of husbands, wives, and nursing infants, a daughter. The villagers described all sorts of unusual phenomena: She danced in the woods with her hair all awry, refused to leave her house, walked in a thick mist when everywhere else the air was clear. Under torture she named two other women, but the court released them. Only she was ordered strangled and burned.[77]

To execute a woman like Marie Lanechin, the court, whether secular or religious, required more than just testimony of "superstitious practices." There must be proof of heresy, of the pact with the Devil, of his worship. Churchmen and local officials readily found women who innocently admitted to having special power and special knowledge. In the sixteenth- and seventeenth-century trials, women spoke of cures, of ointments, of potions, but not of rituals and worship of Satan. Katherine Kepler, the astronomer's mother, spent fourteen months in prison denying such charges. On the other hand, some confessed too readily, and the courts doubted their testimony as well. By the sixteenth century the basic facts of how the Devil looked, what he promised, the fact of intercourse, the power of his charms and potions, had become common knowledge in the small village communities. The inquisitors and judges believed that women might say anything.

In the past men would have used the ritual of the ordeal to determine the truth of the accused's confession, the innocence or guilt of the woman. They would have left it to God to mark her or leave her unmarked. But this was a different age, a new era. The learned no longer called for what they considered divine intervention as in the ordeal. Instead careful procedures of law and scientific evidence had evolved. But this was an age of law not yet just, when judges ordered torture, of science not yet scientific, when learned men believed in astrology and alchemy. As a result, the courts, whether run by churchmen or laymen, became places of extraordinary contradictions with the rational and the irrational jumbled together.

On the one hand, the Catholic courts of the Inquisition and the Protestant secular courts followed strict rules of procedure. As Sprenger and Kramer had explained in the *Malleus Maleficarum* "the accused shall as far as possible be given the benefit of every doubt."[78] The accused might have an advocate, a defense lawyer. There were printed indictment forms listing the accusations, with blanks left for the name of the accused and the date of the trial; there were set questions to be used before and after torture. Some regions allowed appeal to a higher court. On the other hand, Sprenger also believed that he had heard voices screaming obscenities, seen animals like monkeys, dogs, and goats at the windows, all during the trials he conducted in northern Germany. He and Kramer warned prosecutors not to let the witch touch them, told them that they "must always carry about them some salt consecrated on Palm Sunday and some Blessed Herbs" to be "worn around the neck."[79] A French judge, Nicolas Rémy, believed that the witches put poison on their hands and then grabbed the clothes of the presiding officer as if asking his mercy.[80]

Most contradictory of all were the kinds of corroboration demanded in the name of law and science. In fact, it was not so much corroboration of charges that was sought as elaboration on them, transformation of simple accusations of harmful deeds into detailed accounts of heretical practices. Just as the authorities had initiated the search for the witches, so they now by their intervention assured admissions of heresy and thus the deaths of the accused.

According to the *Malleus Maleficarum,* the best evidence of all was a confession, for "common justice demands that a witch should not be condemned to death unless she is convicted by her own confession."[81] But how can a woman in the power of the Devil be made to confess? The learned believed that torture would free her from the Devil's power. Marie Lanechin of Bazeul, in the Cambrésis region of Namur, confessed to relations with the Devil—once she had been sent to a higher court and tortured. Commonly in both Catholic and Protestant religious and secular courts, judges ordered use of the strapado, hands tied behind the back, the rope thrown over a beam and pulled so that the accused hung just off the floor, the arms in the air, at the least the shoulders dislocated. Courts that followed the Carolina (the law of the Holy Roman Empire), or the Inquisition, regulated how many times it could be inflicted—in the Jura, for instance, on three occasions on three separate days. In Offenberg in 1627 the court used a metal spiked chair that could be heated from underneath. The English judge, Matthew Hopkins, used neither rack nor strapado, instead he broke the women by not letting them sleep for hours or for days.[82]

The religious and secular authorities asked the woman under torture a series of questions, most requiring only a "yes" or "no" answer. Drawing on men's pornographic fantasies, they reveal the hysterical and fantastic world of the witch hunters: What is the devil made of when he is having intercourse? Can you see him? What is the source of his semen? Is there a time he favors for the act? Does he choose special women? Was the sex act more or less pleasurable with him?[83] Many women answered as their questioners wished, repeating after their ordeal information about initiation ceremonies, confessing to fornication with the Devil, with animals, with imaginary beasts, speaking of dead infants, toads in fancy clothes, desecration of the host, transformation, salves made of excrement, flying, all the mythology of the heresy.[84]

Jeanette Clerc, tortured and executed in 1539, gave her Geneva judges all they could hope for: a devil named Simon with whom she rode on a stick to the "synagogue"; unnatural sex, icy cold semen, and a mark from his bite on her face; fancy food of white bread, apples, and white wine at the Sabbath, and meetings on Thursdays and Fridays. They had found Jeanette Clerc because her neighbor's cow had died.[85] In 1587 Walpurga Hausmännin, the midwife of Dillingen, "having been accused persistently, uniformly and justly, has been interrogated gently and under torture and has now confessed her witchery."[86] She said that her association with the Devil began with sexual intercourse in her own house. It had frightened her when she had seen his goatlike foot and his wooden hand, but he promised to help her, gave her salves against her enemies, and marked her as one of his own under her left arm. In return she killed babies for him, just at birth before baptism when the holy water did not yet protect them (431 infant deaths were attributed to her). She rode on a pitchfork to the Sabbath celebration, where she paid homage to a Devil enthroned, watched rituals mocking the host, copulated, and gorged on roast pig and babies. The Devil had slapped her when she mentioned Jesus.[87]

Sometimes the torture did not have the desired effect. A woman like Suzanne Gaudry, questioned in Rieux in 1652, initially described meeting Satan, his skin like hide, wearing black breeches and a "flat hat." She admitted they had been lovers for twenty-five to twenty-six years. But without the prompting of the questions like those composed by Sprenger and Kramer, she had little to say of the Sabbath; she spoke of a "guitarist," "some whistlers," and she would admit to no *maleficia* (acts harming others or their possessions). On the third day she said she had killed Philippe Cornie's horse, but later explained that she had offered this *maleficium* because she felt she must say something. Stretched on the rack, she denied everything; tortured again, she refused to speak at all.

In such a situation, the authorities looked to other kinds of evidence to corroborate their assumptions of guilt. Each court had its favored "scientific" proof of the pact with the Devil. In England "swimming" might prove the woman a witch. If she floated, the water had rejected her. In Denmark, they tied her before throwing her in. If she faltered in saying the Lord's Prayer in a Scottish court, she showed herself to be in the Devil's power. In Germany, in France, in England and Scotland, officials stripped and shaved the woman looking for the "Devil's mark," the spot he had touched after their intercourse that neither bled nor pained when pricked with a needle. The village barber surgeon was called in to make the test, first on her breasts and genitals. The French judges found such a mark on Suzanne Gaudry, that and the fact that she did not cry proved her an agent of Satan. English courts also looked for suspicious lumps, believed to be the nipple for the witch's familiar to suck.[88]

By the end of the 1500s in some regions of Europe—southwest Germany, for example—the accusations and the hunting were out of control. No one could say who was a witch and who was not. In Rottenburg in 1602 even the chief prosecutor, Hans Georg Hallmayer, had confessed and had to be executed. The only way to end the random victimization seemed to be to end the whole process. And just as the learned and powerful had begun the panic, so they ended it. In southwestern Germany, in France, in Belgium, in the Netherlands, beginning as early as 1613, authorities refused to hear new accusations, made the rules of evidence stricter, stopped using torture, did not carry out executions, reversed sentences on appeal, and set fines for false accusations. Just as they had given the power and sown fear of those who held it, so now they began to deny it and to discredit those who claimed it. All still observed the strange behavior of the village wisewomen, but instead of calling them heretics united with the Devil, they began to describe them as "possessed," unwilling souls in need of exorcism and prayer, not execution. Others called the women deluded.[89] By the 1730s in England, to claim one could use magic constituted "fraud."

Events in Spain from 1611 to 1613 heralded the changes in attitude and approach that would come for the rest of Europe in the course of the seventeenth century. In 1611, given a Year of Grace, an opportunity to confess without punishment, over 1,800 people of the Logrono region in Spain (most of them children, ages seven to fourteen) admitted to making people sick, to changing form, to ruining crops.[90] With so many confessions the authorities called for a special inquisitor to examine this outbreak of heresy. From May 1611 to January 1612 Alonso de Salazar y Frias traveled the countryside and questioned and cross-questioned those who had confessed. By October 1613 he had rejected most of the testimony. He wrote: "Indeed, these claims go beyond all human reason and many even pass the limits permitted the Devil."[91] His report published in 1614 by the Supreme Tribunal of the Spanish Inquisition concluded, "I have come to believe, and shall continue to do so, that none of the acts which have been attested in this case, really or physically occurred at all."[92] Nothing had happened, so none were guilty: The 1,384 children were absolved; of those fourteen and over, 290 people were reconciled, forty-one confessed witches received absolution, eighty-one others took back their confessions.[93]

In a time of political, economic, and religious uncertainty, men of Europe's elite had turned for security to the persecution of women and the extirpation of an imagined conspiracy of the Devil. The witchcraft persecutions remain the most hideous example of misogyny in European history. Erupting so suddenly in the mid-sixteenth century, the persecutions declined and the fear died equally dramatically in the mid-seventeenth century. Both the religious and secular authorities came to scorn the very beliefs they had honored and that had inspired the hysterical accusations and killing. By the early eighteenth century, the learned and powerful looked to new kinds of order, to new explanations for their universe, now transformed into a rational, scientific place, a place without need of religion or magic.

But while the learned and the powerful stopped their persecutions and moved on to a new world, one without magic, the women and the men of the villages did not abandon the supernatural so readily. There was no heresy, but left to themselves once again, peasant women and men went back to the old beliefs in magic and the spirits, for no one had proved to them that the world was not filled with the irrational, or that the cunning folk had lost their power. Into the twentieth century countrywomen and men continued to rely on the traditional rituals, the combinations of the old religions and Christianity, to guarantee white magic for their families and their property.[94] In eighteenth-century Serbia, peasant women and men thought of the plague as an old woman and had special rites to make her leave the village alone. In nineteenth-century Russia, when the cattle sickened, the men enclosed the herds, went into hiding while the women went out into the fields at midnight and screamed abuse to drive death away. (The oldest woman in the community was tied to the plow and made a furrow three times around the village as another part of the ritual.) In 1910 in Sicily a woman would not comb her hair on Friday for fear it would mean a curse.[95] Into the twentieth century Englishwomen believed that saving hot cross buns from one Good Friday to the next would keep away whooping cough. Irish families in Lough in County Clare in the 1930s left the west side of the land free of outbuildings, because this was the fairies' side. Men could not go near butter as it was being churned or it "won't come."[96]

Throughout Europe old women were still thought to have mysterious ties to the supernatural. They might be midwives. They might have the power of the evil eye. To ward it off, a peasant woman in Greece in the 1950s wrapped her new baby with a special blue bead on a band around its waist, ritually crossed the child, and spat three times when she swaddled it.[97] Many of these old beliefs, these old ways of the peasant woman's world remained intact even as so much else changed with the economic, social, and political transformation of Europe in the nineteenth and twentieth centuries.

4

WHAT REMAINS OF THE PEASANT WOMAN'S WORLD

INDUSTRIALIZATION AND urbanization transformed the economic, political, and cultural patterns of nineteenth- and twentieth-century Europe. Familiar institutions and ways disappeared as governments and social service agencies assumed powers and resources unimaginable in earlier centuries. Peasant women and men learned new relationships to the land, new methods of cultivation. After World War II, modernization and its technology altered the basic routines of their lives. Yet for all of the transformation, many traditional attitudes and patterns prevailed, proving resistant to even such powerful forces for change.

At the beginning of the 1980s in Italy, France, the Soviet Union, and much of Eastern Europe and the Balkans large percentages of the population still worked in agriculture. Women comprised more than fifty percent of the rural population in parts of Eastern Europe.[1] Despite peasant women's contribution to the economy as a whole and to the agricultural sector in particular, the old patterns of devaluing and designating their work had not changed. Many tasks, such as care of the household and children, continued to be unpaid labor, not even valued enough to be counted in national economic statistics of goods and services. Such tasks remained part of women's traditional "double burden"—of work in the home as a wife and mother and outside as a wage earner.[2] When paid wages for their work, peasant women, as in centuries past, continued to earn one half to three quarters as much as men.[3]

Another ancient tradition regarding countrywomen's work proved equally resistant to change. In comparison with countrymen, Europe's rural women still performed the least skilled tasks, and had less access to mechanization and the technological education needed to make use of it. In 1980 the Soviet magazine for peasant women, *Krest'yanka,* published a woman sugar beet grower's account of her work:

> Altogether up to 400,000 seedlings are pulled up from each hectare. And you have to bend over every one of them, have a long look at some of them; you don't choose immediately as if one were as good as another. And from the very first the weeds have to be pulled up. Your back aches and your feet grow heavy. That's what it's like, farming the crop.[4]

In addition, as Soviet Premier Khrushchev explained to a conference of farmers in 1961, "It turns out that it is the men who do the administering and the women who do the work."[5] As late as 1975 women represented only 1.5 percent of the administrative personnel on collective farms.[6] Recent improvements, like using two shifts of women workers at harvest instead of demanding twelve or more hours of labor from one shift, have eased peasant women's lives. Even so, they remained disadvantaged members of the Soviet economy. This inequity was not unique to the Soviet Union and continued to exist in the rest of Europe as well.

The expansion of the role of government and the services it provided, which evolved in the twentieth century, changed some of the ways in which countrywomen dealt with their own and their families' basic needs. In the 1960s in southern Italy and in Sicily, peasant women learned to use the bureaucracy to generate benefits as they had once used private and religious charity. They procured health insurance for a son's operation, a housing subsidy for a new apartment. Still, when rural women needed extra sources of income, they depended on the same kinds of tasks as they always had. Italian peasant women baked bread, raised, slaughtered, and cured pigs, cut wood, and cleaned for the *signora,* the wife of the local landlord. Russian women sold the produce of their household gardens.[7]

The advantages of modernization, the amenities that improved the quality of a family's comforts, came to the women in the countryside but more slowly than to women living in the cities of the twentieth century. In the post-World War II apartments of southern Italian villages, peasant women had a gas burner, but with only two rings, running water but only for two hours a day. Studies done in France in the mid-1960s showed that changes that had become standard in the cities were still rarities in the countryside: 84 percent of peasant homes had no indoor toilet, compared to 48 percent of urban homes; 42 percent had no running water, compared to 12 percent of urban homes; 17 percent of peasant homes had a vacuum cleaner compared to 35 percent of the urban homes; 16 percent had a television set, compared to 27 percent in the cities.[8] By the 1980s in the Soviet Union, nine out of ten rural households had electricity, half had natural gas for cooking. In the mid-1970s, however, only one third of the country families had running water, central heating, or access to sanitation mains.[9]

Though by 1980 the vast majority of Europe's rural women had all of the advantages and services of modern health care for themselves and their families, in some circumstances they felt the impersonality of the new technological medical facilities. In the southern Italian towns, peasant women went through the last stages of labor together, on "thrones" arranged in a circle around the delivery room. Their legs in leather slings, their abdomens draped, their backs propped up, vaginas uncovered, the women of Torregreca shoved the white towel they had brought from home in their mouths, bit hard, and prayed that they would not disgrace themselves by crying out. Where before her friends and family had been there to encourage and comfort, now each woman looked away so that she would not see or be seen.

When hard times came in the modern age, peasant women continued to take solace in their faith and to accept responsibility for their family's religious well-being. European men stopped going to church long before women.[10] In Spain, early twentieth-century Santander was no different from sixteenth-century New Castile with its yearly processions of women, men, children, and churchmen carrying the image of the local saint to the church and then to the shrine. But after 1940 in Santander, the men no longer went up the mountain to the shrine. They left it to the wives while they waited halfway up the path. A shopkeeper's widow explained the women's point of view: "Praying is a job, an obligation, like washing the dishes."[11]

Even with all of peasant women's contributions to their families and their society, traditional attitudes denigrating the female continued to affect their lives adversely. Declarations of women's equality in constitutions and laws did not end misogyny. In the 1970s a Sicilian woman explained:

> Here the woman is not equal to the man. Among the young perhaps there is a dialogue, but not between adults. The woman always feels herself inferior. . . . You're not emancipated just by work. The man continues to treat the woman like an animal, capable of working, but not of thinking.[12]

Neither the states nor the churches of the twentieth century have encouraged peasant women to oppose such attitudes or to seek less traditional roles. The national report of the Austrian government to the United Nations in 1980 gave the goals of the Ministry of Agriculture and Forestry. The ministry described peasant women in traditional terms: within the family, married, defined by their relation to men. By its efforts the Austrian government hoped "to enrich the personality of farmers' wives" (not of farm *women*). They hoped to make the "wives" better able to serve in their traditional capacities. The teacher in a southern Italian village of the 1950s expressed these attitudes much as her counterparts had in earlier centuries. She explained why the girls had only a book about the Madonna: "It wasn't a sermon, but it was right for these girls. After all, why do they need to read and write? . . . What they need to know is how they are supposed to act."[13]

Throughout the twentieth century, as in previous ages, countrywomen have always been the last to learn to read. The 1938 League of Nations *Yearbook* gave percentages of illiterates: in Yugoslavia 57.1 percent women to 32.7 percent men; in Bulgaria 43.3 percent women, 19.5 percent men; in Italy 25.2 percent women, 17.8 percent men; in France 4.2 percent women, 3.4 percent men.[14] Vuka from a village in Serbia in the 1930s remembered how excited she had been after her brother had taught her to read. Her mother left her to work and then "I would take a book on my lap and knit blind, not looking at the knitting but reading under it."[15] Her favorites were the love stories of women betrayed. Betrayal in love led her to think of other kinds of betrayal: "I was sorry for those unfortunate girls. I was sorry for myself, too, because I felt unhappy since I could not do what I wanted." What Vuka wanted was to "have a school education and learn and know everything."[16]

In the twentieth century, to have such opportunity a girl would have to leave the land and go to the cities. By the 1920s peasant women began to leave the countryside in massive numbers.[17] The trend continued after World War II. Parents sometimes encouraged their daughters to seek new kinds of employment away from the farms, like these Soviet parents interviewed in the 1980s:

> Her father and I have spent our whole lives in muck and filth, let Zina do some other work. There's nothing for her to do in the country. . . . [In the city] she can sit in an office, she can learn bookkeeping by all means.[18]

Into the last decades of the twentieth century those peasant women who remained on the land drew on traditional strengths to reconcile themselves to their difficulties. Countrywomen across Europe continued to find pleasure in their lives, however circumscribed, and to take pride in their tasks. Many passed these values on to their daughters and granddaughters. Some rural women of the twentieth century invested the traditional tasks with new importance. Asked her "profession" in the 1960s, a young French peasant woman responded proudly, "*Cultivatrice.*" Her mother had answered "none" when asked the same question.[19] Some rural women laughed at the ways in which husbands and fathers asserted their supposed superiority. An elderly woman from a village in the south of France sat with her neighbors and looked at the men free to meet in the public square. She, a female, as in the past, stayed at home. The old woman turned the customary restriction into a welcome choice. "Up there," she explained, "it's windy and chilly, not as comfortable as on my own front stoop." Her neighbors nodded agreement.[20]

Anna Cecchi from an Umbrian village of northern Italy, born in 1913, had been an orphan since the age of six months. Her grandmother raised her and made the linens for her dowry: sheets, towels, and twelve nightgowns. By the 1980s everything was gone, all but one nightgown which somehow had been spared and kept. This the peasant woman showed to her own granddaughter. Both admired the soft hand-woven cloth, the embroidered pink flowers, the delicate initial on the bodice, and the edges scalloped in lace.[21] The garment represented the peasant woman's legacy of the beautiful and the practical, the heritage of painstaking labor.

The French woman's answer to a government questionnaire, the old women's laughter, the Italian grandmother's moment with her granddaughter exemplify the special qualities and strengths that have characterized Europe's peasant women through the centuries. Pride in occupation, in place and custom, the ability to give value to the most basic realities and to enhance the necessary tasks of life, these sustained them in every kind of circumstance. In addition, with these qualities and strengths, Europe's peasant women sustained generation after generation of country families.

III

WOMEN OF THE CHURCHES

•

THE POWER OF THE FAITHFUL

1

THE PATTERNS OF POWER AND LIMITATION: THE TENTH TO THE SEVENTEENTH CENTURIES

THE PEASANT WOMAN'S FAITH in the Church, in the saints, in the power of the priest, gave her solace and a sense of control over the unpredictable nature of her experiences. Most Christian women and men, whatever their estate in European society, held this simple faith in goodness and sin, and followed this simple path of transgression acknowledged, forgiveness granted, and faith rewarded. Rituals marked their passage from birth and baptism to final communion and the ultimate promise of "life everlasting" in the Christian heaven. But there were other levels of piety, other levels of religious experience.

European women inherited from the early centuries of Christianity traditions that would enable them to claim much more. In the first three centuries of the Christian era, by her religious enthusiasm, a woman, even more than a man, could break the rules of her society. She could overcome what the ancient Greek, Roman, and Hebrew cultures had decreed as the innate flaws and disabilities of her body and nature. Under the protective mantle of the early Church, women had the possibility of roles and functions different from those usually designated for them. They could by their faith claim a life outside the family and marriage.

The first centuries of Christianity had preached "the equality of all believers," making women through their faith equal in potential to men, equal in the eyes of God, powerful by divine authority. Equal and empowered, a woman could reject her traditional function of childbearer, her life defined in relation to men as daughter, wife, and mother. In the world of the Church, there was no separate female or male role; instead, both sexes defined their lives in relation to God, a tie that superseded obligations to all others. In the first centuries of Christianity, zealous women—Mary Magdalen, Prisca, Saint Perpetua, Saint Paula, Saint Melania the Younger, Empress Pulcheria—had performed many traditionally male functions. They had preached, studied

Scripture, prophesied, converted others, and died in the name of the faith.

Twice in European history in the centuries after 800, women would again have these opportunities. In the twelfth and thirteenth centuries and again in the sixteenth and seventeenth centuries, women knew the exhilaration of "the equality of all believers," could forget distinctions of nature and function, and embrace forbidden roles and activities in the name of the revitalized Christian faith. In the religious enthusiasm of the twelfth-century Renaissance and of the sixteenth-century Protestant Reformation, women again became rebels and zealots, seizing opportunities with a fervor as intense as that which had motivated the believers and proselytizers of the early Church. They protested, they fought and died as martyrs. They founded new orders, they reformed the old. They studied, they preached, they converted others. For some, God spoke through their visions and thus gave them the authority to criticize and to prophesy.

But just as the traditions inherited from the early Church gave legitimacy to rebellion and unorthodoxy in the name of faith, so its inherited traditions took it away. The power granted by "the equality of all believers" had come to be limited and circumscribed in the early Church, and so it would be again by the end of the thirteenth century, by the end of the seventeenth century. In each age redefinitions of belief and newly formalized institutions monopolized all access to religious authority in the hands of a male hierarchy. Armed with new dogmas, the ecclesiastics of all Christian sects contained and controlled the faithful of both sexes. Those who did not conform found themselves cast out as heretics and schismatics.

In these successive ages of definition and conformity, most European women of special piety accepted the limitations. They gathered in convents and in congregations under the auspices of the male leaders of their churches. They accepted the religious roles of prayer and charitable works left open to them. Most did not realize that with each cycle of enthusiasm and redefinition, the limits had tightened, the range of activities had contracted. Women could still justify an alternative life outside marriage with their Christian faith, but this was not what their priests and ministers praised. Instead, the male ecclesiastics of the later centuries had etched the disabling and denigrating images of the female's nature and biological function more finely—images which reestablished female subordination, not equality, as the ideal. By inference for Catholic women, by direct commandment for Protestants, their Church leaders had glorified the Christian family, a world dominated and demarcated by male authority. By the end of the seventeenth century the preferred life offered to even the most pious and devout of European women lay within this narrow sphere, a life of service to the new hero of society: the husband, father, provider—the patriarch.[1]

2

AUTHORITY WITHIN THE INSTITUTIONAL CHURCH

The Great Abbesses and Learned Holy Communities

By the sixth and seventh centuries, the creation of bishoprics and monastic orders and the definition of Scripture and holy customs by the authority of popes and Church councils had contained and controlled the enthusiasm of the early Church. Within this new institutional faith, European women could still make a life outside marriage and the family. The Christian Church offered the religious life: membership in communities of women and men pledged to obedience, to poverty and chastity. Within these monasteries, convents, abbeys, and priories, individuals of faith and intelligence could exercise their intellectual and spiritual capacities to the full. No longer a wife, a mother, a daughter, a woman, like a man, could dedicate herself to study and to prayer. Though not equal in nature or power to a monk, a nun could nonetheless share equal access to divine favor, to knowledge, and to spiritual authority on earth.

The cloistered life of the monastery and convent evolved from the earliest centuries of the Church, when lay women and men had grouped themselves together, adopting a common rule to live and pray by. These rules of the fourth, fifth, and sixth centuries became formalized and the future of the group assured by the endowment of lands for them to live from. Between the seventh and twelfth centuries, royal and noble women, the daughters and wives of the propertied, founded such groups, often joining the monastery or convent themselves and presiding as abbess.

From the seventh to the tenth centuries, privileged foundresses and abbesses could assume powers usually reserved to bishops, abbots, and the ordained clergy. Many of these communities consisted of adjacent foundations for women and men, which historians have called "double monasteries."[1] Women often ruled these communities. As abbess such a woman exercised both religious and secular power. Because of the lands held in the name of the order, she was responsible for fulfilling the feudal obligations of a vassal and responsible for the administration of the manors and fields upon which the maintenance of the pious group depended. She also supervised the religious life of those living on the monastery lands, the collection of tithes, and the choice of the village clergy. As abbess she assumed responsibility for the religious life of the nuns and monks of her order.

In the exercise of these powers, some women acted no differently from their male contemporaries. In the mid-seventh century Saint Salaberga of Laon in France founded seven churches and took responsibility for 300 nuns.[2] Her contemporary Saint Fara (consecrated as a child by the Irish missionary Saint Columban) founded a joint community at Brie in the north of France, ruled as abbess, and assumed priestly and episcopal powers, hearing confessions and excommunicating members. The abbess of Jouarre in France obtained a papal dispensation in order to wear insignia usually reserved to bishops. Into the twelfth century at the Spanish abbey of Las Huelgas, the nuns appointed their own confessors. As late as 1230 the abbess Dona Sanchia Garcia blessed the novices like a priest and presided at chapter meetings for the twelve other monasteries under her authority.[3]

Benefactors and ecclesiastics expected that the principal duty and activity of the holy women would be prayer: prayer to exculpate the sins of their supporters and themselves. Men in religious orders, monks, could also preach and administer the sacraments when they had been ordained as priests; without ordination, women, in theory, could only pray. Into the twelfth century, however, the powerful female leaders of these holy communities—whether monasteries, joint communities, or convents for women alone—defined the lives and functions of their nuns more broadly. Prayer, yes, but also education. Literacy in these early centuries meant the ability to read Latin, and this they wanted for their charges, the ability to study the important Latin texts of antiquity and the early Church. By 694 Repton in Derbyshire, England, had become known as a center for education through the efforts of Abbess Aelfthrith. By the eighth century Chelles in France had achieved a similar reputation. In ninth-century Saxony in Germany, the nobility sent their daughters to Quedlinberg or Gandersheim to be educated.

From the seventh to the eleventh centuries, male leaders of the Church encouraged this learning. Saint Columban and the other missionaries of the seventh-century Irish Church emphasized education as they traveled through England, France, and Germany. Contemporary theologians in their treatises on virginity extolled the value of reading the Bible and the Church Fathers. In the ninth century some asserted that through such a "masculine" education a woman could become literally more "virile," more like a man, and thus more holy.[4]

Under the direction of abbesses of particular energy and intelligence, these religious centers were unique, the equivalent of small endowed universities. Saint Hilda (616–680), the grandniece of the seventh-century English king, Edwin of Northumbria, founded one and supervised two other monasteries in the course of her life. She made Whitby, a settlement of female and male anchorites, into a center of learning. Encouraged and advised by Saint Aidan, the bishop of Lindisfarne, she set the regular schedule of prayers and holy days and then turned her attention to the education of her charges. Five of her community became bishops. When a reeve (the layman given responsibility for supervision of one of the monastic estates) brought to her Caedmon, a servant on the manor, she recognized the beauty of his verses. Through her encouragement he joined the order and turned to holy subjects for his poetry.

Women from other English monasteries and convents—from Wimborne in Dorset, for example—went with missionaries to the Continent and stayed to administer religious communities and centers of learning established there. Lioba (d. 782) went to join Saint Boniface on his mission to the Saxons. As his particular friend and adviser, she presided over the double monastery of Bischofsheim and supervised the organization of other communities. Women from her monastery left to become teachers in establishments made by other male emissaries of the Pope. Because of her learned reputation, other members of the Church hierarchy, and Hildegard, wife of the Emperor Charlemagne, sought her advice.

Lioba's mother had dedicated her to the Church as a child, reasoning that the religious life would give her daughter "freedom." Thus it must have felt for Lioba and other women of the ninth, tenth, and eleventh centuries. For within these protected and sanctified cloisters women could be free of the intellectual disabilities placed on their sex. In this regard, they could enjoy opportunities usually reserved for men. They could use their minds; they could read the great texts; they could write their own words and thoughts. Lioba's biographer explained that the abbess had been so eager "to attain perfect knowledge of religion" that even as a little girl "she never laid aside her book." In her adulthood, more moderate though no less dedicated, she rested after the midday meal "for she maintained the mind is keener for study after sleep."[5]

In each century churchwomen explored the different literary forms. Baudonivia wrote the official biography of the sixth-century abbess, Saint Radegund of Poitiers, emphasizing "womanly" strengths such as her piety, her motherly care of the community, and her ability to keep peace. The seventh-century abbess, Gertrud of Nivelles, journeyed to Rome for books and enlisted Irish monks to teach. One of her own canonesses wrote her life. (A canoness took partial vows—obedience and chastity, but not poverty.) In the eighth century, the abbess of Heidenheim is considered to have begun the tradition of English travel books by recording for the bishop of Eichstatt his recollections of his trip to the Near East.

In this sheltered academic environment, one woman went beyond even her male contemporaries. Hrotsvit of Gandersheim (c. 930–c. 990, popularly known by a variety of simplifications of her Saxon name, i.e. Hroswitha, Roswitha), alone among the learned of Saxony, wrote verse, history, and the only dramas composed in all of Europe from the fourth to the eleventh centuries. Founded in the 850s, by the tenth century, Gandersheim had become a rich, independent nunnery, favored by the Holy Roman Emperors. Otto I allowed the abbess her own court, her own knights, the right to coin money and to sit at the meetings of the Diet. Hrotsvit entered the monastery as a child, probably dedicated to the Church by her noble family. Rikkardis, "our learned and kindly teacher" as she called her, taught her Latin; the abbess Gerberga may have given her Greek.[6] Hrotsvit's works show that she was familiar with the great classical and religious texts. (By her time, the tenth century, a good monastic library would include Virgil, Horace, Lucan, Cicero, Seneca, Pliny the Elder, Tacitus and the historians of the later years of the Empire, Ovid, Juvenal, Terence and Plautus, the philosopher Boethius, the Church Fathers, Fortunatus, Alcuin, Bede, Isidore of Seville, Legends of the Saints, the Vulgate, Psalms, and the apocryphal books of the New Testament.)[7] When she came to write, Hrotsvit, like an educated man of her circumstances, composed in Latin: legends of the saints, epics, a poem and her plays.[8] Each demonstrates her inventiveness, her originality with the language, her ability with meter. She created her pen name in the same spirit; it is a word play from Saxon, *hroth* meaning "sound" and *swith* "loud" or "strong." Thus she gave herself the title "the strong voice of Gandersheim."[9] In her poem she described her flounderings as a novice author working alone: "Sometimes I composed with great effort, again I destroyed what I had poorly written," working on so "that the slight talent . . . given me by Heaven should not lie idle in the dark recesses of the mind and thus be destroyed by the rust of neglect."[10]

Hrotsvit wrote her legends of exemplary saints and her plays to counter the images of weak corruptible women that she found in the secular and religious literature. She consciously chose to borrow her plots from the "pagan" playwright Terence, and enjoyed the irony of transforming his stories of deflowered maidens and prostitutes, tales of "the shameless acts of licentious women," into dramas showing "the laudable chastity of Christian virgins." She demonstrated over and over again the victory of the "fragile woman" over the "strong man . . . routed with confusion."[11]

Hrotsvit shared her plays with a few churchmen, who encouraged her in her efforts, but they were primarily meant for the nuns and canonesses to read

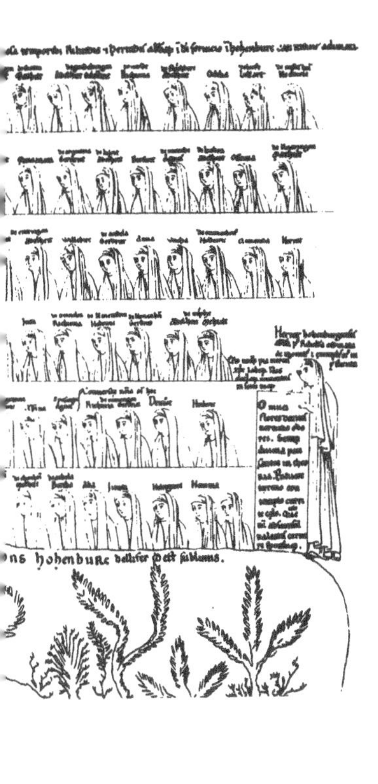

1. The nuns of the convent of the learned abbess Herrad of Landsberg, portrayed in her manuscript, *Hortus Deliciarum* c. 1160.

2. The mystic Hildegard of Bingen's "Awakening: A Self-Portrait" from her *Scivias,* mid-twelfth century.

3. St. Anne teaching the Virgin Mary to read. From a Burgundian prayer book, early fifteenth century.

4. Relief of a group of nuns from a French tombstor
thirteenth century.

5. Nuns tending the sick at the Hôtel Dieu in Paris. From a French manuscript, fifteenth century.

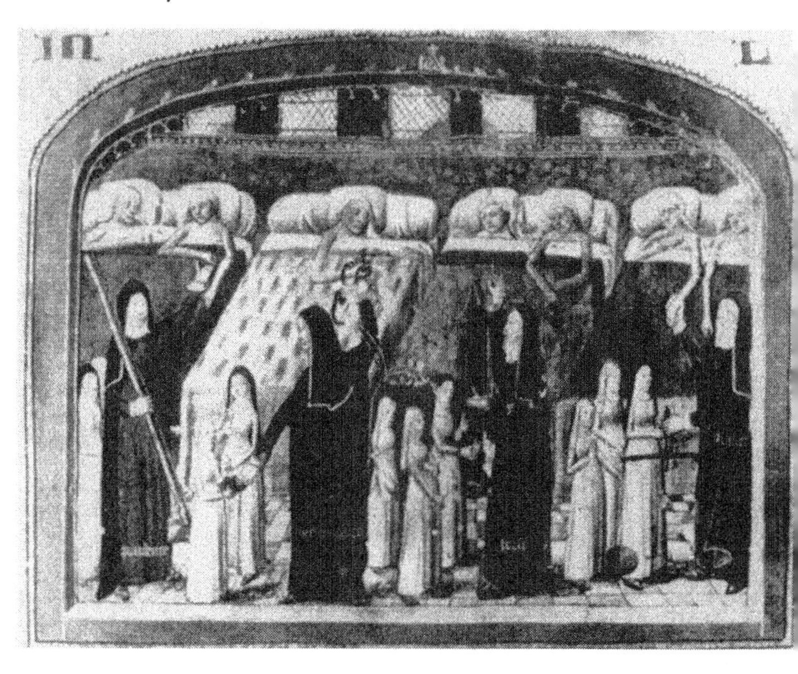

6. Drowning of the Anabaptist, Maria van Monjou. From a Dutch engraving, seventeenth century.

7. The mystic, religious reformer and Doctor of the Church, St. Teresa of Avila. Portrait by Fray Juan de la Miseria, 1576.

8. Mother of the artist Rembrandt reading a lectionary (selected biblical texts). Portrait by Gerard Dou, seventeenth century.

9. Family of the Protestant minister of Hjerm, Denmark c. 1660.

on feast days or for visits of Imperial or Church officials. Aspects of the dramas are weak: abrupt shifts of scene, wooden male characters, long monologues to show erudition. But there is also pathos, freshness, and humor in the situations and characterizations of her heroines. Faith, the heroic virgin of *Sapientia,* taunts the Emperor Hadrian from her pot of boiling pitch; Thais of *Paphnutius,* a repentant prostitute, balks at living in the small cell shown to her. "It will soon be uninhabitable . . ." she explains.[12] *Abraham,* probably her fourth play, has her best characters and most effective dialogue. Maria, trained to the holy life by Abraham, is seduced by a "false monk," and flees to the city where she becomes a prostitute. Abraham finds her and the reconciliation scene gives words to their love and piety. She could not return she explains, "once fallen into sinfulness I dared not face you, who are holy." Abraham replies: "But is there any one entirely faultless, except the Virgin's Son?" "Nay, no one," she answers. Thus assured of forgiveness, "divine grace overflows, and overflowing washed out the horrors of wrongdoing," and Maria returns with him to her pious life.[13]

None of Hrotsvit's writings had wide circulation, though their very existence shows the possibilities offered by the Church to an intelligent woman.[14] Few knew of the work of another especially learned woman, Herrad of Landsberg, abbess of Hohenberg in Alsace (1167–1195); her writings also demonstrate the intellectual heights open to a religious woman. A leader of unusual energy and learning, Herrad founded a community of canons, another of nuns, and a hospital. Under her supervision, Hohenberg embarked upon the creation of an encyclopedia, the *Hortus deliciarum (The Garden of Delights).* Between 1160 and 1170 she and the sixty nuns of her convent worked on what would be a 324-page manuscript, a compendium of all extant knowledge and a history of the world.[15] She described her efforts: "like a bee . . . I drew from many flowers . . . and I have put it together . . . as though into a sweet honeycomb." She filled this "honeycomb" with sacred and philosophic writings, her own lyrics and songs, popular exemplary tales, and with illustrations scattered through the text, sometimes taking the whole page.[16] Her knowledge of the Church Fathers and classical authors is evident in her language and use of traditional allegories. Yet the work also reveals her own imaginative point of view. Herrad's "ladder of virtue" shows souls unable to resist the temptations of the earthly world; only one woman calmly and quietly reaches the top. The last page of the manuscript portrays the nuns of the convent, grouped together with scenes of their daily life. Herrad carefully captioned all the illustrations in both German and Latin so that the novices could begin to learn the language of the Church. (Unfortunately, the original manuscript was destroyed in a fire in the Franco-Prussian War of 1870. The only extant copy of the *Hortus deliciarum* is a tracing.)

Of all the abbesses of these centuries, it was Hildegard of Bingen (1098–1179), founder of the convent of Rupertsberg, who, both in her learning and the authority she exercised, most fully used the power open to women of the Church. Hildegard was unique among the women and men of the learned monastic enclaves both by the breadth of her knowledge (which ranged from the scientific to the musical to the theological) and for the recognition given her in her lifetime. Popes and emperors accepted her scientific treatises. Popes and emperors believed her to be a prophet, a woman receiving divine revelation, recording and interpreting it for her contemporaries just as Deborah and Isaiah had for the ancient Hebrews. (The process of canonization was begun in 1227 but never completed.)

Though she was extraordinary in her accomplishments and reputation, there was little to distinguish the way in which Hildegard came to the Church. She was the tenth child of the Count of Spanheim from southwest Germany. Members of her family had already risen within religious orders; her aunt was an abbess, her brother an abbot. At eight she became the serving maid to an anchoress living near her family's favored monastery, the Benedictine abbey of Disibodenberg. At fourteen or fifteen she made her vow of virginity and was professed. Jutta of Sponheim educated her and in 1136 on Jutta's death, Hildegard succeeded her as abbess. At forty-nine, as a result of one of her visions, she left Disibodenberg with eighteen of her nuns to found a new convent, Rupertsberg near Bingen.[17]

Hildegard remembered her first vision at three, but, she explained, "I kept it hidden until God in His grace willed to have it made manifest."[18] When she was forty, she believed that God had commanded her to write. When she resisted, she fell ill. So, she began. From 1141–1151 she worked with her nuns to produce *Scivias (Scito vias domini,* or *Know the Ways of the Lord),* the source of her authority as a prophet and an interpreter of God's "mysteries." She "saw," she "heard" images and their meanings, twenty-six visions in all. She told of creation, an event similar to the uniting of woman and man. She envisioned the Creator as feminine, nurturing and sustaining the fruits of the union. She saw Jesus and described how humanity could be saved on a Day of Judgment—an endless day when the movements of the sun and stars ceased, when fire would not burn, when air would be thin and water forever calm.

Her close friend, Richardis von Stade, helped her translate and refine her Latin prose. Other nuns of the community did the manuscript, the illuminations, and the illustrations that she had designed. When Richardis left in 1151 to become abbess of another convent, Hildegard found others to act as secretary. From 1163 to 1173, her second book of revelations, *Liber divinorum operum (Book of Divine Works),* was completed in the same way. Visions and their interpretations came to her again as God's word—visions as if on a mirror, "the Shadow of the living light" as she described them.[19] There were visions in which she saw the harmony of the universe, the interrelationships and interactions between human beings and the cosmos. The last group of images gave her the whole sweep of human history from Adam to the Apocalypse, which she then interpreted for her contemporaries.

This second book of revelations included in the explanation of the third and fourth visions a long analysis of the function of the body. Thus Hildegard demonstrated another aspect of her remarkable intelligence. For in addition to the two visionary works, she also wrote between 1151 and 1161 on medicine and natural science. In the *Physica (Liber simplicis medicinae* or *Book of Simple Medicine)* she listed almost 300 herbs, telling when to pick them and giving their medicinal uses. She described animals, plants, and rocks. In the *Causae et curae (Liber compositae medicinae* or *Book of Medicine Carefully Arranged)* she catalogued forty-seven separate diseases, speculating on their causes and possible cures. Both books show her knowledge of classical authors like Pliny, and contemporary twelfth-century sources like the medical texts from Salerno, Italy. She explained that disease came from disruptions to the body's equilibrium and suggested physiological insights centuries ahead of her time, including the circulation of the blood, the ties between sugar and diabetes, nerve action to the brain, and contagion. She commented on the development of the female reproductive system, including descriptions of adolescence from twelve to fifteen and noticed the tendency for women to miscarry or produce defective infants if they conceived before twenty or after fifty.

While finishing these manuscripts, she also worked on the *Liber vitae meritorum (Book of Life's Rewards),* her allegorical dialogues between the vices and virtues, and her symphony, *Symphonia armonie celestium revelationum (Symphony of the Harmony of the Heavenly Revelation),* seventy-seven songs arranged as a cycle of devotional music for performance on feast days of the Virgin and of local saints.[20] Here too Hildegard proved her gifts, for the verses and notes are freer in form and image, more original than the usual Gregorian chants.[21]

Pope Eugenius first heard of her revelations in 1146. Hildegard wrote to Bernard of Clairvaux, the leader of the Cistercians, and he encouraged the Pope to designate her a prophet. In 1148, a group of churchmen, gathered together under the auspices of the Pope, confirmed that God was the source of her visions and that she did have the gift of prophesy. The learned read *Scivias.* Men of power sought her advice. Hildegard willingly took the authority thus given. She made tours of towns like Trier, Mainz, and Cologne, preaching. She wrote a tract against the Cathar heresy. She admonished religious and secular leaders, accusing them of corruption and of bad governance. She warned Pope Anastasius: "You neglect Justice, . . . you permit her to lie prostrate on the earth, her diadem smashed, her tunic torn. . . ."[22] When the Holy Roman Emperor, Frederick Barbarossa, did not answer her letter, she told him God's message: "Hear this, king, if you would live—else my sword will pierce through you!"[23]

In 1178, the last year of her life, Hildegard of Bingen found herself in opposition to the local Church officials of Mainz over the burial of a nobleman in the abbey cemetery. The abbey was placed under interdict, and it was only through her appeal to the archbishop in Rome and by references to visions attesting to the rightness of her actions that the condemnation was lifted, and she and her nuns allowed the sacraments. Only Hildegard's reputation and her acknowledged ties to the divine gave her the authority to oppose the male ecclesiastical hierarchy.

But the world of the great abbesses was disappearing. By the end of the next century, the circumstances and attitudes that had favored learned establishments and had allowed holy women like Hildegard intellectual achievement and spiritual power had gone forever.

How did this happen? How was their protected world dismantled and the sources of their authority vitiated? First, they lost their lands. Many of the great nunneries and monasteries over which abbesses presided never recovered from the attacks of the Viking and Arab raiders of the ninth, tenth, and eleventh centuries. Those joint houses not destroyed were disbanded in the reforming zeal that decreed segregated establishments for women and men.

Then, they lost their independence. Just as the officials of Mainz meant Hildegard to bow to their authority, so popes, bishops, and generals of orders would come to expect obedience from their abbesses. The rules of convents and their nuns fell prey to the centralizing tendencies of the eleventh, twelfth, and thirteenth centuries, victims in the struggles between male secular and religious leaders over spiritual and temporal supremacy. (The investiture conflict between Pope Gregory VII and the Holy Roman Emperor Henry IV from 1073 to 1122 is but the best known of these conflicts.) Both new foundations for women and the old establishments now came under the direct supervision of male ecclesiastics. Property, even the endowment of a nunnery, brought temporal power, power too important to leave in the sole care of a woman.

Last, the abbesses and nuns lost their claims to privileges and powers usually reserved to the male, ordained clergy. Restrictions against women avowing novices, hearing confession, preaching, and singing the Gospel often cited over the centuries found uniform enforcement when stated again by Pope Innocent III at the beginning of the thirteenth century. Renewed prohibitions had begun as early as the eighth century, when Saint Boniface imposed conformity on his followers, insisting that women's communities follow the Benedictine rule. The Emperor Charlemagne upheld in his decrees the superior authority of bishops over abbesses and forbade the women even the privilege of assisting in the administration of the sacraments.[24] A few great houses managed to retain special rights into the seventeenth and eighteenth centuries by avoiding the supervision of the local bishop, or reporting to no one but the Pope. Into the fourteenth century, the English abbesses of Romsey, Barking, Wherwell, and St. Mary's Clerkenwell acted more or less independently. In France at the beginning of the thirteenth century, the abbess Agnes II gained similar privileges from Pope Innocent III for the abbey of Jouarre. Bossuet, acting for the French king in the seventeenth century, broke this independent authority in his efforts to centralize the power of the monarchy. Most famous of all the independent convents was Las Huelgas in Spain, established by King Alfonso VIII of Castile and his wife Queen Lenore, the daughter of Eleanor of Aquitaine and Henry II of England. Until 1874 it was exempt from all control by the bishop and answered only to its own order and to the Pope.

Having lost access to authority through land and office, religious women also lost access to learning. In the twelfth and thirteenth centuries the excitement of study, of disputation and explication passed from the great monastic centers, whether female or male, to the exclusively male enclaves of the episcopal schools of bishops in their cathedral chapters. These evolved into the universities like Paris, Poitiers, and Oxford, where curricula became fixed and the opportunity to study limited. Just like the guilds of craftsmen, the university charters closed participation to all but a select few. Masters and doctors of theology alone could learn and teach. A prerequisite to study became ordination, and by the thirteenth century this sacrament and the priesthood had been officially closed to women.

Gratian expressed it baldly in his twelfth-century codification of the canon law, the *Decretum* (c. 1140). Only the baptized male could "validly" receive ordination.[25] In seeking justifications and proofs of this tradition, now elevated to custom and precedent by the Church, theologians drew on all of the old fears and prejudices against women. Classical authors, the Church Fathers, and commentaries on the Gospels supplied the essential arguments now embellished with the logic of the Scholastics and the enthusiasm of those who sought to reform and purify the clergy. Polemics condemning women because of their nature and their past actions filled the discourses and the sermons of those who insisted women could never function in male roles within the Church.

Saint Thomas Aquinas explained that only the superior male essence could receive the priestly authority. The woman was in a sense immune, for the power of grace could never be transmitted to the female creature. In addition, as a priest, she would have dominion over men, a logical and natural impossibility according to the great Scholastic. In his *Summa Theologica,* the fact of being a woman came first among the disabilities that automatically disqualified individuals from the priesthood, before the lack of reason, being enslaved, committing murder, being born illegitimate, or missing a part of the body. Not only was a female in authority beyond reason, it was dangerous. Gratian remembered Eve and the Fall; the woman forbidden by God to teach or judge was an ever-present image. He explained, "Adam was beguiled by Eve, not she by him," therefore, she must be subject to him "so that he may not fail a second time through female levity."[26]

With the fear of Eve came all of the other traditional nostrums about the potential evil of the female, whether inside the Church or outside it. Women had led man into "wrongdoing," always she had the greater propensity to sin. And there was the matter of her body. The great ecclesiastics of the twelfth and thirteenth century spoke of menstruation as a source of contamination. The menstruating woman, even as a nun or an abbess, could sully the holy places and must be excluded from the mass. Worst of all, her body might tempt the pious males who saw or spoke with her. This became the most damning and confining of the ancient traditions revived and restated—that even under the protective vows of a religious order, the woman's sexuality could not be contained or resisted. The pious woman must be protected from herself, pious men must be protected from her. Women of faith must be kept separate and subordinate. For Popes Gregory VII and Innocent III such considerations became key. In their disputes with kings and emperors, they insisted that their superior temporal power came from the spiritual purity and moral authority of the Church and its agents. Association with women threatened this fragile claim.

By the thirteenth century, attitudes toward women's roles in the Church had crystallized around these male concerns, this medley of traditional fears of the female as a sexual being. Yes, allow the woman a life devoted to religion, but, given the dangers inherent in her being, this life must be closely restricted and guarded. Nuns must be kept separate from men, even male religious. Prohibitions against joint houses, against nuns teaching boys, against any but the most necessary contacts between the women and their male confessors, gained wide acceptance. Even more important, women must be kept strictly cloistered. They must not leave the convent, they must have no contact with anyone from outside its walls. The English Council of York in 1195 described nuns' pilgrimages as "the opportunity of wandering"; any dealings with "secular folk" troubled "the quiet and contemplation" of the sisters, according to a Cardinal Legate in the 1260s.[27]

In 1293 the Bull Periculoso of Pope Boniface VIII gave the justification and set the narrow perimeters of the life allowed women within the Church: Desiring to provide for the perilous and detestable state of certain nuns, who, having slackened the reins of decency and having shamelessly cast aside the modesty of their order and of their sex, he decreed: "that all and sundry nuns, present and future, to whatever order they belong . . . shall henceforth remain perpetually enclosed . . . ; so that no nun . . . shall henceforth have or be able to have the power of going out of those monasteries for whatever reason or excuse. . . ." They shall remain "altogether withdrawn from public and mundane sights."[28]

Not only did the Church restrict and confine women, but the facts of their past achievements and authority were ridiculed. The woman remembered for exceptional learning, piety, and power was not Lioba, Saint Hilda, Herrad of Landsberg, or Hildegard of Bingen. Instead the fantasy of a thirteenth-century French Dominican, Steven of Bourbon, was passed on from generation to generation, the tale of the fictitious Pope Joan. In the story, Joan studied, disguised as a man. Brilliant and devout, she advanced quickly within the order, went to Rome, became a cardinal, and then was elected Pope. But in this life of achievement, this life without restriction, her woman's nature betrayed her. In Rome she lusted for a young man, she fornicated, and she met a fitting end. On the day of her installation as pope, she died in childbirth during the inaugural procession through the streets, death in the gutters of Rome like a common whore. A salacious, cautionary fable displaced the memory of the lives and writings of the great abbesses and their women scholars.

Reformed Religious Orders: The Twelfth and the Thirteenth Centuries

Despite the restatement of negative traditions, the restrictive customs and decrees of the Church during the twelfth and thirteenth centuries, privileged women still joined its convents, priories, and abbeys. They, like men, wanted to participate in this renaissance and revitalization of the Christian faith in Europe. Ironically, the writings of the twelfth- and thirteenth-century theologians encouraged them. On the one hand the churchmen condemned women as inferior by nature, unfit for the priesthood, unable to study at the universities, and of necessity subordinate to male ecclesiastical authority. On the other hand, they revived and made dogma other positive traditions of the early Church which empowered women. They wrote of the transformational power of chastity and the superior status of the virgin. Like the Church Fathers, thirteenth-century theologians like Saint Thomas Aquinas promised that the taking of vows would be the means to escape the disabilities of a woman's sex and the horrors of pregnancy and childbirth.[29]

As nuns, Aquinas explained, women "are promoted to the dignity of men whereby they are liberated from the subjection of men."[30] Rejecting the role of wife and mother, these women came to be seen as the literal and spiritual brides of Jesus, in a marriage described by *The Ancren Riwle* (a guide for women of special piety) with all the erotic joy of a worldly wedding night:

> After the kiss of peace in the mass . . . forget there all the world, and there be entirely out of the body, there in glowing love embrace your beloved spouse Christ, who is come down from heaven into the bower of your breast, and hold him fast till he have granted all you wish.[31]

The life of a nun in a convent much like the one imagined by Aquinas and his contemporaries could bring a woman dignity and honor. It did so even for Heloise (1101–1163), the woman notorious in her own day for her seduction by and affair with Abelard, the learned Schoolman of twelfth-century Paris. Heloise had begun her life as a gifted student at the convent of Argenteuil. Her uncle Fulbert, a canon of Nôtre Dame cathedral, brought her to Paris and arranged for Abelard to tutor her when she was just seventeen years old. The drama of their affair lasted eighteen months, with secret meetings, her pregnancy and the delivery of their son, Astralabe, in Brittany with Abelard's family, and a clandestine marriage at Abelard's insistence to placate Fulbert. Her uncle refused, however, to keep silent about the illicit union of the cleric and his niece, especially when he feared to lose Heloise altogether because Abelard began to consider holy orders for her. It was then that Fulbert sought revenge and had Abelard attacked and castrated.

At nineteen, Heloise took vows at Argenteuil to please Abelard. Her letters to him, written eighteen years later in 1135, reveal how she mourned and suffered the separation. Images of the "delights of lovers" invaded her prayers, and "when I ought to lament for what I have done I sigh rather for what I have had to forgo."[32] Even so, she rose quickly within the monastery and soon after accepted Abelard's offer of an independent foundation in association with the male order of the Paracletes. In 1131 Pope Innocent II confirmed the charter for their lands and thus sanctioned her administration as prioress over the group of nuns who had joined her.

Despite all these accomplishments, however, she explained in her letters that she feared her pious reputation was a sham. Abelard in response tried to reassure her. He saw purpose in all they had experienced together, and explained that now they had been freed from the burden of concupiscence. Using the images Saint Jerome painted for Eustochium, Abelard glorified her chaste life: "Formerly the wife of a wretched man thou art now exalted to the bed of the King of Kings."[33] She could now as prioress, through her tutelege, birth not just a few children in filth and pain but be "delivered in exultation of numerous progeny," the nuns under her care.[34]

Abelard was but the first of her contemporaries to praise "her piety, her prudence, and in all things the incomparable meekness of her patience."[35] Peter the Venerable, the abbot of Cluny, compared her to Deborah, leading her nuns to holiness, a model of humility who had chosen the life of the contemplative over that of the scholar. In the popular mind, she had become like Mary Magdalene, the woman fallen into lechery, then risen through repentance to a state of virtue. Historians assume that she became reconciled to her vocation and knew peace. The only other letter she wrote concerns her son. She asked Peter the Venerable: "Would you for the love of God, keep also in mind my son Astralabe, and win for him office in the Church as Prebend, from the bishop of Paris or some other prelate?"[36] The model of piety could not forget everything from her other life.

Just as Heloise and a few of her nuns had been able to join together, choose a rule to live by, and then gain approval from the pope, so other women in the first centuries of the Church had simply found land, decided to live according to holy vows, and, with the assistance of the local bishop or abbot of one of the established male orders, set up a new convent. In the twelfth and thirteenth centuries, well-born and privileged women read and listened to the new leaders like Bernard of Clairvaux, Robert of Arbrissel, Francis of Assisi. Like their male contemporaries, women of exceptional piety and strength responded to their calls for new religious orders, following a simpler, reformed rule. They too wanted access to this uncorrupted life within the Church, though circumstances would limit their numbers. Historians have not yet completed the work of investigation and compilation that will give the number of convents—both old and new orders—in Europe and how many followed each rule. The records are scattered throughout the Continent and throughout the centuries. Those for England have been the most extensively studied, showing that between the thirteenth and the sixteenth centuries, about half of the establishments were Benedictine, a quarter Cistercian, and the rest divided among the Cluniac, Augustinian, Dominican, and others.[37]

The disparity in England between the number of women's and men's foundations is noticeable. From the tenth to the seventeenth centuries the female orders tended to be fewer, with fewer members, less well endowed.

Historian Eileen Power's study of the 1535 *Valor ecclesiasticus,* ordered by King Henry VIII for purposes of assessment, reveals that the richest and the biggest nunneries were in the south of England, with thirty to thirty-five members each and histories dating back to endowment in the Anglo-Saxon period. Many existed in the north and east Midlands, but they tended to be the poorest, most with fewer than ten members. Suzanne Fonay Wemple, the historian, estimates that in Italy women's houses made up one quarter to one fifth of the houses in the centuries before 1500. In the thirteenth century, parts of Belgium and Germany appear to have been unique in that the number of women's establishments was equal to that for men, but the overall pattern of the Cistercian records for Germany reflects that of England and Italy: by 1500, 654 nunneries and 742 establishments for monks.[38]

Though many churchmen insisted that Bernard of Clairvaux's Cistercian order was too harsh and strenuous for the female, it inspired religious women all over Europe. In France, the older Benedictine houses just took on the rule. At the beginning of the thirteenth century Saint Hedwig founded a Cistercian house against the will of the male abbot at Trebnitz. In some areas, the zeal of the women for Saint Bernard's purified, simpler life surpassed that of the men. It was Ulvhild, the twelfth-century Queen of Sweden, who prevailed upon her husband and then her son to found a monastery and then a convent. In Flanders, in Brabant, and throughout Belgium the number of Cistercian foundations for women exceeded that for men.[39]

Other new orders appealed to propertied women. At the end of the eleventh and the beginning of the twelfth century, Hersende of Champagne and Eleanor of Aquitaine, inspired by the preaching of Robert of Arbrissel, founded or supported female houses with male members obedient to an abbess.[40] The French Premonstratensians of Norbert of Xanten allowed women to join. In 1206 the female followers of Saint Dominic in Prouille, France, gained his approval for a religious community. Women in Madrid and Bologna created Dominican foundations in 1218 and 1219. In Assisi in 1212, Clare, the eighteen-year-old daughter of the Count of Sasso-Rosso, defied her family to gain the approval of Saint Francis of Assisi for her group of pious women at the church of St. Damian. With a rule he had written for them, she hoped to follow his example, to live like Franciscans according to strict vows of poverty and like the friars to have no endowment, only alms collected by the sisters.

Just as the leaders of the older Benedictine and Augustinian orders had come to fear the women they had sponsored, so the leaders of the new foundations indulged in the traditional prejudices and terrors. St. Bernard of Clairvaux opposed any contact for his followers with women, and in general the Cistercians refused to supervise the women's priories. When in the late

1200s the Cistercians, the Cluniacs, and the Carmelites officially accepted female establishments in their name, they insisted upon strict enclosure and supervision by a male prior. Even those who had established "joint communities" did so in a different spirit from the ninth and tenth centuries. Robert of Arbrissel gave women responsibility for administering the abbey only while he traveled the countryside preaching. Norbert of Xanten kept the nuns strictly cloistered and would not allow them to participate in the services. On his death in 1134 his successors expelled them altogether.

Women who wished to become followers of St. Dominic and St. Francis and to establish houses in their names had the most difficult time winning acceptance. For these religious enthusiasts the contradictions between what men preached to all souls and what they allowed to women limited them over and over again. St. Dominic and St. Francis preached "the apostolic life"—a vocation outside the cloister of service to others, not just of prayer and masses, a vocation of absolute poverty with no endowment, only the fruits of the alms bowl to sustain the group. Neither St. Francis nor St. Dominic would permit such a life for women. Both men fell prey to the traditional prejudices and fears about women. St. Francis acted as priest to St. Clare and her sisters, but refused responsibility for any other groups founded in his name. St. Dominic did not wish his friars to service the spiritual needs of the women attempting to follow his rule. Both men insisted on strict enclosure and balked at the idea of females begging. St. Clare and her group survived by spinning and weaving linen for the altar cloths of Assisi and even this had required a papal dispensation. Between 1227 and 1250, popes had to force acceptance of the female establishments by decree, by reversing decisions of the chapter of the orders. But only in this respect did the papacy prove an ally. The Fourth Lateran Council of 1215 forbade the founding of new orders, so women had no choice but to follow the men's rule with their newly stated restrictions. Once again their spiritual lives were limited by the traditional prejudices.

St. Clare and her followers did almost succeed in creating their own new rule with no allegiance owed to a male order. Later in the thirteenth century when women like Agnes of Bohemia, the betrothed of the Holy Roman Emperor Frederick II, and Margaret, wife of King Louis IX of France, founded establishments they called themselves Poor Clares, not Franciscans. They did not acknowledge either St. Francis or St. Benedict as the inspiration for their rule. Still they needed papal permission to join together without an endowment, to live from their own earnings. When in 1263 Pope Urban IV insisted that cloistering be an official part of their rule, they became like the other groups, and the chances of an alternative life, the apostolic life of the men's orders, was completely closed to women. In every instance the inherited traditions of the early centuries of Christianity were evident: Pious women active in the first decades of the new orders were excluded or restricted when institutionalization replaced enthusiasm.[41]

The Cloistered Life of the Nun

From the twelfth to the seventeenth centuries privileged Christian women all across Europe entered this spiritual world of the nun even with its restrictions. For the taking of vows, the consecration, the giving over of herself to the convent meant that a woman had made a choice. In deciding not to follow the traditional life of a wife and mother, such women often had a sense of taking charge of their own future, of the power to choose one alternative over another, however limited.[42]

When it came to making the decision, women entered convents for a variety of reasons, subject to a variety of pressures. Like their male counterparts, the most devout came for spiritual reasons. A few exceptional women challenged their families in pursuing their enthusiasm for the faith. At the beginning of the twelfth century the young Englishwoman, Christina of Markyate, defied her parents, who wished her to marry, and escaped secretly, riding off to the protection of her spiritual mentor dressed in men's clothing. Jacqueline de Sainte, sister of the French mathematician and philosopher, Blaise Pascal, in the midseventeenth century opposed her brother's prohibitions and gained admittance to the order of her choice even without the usual dowry.

Others came for more practical considerations. Commonly, an older woman, usually a propertied widow, would retire to the convent she had patronized during her lifetime. She might take the full vows or she might live as a canoness of the Augustinian order in chaste, simple ways but without the habit and without the restrictions of claustration. In the tenth century, the Holy Roman Empress Theophano had used the nunnery as a refuge when she was ill and when she gave birth. Queen Eleanor of Aquitaine at the end of the twelfth century used Fontevrault, the joint monastery she supported, as the place of her retirement. In the fourteenth century Queen Isabel of Portugal retired to the convent of Poor Clares that she had founded.[43]

From the earliest days of the Church in Europe, parents arranged for their daughters to become nuns, either to insure their own access to salvation or to provide for those daughters who remained unmarried. Elfled, the daughter of the seventh-century English King Oswy of Northumbria, was pledged by her father to a nunnery in hopes that this gift would ensure his victory in battle. The tenth-century Holy Roman Empress Theophano's two daughters, Adelheid and Sophia, became nuns and rose to be abbesses at the

German establishments of Quedlinburg and Gandersheim. In fourteenth-century Sweden, Saint Bridget and her husband gave their daughter Ingeborg to the Church so that she might "fight for God" on their behalf.[44]

For propertied lords and townsmen like the early fourteenth-century English knight Sir John le Blund and the fifteenth-century London alderman Sir John Stocton, the convent was the alternative protected life for a daughter who did not marry, a "devout place" where she could "lead her lyfe in cleanness."[45] The Dowry Fund established in fifteenth-century Florence gave the fathers who contributed to it the right to withdraw money either for marriage or as an annuity for a nunnery. Marie-Blanche, the granddaughter of the seventeenth-century French noblewoman, Mme. de Sévigné, went to the convent when she was only five and a half. Her father had two daughters by a previous marriage and did not wish responsibility for another suitable dowry.

Into the seventeenth century, privileged fathers and mothers also used the Church as a sanctuary for other daughters who might never marry, those born of an illicit union. In the sixteenth century, the Florentine Piccio Velluti managed to marry off his illegitimate daughter Angola, but when her husband died, he arranged for her to join an order "to protect my honor and to assist her in her need."[46] Galileo, Garcia Lorca, Lope de Vega, even the Duchess of Cleveland, once mistress to King Charles II of England, made such arrangements for their illegitimate female offspring.

Whether they had chosen to take vows themselves, or came to accept what they could not change, within the protection of the convent a woman could have a full and rich life. The best endowed establishments were sizable: Ronceary in western France in the eleventh century numbered seventy-three. Most of the nuns were professed virgins, but there were also twenty-two wives and widows, most from noble country families. Four were from the towns, three daughters and a widow of a well-to-do guildsman.[47] Similarly, Fontevrault became a sanctuary for French noblewomen, with a main house for the nuns, a smaller unit for the male canons, a third for widows and repentant prostitutes, a hospital, a school, and a hostel for pilgrims. By the fifteenth century Poissy was the favored nunnery of the French royal family and had become a community of 200 women.[48]

Most houses would be less grand, having closer to twenty than 200 members. All would be women of good families, daughters, or former wives, of men with prestige or property in the community. Those of lower caste might become lay sisters, or might be employed like the peasant women who worked the nuns' fields, or who laundered and cooked. A young woman brought to the nunnery as a child usually made her final profession at sixteen or after (sixteen was the age of consent designated by the Church), in a ceremony performed in the abbey chapel presided over by the bishop and the prioress, witnessed by the other nuns, by family and visitors. Candles, singing, and ritual marked this occasion, when the novice became *sponsa dei,* literally and figuratively the bride of Jesus. From that day forward she wore a wedding band and the habit of her order, and lived according to the convent's rule, subject to admonition and punishment for infractions.

Because of the insistence on enclosure and the assumption by male leaders that women could not endure the same austerities as men, there was not the same variety of religious life offered to them—no wide range from the most sequestered Cistercian to the service and preaching of a Franciscan, to the power of a Dominican advising a king. Just as the late thirteenth-century Bull Periculoso envisaged, life in the female Benedictine, Cistercian, Dominican, or Poor Clare foundations meant days of prayer for oneself, for others, an enclosed existence within the confines of a spiritual community. There would be the seven occasions for prayer, from early in the morning to the time of retiring, times when the nuns would gather and walk silently to the chapel. In the thirteenth century, Odo Rigaud, archbishop of Rouen, ruled that the nuns within his jurisdiction could confess and have the privilege of communion from six to seven times a year.[49] Otherwise, their participation in the mass would usually be restricted, and they would chant responses only when alone at their own services.

As they had in the ninth, tenth, and eleventh centuries in the great abbeys, religious women found ways to stretch the limits of their regulated days, months, and years. Like Hildegard of Bingen, they made music for each other into the eighteenth century. They wrote chants, plainsong, and polyphonic music for the intervals in the mass, for special holidays, and for the ritual occasions of the order. In the fifteenth century, Barking Abbey in southern England had six soloists to sing on feast days and a cantrix to supervise the music, and the choir, and to choose the pitch and the rhythm. At Las Huelgas in Spain the nuns chanted original music for the liturgy of Saint Agnes celebrated at the consecration of a new member of the order. In the seventeenth century, convents in Ferrara and in Milan became famous for their music, their musicians, their singers, and their composers. By 1700, at the age of eighty, Isabella Leonarda (1620–1704) of the Collegio de Sant'Orsola in Novara had composed twenty books of music, motets, music for four voices for the services, and the first trio sonatas for violins and basso continuo ever written by a woman.[50]

The women of the convent prayed together, but they also had time for solitary meditation and reading in the privacy of their cells. From the thirteenth to the sixteenth centuries, especially in the Benedictine and Dominican houses, the women would also be expected to study as part of their religious duties. All of the seventy-four Dominican convents across Europe had libraries in 1300.[51] The one in Nuremberg in its inventory for the years 1456–1469 showed fifty books, many of the manuscripts perhaps made by the nuns.[52] Though the women would no longer read the classical texts as Hrotsvit and Herrad of Landsberg had, they studied books of moral instruction written especially for them in the vernacular and meditated on the writings of mystics as an inspiration to their own devotions.

The nuns had other activities that they performed when together. They read works of piety aloud during their meals in the refectory and in the common room while they sat and embroidered decorative cloths for the churches and vestments for male officials. They cooperated on hangings like that designed to circle the chapel at Bayeux. This imaginative rendering of the eleventh-century Battle of Hastings was worked in woolen thread on linen by the nuns of Normandy to commemorate their duke's conquest of England. The Bayeux Tapestry is like a manuscript, with the whole story shown in the pictures, from the comet portending the death of Edward the Confessor to the arrow in Harold Godwinson's neck. Multicolored horses look over the sides of the Viking boats; Norman and Saxon warriors go hunting; sailors load the ships in their bare feet. The dead decorate the margins along with fanciful beasts and erotic scenes.

The names of a few artists and designers from the seventh and eighth centuries have survived: Herlind and Reinhild at Maasryck in the Low Countries; the Abbess Agnes at Quedlinburg for her curtain showing the marriage of Philology to Mercury. No names of artists from the later centuries remain, but the sacristies of every great cathedral of Europe still display the jewel-embossed robes woven and embroidered in gold and silver thread by the women of the convents. Their lace still graces the altar cloths from which the priests administer the sacraments.

Women expressed their artistic talents within the context of their religious life in other ways. Herrad of Landsberg's *Hortus deliciarum* and Hildegard of Bingen's *Scivias* were famous for their illustrations as well as their theology. Just like a male establishment, a convent might have a scriptorium, a place for copying and illuminating books for use by the nuns or for a patron of the house. Although they never had the resources to do big works and confined themselves to the calligraphy, to illuminating the borders, and to decorating capital letters, some of the artists of the eleventh to the fifteenth centuries have been remembered, particularly from German and Italian houses. Diemud (c. 1057–1130) of the convent of Wessobrun has left the greatest number of works: forty-five books including a Bible in two volumes, which her monastery used to acquire a manor. At Mallersdorf, in Frankfort, at Osnabruck, at the fifteenth-century Dominican houses in southern Germany, in Siena and Pavia in Italy, there are records of women copyists and illustrators.[53]

With the coming of painting in fresco and oil, the nuns of Italy experimented with these new forms. Single works in fresco and oil done for the churches survive in Brescia, in Siena, in Florence and Rome. In the seventeenth century Lucrina Fetti (c. 1614–1651), sister to a painter and an Ursuline nun, came under the patronage of the Duke of Mantua. She did religious scenes for the convent of Sant'Orsola and portraits for the Gonzaga family.

The Bull Periculoso had envisaged religious women living like Jesus' follower Mary, thinking only of prayer, their establishments so generously endowed that servants and laborers would tend to their simple needs, to the supervision of the estates on which their livelihood depended. But the very nature of their situation, whether in a large or a modest foundation, made the pious enclosed life virtually impossible for women. Circumstances forced them into conflict with their male superiors. The simple administration of the convent necessitated contact with the secular world, and rarely would the community be financially viable, especially after the economic changes of the twelfth through the fifteenth centuries, when families found property too valuable to give to daughters or the Church, when more had to be procured with money, not with exchanges of goods or services generated by the convent's estates.

A sense of the problems and contradictions comes from the records of one nunnery and its actual needs. The Benedictine order of Saint Radegund in Cambridge, England, founded in the twelfth century, had only twelve nuns by the second half of the fifteenth century. The account records for 1449–1450 show that they needed the fairs of Stourbridge and Ely for fish, salt, pepper, timber, beds, a churn, pewter pots, a horse, soap, and pitchfork staves. In 1481–1482 they had to buy iron nails, paper, and parchment.[54] The community employed seventeen people to serve its needs and to do the work on the manors belonging to the convent.[55] In a bigger house, like Sion in England or Poissy in France, the needs would be even greater, the life even more compromised when the powerful families whose daughters had taken vows came to visit and expected entertainments and comforts, and otherwise made the simple cloistered existence impossible.

In theory the money for all the needs of the convent, whether large or small, would be generated from the property given at the founding of the order and from the "marriage gifts" brought by each nun at the time of her novitiate. Until the nineteenth century the need for this dowry kept all but those with access to land or other wealth from gaining admission to the contemplative religious houses. The management of these dowries, this endowment, represented part of the activities of the community, and was the responsibility of the mother superior, the abbess or prioress, elected by the nuns with the approval of the bishop and perhaps of the patron of the establishment.

The prioress or abbess would be the one most often forced to act against her vows, which assumed isolation from the secular world. Just like a well-to-do woman in the town or the countryside, she would have to supervise the property—the farms, stables, fields. She would be responsible for providing for the group and for the obligations owed to the lord from whom the land had originally been granted. The lands would mean knights or knights' fees, services owed, rents, fines to be paid to their lord, to be collected from their vassals and serfs. Some rarely came into conflict with their vows. They would merely serve their sisters: the chantress to supervise the service, a sacrist to watch over the chapel, a fratress to deal with the refectory and the lavatory, an almoness to supervise the almsgiving of the community, a chambress to see to their clothing, a kitcheness to supervise the cooking, a mistress of novices to teach the postulants and to watch over the professed nuns' behavior. Others with responsibilities for the group would, like the abbess, often find themselves open to criticism for not following the Bull Periculoso to the letter. The cellaress supervised the servants and the farm. She sold surplus produce. When there was no money for laborers, she did the work herself. At the Gracedieu priory, Margaret Belers, the fifteenth-century cellaress, "goes out to work in autumn alone with Sir Henry (the chaplain), he reaping the harvest and she binding the sheaves, and at evening she comes riding behind him on the same horse. . . ." went the report criticizing her behavior.[56]

Given the economic realities, these communities had to find a variety of ways to earn the money they needed to maintain themselves. Records tell of German and English abbesses lowering rents to lure tenants back to manors they had deserted, hiring lawyers to take dishonest estate managers to court. They took in boarders. Their nuns did spinning, sewing, weaving, knitting, embroidery. The fifteenth-century Dominican convent, St. Catherine's of Nuremberg, became famous for its artists and the small carpets they produced, much like a craftsman's shop.

In the earlier centuries, the nuns had taught girls and boys to train them for the order, or to teach them basic literacy, some arithmetic, and their prayers. In the eleventh century a Russian princess who founded a school for girls in Kiev was but one of many patronesses of such institutions. Matilda, queen of King Henry I of England, later known for her piety and learning, went to the monastery of Wilton to study at thirteen. She remembered her instructress with special hatred, a teacher who scolded and slapped; "as soon as I was able to escape out of her sight, I tore it [veil] off and threw it on the ground and trampled on it. . . ."[57] With the reforms of the thirteenth century and the insistence on strict enclosure, the nuns were forbidden to teach. But the orders and the families ignored the prohibitions. Nuns stopped admitting boys, but they continued to teach the daughters of their friends and families and to use the revenues generated. The Benedictine house of St. Michael's at Stamford had twelve nuns and seven to eight pupils and was typical.[58] Houses all over Europe continued the practice.[59]

Despite its inherent practical contradictions, this ordered, cloistered life of the nun remains unchanged in the twentieth-century convents of the contemplative orders. This role offered by the Catholic Church gave devout Christian women a haven, an accepted model of piety, prayer, and communal activities with the promise of spiritual rewards beyond anything in the secular world.

Mystics: The Ecstatic Life

This life of masses, of reading, sewing, teaching satisfied many women in both their wordly and spiritual aspirations. But there was another path to Christian holiness, one which women had followed since the earliest days of the Church. This was the way of the mystic, whose affinity to God came not from books and pious works, but from meditation and prayer. Direct visions from God—images and words of supposedly divine origin—continued to have authority within the Church, even as it also embraced the rationalism and logic of the university and the Scholastics. The Cistercians and the Franciscans particularly honored such "experiences" of God, rather than "knowledge" of his powers, describing them as "interior" rather than "exterior" manifestations of faith, a passive joining with the deity through self-abnegation.

Traditionally, mysticism had been a path open to women equally with men, one they were encouraged to seek. Ailred of Rievaulx, the twelfth-century English Cistercian, used much the same language in his directions on meditation to his sister that Saint Jerome had for his female pupil Eustochium in the fourth century: "Run, I beg, run and take part in such joy, prostrate yourself. . . ." Give over all feeling to God, he explained, and even lust would become holy, for her "whole affection" would be for "the attractions of the Lord's flesh" and nothing else.[60] Scholastics like Saint Thomas Aquinas even suggested that a woman, with her less rational, more credulous nature, was more open and receptive to this kind of contact with God. Others explained that her humility and patience suited her better than the male to this kind of spirituality.

Thus, by tradition and by default, women of the cloisters retained a means to gain status, recognition, and authority that the reformed Church of the thirteenth century could not reason away with classical citations and would not supervise or decree away. Once the divine source of her experience had been recognized and accepted, a female mystic held power almost as potent as that of the learned abbesses. She served herself and her sisters. In addition, she served the Church as a whole, for the women mystics of the twelfth to the fifteenth centuries kept alive the prophetic and contemplative side of Christianity. Their descriptions of the discovery of self through union with God, "interior monologues" as Saint Teresa of Avila (1515–1582) called them, became models of prayer for the devout both female and male.[61]

To place one's self in a situation where mystical experiences and the reward of union with God could happen required a special kind of courage, strength, and faith on the part of the pious. The women who achieved the honor of the title "mystic" rebelled against their families, subjected themselves to extremes of physical privation and to endless spiritual exercises. Some even had to fight those within the Church meant to give them solace and protection, confessors who questioned the source of their visions. The English holy women, Christina of Markyate and Julian of Norwich, found the path by living the life of the recluse; others like Gertrude the Great and Saint Teresa of Avila found it within the community of the convent.

Christina of Markyate (c. 1123), daughter of a well-born twelfth-century Anglo-Saxon family, made her private vow of virginity as a child, but this ran counter to her parents' plan for her, a useful alliance through her marriage. Christina refused, insisting that she wanted to retire from the world and live as a religious recluse like Hrotsvit's heroines. Her mother beat her and encouraged her betrothed to seduce her. Only a clever hiding place protected her when he broke into her room planning to rape her. (She hid behind the cloaks hung on the wall.) Her father threatened to turn her out: "So come and go as I do, and live your life as you please. But don't expect any comfort or help from me."[62] Even the men of the Church would not help her. Her parents bribed the bishop of Lincoln to favor the betrothal, and he supported their wishes. Roger, the hermit who later became her spiritual mentor, rejected her because of the betrothal already pledged. In the end she made all of her own decisions and took all of her actions alone. She arranged for a horse with one of the family servants, dressed as a young man, and ran away. Then she spent six years in hiding to insure that she could have the life of faith she craved.

Having won the battle with those of the world, Christina battled within herself. She never recovered her health from the time spent in hiding and the austerities to which she subjected herself. She had spent years in a corner of Roger's cell, carefully blocked off so that no one might suspect her presence, in a space so small she could not move or even wear enough layers of clothing in the winter. She held still and quiet, never summoning him for "how could she do this. . . . For she was afraid that someone else besides Roger might be near and, hearing her breathing, would discover her hiding place."[63] This harsh discipline seemed to her the way to God. There in this self-imposed solitude she fasted, she recited the Scriptures, she allowed herself no solace, not even holy books. Ultimately the voices and the images came.

Julian of Norwich (c. 1343–1416?), a fourteenth-century English recluse, did not have to subject herself to such extreme deprivation, but she, like Christina, received the experience of direct communication with her God. Julian was granted her first vision, the first "showing" as she called it, while suffering from an almost fatal illness at thirty. The vision began, she remembered, about four in the morning and lasted into the early afternoon. It had come from her prayers asking for knowledge of Jesus' Passion, for an illness to free her of her body, and for "God's gift—the three wounds."[64] She too arranged to live in a special space, but her life was neither so confined nor so spartan as Christina's. Her little room was attached to her parish church, with one window facing out for visitors and one opening into the church so that she could hear mass. A servant ministered to her basic needs. There she wrote, read the Vulgate, some of the Church Fathers, the lives of other mystics and saints, and directions for meditation for nuns like *Hali Maidenhad* and *The Ancren Riwle.* Later, she wrote again, recording first in a short draft and then twenty years later in a longer version complete with explanations and revision, what she had experienced, sixteen "showings" in all. She divided life into three parts: "being," "growth," and "fulfillment."[65] For her, the period of growth meant struggling and visions; her first one gave her a sense of "the ebbing away of my life" until she saw the crucifix and experienced a new kind of faith.[66] Then she saw the devil who taunted her, took her around the throat with his paws, and put his ugly smelly face close to hers.

Other women visionaries sometimes lived together, seeking the protection or the encouragement of an established community of religious women. There were clusters of mystics in the Low Countries and Germany in the late twelfth and thirteenth centuries.[67] The most famous of these groups of religious women were the nuns of the convent of Helfta in Germany, three in particular: Mechthild of Magdeburg, Gertrude the Great, and Mechtild of Hackeborn. At Helfta, founded at the beginning of the thirteenth century by Count Burchard von Mansfeld and his wife, Elisabeth, for their two daughters, the nuns followed the Cistercian rule, without being formally accepted by the order. The nuns moved to Helfta in Saxony in 1258 and were supervised by Dominican confessors. Gertrude's brother had given the property. With 100 nuns by the 1290s, the convent became renowned for its wealth, piety, and learning. The count's daughters, Sophie and Elisabeth von Mansfeld, did miniatures; other nuns worked as scribes and illuminators.

When contemporaries criticized the well-born mystic Mechthild of Magdeburg (c. 1210–c. 1282), calling her visions suspect, she sought refuge at Helfta, where in her early sixties she became a professed nun. She had experienced her first vision, or "greeting," as she called it, as a young girl of ten and knew many more in the course of her life, one a day in one period of thirty-one years.[68] Like Christina, she subjected herself to harsh penances, scourgings, and ascetism to the point of illness, first under the modified vows of a Beguine, then as a Dominican Tertiary. She suffered from her separation from God wrought by her "small sins" when "my soul becomes so dark, and my mind so dull, and my heart so cold, that I am constrained to pray heartily and long, and humbly to make confession."[69] She had a "holy desire," for union in which her soul fought its way to the "Divine lover."[70]

Mechtild of Hackeborn (c. 1241–1299) and Gertrude the Great (1256–1302) were little girls of five, perhaps awed by the presence of Mechthild of Magdeburg, the renowned old holy woman, when they first came to Helfta. At seven Mechtild begged to be a nun. In 1291 she began to have visions. Two of the sisters took them down as she spoke, and they became *The Book of Special Grace (Liber specialis gratiae).* Mechtild ran the school for the young propertied girls who came, as she had come, to the convent for their education. With her visions, she saw herself as an adviser to the nuns as well. Gertrude the Great rose to become the abbess of the convent. Her visions began in 1281. Along with prayers and meditations in Latin that she wrote, she herself described her mystical experiences, gathered together as *The Herald of Divine Love* or *Revelations (Legatus divinae pietatis).*

Most honored and influential of these mystics within the established orders of the Church was the Spaniard, Saint Teresa of Avila (1515–1582), who struggled at every stage of her spiritual journey, with her father, with her confessors, with bishops and heads of orders, and most painfully with her own doubts. She came from the new mercantile world of the walled towns, from a wealthy Toledo merchant family of Spain. Her grandfather had been a Jew who converted. As children she and her brother played with religion, fantasizing lives as martyrs and hermits. At fourteen her mother died, and she turned to the regular pursuits of a young woman of her rank. In 1531, with her sister married, she was unhappy with her father's decision to send her to the chaperoned care of the Augustinian nuns of Our Lady of Grace in Avila.

It was there in 1537, influenced by the teaching nuns and by an uncle at whose recommendation she read Saint Jerome's letters, that she decided on a religious life. Even so she chose the Carmelites not out of special vocation, but because her friend Juana Suarez was there. It had been visions of hell that brought her to contemplate the retreat, even over her father's objections. She described the life of a nun as "the best and safest state." She planned to "force" herself to accept "the trials and hardships of being a nun," for "afterward I would go directly to heaven."[71] Once in the convent, she experienced illness and pain; she was sick for two years, so severely that on one occasion they thought her dead and placed the wax on her eyes to hold the lids closed. She spent another three years as a semi-invalid suffering periodic paralysis. Images of hell brought bodily pain worse than anything she had known in her illnesses. These, however, were nothing, as she wrote in her autobiography, next to the soul's agonizing: "a constriction, a suffocation, . . . a despairing and tormenting unhappiness . . . as though the soul was continually being wrested from the body . . ." as if, she explained "the soul itself tears itself to pieces."[72]

Relief came only with her mystical joining with God. She remembered the first time, her "conversion" as she called it. It was Lent of 1554, when she was thirty-nine. She was kneeling in the oratory before a statue of the "wounded Christ." Tears came as she thought of Mary Magdalene and her conversion. She could not pray. She imagined Jesus in the garden, Saint Augustine at the moment of his conversion, and then she received a sense "He was within me, or I totally immersed in Him."[73] But her struggling was not over. The visions came, but her confessors denied them and told her to close out the images that gave her such pleasure. These, they insisted, were not gifts from God but messages of the devil. For St. Teresa "among the very severe trials I suffered in my life, this was one of the most severe."[74] Almost ten years were spent like this, doubting the images and words, a soul "alone in the midst of so many dangers," periods when "I often turned back and was even completely lost."[75] Had it not been for a visit to the Franciscan Peter de Alcantara, himself a mystic, and his advocacy with her confessors, this last kind of torment would have continued.

Once recognized by the Church as divinely inspired, these women could experience their ecstasies of faith without doubts or fear of ecclesiastical displeasure. When they wrote of their visions, although all owed debts to male mystics like Saint Bernard and Joachim of Fiorli, together they established a unique visionary voice for women, in which they described the process of acquiring this special access to God, created their own images of the deity, and their own phrasing for the ultimate gift of union.[76]

All of the women mystics described their inner battle to conquer the will in order to be receptive to the spiritual experience awaiting them. Julian of Norwich remembered the "frequent changing of experiences between feelings of well-being and woe," the woe of temptation.[77] Mechthild of Magdeburg wrote that,

> I looked at my body—
> it was mightily armed
> against my poor soul
> with much might
> and with the full power of nature.
> I saw that it was my enemy
> and saw also that if I wished to escape eternal death I
> would have to conquer myself
> and that would be quite a battle.[78]

The agony of the separation from their God, the frustration of striving for complete surrender filled their hours. St. Teresa described this period as "the most painful of lives . . . for neither did I enjoy God nor did I find happiness in the world."[79] The means of conquest lay in complete humility; "I must open myself," were the instructions given to Mechthild of Magdeburg.[80] In the late thirteenth century Marguerite of Oingt, a prioress of the new Carthusian order (d. 1310), described the self-abnegation and the devotion required:

> You know, my sweet Lord, that if I had a thousand worlds and could bend them to my will, I would give them all up for you . . . for you are the life of my soul. Nor do I have father or mother besides you nor do I wish to have.[81]

For the women, more often than the men, it was through contemplation of the Passion, of Jesus' pain in his last hours, that they found their way to ecstatic union. In their prayers they dwelt on Jesus' wounds. For them the mass, the receiving of the sacrament of holy communion, with its symbolic re-creation of the spilling of the Savior's blood took on special significance. They glorified the eucharist, the sacrament of communion, depriving themselves of all other food and drink much as men deprived themselves of sex.[82] It was in the thirteenth century that female mystics made popular the feast of Corpus Christi, the holy day commemorating the eucharist. Mechtild of Hackeborn and Gertrude the Great saw this ritual, communion, the taking of the bread and the wine, as the spiritual and literal union with the deity, the way to participate in both his humanity and his divinity.

Unlike male visionaries, a few female mystics like Julian of Norwich focused on the feminine aspects of the divine, of God in her image, the Mother as the second part of the Trinity. As the Father is the "noble strength," so the Mother is the "depth of Wisdom," Julian wrote.[83] She described "three ways of seeing motherhood in God": as creator, in giving the "sensual" to our natures, and in "the activity of motherhood," the nurturing of our souls.[84] Julian also had a more literal image of Jesus as the mother painfully giving birth. By his death he gave "eternal life," and ever after like a suckling mother he "feeds us with Himself."[85]

More often, however, women mystics, like the male theologians who encouraged their meditations and like the male mystics, saw God as the Holy Bridegroom and union like that with a lover. Christina of Markyate described her relationship with Jesus in these terms:

> So the maiden took Him in her hands, gave thanks, and pressed Him to her bosom. And with immeasurable delight she held Him at one moment to her virginal breast, at another she felt His presence within her even through the barrier of her flesh. . . . From that moment the fire of lust was so completely extinguished that never afterwards could it be revived.[86]

Some mystics like Mechthild of Magdeburg used the images and language of courtly love to describe their Lord and the soul's encounter with him. Mechthild imagined the Lord inviting the soul to dance, the soul disrobing and approaching a bridal bed, for it was also a wedding night. She envisaged a deity "overflowing" with love into which the soul would be "melting in union," both lost in the power of the embrace.[87] The soul crying out, "Fling me down, under Thee, gladly, would I be ravished;" the Lord feasting on the soul's beauty:

> You are my resting place, my love, my secret place, my deepest longing, my highest honor. You are a delight of my Godhead, a comfort of my manhood, a cooling stream for my ardor.[88]

Others found different images to express the rapture, and emphasized the sublime nature of the mystical experience. Hildegard of Bingen had seen it as light, "far, far brighter than a cloud that bears the sun in it" and spoke of the peace it gave her "when I do see it, every sadness and every anxiety are lifted from me, so that I feel like a simple maiden and not like an old woman."[89] Mechtild of Hackeborn envisioned the Deity as a crystal bowl into which she was poured like the purest water. She remembered bright light, trees, a river, vineyards.

Saint Teresa left the most complete description of the successive steps to mystical union with God. She turned the path to the final union into a metaphor of a garden and its gardener. The garden became the symbol of the soul, the four degrees of prayer the ways in which God watered and nurtured it. Each state required less striving and struggle and more abnegation of self, more receptivity until one knew complete humility and passivity. At first the "water" had to be pulled up from a well; then a Moorish windless drew it up; then a stream ran through the garden and at last, in "rapture" the rain fell naturally, creating "the loving exchange that takes place between the soul and God. . . ."[90] The final state of ultimate receptivity gave the visionary the ecstatic "pain of love" with flame, fire, arrows, a "transpiercing of the soul," St. Teresa called it.[91] This was the moment that Bernini tried to capture in his famous sculpture of her. An angel, its face all aflame, came beside her and sent a flaming dart into her heart: "When he drew it out, I thought he was carrying off with him the deepest part of me, and he left me all on fire with the great love of God."[92]

Visions did more for these women than just give them spiritual rewards. Having been chosen by their God, they felt a confidence that empowered them to help others with their prayers and to participate in the reforming zeal of their eras. They decried the corruption of the Church and its hierarchy and worked to purify the existing religious orders. Hildegard of Bingen had set the pattern in the twelfth century, educating her nuns and speaking out, even admonishing emperors and popes. When her community was placed under interdict, she warned the churchman involved that Satan might be using him to keep her nuns from God. Mechthild of Magdeburg described the Church as a "maiden" become "filthy, for she is unclean and unchaste."[93] She described the horrors of hell and purgatory for the sinners she saw all around and wrote letters of advice and warning to religious and secular leaders. When her writings were criticized, she appealed to God—"You yourself told me to write it"—and felt reassured that in her book she held the truth. "He who would take it from My hand must be stronger than I!" she declared.[94]

St. Teresa, honored for her actions and writings in a later century with the title "Doctor of the Church," felt driven to help others through her prayers and teaching. Her writings inspired others to the spiritual life. Her books, verses, and letters influenced male theologians (like the Dominicans Garcia de Toledo and Pedro Ibanez) to seek out lives of prayer and mystical union with God.[95] St. Teresa admired the reformed foundation of King Philip II of Spain's sister, Princess Juana, in Avila. She too wanted to return to the austerity of the early Carmelite rule. With twelve nuns and a prioress, she established a community of Discalced Carmelites (the reformed branch of the order). When she met criticism for refusing to accept dowries, insisting that the nuns live on alms, the Virgin and Saint Clare appeared to her. Other visions comforted her when subsequent actions met with disapproval—the austerity of the life she set, her decision to found male as well as female houses. She wrote: "I was startled by what the devil stirred up against a few poor little women."[96] With special dispensation from the Pope and approval of the general of the Carmelite order, she went on to found other establishments according to her strict rule, thirty by the time of her death. It was in one of her friaries, at Duruelo, that the well-known mystic, Saint John of the Cross, lived.

Given the Church's traditional views of women, the ecclesiastical criticism of St. Teresa was predictable. Hildegard of Bingen had had her encounter with episcopal authority, Mechthild of Magdeburg took refuge in the monastery of Helfta to escape accusations of wrong belief. They and the other women mystics of these centuries survived and continued to exercise their authority because what they preached and advocated coincided with the concerns of their male ecclesiastical contemporaries. What latitude they were allowed came because in the last resort they always acknowledged and accepted the traditional limits. They never refused obedience to the Church.

Powerful churchmen had listened to Hildegard of Bingen and declared her pronouncements prophesy, because she articulated their fears and admonitions. By 1220 Flemish churchmen had compiled her revelations and used them as evidence in their own warnings of the consequences of corruption. St. Teresa's efforts coincided with the Catholic Reformation and the reign of the devout King Philip II of Spain. She corresponded with Saint Ignatius Loyola, founder of the Jesuits. The king encouraged and supported her foundations and added an autographed copy of her life to his personal library in the palace of the Escorial.

In every way short of denying their visions, these women were obedient to the Church. They claimed no power in their own person and always accepted their need for guidance. Never did they even claim any part in the interpretation of what they had experienced. Mechthild of Magdeburg called any comments her "reaction" to her visions; St. Teresa cloaked her descriptions and ideas about her experiences, in the phrase "it seemed to me," and disclaimed any intelligence for such an exercise, "For God didn't give me talent for discursive thought or for a profitable use of the imagination."[97] Not one of these female mystics acted without the support, advice, and approval of a male confessor. Even the women's special emphasis on the eucharist enhanced the power and glory of their ecclesiastical mentors, for only the male clergy could administer the sacrament.

Even the unorthodox St. Teresa proved her willingness to obey the male ecclesiastical hierarchy many times. In 1576 she wrote a statement for the inquisitor of Seville, explaining:

> She never did anything based on what she understood in prayer. Rather, if her confessors told her to do the contrary, she did it immediately, and always informed them about everything.[98]

When confessors had told her her visions came from the devil, she tried to resist them. When they said to write her life, she did. When one insisted, she revised her account of prayer, the *Interior Castle.* The confessor in Segovia told her a woman's commentary on the Song of Songs was improper and so St. Teresa burned it. When she came to found her own religious communities, she was as strict as the most conservative of churchmen. Though she herself traveled on mule back, her nuns rode in a coach with curtains drawn so that they would not be disturbed at their prayers, and lived a life of complete enclosure.

If a woman accepted the absolute authority of the male ecclesiastical leadership and acted within the bounds they set, she might attain sanctity within the Catholic Church into the twentieth century. Women visionaries continued to appear like Saint Margaret Mary Alacoque (1673–1675), a sister in the Parisian Order of the Visitation, who popularized the image of the Sacred Heart. By the nineteenth century, however, the Church had changed its attitude toward women who claimed mystical experience. Significantly, the churchmen made no statements about the authenticity of the visions of later mystics and focused instead on other aspects of their lives. In this way the Church rejected the potential authority of its female visionaries, honoring them for their accepting childlike innocence, not for their special ties to the deity. Even Joan of Arc, the image of vital aggressive spirituality, was transformed in the wake of these Church policies. When canonized in the twentieth century, she too became a saint not because of her visions and declarations in the name of God, but for her "exemplary life," for her virtue and pious devotions.

3

AUTHORITY OUTSIDE THE INSTITUTIONAL CHURCH

Challenges to Established Dogma and Established Orders

The religious enthusiasm of the twelfth and thirteenth centuries worked changes within the institutions of the Catholic Church. To contain its zealous women and men, the hierarchy of popes, councils, bishops, and captain generals of orders revived old doctrines, modified traditional ones, and created new definitions of the religious life for those wishing to live under the Church's auspices. With the Bull Periculoso of 1293, the Church alone defined the way nuns would live and the premises about women's potential weakness and danger to themselves and to men. Only the Church had the right to nod and ignore necessary lapses in the rule. The Church alone determined when pious individuals like the mystics had gained access to the divine and thus access to power and authority among the faithful. Only the Church could validate the divine origins of the cloistered visionary. These decisions and actions brought order and uniformity of belief and practice for a time. The Catholic faith maintained its hegemony into the sixteenth century by such measures.

To continue to maintain order, peace, and conformity outside the traditional institutional framework, the Church from the twelfth to the fifteenth century would have to make other kinds of accommodations as well. The new religious enthusiasm fired hundreds of thousands of Europeans, both women and men, with such force that it could not be contained in the sheltered environment over which the Church had jurisdiction. In these unusual circumstances, in the face of this fervor, the Church had to adapt and accommodate or lose the faithful.

Women and the woman's voice posed the most difficult challenges. For the first time, in these centuries of popular piety, the Virgin Mary emerged as a religious figure in her own right: an idealized woman, almost divine, who could not be ignored. Women visionaries came to the fore and gained followings with no affiliation to any established order. In addition, women in the thousands, numbers far beyond what both the old and new orders could admit, responded with outpourings of religious zeal. Calling themselves "Beguines," they insisted on fashioning their own life of piety not only outside the authority of the family, but also outside the institutional Church.

By a series of leaps of faith and logic, the Catholic Church renamed and refined the worship of the Virgin and incorporated the "veneration" of Mary into its dogma, thus gaining control of a female image so powerful that she at times threatened to become a deity in her own right. Similarly, the Church found ways to accommodate devout, unorthodox individual women. The Beguines of northern Europe remained a more complicated problem. Given their numbers in the twelfth and thirteenth centuries, the Church had to accept their informal vows and allegiance without question. By the end of the 1400s it could afford to be more restrictive. In each instance the ecclesiastical leaders made the same choices, the same compromise. Allow some latitude, occasionally allow the unorthodox as long as believers accepted the ultimate authority of the Church and its absolute right to define and control access to God.

Until the sixteenth century the Church's authority stood reaffirmed; the majority of devout women, whether worshippers of the Virgin Mary, visionaries, or Beguines, accepted its dictates. The Church condemned as heretics and cast out those who did not: the Albigensians of southern France, the followers of the Free Spirit of northern Europe, the Lollards of England. Then in the 1500s, in a similar resurgence of religious enthusiasm, the Catholic prelates and theologians lost their ability to accommodate, and thus lost control altogether. Christian women who sought different images and different roles left Catholicism and, along with the men of their families, found protection as the followers of Luther and Calvin. They rejected the old faith and joined the myriad of new Protestant sects.

The Virgin Mary

From the twelfth century on, the Church gave authority and validity to the overwhelming popular responses to the image of the Virgin Mary that arose spontaneously throughout Europe. Theologians and popes gradually made dogma of her veneration, accepted the hypothetical re-creations of Mary's life, endorsed the celebrations of her festivals and thus made a place for the popular need for a female aspect of the faith.[1]

Before the 1100s, popular faith had little to do with the Church's images of and pronouncements about Mary. The Gospels of the authorized version of the New Testament, the Vulgate, scarcely mention her. The Church Fathers and the Church councils of the early centuries never considered her as a significant individual in her own right; rather they used Mary in their theological debates to define and then to prove the special dual nature of her son, Jesus.[2] By the twelfth and thirteenth centuries, however, churchmen began to write more about Mary. They rhapsodized over what it must have been like to have been literally so close to God, both in the impregnation by the Holy Ghost and in the carrying and suckling of the Son, the child. In their sermons and meditations they made allegories of these experiences. To Saint Bernard, Mary's union with the Holy Ghost recalled the sexual imagery of the Song of Songs and became symbolic of the soul's union with the Savior. Mary is Israel, the new Church, the Bride of God. Churchmen wrote of her as the "miraculous container," her body compared to the tabernacle kept behind the altar that held the host, both containers of Jesus Christ.

Popular images of Mary and stories of her life gradually influenced official Church dogma. By the thirteenth century the learned discourses and the meditations of her advocates within the Church had given validity to the accounts from the Apocrypha, especially the Protevangelium of James (dated from the second or third century). In addition, popular piety and enthusiasm brought the Church to adopt one by one each of the festivals that had arisen more or less spontaneously to celebrate the events of her life.

Already by the tenth century, Hrotsvit, the learned Saxon canoness, incorporated many of the popular stories of Mary's life in a verse legend about the Virgin. Everything about the events and Mary's behavior demonstrated her perfection. She herself was conceived without sin (the "immaculate conception"); at three she presented herself at the temple, there eating nothing but manna brought once a day by the angels—the only substance pure enough for such a pure being. Mary vowed virginity, she refused marriage until God designated her spouse Joseph with a miraculous flowering of his staff. Hrotsvit portrayed the young virgin as the perfect spirit in the perfect body. "Thus Mary lived, prudent, humble, inflamed with charity of soul, affable to all, resplendent with every virtue. No human ear ever heard her speak in manner unkindly; no one knew her to offend in any way . . . a noble exemplar of universal righteousness."[3] Hrotsvit had only superlatives to describe her appearance: "It is said that Mary's face beamed with snowy whiteness, and that her beauty excelled the dazzling splendor of the sun and utterly surpassed in its loveliness every human countenance."[4] Subsequent popular portrayals continued Mary's glorification, especially in the scene of her encounter with the archangel Gabriel, the Annunciation. Writers portrayed her spinning thread; artists imagined her reading, at her prayers—the humble, innocent maiden chastely receiving the word of God.

The fruits of this coincidence of theology and popular piety about a female figure are without parallel for the women of Europe. Men built cathedrals to Mary, the Queen of Heaven—Notre Dame of Paris, of Chartres, of Rheims, of Amiens, Laon, and Senlis in France alone. Christine de Pizan, the writer of the late fourteenth and early fifteenth century Burgundian and French courts, explained the Holy Mother's significance for women; she chose the Virgin as the "defender, protector, and guard" of her City of Ladies, the redeemer of women:

> Temple of God, Cell and Cloister of the Holy Spirit, Vessel of the Trinity, Joy of the Angels, Star and Guide to those who have gone astray, Hope of the True Creation. My Lady, what man is so brazen to dare think or say that the feminine sex is vile in beholding your dignity. For if all other women were bad, the light of your goodness so surpasses and transcends them that any remaining evil would vanish.[5]

Like others, both within and outside the institutional Church, Pizan believed that Mary must enhance other women with the reflections from her glory. Such a perfect female being could, in theory, counter the negative images of women inherited from previous centuries.

Exemplars, Tertiaries, and Beguines

The Virgin Mary had other significance for Christian women. By her experiences she made real a religious alternative to the cloistered life of the nun. Like Mary, a young woman could marry, bear, and nurture a child, live as a wife and mother, yet still vow chastity and, by her devotions and her actions for others, find protection and gain honor and even sanctity from the Church. By identification with Mary, a well-born woman could find her way to God outside the family. Male leaders of the Church made this identification with Mary as well, thus giving the experience even more potential authority.

Also powerful was the image of the grieving mother, the Mater Dolorosa at the foot of the cross, or holding Jesus' lifeless body across her lap (the pietà). Contemplation of Mary's pain, explained men like the English Cistercian Ailred of Rievaulx to his sister, a recluse, could lead a devout woman to Jesus' pain, to his sacrifice, and thus to her own salvation. The religious extremists of the fourteenth century, the bands of individuals called the flagellants, sang the "Stabat Mater," a hymn to Mary standing at the foot of the cross. The believer weeps at her torment, asks that her grief and her pain be transferred to the suppliant: "Holy mother, do this for me: fix the wounds of the crucified deeply in my heart. . . . Grant me to be wounded by his wounds, to be inebriated by his cross, for love of thy son."[6] Not only did contemplation and emulation of the Virgin bring choices and salvation, it also gave authority. A devout woman's actions, her self-abnegation, her devotion, gained the respect and support of churchmen, especially the leaders of the newer orders, the Dominicans and the Franciscans. This life of the "exemplar," seemingly so different from that of her enclosed sisters, brought visions and mystical experiences that the Church also came to acknowledge as divinely revealed. This gave authority to these women just as it did to those avowed and part of the regular orders.

At the beginning of the thirteenth century Saint Elizabeth of Hungary (1207–1231) became the model exemplar, living an uncloistered life with all that it could offer the extraordinarily pious woman. Aside from popular images of the Virgin, the young Queen had the stories of other well-born women who had chosen this secular path to sanctity: women like Radegund, the sixth-century Queen of the Franks, or Etheldreda, wife of the eighth-century King of Northumbria, who had, according to her contemporary, the historian Bede, even thanked God for the cancer that led to her death, a tumor on her neck, so that she might "be absolved from the guilt of my needless vanity."[7]

St. Elizabeth, however, went to new extremes of penance and contrition. In her own lifetime she came to exemplify all the qualities that churchmen and the popular imagination sought in their pious heroines. Out of obedience to her father, she married at fourteen. She fulfilled what she described as her duty by bearing three children to her husband, Ludwig IV, Landgrave of Thuringia. On his early death, she refused to remarry, took vows of chastity, and became a Franciscan Tertiary, a member of a lay order that allowed pious women to live uncloistered. She devoted herself to prayer and to care of the diseased at the hospital of Marburg. Even when married, she had influenced her husband to help her found a hospital and infirmary for the poor. Her mentor and confessor, the Franciscan Master Conrad, encouraged her in extremes of devotion. She imposed fasts and prayer vigils upon herself. He had her whipped for missing a sermon. Stories of miracles surrounding her actions began even before her death at twenty-four. Chroniclers reported that crowds tore at her shroud, at her body, as she lay in state and as she was carried to her grave, crazed with the desire to have her relics. By canonizing her only four years after her death, the Church had bowed to the power of Elizabeth's faith and her popular veneration.

Saint Bridget of Sweden (1303–1373), also a noblewoman, made many of the same choices as Saint Elizabeth. She testified that the Virgin inspired her both by her example and by her appearances to her. The daughter of a "lawman"—the equivalent of a judge in her Swedish community—she was already betrothed at eleven when her mother died and she had her first vision, a voice speaking from the crucifix in her room. She married Ulf Gudmarsson, also a lawman, bore him eight children, supervised his household, learned Latin with her sons' tutor, and served the king, Magnus, becoming teacher to his wife Blanche of Namur. As a dutiful wife, Saint Bridget performed the acts of charity traditional to a well-born woman of the fourteenth century, but her visions continued—of Jesus, John the Baptist, Saint Agnes, and the Virgin. Mary had first appeared to her with her eighth delivery, the labor so difficult it seemed she would not be able to give birth. She believed the Virgin's touch took away the pain and brought the child. Mary came again to comfort her at the death of her favorite son, giving her peace by assuring the grieving mother that the young man would know salvation because of her goodness. Saint Bridget adopted the dress of a Franciscan Tertiary and set penances for herself. From her husband's death in 1344 she pledged herself to a life of austerity and piety.

With the blessing of her spiritual mentor, Master Matthias, the canon of Linköping cathedral, she retired to the Cistercian convent of Alvastra. There between 1345 and 1349, the visions flooded her consciousness. Petrus Olai, the subprior, acted as her secretary, taking down and translating into Latin the ecstasies as she experienced them: the Virgin and Jesus calling for an end to war, for the return of the Pope from Avignon to Rome, and her designation by Jesus as his "new bride" with promises of all the glories of the mystical union:

> Thou shalt rest in the arms of my Godhead, where there is no carnal pleasure but the joy and delight of the spirit. . . . Love also me alone, and thou shalt have all this is needful for thee, and thou shalt have it to overflowing. . . .[8]

The ecstasies Saint Bridget experienced formed the first books of her *Revelations,* which became guides and sources of meditations for both pious women and men in the next centuries of the Church. The calls to action of her visions inspired and empowered her just as they had her cloistered sisters. She established a nunnery at Vadstena. Originally she planned it as a joint community, but in the end, the thirteen priests to service the spiritual needs of the sixty nuns lived very separately, the nuns even watching the service from a choir upstairs above the main part of the chapel. She hoped to create a new order, combining the dress and begging of the Poor Clares, the fasting and austerity of the Cistercians, and the strict claustration and subordination to the confessor required by the Bull Periculoso. Given the chaos of the papacy and perhaps the imperfections of her Latin, it was not until 1370 that Pope Urban V granted her an audience. He gave her no right to a new rule, but with modifications accepted her community as an adaptation of the Augustinian pattern.[9]

Like Hildegard of Bingen in the twelfth century, Saint Bridget's visions exhorted her to another kind of action. She threatened Pope Clement VI with Jesus' anger:

> Take heed to the days of your life, when you stirred me to wrath and did what you would and what you should. But now my hour will soon be at hand when I will call you to account for all your forgetfulness. And as I have let you rise high above all, so shall I also let you descend to fearful torment of body and soul. . . .[10]

Such visions were but one of the rewards for a woman like Saint Bridget, who adopted the life and accepted the obligations of a tertiary. Under the protective cloak of her piety, her widowhood, and her visions she traveled to Rome, to Naples, to Cyprus, and finally in 1372 on a pilgrimage to Jerusalem. She condemned the vanities of women, the laxity of archbishops, cursing queens and kings for what she viewed as their sexual depravity. She saw herself like Samson letting "none of Thy enemies escape until they beg for forgiveness."[11] When contemporaries balked at the extremes of vilification, she believed that she had the authority of the voice of God who assured her that she was but his instrument: "I speak to thee, because it pleases Me, so that all may know how sin must be atoned for, punishment lessened, and the crown increased."[12]

The visions of the Virgin filled the last years of her life, coming with special vividness in her old age when she journeyed to Jerusalem. There, she saw Mary at the Nativity, "an exceedingly fair maiden . . . clad in a white cloak and a robe," and at the Crucifixion: the grieving, sorrowing mother.[13] Saint Bridget felt that she became one with the Virgin as she watched Jesus taken from the cross and experienced Mary holding him in her lap, her eyes "glazed and full of blood," "His face shrunken, His arms . . . grown so stiff that they could not be bent further down to the navel." Mary closed his eyes and mouth; others wrapped him in the linen shroud, and

> the good John came and took me home. Behind us we left the cross, standing high and alone.[14]

In addition to these royal and noble wives and mothers, some propertied women of the new walled towns also gained authority within the Church as exemplars. They too lived chastely and simply outside the cloister, receiving visions that enabled them to lead unorthodox lives. Some took modified vows like Saint Bridget of Sweden. The most famous and influential of these women was Catherine Benincasa (c. 1347–1380), who as Saint Catherine of Siena became the patron saint of Italy, honored as a Doctor of the Church. Born to a wealthy Siennese dyer, she made a private vow of virginity, refused marriage, even cutting her hair in protest, and at eighteen became a Dominican Tertiary. For three years she lived at home as a recluse under the supervision of her confessor, rewarded for her austerities and prayers not only with visions, but with the experience of the stigmata, feeling the pain of Jesus' wounds. At the age of twenty-three, with her father dead and her brothers in exile, she obeyed her visions' command "to go abroad in the world to win souls."[15] With a small group of followers she gained a reputation for piety and exceptional sanctity like Saint Elizabeth, as a result of her efforts nursing the afflicted in the plague of 1374. Stories told of how she had given her cloak to a beggar who asked for alms, explaining, "I prefer to be without a cloak rather than without charity."[16]

In the 1370s, her revelations now accepted by men as influential as Raymond of Capua, the provincial of the Dominicans (who had become her confessor), she began to act with the special confidence and power of the acknowledged visionary. She described the feeling to Giovanna of Anjou, the Queen of Naples: "After our soul has been set free by fear, it rushes along with perfect solicitude and chases away every sin and defect." In a sense, "fear clears the dishes" just as a servant would in washing them.[17] With the help of her secretary, the poet Neri de Landoccio dei Pagliaresi, she wrote numerous letters: reconciling Siennese families, admonishing cardinals and advising popes.[18] Her reputation as a zealous reformer rests on these letters. It was, however, her exemplary life caring for the poor and her mystical experiences that earned her canonization in the fifteenth century.

St. Catherine of Siena never joined an order and described herself to Pope Urban VI as one of "the sheep that are outside of the sheepfold."[19] Another townswoman, Margery Kempe of Lynne in Norfolk, England (c. 1373–after 1438?), also rejected her traditional life and followed a religious calling outside the institutional Church. The daughter of the mayor of Lynne, she married a successful townsman at twenty, but from the time of the difficult birth of her first child she was disquieted about her salvation. Jesus then appeared to her, dressed in purple silk, the "most seemly, most beauteous and most amiable." He sat at her bedside and thus was she first "calmed in her wits and reason."[20] She bore thirteen more children, but continued to crave the solace of Jesus' appearances. He reassured her that though she was no longer a maiden, "I love thee with all My heart, and I may not forget thy love."[21] Reluctantly her husband allowed her both a vow of chastity and to make pilgrimages, first in England, then to Rome and to Jerusalem, to Compostela in Spain, to Danzig in northern Poland.

With confidence in the authenticity of her visions, her "cryings" as she called them, Margery Kempe paid little attention to the proprieties of the normal townswoman's life. Most contemporaries believed her zeal excessive, found her sobbing throughout church services a nuisance, thought her fasting and extravagant gestures of charity—like giving all of her money away in Rome—the actions of a hysteric, not a religious. Churchmen and secular leaders reacted in every way to her: a sympathetic priest read with her for seven to eight years, another helped her inscribe her memoirs (begun in the late 1420s or early 1430s); others ridiculed her. The mayor of Leicester imprisoned her as a heretic; the archbishop of York accused her of preaching. The people of Lynne tolerated her eccentricities.

Even with the dictation of her memoirs and the relative acceptance of her outpourings, Margery Kempe remained an anomaly among the faithful. Only in Spain would women like her—visionaries living alone without formal vows, just informal ties to the Church through a sympathetic confessor—continue to be accepted. The *beatas* of Spain exercised prophetic authority among believers into the sixteenth century. The most famous, like Maria Díaz and Francisca Hernandez, came from peasant families, and part of their credibility came from their childlike sincerity and simplicity.[22]

The preaching of the Franciscans and the Dominicans, the encouragement of sympathetic confessors, touched other townswomen. These women, known to contemporaries as the Beguines, responded to the call for austerity, for chastity, for service to others. In the enthusiasm of the popular piety of the thirteenth century in the towns of northern France, Belgium, and western Germany, they chose to leave their families, to give up their marriages, their homes and material possessions, to come together in a life of prayer and work. The regular orders could not have accommodated their numbers, and many would not have had the endowment needed to become a nun, yet they possessed both the religious zeal and the desire for an alternative life. So in towns all across northern Europe from the thirteenth to the fifteenth century women acted for themselves, establishing their Beguinages or *Götteshäuser,* and living a pious life, separate from the Church yet under the protective mantle of the faith, honored like the *beatas* for the sincerity and the simplicity of their devotions. Historians have only sporadic records of the numbers of Beguinages founded. Estimates from property records in towns give only an indication of the extent of the movement: houses for 600 women in Strasbourg between 1250 and 1350; 169 groups established for 1,500 women in Cologne between 1270 and 1400.[23]

These women sold whatever goods they had and bought the house they lived in, or like the women in Liège and Ghent had it willed to them by their relatives. Initially the Beguines earned respect wherever they established their groups. The first groups so identified, in fact, were the followers and companions of women accepted by churchmen as especially holy; some were believed to have mystical ties with God. Throughout the thirteenth century, towns of this part of northwestern Europe boasted of these holy women: Ivetta of Huy, renowned for giving away her inheritance and establishing a hostel for pilgrims, and for serving a leper colony; Christina of St. Trand, whose body believers saw rise to the church ceiling at the service after her death; Hadewijch of Flanders whose visions became religious lyrics; and most famous of all, Mary of Oignies (d. 1213), in many ways the model of lay piety for all other Beguines.[24]

The daughter of a privileged family, Mary of Oignies had married at fourteen, but, inspired by the religious enthusiasm of the era, she convinced her husband that they should leave their life and join those caring for lepers near Nivelles in Belgium. Driven to even more extreme acts of piety, she sought refuge with the Augustinian canons at Oignies, living alone, fasting, praying, and spinning to support her few needs. Her devotions inspired both her brothers to join the Cistercians and gained her the protection of Jacques de Vitry, a member of the order that sheltered her. De Vitry believed that she had been his inspiration when he had to preach to the heretical Albigensians. He wrote her biography and after her death kept her finger in a silver case. In 1216 he gained approval from Pope Honorius for a life such as hers and for the religious communities of women that she inspired. In this way he saved these early Beguines from the Church's prohibition against new orders and from being forced to join the established convents.

Into the thirteenth century, churchmen applauded the Beguines and their piety. They spoke of their devotions under the auspices of a confessor as the way to keep untutored women from heresy, women who, as Vitry suggested, had the vocation but not the financial means to join the regular orders. Robert Grosseteste, bishop of Lincoln, thought their way of life and their work to support themselves exemplary because they kept the women from being a burden to others. Pope Gregory IX's bull glorifying virginity seemed a formal blessing upon them.

The nobility and royalty like Jeanne of Flanders, the Counts of Holland, and King Louis IX of France helped support Beguinages. Townswomen and men gave them their respect. None of the stories of abuses that colored views of the convents and their nuns collected around the women of the Beguinages. In the fourteenth century in his instructions to his young wife, the Ménagier of Paris gave over the education of his bride to Dame Agnes, the Beguine. She was to hire the servants, to give them their instructions, and she will "teach you wise and ripe behavior and to serve and lesson you."[25]

To contemporaries, the Beguines' reputation came not only from their simple piety but also from their labors. Like Martha, the sister who prepared the meals when Jesus visited her and her sister Mary, all groups expected members to spend their days in alternate periods of prayer and work. To honor this aspect of their lives, the Beguines of Strasbourg called the leader of their house the Martha. In Bruges, the women thought of labor as a form of penance, and a way to avoid sin because of idleness. They might set up a hospital, establish a shelter, and minister to the poor. More often they wove, spun, or embroidered, because in most instances it was their labor that supported them. Few came from the very privileged. Rather women joined out of zeal, willingness to live together in a pious community, commitment to prayer and work. Unlike the nuns in their cloisters, these women followed no established rule. They took vows of chastity but could leave to marry, to return to their family, or to enter one of the regular orders. They might join at any point in their lives, as young girls, as widows, as married women who brought their children along to live with the group. They might, like the Beguines of Ghent, wear gray like a habit, or more commonly, dress simply and avoid bright colors.

Heresy: The Limits of Power Through Faith

By the end of the fourteenth century, however, circumstances no longer favored the unorthodox. From the 1260s to the beginning of the fifteenth century, religious and secular authorities turned against the Beguines and eventually brought about the dissolution of the movement. Once again popes and kings wanted the order and loyalty that came from uniformity of religious doctrine and practice. Women living without endowment, with only the loosest supervision of a male confessor, reading the Scriptures together, confessing to each other, able to sustain and supervise a community outside the framework of a regular order, appeared to usurp the functions of the institutional Church and its priests. Such a life and such confidence in one's devotions seemed like heresy.[26]

Most townswomen and men acquiesced, as churchmen made associations between the women of the pious communities and the most feared heretics of the day: the Beghards, adherents of doctrines identified by the Church as the heresy of the Free Spirit. (The concept of the Free Spirit came from mysticism, the moment of union with God that freed the soul from all concern for the material and left it filled with the deity. Once so filled, followers of the Free Spirit believed themselves unable to sin again and were thus without need of sacraments, priests, and Church.) One woman, Marguerite Porete (?–1310), identified as a Beguine at her trial, did all that the Church feared from those outside its established orders and died condemned a heretic. She preached in Valenciennes, in Châlons, in Tournai, in Paris. She wrote a book of mystical devotion between 1285 and 1295, and she defied the judgment of the Church about her religious message. In 1306, religious officials in Valenciennes burned her book publicly. According to her accusers, she subsequently sent it to the bishop of Châlons-sur-Marne, and to other theologians. She was in prison for a year and a half without a trial. In the spring of 1310 the University of Paris studied her text and declared it heretical. In May, on the basis of her theoretical renunciation at the time her book had been burned in Valenciennes, she was declared a relapsed heretic, and on the first of June was burned to death by the secular authorities.

In the hands of hostile churchmen, Porete's reasoned piety, her logical, clear language devoid of the usual feminine disclaimers, her emphasis on a "religion of inwardness" suggested a faith without need of theologians and rituals. She wrote that her book, *The Mirror of Simple Souls,* had been inspired by "Love" and admonished the theologians "so learned," so hampered by "Reason" that they would be unable to receive the gift of Love or to be "illumined by Faith."[27] She pictured a community of believers by implication superior to the prelates of the institutional Church.[28] Porete outlined the seven steps by which the soul could be led to union with God, to membership in this group of believers, and then to its ultimate goal, the rise into heaven. To achieve this union, Porete saw the soul as a passive agent, needing no rituals or good works, no intermediary, just the ability to surrender to the deity, "Love," as she called it. Once joined in a kind of mystical nihilism, the soul literally had no will, nothing but "the will of God, which causes her to will all that she should will" and thus be incapable of sin.[29] The regents of the University of Paris interpreted such assertions of the power of the individual's faith as denial of the need for communion, confession, and the priesthood. Such belief in the purity of the loved soul outside the Church's guidance could be seen as a glorification of a life out of control, even a call to unlicensed sexuality. Given these attitudes, the churchmen turned Porete's meditations into fifteen charges of heresy, all that they accused the preachers of the Free Spirit of advocating.[30]

By the beginning of the fourteenth century when Marguerite Porete first spoke publicly of her spiritual vision, the local ecclesiastics had already begun to suppress the Beghards, the men of the Free Spirit, and the Beguine communities had already come under attack by association. As early as 1261, hoping to ensure the Beguines' chastity, the Council of Mainz had decreed that a woman had to be forty to join a Beguinage. In the same year the synod of Magdeburg, fearing the Beguines' independence, insisted that they must obey their local priest. In 1273 Bishop Bruno of Olmütz wrote an official letter disapproving of their lack of formal orders and their refusal to marry. Then two general Church councils at Lyons in 1274 and at Vienne in 1312 reaffirmed that there could be no new orders, assumed that these women could not continue without endowments, and condemned them for their independent devotions:

> We have been told that certain women commonly called Beguines, afflicted by a kind of madness, discuss the Holy Trinity and the divine essence, and express opinions on matters of faith and sacraments contrary to the Catholic faith, deceiving many simple people. Since these women profess no obedience to anyone and do not renounce their property or profess an approved Rule, they are certainly not "religious."[31]

In 1317 Pope John XXII published the condemnation, and the churchmen of northern Europe used it to seek out the unorthodox and to prove their own zeal by purifying their dioceses. The bishop of Breslau found fifty heretics among the Beguines. The archbishop of Cologne ordered the Beguinages disbanded. In this new atmosphere, such groups as the sixteen followers of Heylwig of Prague, who supported themselves, supervised, and practiced their own mortifications, found themselves condemned by a papal inquisition in 1332. Dominicans and Franciscans vied with each other, proving their orthodoxy by condemning the very women their orders had once protected. Thus the Dominican John Mulberg persecuted both Beguines and Franciscan Tertiaries in Basel and Mainz in 1405 and 1406.

Questioned by the Church, living in a world in which kings and lords crusaded against centers of heresy like the Albigensians in southern France, the Lollards in England, and the Waldensians of northern Italy, some Beguines like Mechthild of Magdeburg joined a convent. Others chose from among the extant rules, and having gained the approval of their bishop and acceptance from the order's general chapter, they transformed their communities into convents, most commonly Benedictine or Augustinian. Some may have found refuge with the Brethren of the Common Life founded by Gerhard Groote at the end of the 1390s, who allowed some female members to work and pray in the communities he created for men in the Netherlands, in Westphalia, and the Rhineland in Germany. But these laywomen, like the *beatas* of Spain, were the exception. To all intents and purposes, by the end of the 1300s women living outside the traditional institutional authority of the Church found themselves numbered with the heretics, vulnerable to persecution and to death.

The Beguines became suspect largely through association and because of the simple unsupervised life they had chosen. Other women openly and enthusiastically embraced doctrines and roles that contradicted all that the institutional Church had come to represent. First in the twelfth and the thirteenth centuries, devout enthusiasts fired with the same desire for salvation and sense of righteousness that had inspired the women of the early centuries of Christianity, joined groups actively hostile to the dogma and hierarchy of the orthodox faith. In particular, they responded to the preaching of the original cry of Christianity, the call for the "equality of all believers." This more than any other assertion gave women a nature like man's and promised them roles outside the family, outside the convent, roles of authority within the faith independent of their relationship to fathers, husbands, and sons.

The enthusiasts of the thirteenth century in southern France became known as Albigensians, or Cathari, rejecting not only the Church and its rituals as the work of the Devil, but also all of the material world. By her baptism and her study, a woman could become the equal of male believers, achieve the ultimate purity, and be called a *perfecta.* French noblewomen made the sect particularly significant. It was the noblewomen who used their castles as refuges for the 1,000–1,500 who claimed this highest spiritual state. Some became *perfecta,* like Furneria, wife of Guillaume-Roger de Mirepoix, and Philippa, Countess of Foix. When the king of France and the Church launched the crusade to extirpate the heresy, Blanche de Laurac and Esclarmonde, sister of the Count of Foix, directed the defense of the castle of Montségur, a refuge for believers and the last fortress to be taken. Besieged from 1243 to 1244, the defenders eventually surrendered and the 200 Albigensians were burned to death.[32] The Lollards of thirteenth- and fourteenth-century England (another group declared heretical) allowed devout female followers to read the Scriptures in English and to preach to the faithful, just like the men. Historians estimate women as one third or more of the sect.[33] Lollards also condemned the corruption and ostentatious wealth of the Church and called for an end to tithes and a simpler worship. The dissenters that survived and reemerged in the early fifteenth century as the Hussites and the Waldensians kept alive the same kinds of protests and promises of different roles and means to salvation. A woman follower of Jan Hus could, like a man, take both the wine and the bread at the mass and her Waldensian priest spoke to her in her own language.

In 1215 the Church responded first with the inquisition—groups of learned ecclesiastics sent to question and to identify those who could not be reclaimed, those who would not recant. They condemned the heretics and with the armies of the lords and kings, like Simon de Montfort who fought the Albigensians in southern France and King Henry IV of England who eradicated the Lollards, conquered and hounded down the resisters. By the end of the fifteenth century the Church's claim to absolute authority and control of dogma stood intact and confirmed.

4

AUTHORITY GIVEN AND TAKEN AWAY: THE PROTESTANT AND CATHOLIC REFORMATIONS

Religious Enthusiasm Reborn: The Sixteenth and the Seventeenth Centuries

The Church's containment of religious enthusiasm was not to last for long. At the beginning of the sixteenth century Martin Luther spoke out and called again for a "true" faith, an uncorrupted clergy, and the ways of early Christianity. Pious Christian women responded to his words, and to the words of the others who came to be known as Protestants, as they had responded in earlier centuries to those who had protested against what appeared to be the restricting and denigrating aspects of the institutionalized Church. The Protestants questioned established dogma and at first appeared to banish the limits on the female believer.

Lutherans and the English Congregationalists (called Separatists or Independents) spoke of a devout woman's individual ties to God through her faith rather than a male priesthood; Calvinists wrote of a "priesthood of all believers" suggesting equal status for the chosen whether female or male; German Baptists asserted that women were the spiritual equals of men. Other sixteenth- and seventeenth-century sects went even further, speaking of feminine aspects even of the deity. Katherine Zell, a Protestant leader and wife of Matthew Zell, commented on the Lord's Prayer:

> He is called not Lord or Judge, but Father. And since through his Son we are born again we may call him grandfather, too; He may be likened also to a mother who has known the pangs of birth and the joy of giving suck.[1]

The female aspects of God led the English Leveller women to refer, in their 1649 petition, to the dual image of Genesis: "We are assured of our Creation in the image of God, and of an Interest in Christ equal unto men."[2] A 1552 English pamphlet insisted that women's disabilities and inadequacies as observed in the past came not from her nature as had always been assumed, but from "the bringing up and training of women's life," "kept as in a prison."[3] All of the new sects translated the Scriptures and advocated literacy, regardless of gender, so that women too could participate in the process of "enlightenment" as the seventeenth-century English Quakers called it.

In response to such claims and such promises from dissenters, the Catholic Church ignored many of its own restrictions on women and called on its female believers to prove their zeal and loyalty on behalf of the "true" faith. The Church honored women's exceptional piety in this crisis and favored women's actions on its behalf even if they were unorthodox and outside the carefully defined traditional worlds of the family and the convent. Thus, for women, whether Protestant or Catholic, the roles opened to them in the early centuries of Christianity opened to them again in the spiritual chaos and competition for souls of the sixteenth and seventeenth centuries. The appeal and the possibilities were the same. Once again women could use their political power to protect their new faiths. Once again women could preach, could prophesy, could die the honored death of the martyr. For the second time in history, religious faith gave Christian women access to authority within their churches and the opportunity to act as the spiritual and actual equals of men.

Queens, Princesses, and Noblewomen

Queens, princesses, and noblewomen played many roles in the dissemination and protection of the "true" faith just as they had in the days of the early Church. As Christianity spread across Europe in the seventh and eighth centuries, a royal or noble marriage had often been the means of conversion; so it was again in the sixteenth century. Katherine of Mecklenberg is credited with converting her husband, Duke Heinrich of Freiberg, to Luther's views. Royal wives, princesses, and noblewomen throughout Europe used their position to advance their beliefs. Ursula of Munsterberg worked for the appointment of a Protestant chaplain and brought Luther's tracts to her husband's court. Elisabeth of Braunschweig (1510–1558) is credited with the duchy's Lutheran role at the Diet of Augsburg in 1530. Converted after hearing pastor Antonius Corvinus preach in 1538, and appointed regent on her husband Duke Erich's death, she changed the official state religion in only five years.[4] In England two of Henry VIII's wives openly espoused more radical Protestantism than the King was willing to accept. Anne Boleyn, his second wife, is supposed to have gained the reformer Hugh Latimer his appointment as bishop. Katherine Parr, the sixth wife, quietly attempted to make the King's Anglican Church less Catholic.[5] Maria of Hungary and

Isabella of Denmark, though sisters of Charles V, Holy Roman Emperor and secular protector of Catholicism, favored the Protestant cause. Queen Maria openly entertained Lutherans and Lutheran ideas at her court and warned their brother Ferdinand of the dangers of persecuting the dissenters: "To tell you the truth, I would not advise you to touch [them] . . . for these people go to martyrdom with great joy."[6] Queen Isabella publicly opposed her brothers and their faith by taking the wine with the bread at communion (a privilege usually denied Catholic laity) at the Diet of Nuremberg. Renée of Ferrara (1510–1575), daughter of Louis XII of France, remained a Protestant militant when she married the Catholic Duke Ercole, and moved to his court in northern Italy. Until 1542 she posed religious and political problems for him by receiving schismatics (Calvin took refuge with her for one month), by trying to intervene to protect condemned heretics, and by refusing to hear mass. In March of 1554, the Ferrara Inquisition wanted to question her. Confined to the palace by her husband, she gave up the more extreme aspects of her behavior, but accepted only the outward forms of Catholicism.[7]

As in the centuries of early Christianity, Europe's well-born women, both Catholic and Protestant, helped to found new orders and became patrons of religious leaders. The Italian noblewoman Caterina Cibo (1501–1557), and her husband, the Duke of Camerino, used their influence with Pope Clement VII to gain acceptance for the Capuchins (a "purified" Franciscan order founded in 1528).[8] Catherine Willoughby, Duchess of Suffolk (1519/20–1580), gave money to Protestant reformers in Germany and Italy for hospitals, and helped over 1,000 religious refugees from Germany. The seventeenth-century Quaker, Anne Conway, opened her house to George Fox and other leaders of the Society of Friends.

When necessary, as in France in the sixteenth century, noble and royal women played significant roles in the wars that arose out of the political and religious conflicts of the sixteenth and seventeenth centuries.[9] Eléonore de Roye, Princess de Condé, in 1562–1563 and Jacqueline de Rohan, Marquise de Rothelin, in 1567–1568 actively mediated on behalf of the Protestants. After the massacre of French Protestants on St. Bartholmew's Eve in 1572, women like Louise de Coligny made their domains refuges for Huguenots (the name given to the French followers of John Calvin). These women never lost their zeal. In 1628 at the age of seventy-seven, Catherine de Parthenay, Madame de Rohan, led the resistance of the besieged Huguenot stronghold of La Rochelle.

Queens holding all the power in their own hands, like Mary Tudor of England (1516–1558) and Jeanne d'Albret of Navarre (1528–1572), were in a unique position. As rulers in the sixteenth century they could do more than encourage and protect; they were able to dictate and enforce belief.

Mary Tudor's devotion to Catholicism and her determination to reinstate it on her accession to the English throne came from a conviction and stubbornness equal to that of the early Christian martyrs. She was fifteen when her father Henry VIII divorced her mother and had Mary declared illegitimate. With the birth of his second daughter, Elizabeth, he tried to force Mary's submission to his Anglican Church and to the renunciation of her title and position. No longer the heir, separated from her mother, her Spanish companions banished from the court, unable to appear in public, unable to flee, she remained adamant in her refusal of all of her father's demands. Fearing for her life, her cousin, the Holy Roman Emperor Charles V, encouraged her to give in. Her mother's death in 1536 broke her spirit. She capitulated, acknowledging Henry as head of the Church, denouncing the Pope, and agreeing that her mother's marriage had been "incestuous and unlawful."[10] She hoped that the Pope would absolve her of her actions, and asked Charles V to write his approval.

On the accession of Henry's son as King Edward VI in 1547, Mary was thirty, stocky, dark-haired, and dark-eyed with an air of strength, but no beauty. She was his sister princess but an anomaly at his Protestant court, with no influence and no prospect of marriage. She sought solace in her Catholic faith but the king's council would not allow her to celebrate mass even in private. While she and Charles V considered her escape, her chaplain was arrested. In May of 1553 when it seemed Edward was about to die, the succession was altered by act of Parliament to eliminate her altogether from any claim to the throne.

When Edward VI died in July, Mary, fired by a sense of divine intervention, gathered support as Henry VIII's rightful heir, marched with her troops in the streets of London and prevented the Protestant nobles from interposing her young cousin, Lady Jane Grey, as queen. Her adviser, the Catholic Cardinal Pole, called her "God's handmaiden" and compared her to the Virgin.[11] She saw herself in much the same way, as God's agent, the means of restoring the "true faith" to England. She released Catholic prisoners, made attendance at the mass mandatory, gave back the Church lands held in the name of the crown, and systematically persecuted those who refused to conform—an estimated 273 were executed during her short reign.[12] (These executions occasioned the Protestants' name for her, Bloody Mary.) Yet Mary knew that such measures would not ensure Catholicism in England. If she did not want all to disappear when she died, she knew that she must have an heir.

By 1554, just a year after her accession, she had already chosen her husband, a twenty-six-year-old Catholic, Prince Philip of Spain, heir to Emperor Charles V. He arrived at Southampton on July 20; they married five days later in Winchester cathedral. She gave him the status of king before her. He stayed just over a year, to act the part of husband. At the end of four months Mary believed—in the language of her day—that she was "quick with child," and allowed the court to celebrate the first movement of the fetus.[13] Six weeks before the estimated delivery date, she retired to Hampton Court to await her confinement. Spring, then summer came, almost a year passed. According to tradition, Mary left tearstains in her breviary (prayer book) on the page with the prayer for a woman's safe delivery in childbirth.

Philip waited out the supposed pregnancy until August of 1555 and then left for Flanders. He had come to dislike the court and the Queen, and now only used her to support his own projects. By the spring of 1557, by then King of Spain, he had persuaded her to make war with him against the French and to contribute £100,000.[14] Tragedies multiplied. The enemy took Calais from the English forces in July. By the fall of 1557 Mary had accepted that the swelling must be illness, not the infant and heir that she had prayed would protect the faith. She appears to have given up. She ceased her official activities, lay in her room, made her will, and planned for her burial. On her death her subjects welcomed her Protestant sister Elizabeth with the same zeal they had welcomed Mary the Catholic.

Elizabeth Tudor (1533–1603) took the throne as Henry VIII's heir just as Mary had. Subjects favored the dynastic line intact over civil war. Unlike Mary, however, she wanted no religious revivals, no calls for conversion. She made her kingdom Anglican more for political than religious reasons. Only as an Anglican was her mother's marriage to Henry VIII valid, and her own birth and thus claim to the throne legitimate. Her parliaments legislated religious conformity, but she asked for no more than outward observance. When she acted against dissenters, either Catholic or Protestant, she did so because their actions constituted treason. Religious faith was never the key motive for Elizabeth I.

In contrast, Jeanne d'Albret, Queen of Navarre in her own right (1528–1572), had felt from her early adolescence a call to the Huguenot beliefs. She had protested a marriage arranged for her with the Catholic Duke of Cleves when she was only fourteen. They had to carry her to the altar, and the couple never had intercourse. When the reasons for the alliance faded, the marriage was annulled. Though headstrong, she feared to mention her religious doubts. As a child she had seen her father strike her mother kneeling at her Protestant prayers. Jeanne d'Albret's second marriage to Antoine de Bourbon was more to her liking, and though he actively opposed her conversion and was unfaithful, after his death she wore mourning for him for the rest of her life.

Only after her parents' deaths did her indecision end and the last restraints on her convictions disappear. In all likelihood it was the preaching of Calvin's successor that gave her the final courage to break with Catholicism. From 1560–1563 she was besieged with pressures to abandon the open avowal of her Huguenot faith. Her husband allied himself with Catherine de' Medici, openly opposing Protestantism. He imprisoned her in her quarters, took their eight-year-old son Henry to Paris for Catholic education at court, and threatened to divorce her. Philip II of Spain tried to have her kidnapped; the papal legate threatened her with Spanish invasion; the Pope excommunicated her and placed her country (Navarre) under interdict. She wrote in one of her poems "Jésus est mon espérance"—"Jesus is my hope"—and this belief sustained her as it had other Christian women before her.[15] Like the legendary Christian heroines, she invoked God for her side when she answered Cardinal Armagnac, the legate sent to examine her: "Your feeble arguments do not dent my tough skull. I am serving God and He knows how to sustain his cause."[16]

Unlike Mary of England, Jeanne d'Albret gradually turned her kingdom to the Huguenots, tempering her enthusiasm for the sect and the protection of her coreligionists with a desire for tolerance and for an end to the bloodshed initiated by the political leaders of both groups. She acted as a link between the Catholic and Protestant sides of the religious war. In 1561 she made the two religions equal in status, issued toleration edicts from 1564 on, legalized marriage in the Huguenot church, forbade a father or husband to force conformity, allowed Huguenot preaching and the replacement of retiring Catholic priests by a Calvinist pastor. When religious rebellions came to Navarre, she gave amnesties and hanged the commander who had allowed his soldiers to rape women.

Though she had to flee her kingdom in 1568 and joined the Huguenot forces at La Rochelle, she continued to act as sovereign until 1570 and to make appeals for peace to Queen Catherine de' Medici and her son, King Charles IX, from the besieged citadel. She spoke of the renown that would come to them "if you will give to one religion as to the other a general repose."[17] She wrote Catherine in December of 1570 that she could not "believe that you would reduce us to the point of having no religion." "With tears and utter affection" she begged for an end to fighting. "Have pity on so much blood already shed which you can staunch with a word."[18] Perhaps influenced by her entreaties, the two sides made peace. When she returned to Navarre in August of 1571, she made the kingdom officially Protestant but allowed the Catholic faith to continue. She made no objection to her son's marriage to the King's Catholic sister, Marguerite of Valois. She died in June of 1572 and thus was spared the Massacre of St. Bartholomew's Eve, when thousands of Protestants were slaughtered in the name of conformity and order.

Royal women inevitably played the roles of both actors and victims in the religious and political upheavals of the sixteenth and seventeenth centuries.

Protesters, Proselytizers, Unorthodox Nuns, and Martyrs

In the religious enthusiasm of the sixteenth and seventeenth centuries, women did not have to be well born to be inspired and active in the name of the faith. In fact, the possibilities created by the division of Western Christianity into a multiplicity of sects made it impossible for men of any religion to curtail the energies of their women. And in the first decades the men did not want to stop them. Popes applauded Queen Mary Tudor, Calvin congratulated Renée of Ferrara. So it would be with women of lesser rank. The male religious leaders in their initial zeal condoned all manner of unorthodox roles, actions, and claims to spiritual authority that they would normally have decried and forbidden.[19]

Protestant Women

In the sixteenth century, women like Petronella van Praet and Elizabeth van Repenbirch joined male dissenters in leading opposition to Spanish rule in the Netherlands. Van Praet even advocated rebellion. Protestant women in Switzerland defied town religious ordinances and preached to nuns, encouraging them to leave their orders; the Catholic women did the reverse, encouraging them to stay and, after Calvinism had been instituted, finding ways to baptize their children in the old faith. Women both Catholic and Protestant in towns in Switzerland, France, Scotland, and England ridiculed the clergy and services of the other sect, disrupted religious processions, demonstrated, and marched in the hundreds, rioting and even killing in Geneva, in Toulouse, in Paniers, in Rouen, in Edinburgh. In seventeenth-century London, Leveller and Quaker women petitioned for the release of those of their sects imprisoned for the faith. Catherine Chidley asserted woman's right to this male prerogative in May of 1649 when she requested from the House of Commons the release of John Lilburne, the leader of the Levellers: "Have we not an equal interest with the men of this nation in those liberties and securities contained in the Petition of Right, and the other good laws of the land?"[20] In July of 1659, 7,000 Quaker women presented a petition to the House of Commons against tithes.[21] After the Restoration of Charles II to the throne in 1660, Quaker women leaders, like Margaret Fell and Elizabeth Taylor Stirredge, went straight to the king.

The thirteenth- and fourteenth-century heresies had left a tradition of women preaching. One of the condemnations of the Albigensians (the Cathari) in the 1190s had been that they allowed their women to preach. The Waldensians had acknowledged some women priests. Among the Lollards, from their beginnings in the fourteenth century into the sixteenth century, there are examples of women from all parts of society who were permitted many functions. Peasantwomen and townswomen alike opened their homes for gatherings, read and taught, spoke out, preached with their husbands and alone, proselytized and made converts. In 1389 Alice Dexter led a congregation in Leicester with her husband and another male leader. Joan and William White traveled the Waveney valley together, converting as they walked. In the persecutions instituted by Henry VIII early in his reign, women like Alice Harding, Alice Atkins, Agnes Edmund, and Isabel Morwin were called to account for speaking against confession and pilgrimages, for having English Bibles, for teaching their family and friends.

Churchmen and kings had crushed these heresies, women had ceased to preach, but once again in the mid-sixteenth century Protestant women spoke for their sects, once again asserting a privilege traditionally reserved to men. Argula von Grumbach (1492–c. 1563), a member of the minor German nobility, found that once she had converted to Lutheranism, she could not be silent even though it meant her husband would lose his livelihood as prefect. She wrote to her cousin:

> I can't help it. God will feed my children as he feeds the birds and will clothe them as the lilies of the field. . . .[22]

As had happened in the first centuries of early Christianity, as had happened with the heresies of the thirteenth and fourteenth century, the needs of the new faiths favored allowing women the unorthodox roles of minister and religious teacher. Some began by helping their husbands. Regina Filipowska in Poland led her husband's congregations with him and wrote hymns for them to sing. Katherine Zell of Germany (1497/8–1562) first spoke beside her husband, Matthew, from the pulpit in 1527 when he had been excommunicated for marrying her. Her two children died as infants, and so she gave all of her energies to her husband and their faith: appealing to the peasants in the revolt, sheltering refugees, writing pamphlets and hymns, always working to reconcile the sects that emerged in Strasbourg, hoping that this town could be the bulwark of "religious liberty."[23]

Perhaps because there was a relatively recent tradition among the Lollards of women active in the faith, it was the English Protestant sects of the seventeenth century that allowed their female believers the most authority; the more radical the sect, the more unorthodox the roles women assumed. Congregationalist and Baptist leaders saw women as independent of their husbands, able to speak without their approval when religion was at stake. As

Separatist John Robinson explained, "in a case extraordinarily, namely where no man will, . . . a woman may reprove the church, rather than suffer it to go on in apparent weakness."[24] Their women preached, whether in exile during Queen Mary's reign in the sixteenth century, or in England in the sixteenth and seventeenth centuries. Lucy Hutchinson, an Independent (the sect of Oliver Cromwell, England's Lord Protector), claimed that her preaching career began when she was six and spoke to the family maidservants. Catherine Chidley, the Leveller, did everything from organizing congregations and petitioners to writing pamphlets and challenging ministers of other sects to debate.

Alone of the major Christian sects, the Society of Friends, or Quakers, of seventeenth- and eighteenth-century England most unqualifiedly endorsed the wide range of activities taken on by their women. The "inner light" made Quaker women the equals of men, and they were embraced as "helpmeets" in the struggle for the new world of "brotherhood." Like Katherine Evans, they preached with their husbands; Elizabeth Leavens gave up her first child to the care of others, believing as did her husband "that shee may continue in the ministry which the desiar of our souldes is."[25] They distributed pamphlets and wrote tracts themselves. At forty-five Barbara Blaugdone fasted six days in prison to prove that "man" does not live by bread alone. Elizabeth Hooton and Elizabeth Williams were the first women Quakers to be flogged, whipped out of town for preaching at Oxford. Women like Anne Austin, Elizabeth Hooton, Barbara Blaugdone, and Mary Fisher left their families and friends to preach in other countries, as far away as Turkey and Barbados.

No one exemplified better the way in which the faith might empower a woman within the Quaker movement than Margaret Fell (1614–1702). George Fox, the leader of the Quakers, himself had brought her "enlightenment" when he came and preached to her household in June of 1652. She had been spiritually troubled for a long time and "often feared I was short of the right way," but when she heard him speak, it was as it must have been for so many other women:

> I saw it was the truth, and I could not deny it; and . . . it was opened to me so clear, that I had never a tittle in my heart against it; but I desired the Lord that I might be kept in it; and then I desired no greater portion.[26]

She could not restrain herself, and at the next service in her parish she began to cry; she stood during the sermon and had to be put out. Her husband returned home to reports that she had been bewitched, but that evening sitting with her she remembered: "The power of the Lord seized upon me; and he was struck with amazement, and knew not what to think. . . . And the children were all quiet and still, and grown sober . . . and all those things made him quiet and still."[27] It was Fox himself who by speaking with her husband spared her the choice she feared as a consequence of her conversion: "that either I must displease my husband, or offend God."[28] In the end, he allowed her to keep her faith, and to use the house for the meetings of the Friends, though he never converted.

With his death in 1658 Margaret Fell became the very center of the new Quaker movement. With Fox and his followers always traveling, often imprisoned, she became the one fixed element, and Quakers called her "the nursing Mother" of the church. Ann Audland wrote her from prison in 1655, "My dear and near and eternal mother by thee I am nourished."[29] John Camm called her comforting letter "full of marrow and fatness . . . it was like balm unto my head."[30] William Caton, part of her household, wrote to her, "Hast thou not often eased me when I was burthened and heavy laden, fed me when my soul hath been hungry, yea clothed me when I was naked, and strengthened me when I was weak and feeble."[31]

She nurtured the movement in more practical ways as well: supervising the finances, finding the money to support the ministries, touring the meetings in 1663 when Fox was imprisoned, even suggesting designs for the houses they would use for worship. After Fox's death in 1691 (they had married in 1669), she tried to reconcile the factions within the sect, decrying the concern of some for outward appearance with their emphasis on gray uniforms. After the Restoration in 1660, she left the running of the house, Swarthmoor Hall, to her daughter and spent more time writing and speaking with the king, Charles II, on behalf of imprisoned Friends, on behalf of Quaker beliefs. She even suggested that members of the royal family convert. While imprisoned herself, she warned Charles II of "wicked councillors" and threatened "the judgement of eternal misery" for him. She gave the remedy:

> The desire of my heart is that you may take these things into consideration betime before it be too late; and set open the prison doors and let the innocent go free, and that will take part of the burden and guilt off you, lest the doors of mercy be shut to you.[32]

His agents did not escape her pleas and admonitions. She had called the local justice of the peace who was harassing her "a caterpiller, which shall be swept out of the way."[33]

The King kept her in prison from 1664 to 1668 and again from 1670 to 1671. Some women took their spinning wheels to prison; Margaret Fell spent the years in Lancaster Castle thinking and writing. In the course of her life, she produced sixteen books and pamphlets. Most important for other Quaker women, she wrote about the equal nature of women and their equal right to preach the gospel. Published as *Womens Speaking* in 1666, the tract began with her assurance that woman had also been created in God's image and had, even though weak, been chosen for his purposes. Those who denied women this role were acting for the Devil, fostering and fulfilling the enmity God had decreed between woman and the serpent. For Fell woman was the symbol of the Church, thus those who "speak against the Woman's speaking, speak against the Church of Christ, and the seed of the Woman, which seed is Christ."[34] She noted all the occasions in the Old and New Testaments when "God made no difference, but gave his good spirit, as it pleased Him, both to Men and Women" and all the occasions when Jesus favored women.[35] Even the traditional arguments of Paul and his follower in the Epistle to Timothy left her undaunted. The churchmen had taken the injunctions that women should be silent, cover their heads, be subject to the man out of context, she explained. Paul meant only the careless, vain, confused women in those congregations, he never meant it for all women and certainly not for those who "have the Power and Spirit of the Lord Jesus poured upon them."[36] In fact, for Margaret Fell, women preaching proved the beginning of the millennium, the end of the era of the "false Church," and evidence for all to see of the arrival of "the True Church."[37]

Catholic Women

The Catholic Church, even in the extreme circumstances of the sixteenth century when all strong voices were needed to counter the zeal of the Protestants, never faltered in its prohibitions against women preaching and ministering to the faithful. The Council of Trent in the 1560s affirmed as dogma the special, separate nature of the priesthood and what appeared its logical corollary, that the clergy must be by definition male. Yet the Church did make concessions to the crisis of faith. In the sixteenth century, popes broke the injunction against new orders of 1215 and established male groups. Similarly in the sixteenth and seventeenth century, the Catholic Church allowed new groups of nuns to be organized: avowed women were allowed to teach, avowed women were allowed to serve the needs of the sick and the poor outside the cloistered sanctuary of the convent.

Angela Merici (1474–1540) became a Franciscan Tertiary in her late fifties, but the life of a lay sister, praying and serving the poor, did not satisfy her sense of divinely inspired mission. Instead she envisaged a new order of nuns, the Ursulines, brought "together for the service of His Divine Majesty," as she described it in the rule she wrote for them. She was no less militant in her enthusiasm than the Jesuits: "let us live so that, like Judith bravely cutting off the head of Holofernes, we may cut short the deceits of the devil and happily enter our heavenly country."[38] The Jesuits taught, so would her nuns, not in the manner of the secularly minded Humanists of her day, but, she explained, in the proper "Christian Spirit." The nuns should know each little girl in their care, for this was a battle for souls. A failure and "God would demand a strict account from you at judgement day."[39] In particular, the nuns must protect their charges from "the trickery of worldly people . . . from heretics," from men, "never mind how spiritual they are," and "idle women" who might be the very ones to lead the gentle innocent to unchastity.[40]

Though Saint Angela Merici described her nuns in the traditional language of the Church as brides of Jesus and potential queens in heaven, the rule she created for them differed from approved orders in other ways. Her followers wore no habit, only simple black or brown linen or serge clothing. They observed the canonical hours, took vows of chastity and submission, but did not remain cloistered. Until her death in 1540 this life of teaching, this apostolic model like the Franciscans, survived. Other devout women, first in her town of Brescia, then in Parma, in Venice, and into northern Europe established houses like hers, and the Ursulines came to be the principal means to educate young Catholic girls in the seventeenth century, the basis for all future Catholic teaching orders to come.[41]

In the seventeenth century other pious Catholic women claimed the right to an unorthodox life in the name of service to the ill and the poor. As with the teaching orders, their versions of the apostolic life, the Order of the Visitation and the Sisters of Charity, conflicted with traditional rules, requiring relaxation of the provisions on dress, hours of prayer, and cloistering. Because of the needs of the time, women were allowed a modified rule and a life of service outside the cloister. Baroness Jeanne-Françoise Fremyot de Chantal (1572–1641), daughter of the President of the parlement of Burgundy, was left a widow at twenty-eight with four surviving children. Retiring to the country, she cared for her family and the people of her husband's estates. In 1604 she met Saint Francis de Sales, and in 1610 she headed his first female house, devoted to teaching and nursing the sick at Annecy in Savoy. By his death in 1622 they had created the Order of the Visitation and had established nineteen houses. By her death in 1641 there were eighty-eight.[42]

Similar needs and dispensations made possible what would become the principal nursing and charitable order of France, the Sisters of Charity of Saint Louise de Marillac le Gras (1591–1660). (From the thirteenth century in Italian and French towns the Church had sometimes ignored regulations on cloistering and had favored having nuns serve in the public hospitals for the poor. For example, from the 1240s Augustinian sisters cared for the female patients and did the housekeeping for the Hôtel Dieu in Paris.) Saint

Louise became the principal agent and supporter for Saint Vincent de Paul in his efforts to minister to the poor of France. A member of the minor nobility, she had been raised by an aunt, a nun at the distinguished convent for the well-born at Poissy. Her desire to be a Capuchin was frustrated when they refused her because of her ill health. She married well at twenty-one, but her husband died in 1625, leaving her with one child. The next year she became part of Saint Vincent's following at court.

De Paul had first inspired well-to-do women in his parish of Châtillon-en-Bresse. As early as 1616 these women had begun to try to fill the glaring gaps in society's care and concern for the indigent and the sick caused by the religious wars and the secularization of institutions in France. Women at the French court like Mme. de Miramion and Charlotte de Ligny used their own houses as orphanages and as centers for feeding the poor. Mme. de Goussault organized visits to the Hôtel Dieu (the Parisian hospital for the poor), Mme. de Lamoignon, visits to the debtors' prison. Louise de Marillac became the spirit and the means behind the expansion and regularization of Vincent's movement. By the time of her death, in 1660, she had helped to establish and had supervised groups like those of the French court, called "Ladies of Charity," throughout France.

No one suggested that these groups constituted an order. The noblewomen chose to dress simply, to meet for prayer and spiritual exercises, and they took no vows. It was because of criticism from fathers and husbands about their going into the street to see the poor that De Paul brought in young peasant women to accompany them and to help them in their work. In 1642 he allowed the first four countrywomen to take a vow and thus began the "Daughters," later the "Sisters of Charity," the most important nursing order of Catholic Europe in the eighteenth and nineteenth centuries. At first the young women lived as "congregations" in the home of one of the Ladies of Charity, promising obedience and wearing simple gray dresses and white headdresses instead of a habit. Vows had to be renewed each year, and this made it possible for De Paul to avoid the papal requirement that professed women be cloistered. When questioned about this unorthodox arrangement, he assured others, "Obedience shall be her enclosure, the fear of God her grate, and modesty her veil."[43] In another way his Sisters of Charity differed from other orders, for from the beginning their service to the poor took precedence over their spiritual life.

From 1655 on the Sisters of Charity functioned separately from the Ladies of Charity with Louise de Marillac acting as their Mother General. Gradually the Sisters took over the visits and activities of the well-born as the court zeal waned. By De Marillac's death, there were 350 Sisters and seventy establishments in France and Poland.[44] They nursed at institutions like St.

Denis in Paris and on the battlefields of the Fronde. They tended those enslaved in the galleys, the foundlings in the orphanages, and taught young children. In the 1650s, requesting Sisters to go to Calais in the midst of the religious wars, Queen Maria Teresa wrote of them and the role that women have performed both in and out of such orders:

> Men go to war to kill one another, and you, sisters, you go to repair the evils which they have done. Oh, what a blessing! Men kill the body, and very often the soul, and you, you go to restore life, or at least by your care to assist in preserving it.[45]

Along with the other groups of nuns, the Sisters of Charity were disbanded in France at the end of the eighteenth century at the time of the Revolution, only to be reconstituted again. They were the first order to be reinstated by Napoleon Bonaparte in 1799. Their numbers and activities continued to increase in the nineteenth century. By the 1850s over 100,000 Sisters nursed in prisons and military wards as well as their own hospitals. In 1841 papal approval had been given for another group of uncloistered nuns, the Sisters of Mercy, an order very similar to the Sisters of Charity. For although the attitude of kings and republics toward the Catholic Church might change, the need for nursing and charity remained.

Martyrs

As in the years of the early Church, the association in the sixteenth and seventeenth century of religious dissent with political disorder, with disloyalty to a secular ruler, meant that women and men suffered persecution. It made no difference what sect ruled, both Catholics and Protestants acted in the name of the "true" faith, to extirpate the "heresy" with violence. For a woman to be among the most outspoken, the most spiritual, one of those seizing opportunities to act for their faith, could involve great risk. It meant death by burning for Agnes Grebill in England in 1512; it meant confiscation of her livestock for the Quaker Loveday Hambly in 1662. It meant prison for the Spanish mystic Isabel de la Cruz, torture for Maria Cazalla when after reading Erasmus she questioned the value of the sacraments. An active religious woman could become the victim of the mass killings: a French Huguenot slaughtered in 1572 on St. Bartholomew's Eve; a German or Swiss Anabaptist, one of hundreds killed in the sixteenth century, herded onto boats, and set adrift to sink in the deepest part of the river.

In their deaths, in this role as martyr, these Catholic and Protestant women could become sources of spiritual strength and renewal just like the sainted heroines of the early Church for those who remained alive. Among the Protestants, in particular, their lives became legends as hallowed as those of the early saints, and with the printing press and increased literacy, everyone could read about them and admire their glorious ends. Jean Crespin wrote his stories of Huguenot deaths in France, John Foxe did the same for the early Lutherans of Mary Tudor's reign and the Dutchman Thieleman van Braght wrote *The Bloody Theater* (also called the *The Martyrs' Mirror*) so that none would forget the deaths of the Anabaptists, the sect universally persecuted.

The traditions of the early Christian martyrs were repeated in the stories chronicling these women's lives. Portrayed as heroines, they showed the same intelligence and simple honesty in answering their accusers and did not falter when tortured. Elizabeth Young's responses in her thirteen examinations in 1558 seemed so clever that one questioner doubted that a female was speaking. In 1556, Joan Waste, a blind young woman, innocently agreed to abjure if the bishop interrogating her would promise to speak for her on the Day of Judgment. He considered the proposal and then refused. Similarly ingenuous, Maria Weber, called to questioning in 1575 for refusing to go to church for two years, excused herself "on the ground that she has to cook for the family."[46] Others brazened it out. The English young woman Rose Allin went on answering her accuser while he burned the back of her hand with the candle and never cried. "He had, she said, more cause to weep than she."[47] Questioned on the sacrament in April of 1556 Katherine Hut responded, "I deny it to be God; because it is a dumb god and made with men's hands."[48] A widow from Tour declared, "I'm a sinner, but I don't need candles to ask God to pardon my faults. You're the ones who walk in darkness."[49]

The Dutch Anabaptist, Elizabeth Dirks, a former nun who had learned Latin in the convent, was arrested in 1559. In her interrogation, she quoted Scripture as she denied the sacrament, the mass, baptism, the power of the priesthood. Torture with thumb and leg screws did not make her recant or name any of her fellow believers. Her response spoke for all the martyred women. The churchmen admonished her: "You speak with a haughty tongue." She replied, "No, my Lords, I speak with a free tongue."[50]

For these women gloried in their new faith and the salvation their persecution assured just as their predecessors had. The Anabaptist Elizabeth Munstdorp embraced the prospect of her death:

> I must now pass through this narrow way which the prophets and martyrs of Christ passed through, and many thousands who put off the mortal clothing, who died here for Christ, and now they wait under the altar till their number shall be fulfilled.[51]

The wife of the French Protestant martyr Langres called out to her husband from the fire, "My friend, if we have been joined in marriage in body, think that this is only like a promise of marriage, for our Lord . . . will marry us the day of our martyrdom."[52] Hellen Stirke wanted to die with her husband and so had to give up her newborn child. She called to him: "Husband, rejoice, for we have lived together many joyful days; but this day, in which we die, ought to be most joyful unto us both because we must have joy forever; therefore I will not bid you good night, for we shall suddenly meet with joy in the kingdom of heaven."[53] Lady Jane Grey, according to Foxe, told her sister Catherine:

> And as touching my death, rejoice as I do, good sister, that I shall be delivered of this corruption, and put on incorruption. For I am assured, that I shall, for losing of a mortal life, win an immortal life.[54]

Elizabeth Munstdorp had been married only half a year and was pregnant when she and her husband were taken in Antwerp. She bore the child in prison and gave her up to her father and stepmother's care with 30 guilders and 1 gold real.[55] From prison she wrote to her daughter, Janneken, then only a month old, commending her to God's protection "in this wicked, evil, perverse world."[56] She encouraged her daughter not to be ashamed of her parents, to keep the letter "for a remembrance."[57] Her greatest fears were for her daughter's soul, that she might not "know the truth when you have attained your understanding"; her greatest hope was that Janneken would know "the glory" of martyrdom. "Do you also follow us, my dear lamb, that you too may come where we shall be, and that we may find one another there."[58] For Elizabeth Munstdorp persecution and death brought union with loved ones and the equality of believers. Women of extraordinary faith could like men die for their religion.

Limiting Roles: The Woman's Proper Place

In the sixteenth century the women of the Anabaptists first discovered what women of the other Protestant sects would soon learn: that these new Christian sects welcomed their enthusiasm and their zealous efforts, but only in the early days. Already in 1529 one of the Anabaptist leaders opposed women teaching or preaching, another condoned beating by a husband, another wanted them silent altogether. The Munster group of the 1530s spoke of death as fit punishment for a disobedient wife and did not allow the women to participate in the choice of their ministers.

Just as in the second and third centuries when Christianity had fought for converts and for recognition, in the seventh and eighth centuries when Rome had wanted its beliefs spread across the Germanic and Celtic kingdoms, so in the sixteenth century women knew a brief interlude of "equality," of roles beyond those allowed by the inherited traditions of European culture. Soon enough, however, the male leaders of the new Protestant sects contended for authority, claimed exclusivity for one truth over another. They called for conformity, for a hierarchy of leadership, and in so doing reasserted the traditional pronouncements about women's proper function and roles. As in the earlier centuries power was given and then taken away.

In the sixteenth and seventeenth centuries, first in one town, then in another, in one kingdom, then another, the Catholics labeled contradictory doctrines and rituals heresies, the Protestant sects called them untruth, the snares of Anti-Christ; what had once been accepted behavior turned into unorthodox and dangerous practices punished by ostracism, exile, excommunication, death. In these new circumstances, women, whether Catholic or Protestant, found that the prerequisites for their inclusion among the faithful were the traditional, narrowly defined, areas of female competence. Once again the old premises emerged. To achieve order, to bring uniformity, meant women in their proper place, clearly subordinate to men. Whatever the sect, whatever the century, Christianity remained the religion of the Fathers, with a woman's nature, proper function, and roles clearly demarcated, defined in relation to a male whether it be within the patriarchal family or under the protective supervision of its ecclesiastical equivalent.

For all of the sixteenth- and seventeenth-century Christian sects, the most evident narrowing and limiting of authority came first with the visionaries, those women believing themselves to have mystical connections to God. Initially the Protestant sects, like the Catholics before them, welcomed women visionaries, women with the gift of prophesy, able, they thought, to communicate directly with the deity. In the desire for confirmation of rebellion and alternative doctrines, the women's revelations gave authority to the new faiths and rituals. Among the most radical of the German and English sects—the Anabaptists, the Fifth Monarchists, and the Quakers—many more women than men came forward claiming that the "Power and Spirit" of God filled their consciousness. Such a sense gave them, like their counterparts in earlier centuries, the courage to speak out, to believe in their ability to forestall disaster, or to heal the sick. Elizabeth Hooton (b. 1600), converted to Quakerism by Fox early in his ministry, braved both the Protector Oliver Cromwell and the English King Charles II. Refusing to kneel when she presented a petition to King Charles, she remembered the effect on the observers:

> And the power of the Lord has risen in me, and the witness of God rose in many that did answer me. And some present made answer, and said, they wished they

> had that spirit. And then they said, they were my disciples. . . . And the power of the Lord was over them all.[59]

Her courage and conviction had a similar effect when she went as a missionary in her sixties to Boston and to Jamaica in her seventies.

Like their Catholic predecessors, these Protestant women prophets found that when they acknowledged the inferiority of their sex, spoke of themselves as vessels of the divine, and echoed the views of the male leaders of the faith, their prophesies gained a hearing, and they avoided persecution. Elizabeth Poole, an Independent like Cromwell, had such an exemplary attitude. She described herself, a woman as "full of imperfection, crooked, weak, sickly, imperfect."[60] God only used her. Anne Wentworth would never have imagined speaking on her own behalf:

> Nor durst I for ten thousand worlds pretend to come in the name of God, or in the pride and forwardness of my own Spirit put myself into this Work, without his express command . . . having learnt . . . how dangerous and desperate an attempt it is, to put the *Commission* and *Authority* of God upon the *Dreams* and Visions of my own Heart . . .[61]

Mary Cary (Rande), though a Fifth Monarchist, outspoken and aggressive in her faith, avoided all of the hazards awaiting a religious woman. Between 1647 and 1653 she wrote and spoke condemning the Pope, tithes, lawyers. She insisted on her ability to prophesy but all with the proper attributions:

> Let them blesse the Lord for it, from whom alone it came: for I am a very weak, and unworthy woman, and have oft been made sensible, that I could do no more herein of my self, than a pencil or pen can do, when no hand guides it . . .[62]

The Puritan Hugh Peter even wrote the preface for her writings, excusing their imperfections by the simple fact of her sex and recommending her goodness and modesty as a model for other women.

In contrast, less orthodox Protestant visionaries found themselves ostracized and ridiculed. In 1654 when Anna Trapnell, the mystical English poet, sang verses against Cromwell, he imprisoned her. Neither Anglicans nor Independents could abide Lady Eleanor Davies, because of her interpretations of Scripture and her predictions. The Protestants kept her in Bedlam, the insane asylum. When in the 1670s Jane Lead (1624–1704) began to write and speak of her trances and her visions, none of the established sects accepted her, finding her beliefs too unorthodox. A follower of the German Jacob Boehme, she envisaged a soul united with Sophia, the feminine Wisdom, a union more awesome than anything promised in other faiths. Much like the Gnostics, she made Eve her heroine—Eve, long dead, now raised to be the mate of Sophia. She ended the leader of her own sect, the Philadelphia Society (later called the Theosophists).

As for the other roles Protestant women assumed, the joining in violence, the ministering, the preaching, all functions that bespoke the spiritual equality of the zealous female believer, the male reformers gradually forbade one and then another. Once official recognition and legitimacy had been established, the efforts of the women became unnecessary and drew condemnation, not praise.

In Lutheran Germany an outspoken anti-Catholic like Argula von Grumbach went to prison twice; when released she was viewed as an aging eccentric. Luther criticized girls and women for trying to speak eloquently in any context: "It suits them much better to stammer or to speak badly."[63] Calvin, like Luther, believed unequivocally that women could not perform any tasks of the ministry. He feared ridicule of his sect: "How will the Baptists and Anabaptists scoff to see us run by women!"[64] In the course of the sixteenth century Calvinist areas of France and Switzerland quietly closed all the options for women. After publication of Marie Dentière's *Defense of Women* in 1538, no more women's writings appeared in that century. The Geneva clergy found her criticism of the pastors disturbing. She had also asserted women's right to comment on Scripture. In France in 1560 according to the Calvinist Pierre Viret, even the purest and most pious of women must be silent in the service; they could teach little girls instead. In 1562 Jean Morely's suggestion of female deacons caused his pamphlet to be condemned in France and Switzerland. By the seventeenth century in some Protestant congregations the women and men sat separately and even took communion separately, the men first.

The radical English sects were no better. The Puritan William Gouge noted that all men had the special qualities of the priesthood and, by implication, certainly not any women. Pamphlets of the Independents attacked congregations with females preaching. The December 1646 law against unorthodox preaching was directed against them. The 1649 women's petition on behalf of the Levellers elicited disdain from the Commons members who told them to "look after their own business and meddle with their housewifery," "to spin or knit and not to meddle with State Affairs."[65] Their husbands must be at fault, Parliament members explained, to have allowed such behavior of their wives: "It is fitter for you to be washing your dishes, and meddle with the wheel and distaffe."[66] Even the Levellers did not give women the vote.

The attitude of the Quakers evolved more gradually, but with the same result, differentiating and restricting women's activities. In the late 1650s in London the four women followers, the "Ranters" as they were deprecatingly called, who spoke out and who called themselves the equals of Fox, were considered in need of "discipline." In the mid-1680s the London Meeting talked of controlling the women prophets and of asking older Friends to train the younger women not to speak so much. By 1700, women were being encouraged not to travel. By 1698 women had been effectively excluded from the administration of the sect and thus from its leadership. This exclusion came gradually. In 1660 Fox had authorized separate Women's Meetings to help in the period of the persecution; at Swarthmoor, in Lancashire, and in London they met to aid those in and out of prison. By 1674 the women of the London Meeting explained that aiding and ministering to the "helpless" had become their chief work.

The woman's role within the sect became defined and delimited. Although the autonomous groups gave women a sense of usefulness akin to that of other Protestant and Catholic women involved in charitable works, they were simply extensions of the traditional wifely functions, not evidence of a new equality. In fact, Quaker men did not minister to the poor and excluded women from meetings that dealt with all other concerns of the sect. In 1698 none were asked to the London Yearly Meeting. In 1701 those who did attend had to give prior notice if they wished to speak.[67]

Most limiting in the Protestants' assessment of women's proper roles within the Church was their blatant hatred for the life of the convent. According to the new sects, not only should women play no public role, they should not follow a separate religious vocation by joining a religious order. In sixteenth-century England the king's agents disbanded the religious communities of women. They sold everything, including the furniture and the vestments, giving the former nuns portions of the proceeds, never very much. In Germany, some women stayed together, some as Lutherans, others as Catholics, until all in the community had died. More aggressive rulers disbanded even these groups; some had to flee in the chaos of the Peasants War.

Perhaps the Protestant leaders like Martin Luther, Zell, Zwingli, and Bucer felt a special dislike for the avowed regular clergy, whether female or male, because they themselves had been monks. They saw nothing but corruption, mocked the pretended celibacy of the Catholic priesthood, and derided the supposed chastity of the Catholic nuns. The sixteenth century was a time of particular anticlericalism, with secular and religious satires echoing the same themes stressed in the declarations of the reformers: seductive nuns ensnaring young men, young women pining for love in the convent against their will, nuns having illicit sex with their confessors.

Visitation records from the thirteenth to the sixteenth centuries were thought by contemporaries to confirm the unfavorable images of the life of the cloistered religious woman. Like the great princes of the Church—the abbots, bishops, and cardinals—the titled and royal women of the richly endowed nunneries lived very different lives from what their founders had anticipated. Princesses and noblewomen followed more lax provisions, with separate apartments, regular meals, the right to converse, to have visitors, and to enter and leave when they chose. English nuns at Sion ate delicacies like peacock, salmon, and sugar, used spices like ginger, nutmeg, saffron, and cloves. According to tradition, the prioress of Sopwell had the time and the inclination to write the *Boke of St. Albans,* a guide to hunting, hawking, and armor, printed in 1486.

The same tendency toward luxury developed in France. Fontevrault had from its founding in the twelfth century been a place for princesses and queens. In 1471 Katherine de la Trémoille paid 200 écus for the dress and trappings she needed for the consecration ceremony and 200 more for the ring, the dinner, and the fee for the bishop when she took her Benedictine vows there.[68] In 1400 Christine de Pizan wrote a poem describing her visit to her daughter at Poissy, a 100-year-old Dominican priory, an exclusive religious establishment patronized by the king, the prioress his aunt. Gaining admission to this safe and leisured world for her dowerless daughter had been a significant accomplishment for the court writer. On her visit she marveled at the party given for the visitors with a lavish lunch served on gold and silver plate. She enjoyed walks through the walled grove of fruit trees, the ponds, the deer wandering the grounds, the pleasant evening spent talking with the nuns.[69]

While Protestant leaders responded to such images with disgust, disdaining any religious orders for women, sixteenth-century Catholic leaders hoped to bring about reform. "Visitations" like those made by the Papal Legate Johann Busch in Saxony led to condemnations of separate living arrangements, private property, entertaining, inappropriate robes, and laxity in the services.[70] Prioresses did not always assent to Busch's orders. At Sonnenburg, one of the wealthiest and most powerful of the Benedictine houses in the Tyrol, an abbess and her seven nuns fought for six years against the papal legate. Both sides resorted to hiring armed men. Finally the group was broken by being excommunicated and so strictly sequestered that they could receive no tithes or provisions. It was starvation or submission. In contrast Busch found the nuns of St. Martin's in Erfurt models of behavior: "very simple and humble, but of good will and ready for all good work; for they applied themselves promptly to obedience and to the observance of their rule."[71]

The Council of Trent in its decrees in the 1560s hoped to enforce this ideal on all female houses. The Catholic prelates made uniform regulations about profession and novitiate—not before the age of sixteen, by free consent with one year in the community before the final vows—and issued decrees on the ownership of property, how the nuns might live, the choice and age of the abbess—a member of the order for eight years and over forty—and regulated observance and contact with the outside world.[72] To insure maintenance of the rule and subordination to the authority of the Church, the Council decreed visitations by a male member of the order and underlined the supremacy of the local bishop on all issues.

By the late sixteenth and seventeenth centuries, with uniformity of regulations imposed on the older Catholic orders, dispensations granted new groups like the Ursulines were gradually rescinded or forgotten. Any individual refusing the authority of the male ecclesiastical hierarchy, claiming a spiritual right to independent action, was reprimanded, controlled, or, if necessary, cast out. While the Church needed the work of the Ursuline sisters as teachers during Saint Angela Merici's lifetime and into the 1580s, the archbishop of Milan gave them his protection. Already by the 1540s, however, the Church had required that they wear a habit. In 1612, under Pope Paul V, the order Merici had written was put away, and the nuns were required to follow the Augustinian rule and to remain cloistered. The teaching continued—it was too valuable for the Church to refuse—but now the nuns remained in the convent, the students came to live with them, to study behind the enclosed walls.

By 1610, when an Englishwoman, Mary Ward (1585–1645), who had become a Poor Clare in France, produced her plan for a teaching order, the Church could make its own terms. Ward saw her "Institute" as the equivalent of a college, her nuns living uncloistered, leading an apostolic life similar to the Jesuits. Like the Jesuits she wished to acknowledge no authority but the Pope. For she explained:

> There is no such difference between men and women that women may not do great things. . . . For what think you of this word, "but women"? As if we were in all things inferior to some other creature which I suppose to be man! . . .[73]

In an age when the Church had just reestablished the power of its hierarchy, no male ecclesiastic supported such plans or such claims. The archpriest of England, William Harrison, thought the women "soft, fickle, deceitful, inconstant, erroneous, always desiring novelty, liable to a thousand dangers."[74] Pope Urban VIII considered Ward's request, met with her, and refused her. When she ignored the order of suppression issued in 1629, she was arrested and sent to a Poor Clare convent, labeled a heretic and a schismatic. The Bull of Suppression in 1631 condemned the "Jesuitesses" who "carried out works by no means suiting the weakness of their sex, womanly modesty, virginal purity. . . ."[75]

In the sixteenth century, Johann Busch in his visitation report complimented the nuns at Ercherde: "These virgins were well obedient, pious and tractable . . . dealing with us and with each other kindly and benignantly by word and deed."[76] These qualities became the ones fostered and glorified by the Catholic Church into the twentieth century. The Sisters of Charity, unorthodox though their lack of cloistering was, continued to win approval, partly because of the great need for their services, but also because nothing in their training or their behavior contradicted the sixteenth-century ideal of the devout nun. Vincent de Paul allowed education for them but, as he explained in his formal instructions, it was "not for the sake of being learned," rather "Learn . . . in order to facilitate the means of observing your Rules . . . that you may be able to keep a correct account of your expenditure, books, receipts; to write to your superiors from distant places . . . to express your respect and dependence upon them; and above all, to teach the poor little girls of the village—in a word, that you may be able to serve God better."[77] Even when caring for their patients, they must not assume knowledge or take independent responsibility for treatment; instead they must transfer their acceptance of the authority of the Church to the absolute authority of the males with whom they worked: "You should act, my sisters, with great respect and obedience towards the doctors, taking great care never to condemn or contradict their orders."[78]

The Catholic women of special piety who came after achieved recognition and even sanctification because they also were obedient and deferential. Barbe Acarie (1566–1618), later known as Marie de l'Incarnation, lived an exemplary life as the wife of a French nobleman, then helped establish Ursuline communities in France and to persuade King Henry IV to allow the order of her heroine, Saint Teresa of Avila, the Discalced (Reformed) Carmelites. Saint Thérèse of Lisieux (1873–1899) came from a very devout French family—four older sisters became nuns. She desired a holy life so strongly that she petitioned first her bishop and then the Pope for permission to enter the Carmelite order at fifteen. Her life there from seventeen to her early death from tuberculosis at twenty-four became the subject of an autobiography *The Story of a Soul,* a life so popular that hundreds of thousands of copies had been sold by the 1920s. Most appealing was her image of the "little flowers" to whom God "shows his infinite greatness." God in his "condescension" turns, she explained, not only to the great Doctors of the Church but to the children, the untutored whose "holiness delights Him."[79]

She and her countrywoman, Saint Bernadette of Lourdes (1844–1879), personified all that the Church wanted of its women. Both were canonized not for their visions, but for their exemplary lives of compliance, obedience, and extreme humility. At fourteen Saint Bernadette, born Marie Soubirous, the daughter of a peasant family, was believed by contemporaries to have had numerous visions of the Virgin Mary. Popular enthusiasm forced the Church to build a shrine and to accept cures at its spring as the work of God. The prelates remained skeptical of Saint Bernadette's mystical experiences, and it was as a nursing Sister of Charity that the Church honored her, for her exemplary care of the sick, her extreme self-abnegation in the face of her tuberculosis, and her obedience to the Church's commands.[80]

With the women it chose to canonize, the Church offered a model of appropriate behavior. Another, more powerful figure was also available to be extolled: Mary, the Virgin Mother of God, agent of the miraculous birth of Jesus, and the Queen of Heaven. In the era of the Counter Reformation and later in response to challenges from the increasingly secular world of the nineteenth and twentieth centuries, these ideal images of a more perfect being took precedence over the more humane images of the mother and her infant. Now Mary's absolute purity, in a sense her inhumanity, was stressed. She emerged in the popular and ecclesiastical imagination as Maria Immaculata, literally, Mary the Immaculate, whose purity stretched back to the point of her own conception. Thus Mary, immaculately conceived, was untainted by the Fall, without sin, without curse, a perpetual virgin who in turn conceived her child without sin. When this belief was formalized as dogma in 1854, Mary became the perfect human being. Thus the Virgin was a contradictory model: the human female honored above all other mortals that women could aspire to, but conversely an unreal being whose unique experiences implied condemnation of women's sexuality and ringed their lives with implied prohibitions in the name of purity.

As with the images of the Virgin in the earlier centuries, a combination of the needs of the faithful and of the ecclesiastics created this new incarnation of Mary. The popular imagination and the logic of the theologians first established Jesus' conception as "immaculate." The mystic Mechthild von Magdeburg's description passed from Germany to England to become the words of a song:

> He came also still
> Where his mother was,
> As dew in April
> That falls on the grass.[81]

Mary's own "immaculate" conception without sin was popularized in the fifteenth century by Nicolette Boylet (1381–1447), a leader of the Poor Clares. The cult was especially popular in France, Germany, and Belgium.

For centuries the papacy neither condemned nor condoned the idea.[82] Finally in 1854, bowing to the needs and the enthusiasm, Mary's special conception became an article of faith by papal decree, and the "immaculate" perfect woman was there to epitomize the Church's ideal. An equally confining picture of the ideal Christian woman would evolve in the imaginations of the Protestant leaders and their believers.

5

TRADITIONAL IMAGES REDRAWN

The Female Nature

With the Protestant and Catholic Reformations, what Christianity gave in one era, it took away in the next. Women would not find lasting equality through their churches. When male leaders of the Christian sects consolidated their beliefs and institutionalized their relationships with their followers in the seventeenth century, the patterns of earlier waves of consolidation and institutionalization were repeated. In the new churches as in the old, authority and conformity replaced independence and diversity. Women in particular had to be dealt with, for woman out of place, acting outside her traditional role, once again symbolized the general disorder of society.

Women out of control were seen as a danger to men—a specter as ancient as European history. To the learned religious male of the sixteenth and seventeenth centuries, whether Protestant or Catholic, the image inspired fear. Though the splintering of the Church gave Christianity many voices and variations in belief and ritual, the fear of women gave the messages from Europe's pulpits an awful resonance and uniformity. It mattered not if the learned ecclesiastic spoke the words of the Evangelists and the Church Fathers, of the Scholastics, the Protestants, or the popes of the Catholic Reformation. Churchmen, divided on so many other issues, united in condemning women who did not remain within the spheres allotted to them. They feared what they assumed to be a woman's latent desire to rule. Ruling she destroyed the male. Even worse, there was her body with its insatiable lust and the lust it inspired. In their efforts to disarm this potentially destructive female nature, men spoke as if with one voice. They declared the woman's nature by definition inferior to man's, by definition, therefore, women must be subordinate. There was no logical progression, no reasoned path by which sixteenth- and seventeenth-century Protestant and Catholic theologians, priests, and pastors arrived at these coincident conclusions. There was no need for logic or reason, they believed such conclusions were self-evident.

Any man who looked could find the divine hand in this basic inferiority of the woman and the need for her to be controlled. By God's design, by the evidence of her physiology, she stood revealed second to the superior male and thus meant to be his subordinate. Explanations and justifications followed one from another like the Scholastics' proofs of the existence of God, the Protestants' treatises on the dual nature of the sacraments and the predestination of souls. In the thirteenth century Thomas Aquinas had argued that:

> For good order would have been wanting in the human family if some were not governed by others wiser than themselves. So by such a kind of subjection woman is naturally subject to man, because naturally in man the discretion of reason predominates.[1]

Borrowing from Aristotle, Aquinas, like Gratian and Abelard before him, saw the consequences of the Fall merely confirming the "natural" order and woman's inferiority. The active, the intellect, the reason is man; the lesser, the passive, the material, the body is woman. Three centuries later, the leaders of the Protestant Reformation reiterated these arguments, citing the same sources to make the same pronouncements. In 1535–36 Luther explained in his lectures on Genesis:

> It is evident therefore that woman is a different animal to man, not only having different members, but also being far weaker in intellect. But although Eve was a most noble creation, like Adam, as regards the image of God, that is, in justice, wisdom and salvation, she was none the less a woman. For as the sun is more splendid than the moon . . . so also woman, although the most beautiful handiwork of God, does not equal the dignity and glory of the male.[2]

John Calvin, like Aquinas, believed her never to have been man's equal, but to have passed from subservience to slavery as her curse after the Fall. Even the Quakers took the Scriptures literally in this regard: "Man and Woman were helps meet . . . before they fell; but after the Fall, in the Transgression, the man was to rule over his Wife."[3] (The Quakers believed Jesus' death restored women's more equal status.)

And if by chance woman did gain authority over man, disaster followed. Eve had taken Adam with her in the Fall, so all women could harm men. Hugh Latimer, the sixteenth-century English Protestant, reasserted the traditional view: "A woman is frail, and proclive unto all evils: a woman is a very weak vessel, any may soon deceive a man and bring him into evil."[4] Calvin and his Scots disciple John Knox had to deal with women who actually did rule and held political power. Calvin had been horrified when Renée of

Ferrara wanted to participate in the Presbyterian Synod. In 1558 in his *First Blast Against the Monstrous Regiment of Women,* Knox called female rulers "visitations of God's anger," a rebellion against nature, a state that defiled, polluted, and blasphemed. The men who accepted their rule were no more than "brute beasts."[5]

The churchmen pondered: With such awful consequences how could women come to rule? Both Catholics and Protestants formulated the same explanation. Man fell before the woman's body, its sexuality, the lust it inspired. The twelfth-century *Ancren Riwle,* a guide for female recluses, issued the warning: that a man "commits mortal sin through you in any way, even . . . with desire for you . . . be quite sure of the judgement: you must pay for the animal because you laid open the pit."[6] Religious women would be reminded of this view every day as they went to mass, for the sculpted and carved reliefs of the great cathedrals portrayed lust as a female.

In the twelfth and thirteenth centuries, male leaders of the reformed orders had tried to protect their men from this ever-present danger, the woman's sexuality. Abelard wanted women's heads covered to prevent their inspiring "lewd desires."[7] Saint Bernard wanted no contact with women at all, not even his own sister, for he believed "to be always with a woman and not to have sexual relations with her is more difficult than to raise the dead."[8] The same fears plagued Saint Ignatius Loyola at the end of the 1520s. He advised a fellow Jesuit:

> I would not enter upon spiritual conversations with women who are going or belong to the lower classes of the people, except in church or in places which are visible to all. For such women are easily won, and through such conversations, rightly or wrongly, evil talk arises.[9]

This view of the woman as sexual temptress became an especially significant concept when both Catholics and Protestants came to define their clergy, the men who represented the strength and purity of the faith at its popular level. In the Catholic dogma of the seventeenth-century Reformation, a priest's ability to resist desire, to remain pure and celibate, became key to setting him apart from and above the laity. From the earliest days of Christianity most theologians had disapproved of intercourse and marriage for ordained men. They had written with horror of a clergy mired in the material world of sexuality, wallowing in lust which they saw as the ultimate act of irrationality.[10] Despite the uniformity of attitude, into the sixteenth century in the Western Church, disparity existed between theory and practice. The highest Catholic ecclesiastics tended to honor the prohibition, setting aside their wives. This was not so for the lesser clergy. Priests continued to take wives as helpmates and to have intercourse both in and out of marriage. Only with the Protestant cries of corruption and laxity, only in the crisis of the Reformation could the Catholic Church enforce celibacy for all its clergy. The canons of the Council of Trent in 1563 forbade marriage, intercourse, and any sexual contact for the ordained male. The clergy must not "live in the filth of impurity and unclean bondage." Only if they remained celibate would the people "revere them the more, that they know them to be more pure of life."[11] Protestant theologians found nothing to contradict in this reasoning. They saw celibacy as an ideal. Calvin called it "a blessing," a "gift from God."[12] In contrast, he wrote of the "dreadful madness of lust," comparing it to epilepsy.[13]

Even so, both Catholics and Protestants acknowledged that all men could not remain celibate. For men the leaders of every Christian sect offered the same remedy, marriage. In matrimony intercourse could take place in controlled circumstances, in a licit manner, for a rational purpose—the procreation of children. Luther spoke of sex in marriage as "medicine" and of having "to make use of this sex in order to avoid sin."[14] Calvin gave similar instructions: "let every man abstain from marriage only so long as he is fit to observe celibacy." When "his power to tame lust fails him," then it must be God's will that he marry.[15] Given these premises about man's imperfect nature, Protestant churchmen expected their clergy to marry.

Even in marriage, however, both Catholic and Protestant ecclesiastics warned that lust must be avoided. Calvin explained: "For even if the honorableness of matrimony covers the baseness of incontinence, it ought not for that reason to be a provocation thereto."[16] There must be no pleasure in the act. In 1584 the catechism by the French Dominican Jean Benedicti echoed the reasoning of Aquinas and the canons of the Council of Trent:

> The husband who, transported by immoderate love, has intercourse with his wife so *ardently* in order to satisfy his passion that even had she not been his wife, he would have wished to have commerce with her, is committing a sin . . . an *uncontrollable* lover of his wife, rather than her *husband,* is an adulterer which means that a man must not use his wife as a *whore,* nor the wife behave toward her husband as a lover, for the holy sacrament of marriage must be treated with all honesty and reverence.[17]

The Christian Family

For the learned Catholic and Protestant churchmen of the sixteenth and seventeenth century, marriage meant more than the means to protect themselves and their believers from the horrible powers of lust and female sexuality. Marriage meant the traditional relationship between women and men and was the way to keep women subordinate and obedient. For with queens and kings contending, parliaments and cortes dissolved or divided, economic changes beyond the individual's ability to comprehend, the end to uniformity of belief and practice, European men looked to the familiar. They wanted limits, certainties, incontrovertible authority in religion, in many aspects of their lives. Women out of place, women unchecked, embodied disorder. Thomas Edwards, the seventeenth-century English Protestant, easily made the connection from females preaching to men losing "command of wives, children, servants."[18]

Edwards' reasoning was typical of his age. The Protestants looked to the Old Testament and in the metaphor of the patriarchal family found the order they sought. The king or the state played the role of father; subjects were like children cared for and subordinate. All were bound together by reciprocal responsibilities and loyalties. All was based on the familiar unit of society, the woman and man joined in marriage. At its essential level, within the relationship of one couple, the male functioned as the patriarch, the woman as the model child willingly loyal to her natural ruler. It was to be a hierarchy of fathers. Order at the lowest level of society could then be mirrored at the highest level with the king, the patriarch of the larger family of all the subjects.

For men this view had the effect of destroying much that had been potentially democratic about the Reformation. For women it was disastrous. Protestants and Catholics saw the man as capable of functioning both within the family and outside it. Not so women. By the end of the seventeenth century, whatever her Christian sect, the preferred role for a woman was to be a married wife and mother. Within the safety and constraints of this society of male-dominated families, more exactly than ever before, a woman found herself defined in relation to a man, given no identity outside that relationship, and judged by her success at prescribed family duties and obligations. The hallowed voices spoke as one, and women were subordinated anew in the service of the Christian family.

From the eleventh century Catholic churchmen like Saint Thomas Aquinas and Saint Bonaventure had written about what in theory would have been a more equal relationship in marriage. At first the views of the Protestant leaders appeared egalitarian. Martin Luther listed living together as the first purpose of marriage even before the traditional reproductive reason. Calvin followed explanations given by the Humanists, thinking of woman as the descendant of Eve, formed from Adam's side and thus designed to be man's companion and complement. Other Protestant leaders used similar phrases and reasoning. The seventeenth-century Puritan, Robert Bolton, called marriage "loving contentment in each other."[19] Some Puritans described the relationship with analogies to the mercantile capitalist world they were experiencing. In his 1627 sermon William Gouge called it "a joint stock company" with the husband making the woman "a joint Governour of the Family with himself."[20] Although the reproductive and sexual reasons for the union continued to be paramount in the writings of the Catholic Reformation, companionship was there as well. The Roman Catechism of 1566 spoke of the "partnership of diverse sexes—sought by natural instinct, and compacted in the hope of mutual help."[21]

With the members of the most radical Protestant sects, the sense of companionship sometimes carried over from doctrine to real life. Matthew Zell agreed that woman was created for man, but believed that man needed marriage to be complete. Katherine Zell, his wife, in turn described herself in the traditional imagery as a "splinter from the rib of that blessed man," but with a difference:

> My husband and I have never had an unpleasant fifteen minutes. We could have no greater honor than to die rejected of men and from two crosses to speak to each other words of comfort.[22]

Such mutual love became almost a part of Anabaptist martyrology. Hendrik Verstralen wrote to his wife before his execution:

> My faithful helper, my loyal friend. I praise God that he gave you to me, you who have sustained me in all my trial.[23]

Of the Protestant leaders, Martin Luther most explicitly described the ideal relationship between a woman and a man and the world it created; he wrote a friend:

> The dearest life is to live with a godly, willing, obedient wife in peace and unity. Union of the flesh does nothing. There must also be union of manners and mind.[24]

"Union of manners and mind" had a very specific meaning to the Protestant leaders, a meaning familiar to all Christian theologians. For there were unspoken traditional premises even to a relationship such as Luther portrayed—premises stronger than any momentary need for a partnership of equals, premises that took precedence over delight in companionship. After the first year of his marriage, Luther wrote of Katherine von Bora: "My wife is compliant, accommodating, and affable beyond anything I dared to hope."[25] Calvin spoke of the wife as a "complement" to the husband, but it was a complementarity of unequals. He used the analogy of the Church Fathers: the man was the head, the woman the body, and her principal obligation was "to please her husband" and "to be faithful whatever happens."[26] Even the otherwise radical Anabaptists came to see a wife's obedience reaching beyond faith. Nothing justified a woman leaving her husband, not even his exclusion from the congregation.[27]

Within this Christian family all ideas of equality fell away. Instead, the Protestant leaders insisted that women should willingly submit to the rule of the fathers and to exclusion from all activities traditionally reserved to men. Luther saw the household as a separate place, a separate sphere from which the man, the husband and father, came and went, but in which the woman, wife and mother, was exclusively confined: "Let the wife make her husband glad to come home, and let him make her sorry to see him leave."[28] Luther described the relationships within this ideal family, the activities appropriate to each of its members:

> The rule remains with the husband, and the wife is compelled to obey him by God's command. He rules the home and the state, wages war, defends his possessions, tills the soil, builds, plants, etc. The woman on the other hand is like a nail driven into the wall . . . so the wife should stay at home and look after the affairs of the household, as one who has been deprived of the ability of administering those affairs that are outside and that concern the state. She does not go beyond her most personal duties.[29]

In their manuals of behavior and faith, the English Puritans and Anglicans praised the wife's willing subordination in this restricted, patriarchal world. Robert Cleaver compared her favorably to man's other helpers in his pamphlet *A Godly Form of Household Government* (1598). The wife was better than the horses, the oxen, the servants, "a fellow comforter of all cares and thoughts" who served not out of fear or for wages but "only by love."[30] Robert Wilkinson in his 1607 wedding sermon compared the good wife to a snail "not only for her silence and continual keeping of her house but also for a certain commendable timourousness of her nature."[31]

In theory this subordination of the woman within marriage was not meant to be absolute. An occasional Catholic or Protestant ecclesiastic noted limits. For example, an early seventeenth-century French churchman said that forcing sex that would end the wife's life constituted "severe and intolerable obedience."[32] A husband could not force her to change her faith. But otherwise all assumed his control of her life, expected willing acquiescence from her and condoned violence if she did not comply. William Whately, the English minister, explained in the 1619 edition of his pamphlet on the duties of wife and husband, the justifications for beating when

> She will raile upon him with most reproachful terms, if she will affront him with bold and impudent resistance, if she will tell him to his teeth, that she cares not for him, and that she will do as she lusts for all him; if she will flie in his face with violence . . . or brake into such unwomanly words or behavior.[33]

In 1713 Antoine Blanchard, the prior of Saint-Mars-les Vendôme, composed questions for the confessional that made it clear that the Catholic Church also believed that a wife brought any beatings upon herself:

> Have you maintained a close union with your husband, suffering his shortcomings with patience and with charity? Have you not given him reason to be angry and have fits of rage on account of your arrogance and your obstinacy? Have you not been insufficiently obliging towards him?[34]

For women this subordinate role of wifely service was but one of the functions of marriage. Even more significant for her was the role of mother and her child-bearing function. Luther explained that, weak and vice-ridden though the woman might be, "one good covers and conceals all of them: the womb and birth."[35] Both Luther and Calvin, like their Catholic predecessors, had a horror of pregnancy and childbirth, yet both valued the experience above all others for women. Luther described the divine plan:

> In procreation and in feeding and nurturing their offspring they are masters. In this way Eve is punished; but . . . it is a gladsome punishment if you consider the hope of eternal life and the honor of motherhood which have been left to her.[36]

Her travails, her groans, all, according to Calvin, symbolized for God her obedience to his will and gave her the means of her salvation. It was not by faith that the woman erased her sin, but by childbirth. Fulfilling her biological function, she came closer to the divine. In her pain, she would come to God. Elizabeth Clinton explained it to her Puritan sisters:

> We have followed *Eve* in transgression, let us follow her in obedience. When God laid the sorrows of conception, of breeding, of bringing forth, and of bringing up her child upon her, and so upon us in her loins, did she reply an word against? Not a word: so I pray you. . . .[37]

Likewise, should the woman refuse any part of this divinely ordained calling, she transgressed. Any impediment to conception was a way of refusing her own salvation. Calvin called coitus interruptus (withdrawal) "monstrous," assuming as had Jerome, the fourth-century Church Father, that she would thus be killing "before he is born the son who was hoped for."[38] In addition, the Puritans insisted that she breast-feed the child herself. If not, she committed a sin against God and rejected the child, asserted Robert Cleaver and William Gouge at the beginning of the seventeenth century. Why else, they wondered, did woman have breasts?

Luther must also have asked this question about woman's function on earth. He had marveled at what appeared to him a divine plan, at all the ways the woman's physiology justified the function and the roles he and his ecclesiastical contemporaries had assigned her: "To me it is often a source of great pleasure and wonderment to see that the entire female body was created for the purpose of nurturing children."[39] Not only that, but obviously: "women . . . have but small and narrow [chests] and broad hips, to the end they should remain at home, sit still, keep house, and bear and bring up children."[40]

In their own marriages the Protestant reformers lived out their ideal relationship between a woman and a man. Exemplary wives created exemplary households, refuges, and safe havens for husbands. It was as the seventeenth-century English Puritan, William Gouge, described it in his pamphlet *Of Domesticall Duties:* The perfect wife looked to him "after a manner a Saviour," "a king to govern and aid her, a Priest to pray with her and for her, a Prophet to teach and instruct her."[41] Subject to their husbands, within this safe confined sphere, these women willingly exercised their considerable energies and talents.

Although Luther wrote of the husband as playing a role in the family, in fact he himself married at forty-two to bring order to his life and left everything to his wife, Katherine von Bora (1499–1550); the twenty-six-year-old former nun even found the means to maintain and sustain the household. Luther had seen no other way to provide for the nine nuns that he had encouraged to escape from the Cistercian convent of Marienthron than to arrange their marriages. Katherine von Bora suggested that he or one of his advisers (Nicholas von Amsdorf) be her husband. Bemused, Luther wrote to a friend: "While I was thinking of other things, God has suddenly brought me to marriage. . . ."[42]

Thus he became the first of the reformers to marry, she the first model wife and mother for the new Protestant sects. She bore him six children between 1525 and 1534: Hans, Elizabeth, Magdelena, Martin, Paul, and Margaretha. She watched Elizabeth die before her first birthday, Margaretha at thirteen. She also raised eight of Luther's nieces and nephews. She turned the "Black Cloister," the Augustinian monastery that Luther had been given, into a home and a source of revenue to shelter, nurture, and sustain her extended family and the many who sought refuge with Luther. She filled the forty rooms with boarders: students from the university, nuns and monks who had abjured their faith. She tended the orchard and garden, she acquired land, she managed their farm in Zulsdorf, seeing to the planting and harvesting, buying and selling the livestock. She stretched these goods and revenues and the 200 guldęn a year from the Duke of Saxony to cover expenses.[43] Meanwhile Luther sat at his dinner table and talked; he preached, he wrote, he traveled. When he died in 1546, she coped as she always had. Ruined twice over when forced to flee Charles V's armies in 1546 and again in 1547, she rebuilt and continued. Suddenly in 1550 she died. The coach horses took fright, she jumped out to try to stop them, but fell into the ditch by the side of the road. The cold and the wet left her ill, and she did not survive this last effort on behalf of her family.

Other German and Swiss reformers expected no less of their wives. Anna Adlischweiler (1504?–1564), in a nunnery since childhood like Katherine von Bora, stayed behind, even after all the other nuns had left, to nurse her mother. Henry Bullinger was her chaplain and proposed marriage, promising to show her his "fear of God, piety, fidelity and love."[44] He wrote her a handbook of proper behavior, *Concerning Female Training and How a Daughter Should Guide Her Conduct and Life,* and perhaps seeing no other course, she agreed to be his wife in 1529 after the death of her mother. She bore him eleven children, and died in 1564 after nursing the household in an epidemic of plague.

Wibrandis Rosenblatt (1504–1564) came from a well-born Basle family and cared for a series of reformers, first the Humanist Keller, her husband for two years, then Johannes Oecolampadius for three years. Then some friends suggested that Wolfgang Capito take on the widow with her four children; five children later she was widowed again by the plague of 1541. The wife of Martin Bucer, the Protestant theologian, sensing the imminence of her own death, asked Wibrandis to take on her family and children. A fourth marriage followed, with two more children. Through the deaths of her children, her exile in England with Bucer, she always coped. When Bucer died in 1551, she collected donations from English colleagues for passage for herself and the children back to the Continent.

Though Luther, Calvin, and their followers disagreed on key points of the faith, nothing marred the uniformity of their thinking about their marriages. Calvin too came to the conclusion that he should wed when urged by his friends. He, like Luther's followers, chose a widow, Idelette de Bures (?–1549). Married in 1540 in Strasbourg, she set up their household in Geneva in 1541, taking in refugees, nursing Calvin through his periodic headaches, trying to bear him children. Five were born, and five died in infancy. She never regained her strength after the death of the first boy in 1542. Her death in April of 1549 left Calvin bereft:

> She passed so quietly so that one scarcely marked the difference between life and death. . . . You know how sensitive I am or rather how weak. If I did not take myself severely in hand I could not stand up. I have lost the best companion of my life. . . . While she lived she was a faithful helper in my ministry. Never did she in any respect ever stand in the way. During her illness she never spoke about herself and never troubled me about her children. . . . Now I am plunging into work to drown my sorrows. . . .[45]

The Presbyterians in England and Scotland expected and found similar solace in their wives. John Knox made the most traditional kinds of marriages; he chose Margery Bowes (1538?–1560), thirty years his junior when they married in 1555. Her mother, Elizabeth Bowes, one of his circle of female believers, arranged the match. In exile in Geneva two years later, Margery cared for him, her mother, and another widow, Anne Locke, who had joined the household. She bore him two sons and died exhausted at twenty-two. Margaret Stuart was just seventeen when Knox, then almost sixty, took her for his second wife in 1564. Elizabeth Bowes, his former mother-in-law, was still part of the household.

Nothing in Protestant theology or practice about women's proper roles and function contradicted Catholic custom or tradition. Though the convent offered an alternative life, increasingly Catholic priests, bishops, and popes spoke less of this as the preferred path for their female believers. Instead, like the Protestants, they too extolled and enshrined the glories of marriage and motherhood for women. Even the Virgin Mary, once the model of the pious solitary female, came to be portrayed as part of the Holy Family.

It was in the fifteenth century that popular and learned believers first gave attention to Joseph, Mary's husband. Writings of earlier ages told little. They showed him a confused old man, needing to be reassured about his pregnant wife's purity, his existence justified by theologians like Saint Bernard as necessary to protect Mary's reputation. Once begun, however, the stories about Joseph multiplied, and he became the husband and father of the Holy Family. Artists showed not just the mother and the infant, but now the three—with Joseph at the manger, leading the ass on the flight into Egypt, at the carpenter's shop working side by side with his divine son. In the images of the sixteenth and seventeenth century, Mary, always a malleable figure, was transformed again. Once the perfect intercessor and the perfect mother, she now became the perfect wife—the Virgin Mary, an obedient, silent spouse, a fit companion to Joseph, her protector, who worked to provide for her and the child.

6

THE LEGACY OF THE PROTESTANT REFORMATION

In terms of attitude and practice toward women, only one of the Protestant changes survived, a change eventually condoned by the Catholics as well. To raise the children, to be the model Christian wife, the woman needed to know how to read. By the end of the seventeenth century, Christianity, whatever the sect, had accepted the ideal of female literacy. All, like Foxe in his *Book of Martyrs,* described illiterate females as "silly women."[1] Luther promised Katherine von Bora, his wife, a reward if she read the Bible all through. In Huguenot France husbands taught wives, the congregations gave classes for women. All favored the establishment of schools for their little girls. Protestants and Catholics taught them how to read and the rudiments of arithmetic—nothing erudite, but literacy nonetheless.

Nothing appeared revolutionary in this favoring of education. In their classrooms, the little girls also learned the household skills traditional to their accepted function and roles. Few saw contradictions in this. Few questioned the prescriptions and tenets of the faith: the views of women's inferiority, her subordinate place in the world of the family. The vast majority of women, like their predecessors expected to marry, to bear children, to live out a life of obedience and service, the dutiful appendage to the man. Elizabeth of Braunschweig, a Protestant noblewoman, advised her daughter on her marriage to obey her husband. The Protestant women active in England started from the same traditional premises. Catherine Chidley, the Leveller, challenged Thomas Edwards to debate, but excused her possible failure as inconsequential "for I am a poor woman and unable to deal with you."[2] With the other women leaders who presented petitions to Parliament, she acknowledged in 1653 that such action might seem "strange and unbeseeming our sex."[3] In 1642 her predecessors had agreed in their petition that women are "frail," "weaker" and insisted that by their action they did not intend *"to equal ourselves to Men, either in Authority or Wisdom."*[4] Anna Maria van

Schurman (1607–78), the seventeenth-century learned German Protestant, despite all of her classically inspired arguments for female education and scholarship, spent most of her youth caring for two aging aunts, writing of the "tight and sweet bond of love."[5] Later in life she became a follower of Jean de Labadie, living in his community in exile until her death. Committed to the foundation of a purer Church and faith, she appeared to have rejected all that her writings suggested and all that she had accomplished with her mind and her reason.

Even though most sixteenth- and seventeenth-century women accepted denigration and subordination, once they had learned to read and had established this privilege for their daughters, women's situation would never be quite the same again. For it was not Anna Maria van Schurman's pious actions and works that other women found in her writings, translated from Latin into the vernacular. Instead they delighted in her unorthodox education, her attendance at the Dutch university (though incognito). Her treatise of 1638, *Whether the Study of Letters Is Fitting to a Christian Woman?*, published in the Netherlands, France, and England, raised questions in their minds about the traditional views of women's limited roles, and by its very existence proved how learned a female could become. Later when Van Schurman acquiesced to all of the criticism of herself and her unusual vocation, she exemplified the contradictions inherent in the messages of Protestantism.

For while preaching to women about their subordinate role, their natural inferiority, and their proper behavior, Protestant ecclesiastics also valued zealous, active piety and success in the world. This they glorified as the true testament to faith. In Protestantism it was not the meek and obedient who inherited the kingdom of heaven, but the quick, the strong, and the defiant. It was as if exemplary women of faith would by definition be denied access to salvation. If allowed to read, a woman would eventually discover this contradiction for herself.

Catholic dogma presented no such glaring contradictions. Within its nunneries there had always been a place for unconventional women who wished to read. In its veneration of humility, acceptance of suffering, and good works the wife and mother was always valued. Her subordinate condition embodied the life of the exemplary Catholic far more than that of her husband or father. The men who had dominance over her were males active in the outside world of temptation, liable to fall into the sins of pride and vainglory.

Perhaps because of this difference in the messages of the faiths, in the eighteenth and nineteenth centuries it was in Protestant nations that groups who worked to give women rights knew their greatest success. Though neither faith favored such changes, Protestantism, more than Catholicism, fostered circumstances and attitudes that allowed women to organize in their own behalf, to assert their place as equals, and to sustain their victories. Unintentionally Protestantism contributed to the long process by which European women began to free themselves from the denigrating and devaluing premises fostered and formalized for so many centuries in the name of religious truth.

IV

WOMEN OF THE CASTLES AND MANORS

•

CUSTODIANS OF LAND AND LINEAGE

1

FROM WARRIOR'S WIFE TO NOBLEWOMAN: THE NINTH TO THE SEVENTEENTH CENTURIES

ELEANOR OF AQUITAINE (1122–1204), the great landholder and queen of the twelfth century, chose to be buried in Anjou at the abbey of Fontevrault, the double monastery that she had favored with her patronage during her lifetime. The painted stone effigy of her sarcophagus suggests a woman younger than Eleanor's eighty-two years. A simple coif covering her hair, held tight and smooth by a band of cloth that goes under her chin, accentuates the roundness of her face. The lips appear to be smiling. She holds a book in her upturned palms as if she had just looked up from her reading. Her blue cloak is loosely pulled across her draped reddish gown seeming to keep her legs warm while she rests. Only a small crown indicates her royal estate.

As if symbolic of Eleanor's place in the warrior culture of twelfth- and thirteenth-century Europe, her sarcophagus now lies between that of her second husband, King Henry II of England, and her favorite son, King Richard the Lion-Heart. For although she held and administered lands equaling one third of present-day France, established her own court, went on Crusade, made her own second marriage, governed as royal regent, and even rebelled against her husband the king of England, all of this power and the masculine roles she assumed came because of her place within a family and only with the acquiescence of the men who headed it. With no male heir, William of Aquitaine gave his daughter the rights to the family's vast lands. Eleanor maintained her claims and authority over her inheritance, not as a woman alone but as the wife of two kings and the mother of two others. For although Eleanor of Aquitaine ruled and fought like a man, she remained physically vulnerable and subject to limitations because she was a woman. Her father arranged her first marriage; she counted on her husband, Louis VII, to assert her claims to Toulouse. When her rebellion against her second husband, Henry II of England, failed, he pardoned his sons but kept her imprisoned for fifteen years. In the last resort, as in the warrior cultures of the centuries before 800, even queens remained dependent upon and ultimately subordinate to their fathers, husbands, and sons.

From the ninth to the seventeenth century Eleanor of Aquitaine's experience—the combination of power and vulnerability within the context of her family—characterized the lives of all women of Europe's warrior and landed elite. They were both active participants and pawns in the wars of their men. They acted as surrogates for their husbands and fathers or, when they held lands in their own right, on their own behalfs. In those eras when warring was the principal occupation of the men of the family, they raised troops, they suffered in seiges, they defended lands and castles. They were captured, ransomed, and killed. In addition, throughout the whole period from the ninth to the seventeenth centuries these women assumed the obligation to supervise the lands on which the family's sustenance and access to political power depended. They maintained the household, they bore the children who continued their own and their husband's lineages.

Eleanor of Aquitaine's effigy portrays her with a book. She like many other "ladies," "gentlewomen," "châtelaines"—as contemporaries called them—could read and write. These women might teach the alphabet to their daughters and sons along with the first prayers. Their own writings, though few, are rich sources for their lives. Noble wives had records of their household accounts on parchment rolls. They left instructions for the bailiffs and seneschals who supervised the work of the manor farms under their administration. They wrote manuals of advice for their children and letters to their husbands to inform them of decisions while they were away. They contributed to the literature of these centuries with songs and narrative poems that they composed to entertain their friends and retainers.

As so few personal documents exist before the fifteenth century, men's literary writings must also serve as sources describing the activities of this warrior and landed elite. The new bride of *Njal's Saga* rode off with her Viking husband and his men to their home. She, like the wives and daughters of other sagas, of the epics and chansons de geste—the principal literary forms of the ninth to the twelfth centuries—served meat and drink to the warrior bands sitting at the long wooden tables around the open hearths of the great halls. The women of the lais and romances of the twelfth and thirteenth centuries, like *Tristan and Isolde,* tended the armor, dressed wounds, gave token favors to their chosen knights and warriors, and watched the ritualized fighting called the joust.

In addition to describing women's activities, men's literary writings from these centuries also reveal their attitudes toward women. The twelfth-century troubadours, or poets, made these noble women the inspiration for and the glorified objects of love. When kings and their followers devised codes of chivalry to govern a warrior knight's behavior, they idealized women as those in need of protection.[1]

From the ninth to the seventeenth century, circumstances gave these women of Europe's elite unusual opportunities and access to power. In certain eras they could exercise the practical power of fighting and ruling like a man. Described in the poetry of the age as the honored and admired arbiters of behavior in the world of knights bound by codes of chivalry, these women acquired a symbolic power as well. But both kinds of power carried vulnerability. The woman who acted as a man would always be an anomaly, easily subordinated in a culture that valued force and physical strength. In addition, while poets reveled in their love of a "lady," they also feared the power that love gave to women. For them, as for men in the cultures before 800, love signified the power to seduce, to lure, to make a man act against his will. Europe's warrior and noble culture maintained the same prejudices about women and the female nature as had the Greeks, the Romans, the Hebrews, and Christianity's early leaders. The ancient views of women and the ambivalence they reflected prevailed into later centuries. When noble fathers of the ninth to the seventeenth centuries considered how to educate their daughters, they chose not to present the image of the independent warrior maiden, but rather the ideal of the chaste, obedient, subordinate wife. The attitudes and images remained as they had been in the warrior cultures of the centuries before 800: the traditional fears, the traditional ideal redrawn for new generations.

2

CONSTANTS OF THE NOBLEWOMAN'S LIFE

War

From the ninth to the seventeenth century warfare defined the world of Europe's noblewoman. Over that period, all of its essential features changed—the occasions for fighting, the weapons, the tactics, and the relationship between the warriors and their leaders. And yet the roles allowed to women of the warrior and landed elite remained remarkably constant. Only in rare circumstances did they exercise political authority in their own right—that is, act as men. More commonly in these centuries noblewomen were the companions and trusted surrogates for their warrior fathers, husbands, or sons.

In the ninth and tenth centuries the patterns of warfare of the Germanic and Celtic peoples prevailed. A man farmed one season and warred the next. He would be numbered among the groups of warriors, the thanes, bound by ties of kinship or by oaths to a chieftain. These warriors fought for booty, for land, or, like the Spanish hero, the Cid, for honor. A successful leader like Charlemagne commanded the allegiance of many such warriors and thus created a vast feudal empire. A noblewoman of the ninth, tenth, or eleventh century would be known as the daughter or wife of such a man.

Women of this elite class heard many stories of the exploits of heroes and of great battles, from jongleurs, singers who entertained at banquets in the high-beamed great halls. In the chanson de geste *Raoul de Cambrai,* a warrior breaks the Emperor's truce by hitting his enemy on the forehead with the thigh bone of the venison the women had served him.[1] In the epics and chansons de geste men battled hand to hand with axes and heavy swords, and streams "ran red" with the blood.[2] "The fields were as white with the linen habits of the dead as they might have been with birds in the autumn," remembered one contemporary after the midninth century battle of Fontenoy-en-Puysaye between claimants to Charlemagne's empire.[3]

In the eleventh, twelfth, and thirteenth centuries fighting changed from an occasional fixed duty owed to a chieftain to the main purpose in life, as noble families fought against each other. In this age of feudalism liege lords had the allegiance of vassals, and thus could call up knights to gain territory and with it the prestige and the means to support and command yet more mounted and armored warriors. In this era of the hegemony of families, Eleanor of Aquitaine became the custodian of her father's land and lineage. In these circumstances she and other women directed warfare in the family's name and ruled in their own.

From the beginning of the twelfth century, men turned from the melée of the battlefield to the more regulated combat of the tournament, the joust, and the challenge, played out before the eyes of their women, and the noblewomen's roles became circumscribed once again. Men of the warring elite had begun to set themselves apart by birth, training, and ritual as a special caste, an aristocracy, and their women, no longer fighting and ruling, became the custodians of this privileged status. The treatment accorded to them designated one man as chivalrous and another as churlish. Noblewomen bore children who carried on the family line and title.[4]

By the beginning of the thirteenth century elite women found the circumstances of the noble warrior's world and the roles they might fulfill altering again. In the wars between families, first one and then another family grew powerful, asserting broader authority. Royal families turned their feudal domains into dynastic holdings and with their new centralized rule kept order not by force and private oaths from warrior followers, but through the codification of customary law, the institution of Roman law, and the passage of statute law. Laws came to be administered and enforced by bureaucracies loyal to the dynasty and by royal courts, not by the manorial and feudal forces of the past. In addition to the lord's "ban," or guarantee of order on his lands, there was the "king's peace." As time passed, women acknowledged this changed reality. The wives of the Paston family in fifteenth-century England dated their letters with the year of the king's reign, signifying a world governed and demarcated by a royal dynasty.[5] From the fourteenth to the seventeenth centuries, some men of the warrior and landed elite continued to fight, but now they became the powerful lieutenants of their monarchs. Kings fought against kings and called on the members of the nobility to gather together companies of retainers bound not by oath but by written terms of service, ranked according to pay. These were wars with massed battles of foot soldiers, artillery barrages, sieges ending with pillage and looting. Just as kings waged such wars, so could queens. As leaders of fourteenth- and fifteenth-century dynasties, as the heirs of royal families, women assumed for the second time in these centuries the traditional male power of ruling and fighting in their own name.[6]

Companion and Surrogate

Amidst the dramatic changes in so many aspects of this world of warriors and kings—political authority passing from clans to families to dynasties, war transformed from a seasonal obligation to a paid profession—the role most often given to the women of Europe's nobility was that of companion and helpmate. Whether a knight's lady or a king's mother, these women tended to the needs of their men in peacetime and acted for them in their absence during war.

The literary sources give vivid pictures of the ways in which Europe's noblewomen aided and comforted their men. Guiburc, the wife in *Guillaume d'Orange,* gave drink, towels, and food to the messenger sent to tell them of danger to her nephew, and after his victory had dinner waiting for 500 knights.[7] In another version of the tale she entertained the warriors with "lais and fables."[8] Kriemhild, one of the heroines of the German epic, the *Nibelungenlied,* makes clothing for the hero Siegfried and his close friends. The women of the castle in the romance of *Lancelot* by Chrétien de Troyes gained the poet's praise for creating "the well-ordered household" with daughters, sons, and retainers carefully instructed so that

> even the least among them prepared to perform his special task while some run to prepare the meal, others light the candles in profusion, still others get a towel and basins, and offer water for the hands. . . . Nothing that was done there seemed to be any trouble or burdensome.[9]

In addition to providing the feasting and entertainment for the returning warrior, the wives and daughters of the warring elite helped in the men's preparations for the fighting and experienced the bloody consequences of the combat. The mother in *Raoul de Cambrai* prepares the clothes to be used by a knight on the occasion of his knighting, his initiation into the rituals of the warrior. Isolde checks Tristan's equipment, including his sword, after he returns from fighting. In *Parzival* the women helped remove the men's armor, and bring the water to wash the rust from their hands and faces.[10] A "maiden" in Chrétien de Troyes' romance of *Erec and Enide* took "excellent care" of the knight's horse.[11]

Women of the noble households also had special knowledge of healing. They tended the injured and the sick. Heroines of the chansons and lais dress wounds, bathe the exhausted heroes, and supervise their diets. The two sisters in Chrétien de Troyes' *Erec et Enide* cleaned the "dead flesh" of the wound away, washed and covered it.[12] Maidens with "soft white hands washed and rubbed away" the hero's bruises in *Parzival.* [13] Nicolete carefully manipulated Aucassin's dislocated shoulder, "And so she handled it with her white hands, and so wrought in her surgery, that by God's will . . . it went back into its place." She then made a poultice of herbs that she bound with "a strip of her smock, and he was all healed."[14] Like Queen Isolde, mother of the heroine of the same name in the romance of *Tristan and Isolde,* women of the warrior and landed elite achieve reputations as "physicians," for "the marvellous efficacy" of their medicines.[15] Beatrice of Savoy, mother-in-law of Louis IX, the thirteenth-century king of France, and Elizabeth Grey, the sixteenth-century English Duchess of Kent, both had books of herbal remedies, and books of "physick" to which they referred in giving care. When their men died, it was the women who ritually grieved and mourned their deaths. In the Spanish epic, *The Cid,* the women screamed and scratched their cheeks to show their despair.

The chronicles of the great warrior families and the dynasties of kings, and the records of the secular and religious courts, tell of another aspect of women's lives: the dangers in times of war. In 1199 Joanna, Eleanor of Aquitaine's daughter, gathered troops to put down the rebellion that arose when her husband Raymond left for a crusade to the Holy Land. Pregnant, she barely escaped when the enemy overran her camp and set it afire. The exertion started her labor and perhaps caused her death soon after the infant was born. In France in 1358 near Soissons, the angry attackers from neighboring districts raped and murdered the wives and daughters of the knights they believed had betrayed them. In the hatred generated by rivalry between noble families, contemporary histories indicate that not even the sanctuary of the nunnery was honored. A St. Alban's chronicler of fourteenth-century England reported that when the abbess refused hospitality to Sir John FitzAlan's men, they systematically raped the inhabitants of the nunnery, beginning with the novices and working their way through the widows, the married women, and the nuns.[16]

Women could be endangered by their own families as well. Warriors, whether kings or knights, used their sisters, wives, and daughters as pawns, part of the paraphernalia of their negotiated truces and settlements. Adelheid, the sister of the tenth-century future Holy Roman Emperor, Otto III, was only seven when the two of them were given as hostages.[17] King Henry I of England allowed one of his vassals to blind the daughter of Eustace de Breteuil in revenge. De Breteuil had put out the eyes of the man's son.[18] In 1340, according to the chronicler Froissart, the English King Edward III's queen, Philippa, stayed behind in Ghent as a "guarantee of his honorable intentions."[19] As late as the fifteenth century in Russia, a husband in signing pledges with the Grand Prince of Moscow used his wife and her property as guarantees of his word.[20]

A woman of the warrior and landed elite was always in danger when she traveled. Both her property and her body could be booty, a source of wealth for an enterprising knight or prince. In the epics *The Cid* and *Nibelungenlied,* noble women travel only with the protection of hundreds of warriors. The epics reflected the real danger. The sixth-century Frankish King Chilperic impoverished himself to provide his daughter, Riguntha, with a suitable dowry for her Spanish marriage. All along the road to her betrothed, her entourage—the men meant to protect her—robbed her. They learned of her father's death and abandoned her near Toulouse, taking all that remained of the fifty wagonloads of goods she had been given for her marriage.[21] Six centuries later the dangers remained unchanged. Five hundred knights came with the son of King Louis VI of France when he went to bring fifteen-year-old Eleanor of Aquitaine north to be his bride.[22] The twelfth-century heiress was a valuable prize, easily abducted and forced into marriage. Into the fifteenth century European noblewomen would still be attacked in this way. René of Anjou's wife went on a pilgrimage, only to have her baggage stolen as she passed through the territory of a lord to whom her husband was in debt.[23]

From the very beginning of the European era, wives, daughters, and mothers of royal and noble families acted as surrogates when their men were absent. Then they exercised a warrior's and lord's authority, defended the family's lands and titles, and were honored as "men" when they succeeded. In the eighth century Pepin, the King of the Franks, first gave official sanction to the authority of his queen by having his wife, Bertha, crowned and anointed just as he had been—thus inaugurating a tradition that lasted into later centuries. Judith (d. 840), the second wife of Louis the Pious, ruled with him. Richilda also ruled jointly with her husband, Charles the Bald, Louis the Pious' successor.

The tradition continued among the next generations of Frankish kings. Engelberge (d. 891), wife of Charlemagne's grandson, Louis II, presided over her husband's court at Ravenna, went on military campaigns with him, negotiated over territory with his uncles, and arranged his reconciliation with the Pope. In the tenth century the Holy Roman Empresses Adelaide of Burgundy (also called Adelheid, c. 931–999) and her daughter-in-law Theophano (d. 991) ruled with their sons and protected their own claims to power. Otto II included Adelaide in his decrees, arriving at decisions "with the advice of my pious mistress and dearest mother."[24] When he died in 983, his wife Theophano assumed the title "imperator augustus" and defended her son Otto's title both from dukes and princes eager to support another claim-

1. Noblewomen defending a castle From a medieval book of chivalry.

2. Noblewoman distributing alms. From a Dutch *Book of Hours* (prayer book) c. 1440.

3. A lady arms her knight while other noblewomen watch from above. From a German manuscript, thirteenth century.

4. Uta of Meissen, patron of Naumburg Cathedral. Sculpted by the Naumburg Master c. 1250. Located in the west choir of the Cathedral.

5. Queen playing chess. From a Dutch manuscript c. 1405–1410.

6. Noblewomen embroidering. From a sixteenth-century tapestry.

7. Isabella of Bavaria, Queen of France and wife to King Charles VI. Fourteenth century.

8. Wedding celebration. From a fifteenth-century manuscript.

ant and from attacks by the Slavs and the Danes. The chronicler Thietmar of Merseburg honored her accomplishment: "She was a woman of discreet and firm character . . . with truly masculine strength she preserved the empire for her son."[25]

The royal women of the Germanic kingdoms in England also acted as surrogate rulers on occasion and fought beside their men. In the early tenth century Aethelfled (d. 918), daughter of Alfred the Great, ruled Mercia after the death of her husband and with her brother realized her father's dream of reclaiming the lands seized by the Vikings. The chronicler of the Anglo-Saxon kingdoms recorded her victories at Bridgenorth, Tamworth, and Stafford.[26] She captured Derby and Leicester and gained the allegiance of the northern nobles just before her death.[27] Viking kings forbade women in their camps, but in England they allowed their wives and mothers to rule for them and their sons. Aelfgifu of Northhampton bore two sons to Canute, the eleventh-century conqueror. On his death she ruled as regent for one as King of Norway and forced brief recognition for the other, Harold Harefoot, as King of England.[28]

From the eleventh to the fourteenth centuries the histories of warring families and emerging dynasties in England and France offer examples of royal and noble wives and mothers who helped and supported their men, holding territories and maintaining claims against ambitious rivals. In part, the English and French families controlled so much land because they relied on their women as trusted surrogates. Matilda, the wife of William the Conqueror, ruled Normandy for him while he attacked England in 1066. Although their son-in-law Stephen of Blois inherited the dukedom, their daughter Adela ruled. When warriors went on the Crusades, women acted in their stead. At the end of the eleventh century, Adelaide of Monserrat ruled for her son Roger II of Normandy. When Philip Augustus, the first of France's strong dynastic kings, joined the Third Crusade in the twelfth century, his mother, Adele of Champagne, retained her title of queen, ruled her own lands, and acted as his regent. In 1338 Agnes, the Countess of Dunbar, defended her castle in a five-month siege by the English king, eager to add Scotland to his domain. In 1433 the French Duchess of Bar contracted with Jean de Vergy to fight the English and to "drive them out of Bar and Lorraine."[29]

Margaret of Anjou (1429/30–1482), wife of Henry VI of England, grew up surrounded by competent women acting for the men of her family. Her mother and grandmother had both ruled their lands. Though only in her twenties in 1453, she proved more aggressive and successful than her husband when his right to the throne was threatened. When attacked by the French king and then by rival families within England, she acted for him. She found money, gathered support, and raised troops until her final defeat in 1471.[30]

In the wars of the sixteenth and seventeenth century in England and France, characterized by struggles, both religious and secular, between the new consolidated monarchies and the old landed families, noblewomen fought, defended, negotiated, and participated on behalf of their men in every way. In the English Civil War wives defended their estates against attacking forces, raised money, and traveled from exile to plead for return of confiscated family property. Anne Murray, Jane Lane, and Judith Coningsby helped to smuggle the two Stuart princes, James and Charles, out of England. Henrietta Maria, queen of Charles I, was indefatigable on behalf of her husband's interests; allowed to leave the country in February of 1642, she sold jewelry in the Netherlands to pay troops, and slipped in and out of England to advise and encourage, even when pregnant with their daughter Henriette.[31]

During the sixteenth- and seventeenth-century French Religious Wars, noblewomen gave equally important assistance to their men. Married at sixteen in 1551 to Louis de Bourbon (Prince de Condé), a leader of the Protestant forces, Eleonore de Roye went to the king to beg for her husband's release when he was imprisoned for a plot against the crown. She raised money and support for him when he later fought the regent, Queen Catherine de' Medici, and once again negotiated his release when he was captured. On her death at twenty-eight, her husband wrote to his sons: "Ordinarily boys are expected to try to grow up to be like their fathers. I would urge you to try to be like your mother."[32] Anne-Geneviève de Bourbon-Condé, Duchesse de Longueville, sister of one of the leaders of the Fronde (the faction opposing the royal family), reigned like a queen while the Fronde held Paris in 1649. When her husband and brother were arrested, it was she who went back to Normandy to raise troops, and she who encouraged an alliance with Spain. Anne-Marie-Louise d'Orléans, Duchesse de Montpensier (known to her contemporaries as La Grand Mademoiselle), acted for the family when her father, the Duke, proved too cowardly. She made the alliance with the leader of the opposition against the monarch; she marched into the town of Orléans and won the assembled town notables with a speech. During the 1652 attack on Paris, when she found her father hiding in his room, she went to the fortress of the Bastille and ordered the men to fire on the royal troops. When the queen, Anne of Austria, and her adviser, Cardinal Mazarin, had put down the Fronde, they acknowledged Montpensier's bravery and initiative by sending her, not her father, into exile.[33]

War filled and shaped the lives of Europe's noblewomen. No matter what the character of the fighting, no matter what the cause, women of the warrior and landed elite were drawn into the conflicts. They comforted and counseled their men. They too suffered from the violence. They acted as surrogates administering conquered territories, defending lands and privileges, negotiating the restitution of lands, the release of prisoners and truces. From the ninth to the seventeenth century, war dictated every kind of role for noblewomen: from endangered observers to active agent on behalf of their families.

Marriage

In this world of warriors, lords, and kings, from its beginnings in the ninth century until its decline in the seventeenth, noblewomen survived and functioned as part of a family. A well-born woman was always defined and identified by her relation to its men: daughter to her father, wife to her husband, widowed mother (dowager) to her son. Except for the few who had the opportunity or the vocation for life as a nun, all women of Europe's warrior and landed elite lived dependent on their male kin.

As in the centuries before 800, marriage remained the most significant event of a woman's life. By her betrothal and wedding she passed from the protection and authority of one family to another, from the tutelege of her father to that of her husband. These two ceremonies also signaled the beginning of a new household and the prospect of a new family, much as they did for peasant women and men. Unlike peasant women, however, a noble or royal bride was usually younger than her husband by five to ten years. Typically, from the thirteenth to the sixteenth century she was twenty or younger, he in his late twenties or early thirties.[34]

Marriage among the well-born always occasioned celebration, marked by ancient customs and rituals. Marriages in the sagas and epics resemble the Germanic weddings of Europe before the ninth century, with feasts, gifts of castles and land given for the young women, dowries (gifts to the groom) of gold and silver, public vows taken between the young maiden and her betrothed. The reality was, if anything, more splendid. Early in the thirteenth century when thirteen-year-old Marguerite of Provence married King Louis IX of France, the festivities started in February and ended at the beginning of June. He spent ten times the normal daily expenditure for his household on the clothing alone.[35]

Sometimes the young woman had to be trained for her new life. In 1385 the Duchess of Brabant taught the young Isabella of Bavaria, betrothed to the French King Charles VI, the ways of the court. Froissart, the chronicler, explained that the duchess had been chosen for the task because she "was very experienced in such things."[36] There was a church ceremony, followed by the banquet. Charles from their first meeting "saw that she was young and beautiful and was filled with a great desire to see her and have her."[37] Perhaps like the royal couples in the romance of *Tristan and Isolde,* they followed the symbolic custom of taking wine together before they retired for the consummation of the marriage. Even then rituals governed the occasion:

> In the evening the ladies put the bride to bed, for that duty belonged to them. Then the King, who so much desired to find her in his bed, came too. They spent that night together in great delight, as you can well believe.[38]

Gradually the church came to play a more and more established role in the rituals surrounding the joining of the couple. In particular, the Church added its authority and guarantee to the vows exchanged by the couple. A history of the twelfth-century Guines family of France described the priest officiating at the taking of vows and then blessing the marriage bed along with the groom's father. Other French noble families of the twelfth century had the priest present for the exchange of gifts, for the giving of the ring, and for the bride's kneeling to acknowledge her husband's authority over her.[39] In 1316 the noblewoman later to be known as Saint Bridget of Sweden went through a church ceremony first. The priest blessed the couples' rings, placed his stole over their joined hands, and intoned the words that described her relationship with her husband: "Let the yoke she is to bear be a yoke of love and peace." In her father's great hall the couple then performed a secular version of the ritual. Her father literally gave her to her husband with these words: "So I give thee this my daughter to honour and wife and to the half of thy bed, of locks and keys, of every third penny and all the right which is hers after the law."[40]

Just as in the early centuries of Christianity, the Church's involvement in the proceedings came from a desire to ensure the consent of the young woman and man to what it viewed as indissoluble vows. In the *Decretals,* the first codification of Church law in the mid-twelfth century, in answer to the question "May a daughter be given in marriage against her will?" the author, Gratian, stated: "By these authorities it is evident that no woman should be coupled to anyone except by her free will."[41] Gratian envisaged a formal betrothal and a wedding ceremony, both blessed by a priest. Although the Church never gained acceptance for its opposition to the exchange of gifts (theologians believed that this likened marriage to enslavement), the need for formal guarantees of consent and the officiation of the priest did prevail. In the mid-sixteenth century at the Council of Trent, marriage was declared a sacrament. The ceremony shifted from a single public exchange of vows to a betrothal announced on successive occasions (usually three for Catholics and Protestants) or written agreement and a wedding ceremony in the presence of a priest and witnesses.[42]

The words spoken in Saint Bridget's ceremony and the gestures made symbolized significant aspects of the union. Most important to Europe's warrior and landed families would be the rings given and received. The rings signified the exchange of gifts: the promise on the part of this man that he would provide for the bride as her father had, and the contribution made on her behalf by her family to the material well-being of the union.[43] The usual provision from the groom's family would be a portion of the land that he had claim to, from one-third to one-half, which she traditionally claimed on his death. An Anglo-Saxon marriage agreement from Kent in the early eleventh century explained that Godwine would give his bride "a pound's weight of gold in return for her acceptance of his suit, and he granted her the land at Street with everything that belongs to it, and 150 acres at Burmarsh and in addition 30 oxen and 20 cows, and 10 horses and 10 slaves."[44] An emperor's gift would go far beyond acres and livestock. The gift of the tenth-century Holy Roman Emperor, Otto II, to his bride, Theophano, included the three provinces of Istria, Walcheren, Wicheln, the city of Pescara, the abbey of Nivelles and four manors.[45] By the thirteenth century it was customary for the woman to receive this, her "dower," at the beginning of the marriage or her "jointure," as a "widow's portion," at the time of her husband's death. She would then have the use of it until her remarriage, when she would pass into the care of another man, or until her death, when it would revert to the heirs. The custom of the bride's contribution to the union had existed in the Roman Empire as the "dowry." It survived after 800 and by the midtwelfth century was the practice all over Europe, families following the pattern of the newly revived Justinian code in setting the amount. In most betrothals the dowry equaled the donation of the husband's family.[46]

The more important the families in the feudal hierarchy, the more significant the negotiations and the less attention given to the wishes of the individuals to be wed. Marriage arrangements became more obviously a matter of business, of hearings before the feudal overlord whose consent was required, or, later in the age of royal justice, a matter of legal documents written by notaries and witnessed by family members.[47]

With the transfer of land so significant, religious and secular authorities rarely questioned the families' arrangements. Even though first Catholic and then Protestant doctrine espoused support of the couple's wishes, when appeals came to their courts, they usually concurred with the parents' contract. Cases heard in English church courts of the fourteenth and fifteenth centuries show that the churchmen rarely supported women claiming forced marriage; only actual beating, not threats of violence or refusals of inheritance, seem to have made them sympathetic to women's pleas. When in the sixteenth and seventeenth centuries the secular authority became concerned with matrimony, the royal statutes and royal courts always upheld the primary right of the parents to give or withold consent to a match. For example, though Danish, Russian, and French laws of the mid and late sixteenth century required periods of betrothal to ensure the willingness of the couple and a Church ceremony, they also affirmed the parents' essential right to consent. Parental authority superseded the couple's wishes. This continued to be standard among Europe's noble and royal families into the eighteenth century.[48]

In addition to the transfer of property, a young woman's marriage meant the transfer of her person, the giving of her reproductive ability. The sexual union of the couple, the children she would bear, made it possible for both sets of parents to continue their lineages, their family names. Prospective brides thought practically of this aspect of marriage. In their song three twelfth-century women poets, the trobairitz (troubadours) Lady Caranza, Lady Alais, and Lady Iselda, jointly advised women to pick a man who will "plant good seed."[49] Royal families expended special care in the choice of potential royal mothers. King Edward II, the early fourteenth-century king of England, sent Bishop Stapeldon to inspect Philippa of Hainaut, a prospective bride for his son. She was only eight but received a good report: "reasonably well shapen; all her limbs are well set and unmaimed; and nought is amiss as far as man may see."[50] Froissart tells of the selection of a wife for Charles VI, King of France, in 1385. The father of the prospective bride, Isabella of Bavaria, agreed to the procedures only after being reassured by the Duchess of Brabant. He disliked the indignity of the process and the idea of his daughter marrying so far from home. The duchess probably supervised the viewing of the young woman, nude, to see that "she is fit and properly formed to bear children."[51]

A daughter's marriage could bring benefits to an enterprising family, especially access to higher status through her joining with a suitable young man. In the twelfth century, the Anglo-Saxon parents of Christina of Markyate (later known as Saint Christina) opposed her religiously inspired wish to remain unmarried for "they feared that they would lose her and all that they could hope to gain through her."[52] As her contemporary biographer explained, her parents wanted her to give up her pious vows and marry well so that "she could have enriched and ennobled not only herself and her family but all her relatives."[53] In the sixteenth and seventeenth centuries land-poor noble families used the marriage of their sons to tap the wealth of the merchants and manufacturers of the walled towns. The townsmen's daughters brought the money to revitalize the estates, the noblemen's sons the lineage to enhance her family's social position.[54]

Royal families could do even more with the marriages of their daughters, using them as the means to maintain, establish, or strengthen their political authority. Control of marriage could neutralize potential enemies. In the ninth century the Emperor Charlemagne feared that, if his daughters married, their husbands would endanger his hegemony. So he allowed them sexual liaisons, provided for their illegitimate sons, but forbade marriage. Others sent their daughters to nunneries for the same reason.[55] Queens and kings usually took a more positive attitude, however. Marriage of daughters could mean alliances to support the dynasty or new lands to increase its holdings, to enlarge the royal domain (demesne). According to the *Anglo-Saxon Chronicle* Henry I of England arranged his daughter Matilda's marriage in the twelfth century "in order to secure the friendship of the count of Anjou, and to secure assistance against William (son of Robert, Duke of Normandy) his nephew."[56] Queen Eleanor of Aquitaine and King Henry II arranged all of their daughters' marriages carefully, placing them with the most powerful feudal overlords in Europe: Matilda married to Henry of Saxony, Eleanor to Alfonso of Castile, and Joan to William of Sicily. At the end of the fifteenth century Isabella of Castile and Ferdinand of Aragon showed exceptional skill in this kind of diplomacy. Having newly consolidated their control of Spain, they married each of their four daughters to one of the great European dynastic families: Juana to the Hapsburg heir Philip the Fair, Catherine to the English heir Arthur (and to his brother Henry when Arthur died), and Isabel (and then after her death, Maria) to Manuel, king of Portugal.[57]

Given the multitude of ways in which a young woman was valuable, all of the older traditions against kidnapping, against raping and then claiming a young woman as bride, survived as crimes in the customary, statute, and ecclesiastical laws of Europe. According to the decrees of kings, emperors, and Church councils, an abductor suffered a wide range of punishments from fines and whipping to exile and death.[58] In fact, however, a family would not protest and would not find the laws useful unless they could appeal to a force or influence as strong or stronger than the abductor. No such countervailing force existed to protect Rogneda, the tenth-century Russian princess. She refused a suitor, Prince Vladimir of Kiev, because of his low birth. He pillaged and burned her father's land and killed him and her brothers. She then became one of his many wives and concubines. In the twelfth century Eleanor of Aquitaine left her lodgings in the middle of the night to escape from Count Thibaut of Blois. She had to ride on again to avoid an ambush set up by another nobleman, Geoffrey of Anjou, brother of the man she was on her way to wed, Henry Plantagenet (the future Henry II of England). Both men planned to kidnap her and to force her, the heiress to so many rich lands, to accept them as husbands.

Protests and objections to such treatment were expected to melt away when fines had been paid, a dowry offered, and the marriage ceremony blessed by a priest. In this, Europeans drew on ancient traditions from the cultures before 800, where marriage of the dishonored woman was more important than punishment of the man. Following this tradition, the midsixteenth-century Council of Trent did not rule out that the woman might in a "safe and free place" consent to marry her persecutor (although they also excommunicated the abductor and made him pay her dowry).[59] Laws of earlier centuries, especially those of the Spanish warrior enclaves like Cuenca and Sepulveda of the tenth to thirteenth centuries, reflected this practical attitude. They punished the theft of their women, but accepted as consensual unions the marriages their citizens made with daughters they had seized from other settlements.[60]

In the last resort, if they wanted a match, fathers, mothers, and feudal overlords would themselves force the daughter to wed. In twelfth-century England, Christina of Markyate's mother beat her, threatened her, and then tried to arrange her rape to force her to break her vow of virginity. In twelfth-century Ireland, Isabella Heron's father had three men drag her to the church, and he beat her to make her go in to take her marriage vows. Katherine Dowdall's mother hit her, and her father threatened to break all her bones to force her into the marriage bed.[61] In fifteenth-century England, the Pastons beat their daughter Elizabeth and kept her locked away to force her to agree to a match with a fifty-year-old widower. She gave in and was spared only because the negotiations fell through.[62]

To discourage opposition and to insure that their plans would come to fruition, parents, especially Europe's kings and queens, arranged marriages for their daughters and sons when they were very young, sometimes before the Church's designated ages of consent (twelve for girls, fourteen for boys at the end of the thirteenth century).[63] At the end of the twelfth century Eleanor of Aquitaine and her husband, Henry II of England, married their nine-year-old daughter Eleanor to the ruler of Castile, a groom of twelve. When a province, the Vexin, was suggested as dowry for the three-year-old Marguerite, princess of France, Henry II seized her, took her to his lands in Normandy, and married her off to his five-year-old son. Another son, Geoffrey, was only eight when his parents betrothed him to the five-year-old heiress to Brittany. The Russian nobility wed their children in similar ways. From the fourteenth to the sixteenth centuries the average age at marriage for children of those holding lands and serfs was twelve for girls and fourteen to fifteen for boys.[64]

Eleanor's contemporary, Philip Augustus, King of France at the end of the twelfth century, abused marriage even beyond the ruthless standards of his day. He unscrupulously used both force and law to increase his lands and his authority. No heiress escaped his attention. He explained to the citizens of Falaise and Caen in northern France: "We will not marry widows and daughters against their will . . . unless they hold of us, in whole or in part, a *fief de haubert.*"[65] A *fief de haubert* meant enough land to support one knight—even such a small amount of territory and a single warrior attracted this king. Philip Augustus treated members of his own family as he did other noblewomen. He sent his sister Alais as a gift of reconciliation to King Henry II. When she returned to France in 1195, unmarried, aged thirty-five, and reputedly the discarded mistress of the English king, Philip married her off to one of his supporters. He treated his own wives just as callously. He took Ingeborg of Denmark as his bride for the 10,000 silver marks of her dowry; one day after the ceremony in 1193 he sent her away. For twenty years he kept her imprisoned, while he lived openly in a bigamous marriage with another woman, despite the interdict of his kingdom declared by the Pope. Only in 1213 after the death of his favorite, Agnes, and the establishment of the legitimacy of their two children, did he release Ingeborg and acknowledge their marriage.[66]

For Ingeborg of Denmark marriage meant helplessness and disgrace, for Agnes, the favored wife of Philip Augustus, protection and provision—what the institution was intended to give a woman. Whether marriage made her the pawn of her family or the leader of it, a noblewoman saw no real alternative. As a daughter, she accepted her family's plans for her. When she reached adulthood, she became a wife. And as a wife, she accepted marriage with its liabilities as well as its compensations.

Land, Faith, and Children

With her marriage the noblewoman, the châtelaine, the lady, assumed responsibilities as significant to the maintenance of feudalism as the days of fighting and the mounted warriors owed by her husband to his liege lord. In the feudal age and in the relatively more peaceful centuries of the centralized monarchies, wives of the lords, castellans, and seigneurs often became the principal supervisors of the lands on which the family depended for food and revenue. These elite women accepted other responsibilities for their families. They, more often than their husbands, saw to the spiritual well-being of the household, offering their prayers and faith for the family's protection. These women bore the children and watched over their early years, thus ensuring the continuation of the family into a new generation.

A noblewoman fulfilled these roles and functions from her husband's fortress, perhaps from a castle like that of the French Countesses and Counts of Matignon. Fort la Latte still sits on the cliffs of Brittany overlooking the sea. Though small, the stronghold appears virtually impregnable. Only a narrow causeway connects it to the mainland, and that is protected by a barbican, a tower with its own drawbridge. Outer walls enclose the inner complex of stone buildings. From the ramparts there is a clear drop to the pebbled beach and the sea thirty feet below.

When attacked, the family took refuge in the round tower keep, a three-story building not more than eighteen feet in diameter with each floor a single room. The roof of the third floor is the simplest brick vaulting. The windows are small, with slits under each to use for defense in wartime. Under seige the family barricaded itself here, surviving on the produce of the household garden, a cleared space beside the keep among the outbuildings where they stabled the animals and stored wood and grain. In the seventeenth century, the Matignons probably thickened the outer walls, bolstered the stone with earthworks, and rounded the towers on the cliff side in an effort to make the fortifications less vulnerable to artillery.

In its simplicity and its size, this Breton castle is far more typical of the kind of dwelling lived in by the wives, daughters, and mothers of feudal nobles than the elaborately crenellated, multitowered buildings drawn in the fifteenth-century Books of Hours or imagined by nineteenth- and twentieth-century writers and artists. By the end of the twelfth century women and men of property lived in stone castles like Fort la Latte, in a central keep roofed with wood, thatch, or shingles surrounded by wooden buildings for the cooking, storage, and livestock. Earthworks, fences, and a moat provided protection for the family and their retainers. A knight's family, members of the lesser nobility or the gentry, lived in a fortified manor, a simple wood and stucco building faced with stone, the doorways made of carved sandstone. Rubble provided the fill to make the thick walls needed for protection. The biggest landholders, like the Guines of Ardres-Flanders in northern France, had three levels in their stone tower keep: the cellar and a first floor were used for storage; the second floor was the great hall with timbered floors strewn with rushes to scent the air and to absorb the offal.[67]

Most activity took place here around the central hearth, where a hole in the roof allowed the smoke to escape, the ceiling blackened from the fires. Windows remained small, shuttered, perhaps the highest ones covered with oiled parchment or horn, to let in some light while keeping out the cold. Despite whitewashed walls, the room, however high the ceiling, would be dark even in daytime. Most women used rushes dipped in animal fat as wicks in metal holders. Mahaut, the fourteenth-century Countess of Burgundy, had a hanging lamp fired by a wick that floated in oil to use inside her enclosed bed.

In the eleventh, twelfth, and thirteenth centuries all was simple and practical. Separate areas for different activities like cooking, sleeping, eating, or giving birth to children did not exist. There was no real privacy. The family, retainers, and serving staff all lived and spent most of their time indoors together in the great hall of the castle. The lady or châtelaine performed the nightly ritual of covering the central hearth fire with a pottery lid, allowing just enough air to keep the embers alive.[68] The household members took their mats from chests along the walls, and arranged them on the floor. The lady and her lord retired to their bed, a wooden framed platform with hangings for privacy. Sometimes the great hall might have a room to the side or above, like a loft, for them.

In these early centuries furnishings were sparse, and everything was portable: the hangings on the wall, stools, trestle tables, straw mattresses, truckle-beds, chests, the wooden tub lined with cloth for bathing, the glass for the windows, even the lady and the lord's bed was a wooden frame easily broken down and reassembled. These were the "movables" described in agreements about women's inheritances and in negotiations for their marriages. All was portable because one of the noble wife's duties was to supervise the moving of the household from dwelling to dwelling for reasons of economy, safety, and administration.

The Lands and the Household

For all but a few of the most powerful families, a castle like that of the Guines filled every need. For the Breton Countess of Matignon, of Fort la Latte, as for other women of this landed culture, it was on the one hand a refuge, at least in theory a sure defense against the attacks of other families. On the other hand it was a base from which her men forayed out to assert their authority, to plunder and to kill. Thus threatened and protected, the noblewoman lived in the world of feudalism, governed by a political system predicated on the assumption that peace and order could be kept by private oath and the prospect and reality of force.

According to feudalism, no family "owned" the land; instead each "held" it from another more powerful man, the head of another more powerful family. To hold the land, or fief, meant to have the authority to use it and its peasant population to support oneself and one's family. In return, a man like the Count of Matignon made his oath of fealty, pledging to fight when called upon and to keep peace on the lands he had been given. The size of the holding corresponded to an equivalent number of warriors; so many acres could support so many fighters. From the twelfth to the seventeenth century, a knight or chevalier, the single mounted and armored warrior, would require the labor of twenty-five to thirty peasant families to supply his family with provisions and to generate the revenues needed to pay for his horse and equipment.[69] By the eleventh and twelfth centuries, the loosely defined obligations that had characterized the earliest centuries of feudalism had been refined to specific duties and payments in kind or service. Feudal customs adjudicated disputes and determined punishments of those who did not meet the terms of their oaths of fealty. When, from the thirteenth century on, money became more and more essential to this system of governance and to the waging of war, the terms of the feudal relationship adjusted to meet the new monetary needs.

Throughout all of the changes and all of the centuries of feudal warfare, the women of the noble family remained the key to the maintenance of the entire system. At the center of feudalism was the castle and the lands that supported the warrior. In his absence it was the lady of the manor, the châtelaine of the castle, who took responsibility for the "living"—the land and its produce. She administered the means by which the warrior could be provisioned and armed, everything from harvesting the oats to feed his horses to selling wheat to pay for his coat of chain mail.

The complexity of her task was directly related to the size of the holding and thus to the size of the household she and her family could support. The châtelaine of Fort la Latte might have worked in the garden herself; she supervised the peasant women and men hired from the lands for specific tasks, like cutting and bundling the grain. Women who lived on larger holdings did no physical labor themselves and instead employed others. Mahaut, the much wealthier Countess of Artois, had a retinue of servants who traveled with her from castle to castle: a shoemaker, laundress, chaplain, physician, secretary, treasurer, steward, and cook.[70] Twice a year, at Easter and All Saints Day, she gave gifts to the servants and to all of her ladies-in-waiting (the young women of noble birth who acted as her companions and personal attendants). The gifts ranged from gold chaplets and gold and silver braid to cloth and furs.[71] The account books of Eleanor de Montfort, sister of England's King Henry III and wife to the successful warrior conqueror, Simon de Montfort, show that in 1265 she had a large staff appropriate to her royal lineage and the lands she and her husband held.[72] In addition to the employees listed for the Countess Mahaut, her contemporary, Eleanor de Montfort, had staff both for the estate and for the household; for example, the estate steward supervised those tending the stables and horses and the groom for the hunting dogs, whereas the household steward dealt with those caring for the plate, preparing the linens and the candles. Tailors made clothing, an almoner distributed food and the occasional pence to the poor, a chaplain and his clerks said mass for the household.[73] When the king came to visit for two weeks at Easter in April of 1265, Eleanor de Montfort was responsible for feeding and tending to his retinue as well.

None of these responsibilities changed into the seventeenth century and the eras of centralized dynastic rule. Rather, more families became part of this elite, and thus more noblewomen came to have larger households requiring provisioning and care. In seventeenth-century England these new members of the elite called themselves gentry. Together with the older noble families they represented 2 to 3 percent of the population.[74] In 1684, for instance, the English baronet Sir Richard Newdigate's wife had seven daughters, all under sixteen, and twenty-eight servants to provide for and look after.[75] Most typical of the gentry women whose records have survived was Margaret Fell, the wife of the local judge, who had a moderately sized household: her children, one female servant, a companion for her son, his tutor, a clerk, and an estate agent.[76]

When Saint Bridget of Sweden married at the beginning of the fourteenth century, her father had spoken of her "locks and keys," the symbol of the responsibilities she assumed as a wife. Perhaps she wore her keys at her waist. As ladies, as châtelaines, Saint Bridget and other women of the warrior and landed elite took charge of everything within the castle. In the thirteenth century these tasks and the supervision of the lands seemed so varied that Robert Grosseteste, the English bishop, wrote a book of household management to guide the widowed Countess of Lincoln, Margaret Lacy. At the beginning of the fifteenth century Christine de Pizan, the French courtier and writer, offered a similar guide in her *Book of Three Virtues (Livre des trois vertus).* Both instructed the noblewoman in duties involving every aspect of the estate: supervising the staff and laborers, calculating the annual revenues and taxes, overseeing the harvest, buying and selling goods, slaughtering and curing livestock. Grosseteste even offered the countess advice on how to rotate use of the earl's different residences, including how to arrange for the moves this entailed.[77]

Women's own records and letters show the reality to have been as complex as the guides suggest. Just a few entries from the Countess of Leicester's account books for the spring and summer months of 1265 (February 19–August 29) indicate the range of her responsibilities and the details involved in providing for a grand household and caring for a large holding. She arranged to stable the new foals, to grease one cart, to repair others, and to buy a new one. She paid the barber to come to bleed a member of the household. She bought wine, shoes, and hose. She hired the messengers to carry her letters to her husband and other noblemen. In the fifteenth century Margaret Paston more or less ran the family's estates all the time. Her husband John served in Parliament, was imprisoned three times, and then died in 1466, leaving her to cope. Her son also left her in charge. She held the property against an armed attack by the Duke of Suffolk in 1465, actively participated in court cases over title to their lands, all in addition to settling disputes among tenants, hiring laborers, supervising the bailiff, marketing the crops, buying weapons, selling timber, and arranging to borrow money. In 1476 Elizabeth Stoner found the duties almost overwhelming in her husband's absence. She explained: "when you wrote me I had no leisure; truly I have been sick or busy, or else I would have written to you before this."[78]

From the tenth to the seventeenth century, all of the noblewoman's skills were called into play to produce "dinner," the main meal, usually served in the middle of the day. Her household would gather in the great hall to eat three or four courses, all her responsibility. She would have overseen the collection of every ingredient—either raised or purchased—supervised the preparation of the food in her kitchens, and the serving of the meal. Dinner began with broth or meat soup followed by a second course of boiled beef, roasted mutton, capon fattened on oats, or pigeons from the manor dovecote. All were eaten on trenchers of bread that sopped up the juices and gravies. Then came fresh or salted fish from the local ponds or herring bought in the market. Hard cheeses or soft ones like Brie came next, followed by fruits or pastries for the dessert. There was beer or watered wine to drink. Rich households served white bread, and those at the manor "high table" ate off plates. The lady of the manor or castle also served two other meals: "break fast" in the morning consisting of bread, meats, and jams and supper in the evening with porridge or broth made from grains, legumes, and roots.[79]

The more lands held, the more revenue generated, so the more variety the nobleman's wife could provide for meals. At markets she would acquire products that she could not produce from the family holdings. Choice and special foods like oranges came as trade expanded and fairs and markets spread throughout western Europe. In the thirteenth century Margaret, the Countess of Lincoln, sent bread and beer to the nearest fair and expected to buy wine, wax, and clothes there. By the thirteenth century, families of property bought spices and fruits. The Countess of Leicester accounted in her household records for purchases of dried fruits, salt, mustard, pepper, ginger, cinnamon, and cloves. In seven months she used 53 pounds of pepper. All were expensive, ranging in price from 10 pence to 2 shillings, at a time when a laborer's daily wage was 1 pence and a cow cost 6 shillings 8 pence.[80] Her household was so large that she needed to buy many of her staples as well. On one Sunday she purchased six sheep, one ox, three calves, 8 pounds of fat, six dozen fowl, and ten geese. One Saturday in May, she bought 600 eggs for 22 ½ pence; on another Saturday, bread, wine, grass, and "oats for 52 horses." She needed 13 pounds of wax for candles to light the chapel.[81]

From the fifteenth century on, wives at all ranks of the warrior and landed elite used the local markets for basic needs as well as for special luxuries. In the fifteenth century, the English gentry family, the Pastons, bought dates and oranges, cloth, weapons, and sugar there. Their contemporary, Queen Catherine, wife of King Henry V, was so dependent on trade that she procured a safe conduct so that merchants could bring goods to her household in the midst of the fighting in France during the Hundred Years War. Into the sixteenth century ladies like Philiberte de Luxembourg sent servants to the Lyons fair for spices, fine clothes, fabrics, laces, threads, and needles. In the seventeenth century the more frugal Sarah Fell (daughter of Margaret Fell) grew and made everything she could on the estates of Swarthmore Hall, including hemp for rope and household cloth. She still needed to sell vegetables and cheese in order to buy wine, sugar, leather gloves, shoes, utensils, and dress fabric.[82]

Faith

The noblewoman's responsibility for the household carried with it responsibility for the spiritual well-being of the family. The Christian religion was as important to the mistress of the manor as it was to the peasant women who worked her fields. Whether Catholic or Protestant, from the ninth to the seventeenth centuries wives of the warrior and landed elite demonstrated their piety in good works that benefited those living on their lands and in gifts to the Church. They believed such acts would help to assure their salvation. The thirteenth-century Countess of Leicester, as the lady of the manor, was the principal source of charity for the local village. She noted 4 pence a day distributed to the poor by her almoner, as well as the leftovers from the meals in the great hall.[83] In thirteenth-century France, Mahaut, Countess of Artois, supervised the giving of blankets, clothing, and shoes to the poor at the beginning of winter. She planned the distribution so carefully that no one received the same gift two years in a row.[84] In the late sixteenth and early seventeenth century some elite women bought grain to distribute in times of shortage. Others like the Englishwomen Lady Alice Lucy, Lady Warwick, and Lady Arundell used their skills to tend the neighborhood's sick.[85] Unfortunately, such efforts would have done little to alleviate the general suffering of the poor.

Noblewomen, like noblemen, made gifts of land and money for use by the Church, for the establishment of needed services, and for prayers both for their own and others' salvation. A Spanish countess articulated the feelings that motivated her charity:

> I fear the pains of hell and I desire to come to the joys of paradise, and for the love of God and his glorious Mother, and for the salvation of my soul and those of my parents, I give to God, St. Mary, and all the saints my whole inheritance in Retoria.[86]

Records for Spain, Italy, France, and Germany from the eighth to the twelfth centuries show noblewomen alone, or noblewomen and their heirs making from 7 to 18 percent of the land gifts to the Church.[87] In Anjou the pattern of gifts received for three religious establishments from 1000 to 1249 showed that of the total of 833 donations, wives alone had given 11.8 percent; together with their husbands they had given 31.4 percent.[88]

Noblewomen founded monasteries, almshouses, and the first hospitals in Europe. In the early eleventh century Agnes of Burgundy spent £1,100 in the foundation of two abbeys: La Trinité and Notre Dames des Saintes.[89] Mahaut, Countess of Artois, was remembered in her lands not only for the gifts of her seasonal charity but also for the hospital she had built at Hesdin. It had an enclosed courtyard and garden—a cloister—a chapel and small rooms for a matron and servants. The patients lay in a 160-foot-long ward with a sixteen-foot-high gabled roof and ten large windows along one wall.[90] Women of Europe's royal families also founded hospitals: Margaret of Bourgogne, Queen of Sicily; Blanche of Castile, Queen of France; Elizabeth of Aragon, Queen of Portugal; Isabella, Queen of Spain; and Margaret Beaufort, mother of King Henry VII.

Some well-born women gave enormous sums of money to have prayers said for others and for themselves. In 1155, Avice, daughter of the sheriff of Nottinghamshire, gave "alms for the refreshment of the monks on the anniversary of my burial so that through their intercession my spirit may be refreshed in the skies with celestial food and drink."[91] Eleanor of Aquitaine gave money to her favored monastery of Fontevrault for masses to bring the family good fortune and then in 1199 for prayers to insure salvation for her dead husband and son (Richard). In a grand gesture of piety, Queen Margrethe of Denmark gave 26,000 marks to establish a chantry (an endowment for masses for the dead) for all of the men who had died in the wars of her country.[92]

Other women of the nobility hoped by special acts of personal devotion to demonstrate their religious zeal. At the end of the fifteenth century Queen Isabella of Spain gave a religious tone to her entire court. She wore the robes of a Franciscan Tertiary, she prayed the canonical hours just like a nun. She and her ladies embroidered altar cloths for the church. The English Puritan, Lady Mary Hoby, in 1599 begrudged walks with her husband as "nothing reading nor profiting myselfe," and asked God to "pardon my omissions and commissions, and give mee spirit to be watchfull to reduce the time." Proper use of time for her meant the fulfillment of her duties: prayer, preparation of meals, the washing, care of the tenants and of the crops, supervision of the Sunday school she had established for the parish children.[93]

Religious zeal might also take the form of encouraging learning: the commissioning of books of piety, the spread of religious texts and their ideas from one part of Europe to another, the support of individual churchmen, and in England the establishment of colleges. In the eleventh century Judith, sister of the Count of Flanders, took the set of Gospels made to celebrate her marriage to the Anglo-Saxon nobleman Tostig with her on the occasion of her second marriage to the Duke of Bavaria. Thus she brought Anglo-Saxon illuminated manuscript to the Continent, where monastic artists could copy the techniques. In the fourteenth century Anne of Bohemia, queen of Richard II of England, owned Bibles in Latin, German, and Czech. It may have been through her spiritual advisers that the writings of the English religious reformer Wycliffe reached his Czech counterpart, Jan Hus.[94] Margaret Beaufort, the mother of England's King Henry VII, supported the churchman John Fisher and encouraged and arranged for the publication of his writings. She also translated part of the mystic Thomas à Kempis' writings. She endowed chairs of divinity and founded two colleges where men might study at Cambridge: Jesus and Christ's College.[95]

Childbearing and Children

In the seventeenth century the English puritan Lady Hoby established a Sunday school on her estates, providing religious instruction for the children of peasant families. Like all noble wives, she also supervised the religious education of her own children. Propertied mothers often taught their offspring to recite their first prayers, to read, to write, and to do simple sums. Famous sons praised their mothers' efforts. The learned churchman of the twelfth century, Peter Damian, remembered the teaching games and alphabet cards he had had to play with.[96] In the sixteenth century Suzanne de la Porte took time from managing the family's properties to educate her son, the future Cardinal Richelieu.

Daughters learned from their mothers and also from the women whose households they joined. Well into the seventeenth century the practice continued of sending little girls to another woman's care. Just like boys who went to train for knighthood under the auspices of a respected warrior, so girls went to learn in the household of an honored lady. In the thirteenth century the Countess of Leicester kept her daughters with her until they were almost thirteen. This was unusual. More commonly a daughter would go to live with another noble family when she was six or seven, even in knights' families. In the fifteenth century the gentry family, the Pastons, had trouble finding places for their daughter. In the end they had to pay Lady Pole 26 shillings 8 pence for boarding Elizabeth.[97] A large noble household, in contrast, could accommodate many girls. The seventeenth-century Duchess of Hamilton took on the children of three other families, giving them the opportunity to learn the manners and skills of the nobility under her tutelage.

More significant in some ways than the care and education of daughters and sons was the fact of childbearing itself. For part of the purpose of marriage, of the joining of the two noble families, was the prospect of the woman's fertility. She brought not only her dowry to the marriage, but also the gift of her body, her ability to carry on the lineage. In the fifteenth century in England a pregnant noblewoman might be pampered with special foods like dates and oranges. Elaborate rituals and customs surrounded birth in a royal household. Queen Elizabeth, wife of the English King Henry VII, paid a religious house for use of the girdle (belt) believed to have been worn by the Virgin to ease her labor pains.[98] Queens like other Christian women went through the ritual of "churching" when the new mother first returned to church to be blessed and thus cleansed by the priest. It was popularly believed that Saint Elizabeth, the thirteenth-century Queen of Hungary, as a special act of devotion went to the altar after the birth of her children just as the Virgin Mary had in simple clothes with the symbolic offering of a candle and a lamb. The other key Christian ritual of birth, the baptism of the infant, might be performed on the natal day or as long as six weeks after the birth.

Neither the passage of the centuries nor the accumulation of more or less property ensured a safe delivery for mother and child. Queens, like commoners, risked their lives each time. On the Crusade in 1270 with her husband Philip III, the pregnant Isabella of Aragon fell from her horse when fording a stream. She miscarried and died a few days later. Jeanne de Bourbon, wife of King Charles V of France, bore four children but died in 1378 from the complications of the last birth.

Historians have tried to compile mortality figures for women of the English nobility in the seventeenth century. Among titled women, 45 percent died before age fifty, of those, one quarter died from childbirth.[99] A strong woman faced many years of childbearing. Seventeenth-century women have left accounts of their experiences. Alice Thornton, a member of the English gentry, recorded the details of each of her confinements. She first conceived seven weeks after her marriage. She had complications with each pregnancy, but still produced a child a year from 1654 to 1657. Her son, born in December of 1657, presented "feete first" and caused her bleeding for weeks afterward.[100] Margaret Fell, another gentrywoman, married in 1632 at seventeen and produced nine children with two miscarriages. In contrast, her daughter Bridget died in childbirth in the winter of 1662–1663. Another daughter, Sarah, wrote her mother of an easier confinement in May of 1684: "I was about six hours in travail, and though it was sharp the Lord was good to me in giving me strength to go through it and endure it and gave me deliverance in his own good time of a sweet babe."[101]

Most brutal and painful was death from the unintentional butchering that occurred when the mother could not deliver the child. In seventeenth-century England, Elizabeth Tufton described her friend Frances Drax's labor. The pains began on Saturday. Not until Monday did the waters break. On Tuesday Mrs. Baker, the midwife, announced that the infant was dead. They called in a surgeon to try to remove the fetus from her body. "Hee continued his endeavors till 10 a Clock at night, and we were forced to give her Cordialls almost every minute . . . the man protested hee tryed all the wayes he could imagine tho in vaine, and shee growing faint and light headed, begged of the Doctor for Christ's sake, to let her dye at rest." Her friends helped her back to bed and sat with her while she bled to death.[102] Not until the nineteenth century would women even of the highest rank have less hazardous experiences of childbearing.

The women of the castles swaddled their babies just as the peasant women did, carefully wrapping each limb and then the infant's whole body in cloth bands. Records from as early as the eleventh century show that noble wives did not always nurse their infants themselves. They paid a peasant woman, usually one whose child had died, to come to the manor as "wet" nurse. In the grand households there were "dry" nurses as well. Eleanor, Countess of Leicester in the thirteenth century, had a nurse for each child. The wealthy seventeenth-century Duchess of Hamilton kept a whole separate staff of attendants for the children: both wet and dry nurses, a chambermaid, and a washerwoman.

Just as death claimed the children of peasant women, so it took the children of the warrior and landed elite. Accidents and disease left roughly half to survive to adulthood. Disease struck often. Alice Thornton's son William died of smallpox soon after birth; another son, Robin, died at four; Christopher lived only three weeks. She prayed "that my soule may be bettered by all these chastisements."[103] Margaret Fell lost two of her children in infancy. Her daughters fared less well. Isabel lost three of her four children, Mary five out of ten, Rachel lost three of her four babies as infants.[104] The very rich had the same experience. Lady Mary Fielding, Duchess of Hamilton, bore six children in eight years. Four died before their seventh birthday. The birth of William in 1638 so weakened her that she died soon after. Her daughter, Anne, proved stronger, producing thirteen children between 1657 and 1671; ten lived to adulthood, a remarkable number in this era.

With so much at risk, with so much subject to change from the ninth to the seventeenth centuries, the constants in the lives of noblewomen remained. Though they rarely held the land in their own right, they gave instructions to bailiffs and seneschals—the overseers of the family manors. They supervised the workings of the estate and the household. To them fell the responsibility for the lineage, since they bore the children to carry on the family name and the claims to the family's lands. They were as essential to the feudal system and the order of the kingdom as their men.

3

POWER AND VULNERABILITY

Wives and Daughters in the World of Feudalism: The Ninth to the Twelfth Centuries

In the world of European feudalism of the ninth to the twelfth centuries, the interests of war were not served by women holders of land. In theory they could not fight; in practice, they were not trained to do so. Therefore they could not be warriors and in theory could not fulfill the basic obligation on which the feudal relationship between lord and vassal depended. But circumstances did not always honor theory. After the death of Charlemagne in the ninth century, a combination of circumstances arose that favored the subversion of this basic premise of a warrior culture. Just as in the Roman Empire and the early Germanic kingdoms, some European women fought on their own behalves and ruled in their own names.

The political circumstances of the eleventh and twelfth centuries resembled those of certain eras before 800. It was a time of transition where one form of political authority evolved into another; the rule of one people, like the Franks or the Anglo-Saxons, gave way to the rule of many, to an era where families contended for power. In this context women acquired land and waged war just like men. Despite the apparent illogic of a female warrior and lord, women played decisive roles in this period of feudalism, not just as surrogates for their fathers or husbands, but as lords and vassals defending their own lands and privileges.

Women of the warrior and landed elite came to have this power because of a series of changes in European feudalism. With Charlemagne's empire divided and centralized authority dissipated, families found ways to weaken the former authority of their chieftains and kings. They favored the right of inheritance. By the end of the eleventh century, land tenure (the right to hold and administer a fief) no longer automatically returned to the bequest of the feudal lord on the death of each holder. Instead families asserted the right of their children to inherit the fief: the right to authority over the lands and the people who worked them, and the right to the produce. From the ninth to the twelfth centuries, the primacy and relative independence of the family came to be established in other ways that would ultimately work to the advantage of women. This period saw the birth of the concept of "lineage." The taking of a common family name and the creation of a line of descent from female and male ancestors replaced identification by clan or by region. By the second half of the eleventh century, individuals born of the same mother and father in some parts of France began to use a common surname. In the absence of sons, a daughter might be the sole repository of this newly established family name and lineage.[1]

In the eleventh and twelfth centuries in Europe, economic change also contributed to the circumstances that allowed women to hold traditionally male powers. In the evolving commercial economy, with money replacing goods and services as the means of exchange, families had to reassess the ways in which their properties passed to the next generation. From the eleventh century on, being a landholder in this increasingly monetary economy required more than the ability to fight and rule. Money had to be found, money owed now as "aids" to the feudal lord, money for hospitality, for display—the trappings of the elite class—and most important, money to equip the mounted and armored knight.[2] Lords and vassals saw their lands as their main source of revenue. If lands passed by right of inheritance into the hands of many sons and daughters, the hard-won wealth would divide into smaller and smaller units. Eventually these units would be too small to sustain the obligations that ensured the continued power and status of the family.

Confronted with these economic realities, the nobility made some of the same choices as the peasantry. They limited their offspring by late marriage and restricted the number of sons who could marry.[3] More significantly for their daughters, families in most regions of Europe abandoned any semblance of equal inheritance and instead designated a portion of the holdings as "the patrimony," giving this to one heir (usually male, who by definition could perform the services owed to the feudal landlord). Lands came to be consolidated into larger holdings. In the eleventh, twelfth, and thirteenth centuries, the greatest of the feudal magnates amassed lands through inheritance, ruling whole provinces as their fiefdoms and using the wealth they generated to stave off the centralizing efforts of kings and emperors.[4]

This combination of political and economic circumstances, these new choices meant to enhance and insure the power of families, worked to the advantage of a few women. The strength and ambition of families had established the right of inheritance and the pattern of consolidated holdings in the hands of a favored heir. The strength and ambition of families meant that in the absence of sons, daughters were identified first by lineage and only secondly by gender. As the ninth-century law of the Thuringians had phrased it, rather than lose the land the inheritance passed from "the spear to the distaff."[5] All across Europe, from Russia to France, from Italy to Norway, men ignored customary prohibitions against female inheritance of the patrimony and allowed a favored daughter to inherit just like a favored son. In this way noble families could sustain the right of inheritance, continue their lineage, and retain or increase the families' properties.[6]

In the absence of sons, daughters became sole heir, held land in their own right, and fulfilled the family obligations to fight and to rule just like men.[7] The economic changes only made this anomaly more possible, for a paid male surrogate or a set amount of money could fulfill the feudal obligation to fight. A woman had only to generate revenue to fulfill the ancient obligation of the warrior, his military service. In the twelfth century two fathers, King Henry I of England and Duke William X of Aquitaine, bequeathed their lands to their daughters. Thus, Matilda became Queen of England and Eleanor of Aquitaine became the most important vassal of the King of France, a landed magnate holding more than one third of the territories under his suzerainty.[8]

Though the Norman kings of England had claimed the throne by right of conquest, their power was so great that they passed it undisputed from father to son. Then the male line failed. In 1120 when his son William died, Henry I recalled his widowed daughter Matilda (1102–1167) from Germany, where she had been married to the Holy Roman Emperor, and planned for her accession. In 1127 he held New Year's court at Windsor, designated her as his heir, and made his tenants-in-chief, his principal vassals, swear fealty to her. To neutralize the powerful Counts of Anjou, he arranged for the twenty-five-year-old Matilda to marry the Anjou heir, the fourteen-year-old Geoffrey Plantagenet. She bore him three sons, and the lineage of William the Conqueror seemed assured. In 1137 Henry I died, and Matilda's right to inherit lands and power stood to be tested. Ironically, her chief rival, her cousin Stephen, claimed the throne through female inheritance as well, through his mother, Adela, Henry I's sister. Many great tenants-in-chief saw these unorthodox claims and this rivalry as their opportunity to assert their independence against this strong centralizing dynasty. Matilda remained in England, while her husband went to claim Normandy. She led her own armies, supported by her Scots uncle, a cousin, the Earl of Northumberland, and her father's two illegitimate sons, the Earls of Gloucester and Cornwall. She won battles, she lost castles. There was no clear victory, and she was never popular. She brought order, but contemporaries described her as arrogant and too direct. In the end it was not her skills as a warrior and ruler that won her the right to the throne, but her female body and fecundity. Her children gave her the advantage over Stephen; Eustace, Stephen's heir, died, and in 1154 it was Matilda's son, Henry Plantagenet, who became king of England as Henry II.

Similar circumstances gave lands and access to power to the daughter of Duke William X of Aquitaine. Eleanor of Aquitaine (1122–1204) came from southern France, where great families had become hereditary overlords of vast territories, paying only cursory attention to the king, their superior in the feudal hierarchy. Like Matilda, Eleanor came to her lands because of a series of deaths. Both her mother and her brother died in 1130 when she was eight.[9] Her father died seven years later on a pilgrimage to Spain. Before leaving on this journey, he had made arrangements to protect her and to keep the lands intact despite his lack of a son. Louis VI, the French king, was to be her guardian. His authority could be used to secure the lands against collateral branches of the lineage. His authority was meant to intimidate other landed magnates who might try to seize the holdings, the daughter in whom all claims lay vested, or both. In return for his protection, the king as her guardian could decide whom she would marry. Eleanor would bring to her marriage lands and revenues, rivaling the power and wealth of his own royal family. She held the provinces of Poitou, Guienne (Aquitaine), and Gascony, almost one third of present-day France. When Eleanor was fifteen, just before his death early in August of 1137, Louis married her to his own seventeen-year-old son, thus securing her lands and the allegiance of her vassals to the royal French dynasty.

Eleanor's young husband, now King Louis VII, looked to others for advice: Abbot Suger of Saint-Denis and his mother, the dowager Queen Adelaide of Maurienne. This did not deter Eleanor. Even at fifteen she acted like a queen and used her own authority as a feudal magnate. She created a court of gaiety and display. She asserted her claim to Toulouse through her mother's lineage. She pressed for the marriage of her sister to Louis' adviser over opposition from her vassals and the Pope. In 1145, Eleanor and the female companions of her court enthusiastically joined the Crusade to win the "Holy Land" from the Turks.

In the Middle East Eleanor raised the question of divorce with Louis, manifesting the independence and courage her heritage gave her. As ally and husband Louis had failed her. He had not reclaimed Toulouse for her and allowed her little influence in his lands. In an age when women and men believed that the father determined the sex of the child, she had borne only a daughter. As the protector of the family's lineage, Eleanor needed sons who could inherit and then govern the lands she had been bequeathed. Without sons the patrimony would be absorbed into the royal holdings and become the king's, to be distributed as prizes to the sons of other families.[10] When Eleanor raised the question of divorce, she came to know the limits of her power. Louis refused, and when she argued, he had her forcibly carried off to be with him on his campaign to regain Jerusalem.[11] His attack failed, and on the voyage home to France, Eleanor prevailed upon him to break the journey in Italy and consult with the Pope. The pontiff personally escorted the couple to bed and a renewal of their marital duties.

She persisted, singlemindedly and patiently, in her quest for dissolution of the unwanted union. The birth of another daughter in 1150 helped her cause, as did the death in 1151 of Suger, one of the most obstinate opponents of the divorce. She prevailed upon Louis to remove his troops gradually from her lands. In March 1152 the Pope granted them an annulment. The last part of Eleanor's planning assured her a new partner and protector, a warrior worthy of an heiress of her stature. In the summer of 1151 she had met Henry Plantagenet, son of Matilda of England. She was thirty, he eighteen. Perhaps they committed adultery.[12] In 1152 Eleanor left her young daughters Marie and Alix in their father's custody (part of the annulment agreement) and rushed south to her capital of Poitiers and marriage with Henry. He was heir to the throne of England and the French provinces of Maine, Touraine, Anjou, and Normandy. Together they ruled more than half of France and, after 1154, all of England. In addition, Henry II's ambition and energy matched hers. And he was young and strong, the kind of man contemporaries believed would give her sons.

The early years of their marriage proved Eleanor's assumptions correct. They ruled as partners. She traveled with him, over 3,500 miles in 1157–58 in the administration of their lands.[13] She presided with Henry at his periodic ceremonial courts, acted as regent in his French territories. She ruled for him in England while he was in France. From 1154 she collected her own tax from Henry's English subjects, known as the queen's gold. Eleanor of Aquitaine was as powerful as any feudal noblewoman could be. In addition, she bore a child almost every year between 1152 and 1166—eight pregnancies in all, with four sons and three daughters surviving to adulthood.[14]

The royal couple maintained this effective partnership until their sons began to come of age. Then all of Eleanor's instincts and ambitions as a feudal magnate in her own right came into conflict with Henry's desire to consolidate and centralize authority in his hands as monarch. Eleanor valued lineage and local hegemony. She wanted to establish each son, bequeathing lands when and as she chose. Henry wanted the lands consolidated to ensure resources vast enough to counter the power of the French king. The birth of a male heir to Louis VII's third wife in 1165 made them both more determined to have their own way.

Eleanor wanted her sons to have title to their own lands and the revenues so that they would be firmly established while this future royal French overlord was still a child. In theory, Henry agreed with her; he allocated lands in January of 1169, but he kept the revenues and the authority in his own hands. At the Christmas court at Chinon in December of 1172, their eldest son Henry, just seventeen, with Eleanor's encouragement confronted his father. When Henry II refused to relinquish any of his power, Eleanor joined the boys and their ally the French king in rebellion. The campaign was disastrous. Her sons barely escaped to take refuge with Louis; she delayed too long in Aquitaine. When Henry turned his troops south to pillage and waste her lands, he captured her. She was riding astride dressed in men's clothes with only a few knights to protect her.

Henry pardoned everyone but his wife. She was too important to kill, too important to leave free. In this world of feudalism she had both too much power and not enough. As a feudal magnate Eleanor commanded the fealty of her vassals, but they had not been able to win her victory. On the other hand, her death at Henry's order would only make them more hostile to his authority. As a solution to his problems Henry may have suggested an annulment and her retirement to the nunnery at Fontevrault, the dual monastery that she patronized.[15] Eleanor refused, and Henry chose what appeared to be his only alternative. He kept her prisoner, isolated from her followers, for fifteen years in Salisbury Castle. As time passed she refused any suggestions of reconciliation. He allowed her the occasional public appearance: for the marriage of their daughters, on the death of their eldest son in 1183. As once before with another king/husband, she waited, and she planned. She encouraged her favorite son, Richard, to claim Toulouse in her name and to rebel again. This time he defeated the exhausted and perhaps disheartened old man in the spring of 1189. Henry II died that same year.

With the accession of Richard, all conflicts between lineage and loyalty ended. All that he claimed, Eleanor acknowledged, all that he ruled, she helped to protect as his mother, the dowager queen. Her first public act was to release herself from custody and to issue an amnesty decree explaining that she knew "by her own experience that prisons were depressing to men and that to be released from them was a most delightful refreshment to the soul."[16] An English chronicler, Richard of Devizes, described her about this time (c. 1190), when she was in her sixties: "incomparable," "beautiful yet virtuous, powerful yet gentle, humble yet keen-witted, qualities which are most rarely found in a woman."[17]

She used all of these qualities, drawing on her intelligence and her feudal authority to ensure Richard's power and then his brother John's, as first one son and then another came to rule. Eleanor now became guardian of both her father's and of her husband's lineage. Into her seventies she helped to secure the family's lands and titles. She fought for Richard when her youngest son John tried to turn his brother's absence on Crusade to his own advantage. She raised the 35 tons of silver needed to ransom Richard in 1173.[18] She reconciled the brothers on Richard's return. In 1199 while she tended her dying son Richard, Louis VII's son, now reigning as King Philip Augustus of France, prepared to take their lands. At seventy-seven Eleanor raised an army, retook Anjou and Maine, and gave John title to Aquitaine. She helped to arrange the truce by which Philip Augustus acknowledged John as King of England.

She stood by her youngest son and fought for him despite his blunders and his defiance of his feudal overlord. In 1202 John seized the betrothed of one of his vassals for his own wife, giving Philip Augustus a reason to attack their lands. Eleanor left Fontevrault again to secure Aquitaine. But she had reached the limits of her physical strength and her powers. Only John's forced march saved her from capture when she was beseiged in her castle of Poitiers in July of 1202. She never recovered her spirit and died in April of 1204.

Eleanor was spared the series of disasters that characterized the rest of her son's reign as he lost most of the lands that she had administered and defended so carefully. Even had John been more circumspect, however, the era of the great feudal magnates and the opportunities it offered aristocratic women was coming to an end.[19] The political and economic changes of the twelfth and thirteenth centuries wore away at their powers. The dynastic monarchs like Philip Augustus of France, not the tenants-in-chief like Eleanor, came to offer the relative order and peace Europe's society demanded. As kings amassed their personal holdings, they commanded the revenues needed to maintain power and authority. From the thirteenth to the seventeenth centuries, only Italy and Germany continued in the pattern of feudal lineage and splintered holdings.

Unusual circumstances had given Eleanor of Aquitaine access to political power well beyond what was normally allowed a woman. Even so, in a world at war, only the strongest personalities, the strongest lords, whether female or male, with clear and protected claims to the lands that made warring possible, could hold power against other vassals, or against the gradually evolving authority of kings. This was no "golden age" for Europe's elite women. However favorable the circumstances for them, they still remained vulnerable and subordinate. Queen Matilda's father arranged her marriages. Eleanor of Aquitaine could be held prisoner by her husband. In an era when landed women assumed men's roles and held vast lands in their own names, it was not as warriors that they best protected their lineage and defended the interests of their families. Instead, as in other ages, in the long term, their traditional function and role proved the most significant. Matilda and Eleanor of Aquitaine bore sons. Those sons inherited the lands and passed them and the authority they represented on to the next generation. It was as mothers and the bearers of the legitimate progeny that even the most enterprising and courageous women best served Europe's warrior and landed elite.

Courtly Love and Codes of Chivalry

The sculptor of Eleanor of Aquitaine's effigy portrayed the Queen holding a book. Eleanor, like queens and noblewomen before her, had acquired a reputation for her reading and for her patronage of a world of peaceful, leisured activities. Into the fourteenth century churchmen chronicling the reigns of their kings praised the traditional learning and intellectual accomplishments of royal daughters and wives. In the ninth century Charlemagne's daughters had been educated with his sons. Alcuin, their tutor, reported Gisela's unusual interest. She liked to study "the stars in the stillness of the night."[20] The thirteenth-century code of Saxon laws, the *Sachsenspiegel*, stated "As only women read books, these are to be inherited by them."[21]

By the late twelfth century many noblewomen's intellectual interests had broadened, and the number of books available to them had increased. Instead of being limited to texts in Latin, those who could read now had works translated into or written in the vernacular. Mahaut, the wife of the Count of Burgundy, had among her manuscripts the usual sacred books but most of them in the vernacular not in Latin. She had her Bible, her prayer books—the psalter, a Book of Hours—the lives of the saints and Church Fathers, the "miracles of Our Lady" and the works of the Roman philosopher Boethius, all in French. In addition, Mahaut owned manuscripts that reflected other than pious concerns: a verse translation of the laws of Normandy, chronicles of the kings of France, and a version of the travels of Marco Polo, reflecting her interest in politics and trade. For her own amusement and entertainment she had "romances," the stories composed for the rich courtly world where women like Eleanor patronized not only churchmen but also poets.[22]

From the tenth to the early fourteenth centuries the interaction of East and West had fostered the creation of these secular interests and this new courtly world. Time spent in the courts of Crusader kings introduced Europe's nobility to leisured and elegant pursuits, many borrowed from their Muslim enemies, many adapted and brought back to their own lands. Trade with the East brought the beginnings of luxury. The clothes of Europe's noblewomen and men, the ways they spent their time, the roles and functions they envisaged for themselves, all changed. The women and men of the warrior and landed elite came to value comfort and peaceful activities. They learned to enjoy reading and writing in their own language. They increasingly did so for pleasure rather than spiritual improvement. They transformed the gatherings in the great halls of their castles. They took the values and rituals of the Crusades and the warrior and turned them to new uses. In their courtly gatherings, honor came to be awarded not only by men for prowess on the field but also by women for success in love. "Homage" came to describe not only the relationship between vassal and lord but that between men and women. The court came to be a place not only for adjudication of manorial and feudal disputes but also a place for the judgment of lovers' claims.

In the late twelfth and early thirteenth centuries, the members of the courts of Henry II of England, of Eleanor of Aquitaine, and particularly of her daughter Marie, Countess of Champagne, initiated this transformation.[23] They sponsored music, singing, and the creation of a new vernacular literature.[24] In the twelfth and thirteenth centuries, the poet singers—the troubadours and jongleurs—traveled from castle to castle. Creators of romances, (long narrative poems) like Chrétien de Troyes entertained their noble patrons. In dedicating *Lancelot* to Eleanor's daughter Marie, he explained that he was but the practitioner, "my lady of Champagne" had given him "the material and the treatment."[25] Courts like these flourished in many parts of Europe well into the fifteenth century: in Aragon, in Andalusia, in Blois, Burgundy, and Flanders. As queens had once spread Christianity by their marriages, so now they brought cultural change. When Anne of Bohemia came to England to be Richard II's bride in 1382, she arrived with an entourage of poets and jongleurs.

In this more luxurious and elegant world, noblewomen and men changed the way they dressed. By the twelfth century the rough wools and linens of their wardrobes were supplemented with silks and satins from the Arab lands in Spain and the Turkish lands of the eastern Mediterranean. Their songs and stories described the new colors, bright purples and scarlets, the gemstones like rubies, diamonds, and sapphires, the jewelry like gold filigree bracelets for a woman's wrist and circlets for her hair, the accessories like mantles and cloaks trimmed with sable and ermine.

Members of the nobility no longer wore the straight, loosely belted long overtunics that had previously given females and males much the same appearance. Instead the man's tunic became fuller at the chest and shoulders and shorter to reveal his legs and his hose in bright greens, blues, and oranges. The woman's underchemise showed the lines of her breasts and was girdled at the waist; the sleeves tapered and could be buttoned to the wrist. The overtunic remained long but might now be fitted and have lined sleeves, a cloak hanging loosely about her shoulders, fastened only at the neck.[26]

Women and men spent their time differently. The images of their writings—of the lyrics, chansons, lais, and romances—show the peace-filled days of "gentlewomen" and their "chevaliers," as contemporaries called ladies and their knights. They rode and hunted together, strolled and talked in the enclosed gardens of espaliered quince and pear trees with nightingales, larks, goldfinches, and thrushes singing in the protected warmth of the afternoon sun. Their songs and stories glorified spring as a time for young women and men to fall in love. Chrétien de Troyes described such a scene, an open meadow:

> maids, knights, and damsels playing at divers games in this pleasant place . . . backgammon and chess or dice . . . most were engaged in such games as these; but the others there were engaged in sports, dancing, singing, tumbling, leaping and wrestling with each other.[27]

A noblewoman had leisure time to work over her embroidery frame, while listening to a lyric or a romance. A thirteenth-century English poet complimented Eleanor of Provence, Henry III's queen, on the embroidered men, birds, animals, and flowers that she had designed and sewed, and especially on "the way she divided her colours."[28] In the fourteenth century Lady Joanna, the daughter of England's King Edward III, bought gold thread and silk to use for the piece she was working on rather than the more common wool and linen.[29]

Into the fifteenth century women used the tent stitch to create the background, sewing the thread diagonally across the weave of the fabric to fill in and around the design. They made the graceful lines of their designs with the split stitch, passing the needle through the fabric and then back again, splitting the previous stitch in a series of small, closed loops.[30] Ladies also did weaving on lightweight vertical looms for large pieces of cloth or on small table looms for the tapes they used to decorate their tunics and cloaks.

Women Poets and "Courtoisie"

The women and men of the courts imagined new roles and functions for themselves, not just wives and husbands fulfilling their obligations, but also lovers with each succumbing to the strength of their passion. Women and men of the courts wrote of these new roles. As early as the eleventh century in the Rhone Valley of Provence, noblewomen were composing songs and poetic narratives—lyrics, chansons, and lais. Of the four hundred known singers and poets, or troubadours, twenty were women, called in the language of Provence "trobairitz."[31] Wives and daughters of landed noble families, they amused themselves, their friends, and their retainers by describing the feelings, the motives, and the actions of lovers. Although they probably did not travel from castle to castle like the male troubadours, they expected their verses to bring them fame and reputation. The trobairitz Castelloza (born c. 1200) from Auvergne told her audience that she entertained to be thought well of, to gain respect and prestige.[32] In style and subject their verses differed from those of their male counterparts. They ignored many of the formal trappings of the men's verses, used less allegory and word play. In the free conversational tone of their chansons and tensons (lyric poems of debate) they idealized neither the lover nor the feelings of love. Instead they were direct, practical, sensual, and spontaneous. They avoided studied reactions and alternately expressed happiness, anger, sorrow, and passion.

The women's chansons celebrate the excitement and the intensity of time with a lover and the pain at separation from his embrace. Tibors (b. c. 1130), the sister of a troubadour and the wife of a powerful Provençal lord and patron, wrote in the mid-twelfth century:

> Sweet handsome friend, I can tell you truly
> that I've never been without desire
> since it pleased you that I have you as my courtly lover;
> nor did a time ever arrive, sweet handsome friend, when
> I didn't want to see you often;
> nor did I ever feel regret,
> nor did it ever come to pass, if you went off angry,
> that I felt joy until you had come back. . . .[33]

Clara d'Anduza of Languedoc (flourished first half of the thirteenth century) wrote that she was so angry and resentful at not seeing her lover that "when I try to sing / I weep and sigh; and I can't make music. . . ."[34] In her tenson Alamanda (flourished second half of the twelfth century) described the spurned lover's rage: "I'm so angry that my body's / all but bursting into flame."[35]

Men's lyrics and chansons praised discretion and delighted in the anguish of anticipation. The women, in contrast, called for forthright declarations, railed at missed opportunities, and delighted in thoughts of the consummation of passion. Clara d'Anduza sang that those who tried to ban her love would be "powerless to help my heart improve / or to increase my love for you, / or my longing, my desire or my need."[36] Beatriz, Countess of Dia (b. c. 1140) has four extant poems, the most of any of the trobairitz. She wrote with anger about her own discreet behavior:

> I have been in great distress
> for a knight for whom I longed;
> I want all future times to know
> how I loved him to excess.

Now I see I am betrayed—
he claims I did not give him love—
such was the mistake I made,
naked in bed, and dressed.[37]

She imagined the ecstasy she had missed:

How I'd long to hold him pressed
naked in my arms one night—
if I could be his pillow once,
would he not know the height of bliss . . .
I am giving my heart, my love
my mind, my life, my eyes.[38]

This imagined ecstasy might be "illicit" sex (as the Church would have called it) before marriage, or adultery with a lover whose "gentle, sparkling face" the poet wanted "in my husband's place."[39]

The trobairitz looked for different qualities in their lovers. While male poets wrote of physical attributes, the women dwelt on the character and behavior of the man—qualities that came to be called "courtoisie." The Countess of Dia explained: "I've picked a fine and noble man, / in whom merit shines and ripens— / generous, upright and wise, / with intelligence and common sense."[40] In an anonymous tenson an older woman instructs a maiden pleading the cause of a knight:

Maiden, if he really wants my love,
he'll have to show high spirits and behave,
be frank and humble, not pick fights with any man,
be courteous with everyone;
for I don't want a man who's proud or bitter,
who'll debase my worth or ruin me,
but one who's frank and noble, loving and discreet. . . .[41]

Most important of all to the trobairitz, the lover must be faithful to her and not humiliate or betray her once she had given him her affection and her favors. Isabella (b. c. 1180) voiced the noblewoman's worst fear: that her troubadour had sung her praises not "out of love / but for the profit [he] might get from it."[42] Others valued obedience in their male lovers. To lie with her love "a single night" the Countess of Dia had asked him to "promise this: / to do all I'd want to do."[43] Trobairitz sent angry verses if the man turned to another. Alamanda, in an imagined dialogue between herself and a troubadour, dismissed the new woman "who next to her's worth nothing, clothed or nude."[44] She wrote that she wanted to jilt him to prove her strength. The Countess of Dia described her rage:

Of things I'd rather keep in silence I must sing:
so bitter do I feel toward him

whom I love more than anything.
With him my mercy and fine manners are in vain,
my beauty, virtue, intelligence.
For I've been tricked and cheated
as if I were completely loathesome.[45]

These women poets mourned love lost and love never returned. The Countess of Dia complained "For all my wit and charms he doesn't care. . . ."

Another now has coaxing words to say,
Another smiles and welcomes you today,
Think of the kissing—dawn of our affair—. . .[46]

Azalais de Porcairages (born c. 1140) confessed in her chanson that "my pain comes from Orange" (possibly Raimbaut d'Orange, the troubadour) and that "My heart is so disordered / that I'm rude to everyone."[47] She used the images of winter to describe her situation:

Now we are come to the cold time
when the ice and the snow and the mud
and the birds' beaks are mute
(for not one inclines to sing):
and the hedge-branches are dry—
no leaf nor bud sprouts up,
nor the cries of the nightingale
whose song awakens me in May.[48]

Just as royal and noble women created their own style and wrote from their own perspective in their short lyrics and songs, so they gave the longer verse form, the romance, a distinctive character. Marie de France (flourished twelfth century), accompanied by her harp, sang of Celtic and Breton heroes—Guigemar, Lanval, Tristan—just like her male contemporaries, Chrétien de Troyes and Thomas. In her hands, however, instead of being like the men's long tales of adventure after adventure, they became briefer episodes giving a picture of one event, one situation, one set of emotions. The story becomes a vignette called a lai narratif. In addition to form, her perspective differentiated her writings from other twelfth-century poets both female and male, and from the later creators of romances like Gottfried von Strassburg (author of *Tristan and Isolde*). She too described beautiful heroines and courageous heroes living in sumptuous surroundings: a baby wrapped in a coverlet "hemmed all around with marten fur," a tent of purple taffeta, a horse with silk trappings.[49] She too wrote of love, lovemaking, and adultery, the themes of the twelfth- and thirteenth-century romances. But when she sang of this world of lovers, she showed its unpredictability, its effects on individuals, not just the moral lessons it might teach. She chose to emphasize character and situation over narrative. In this she was unique, and perhaps it was this choice that made her verses so widely enjoyed. Each of Marie de France's lais exists in a number of manuscript copies, in Middle English, Middle High German, French, Italian, and thirteenth- and fourteenth-century Old Norse. Such variety in an age before printing attests to the continuing popularity of her writings.

Scholars know little about Marie de France. Probably the daughter of a landed Norman family, she lived at the court of Henry II of England, to whom she dedicated her lais as her "brave, courteous" patron.[50] Between 1160 and 1215 she translated from Latin, and perhaps from English for the pleasure and edification of the king and the women and men of his circle.[51] She explained that she decided to compose lais rather than do translations because they were harder to write. In all her work she showed the same pride in her skill and kept to the same purpose—to entertain and gently to instruct. In the lai *Milun* she began:

> Whoever wants to tell a variety of stories
> ought to have a variety of beginnings,
> and speak so intelligently
> that people will enjoy listening.[52]

With the Celtic and Breton tales she turned to verse. She wanted to do justice to the stories for, "Whoever deals with good material/feels pain if it's treated improperly."[53]

To treat the material properly she gave her listeners women and men in their familiar roles: as mothers and ladies, as lovers and warriors, but each with a realistic touch. The knight in *Eliduc* teaches a young princess to play chess; the foster mother of *Milun* urges her nephew "to do good deeds" and more practically "also gave him plenty of money."[54] She portrayed lovers and sometimes justified their adultery, blaming magic (like her male contemporaries) or the possessive and restrictive behavior of the husband.[55]

Unlike her male contemporaries, however, Marie de France depicted illicit loving with imaginative and original turns of plot and characterization. In both *Yonec* and *Guigemar* the villainous husband fears his wife's beauty and locks her up. In both tales, the passion of the young lovers is the means for them to escape the injustice and unpleasantness of their world. In *Yonec* the woman in despair jumps from her window, but instead of dying she is transported to a silver city; she bears her lover's son, who returns as an adult to kill the evil husband. In the story of *Lanval,* the lady, not the knight, is the rescuer. He rests in a meadow, having been forgotten in King Arthur's distribution of "wives and lands." As if in a dream, two maidens appear and bring him to a magnificent tent where

a lady promises him "If you are brave and courtly,
no emperor or count or king
will ever have known such joy or good
for I love you more than anything."[56]

She makes him swear to keep their love a secret, but when he is falsely accused by the Queen of making overtures, he offers as his defense his love for the mysterious lady. Even though he has broken his oath, she appears, dazzles the court with her beauty, and clears his name.

Marie de France takes time in her lais to show the enjoyment women and men took in each other. The lady who dazzled King Arthur's court in *Lanval* is the ideal of female beauty: "no one who looked at her / was not warmed with joy."[57]

She was dressed
in a white linen shift
that revealed both her sides
since the lacing was along the side.
Her body was elegant, her hips slim,
her neck whiter than snow on a branch,
her eyes bright, her face white,
a beautiful mouth, a well-set nose,
dark eyebrows and an elegant forehead,
her hair curly and rather blond[58]

When Lanval first saw her, she was partially uncovered "her face, her neck and her bosom; / she was whiter than the hawthorn flower."[59] After this first meeting, from then on "he was impatient to hold his love, / to kiss and embrace and touch her,"[60] Marie de France's heroines had similar reactions to the men. Guilliadun takes Eliduc "by the hand" and

they sat on a bed,
they spoke of many things,
She looked at him intently,
at his face, his body, his appearance;
she said to herself there was nothing unpleasant about him.
She greatly admired him in her heart.[61]

The young wife of *Yonec* loved her knight on sight, "For he was of great beauty," and after she had laid "beside her love" and they had "laughed and played and [spoke] intimately"

She wanted to see her love all the time
and enjoy herself with him.
As soon as her lord departed,
night or day, early or late,

she had him all to her pleasure.
God, let their joy endure![62]

Marie demanded the same devotion and qualities of character in her lovers as the trobairitz. For her it was not the passion or the illicit love that constituted sin, but rather the denial of love's force or the man's betrayal and inconstancy. Even willing and eager heroines give themselves only after heroes have declared their devotion. Lanval promises:

I shall obey your command;
for you, I shall abandon everyone.
I want never to leave you.
That is what I most desire.

The heroine in turn then "granted him her love and her body."[63] In Marie de France's writings, love was a bond that surpassed all other obligations or laws. In *Milun* the faithfulness of the lovers is rewarded; they manage to send messages to each other over a twenty-year separation and then on the death of her husband, "In great happiness and well-being / they lived happily ever after."[64] Marie punished adultery in only one of her extant lais, *Equitan*. This, however, was love between social unequals, between a seneschal's wife and a king, indulged in "without sense or moderation."[65]

Marie de France retold and adapted familiar tales with special force. *Laüstic* exists in many versions, ending usually with the death of the lover or the husband. She focuses instead on the emotions of the three characters: the lady rising from her bed at night, going to her window wrapped in her cloak, knowing that her lover too "was leading the same life, / awake most of the night," taking pleasure in seeing each other "since they could have nothing more."[66] In the vignette death comes not to the lover or the husband but to the nightingale that has become the symbol of the couple's love. Marie describes how the husband

killed it out of spite,
he broke its neck in his hands . . .
and threw the body on the lady;
her shift was stained with blood
a little, on her breast.[67]

Eliduc is similar to the familiar tale of Isolde and Tristan, but Marie de France uses only the end of the story when Tristan is apart from Isolde and is tempted by the love of another woman. Even more originally, she tells the tale from the point of view of the women. In her lai Tristan becomes Eliduc, the "other Isolde" Guilliadun, daughter of the king he has come to serve when banished from his own lands. Throughout the tale, the women, not the men, are the active agents. Guilliadun first declares her love; Guildeluec (Tristan's wife) resolves the conflict. She saves Guilliadun's life and rejoices that she can give her to Eliduc. "I want to leave him completely free," she explains.[68] Then she graciously prepares to become a nun. Marie makes what she calls "a very good end"; all three join religious orders.

That the three principals of one of the great love tales of the era all go to monasteries has a certain comical quality, and perhaps this was intended. Marie de France, unlike her male contemporaries, made fun of traditional images and actions. At the end of *Lanval,* having saved the hero, the heroine starts to ride away on her horse. He is the one to leap up behind so that they can ride off to live happily ever after. The wife of *Laüstic* is satirized for setting a "wonderfully high value on herself," and for loving the knight not so much "for the good that she had heard of him / as because he was close by."[69] The beautiful educated heroine of *Chaitivel* could not choose which of four knights to honor with her token in the joust. "She didn't want to lose / three in order to have one."[70] So she gave tokens to all four and, when three died, mourned them, planned to write a lai in their honor, and ignored the survivor. Marie retells a familiar folk theme with a particularly satiric twist. In *Les Deus Amanz* the king keeps his daughter for himself by setting impossible tasks for her suitors, for example: to carry her up a mountain. Undaunted, she fasts, and her lover goes to Salerno for a strengthening potion for himself. Both end up dead, he from exhaustion because he never stopped long enough to drink his potion, she in a swoon on discovering that he had died.

Chivalry

Marie de France might mock the adventures and behavior of ladies and their knights, but by the end of the twelfth century the attitudes and activities she referred to reflected new ideals of conduct for the men of the warrior and landed nobility. The concept of "chivalry" had evolved. The romances both influenced and reflected this new definition of knightly behavior and the ways in which it affected women. In the peaceful world of courtly love, men learned "courtoisie" and were embraced or shunned by women for their traits of character and their constancy. When men went to war, their treatment of women became equally significant with the creation of the ideals of chivalry.

The twelfth- and thirteenth-century lyrics, lais, and romances portrayed a very different kind of hero from the earlier epics and chansons de geste. In *Njal's Saga,* first sung in the settlements of Iceland in the late tenth century, Gunnar, the leading warrior, was tall, a "powerful man" who could use his sword with either hand.[71] Rainoart, the hero of *Aliscons,* one of the twenty-four epics glorifying Guillaume of Orange and his family, was known for his "swiftness, his size," and having the "look of a boar."[72]

In contrast the idealized warriors of the romances are described as "noble" with "fair and curling hair" and a mouth that "shone in redness like a ruby."[73] Tristan is beardless, graceful and admired for his hands: "soft and smooth, fine and slender and dazzling white."[74] These heroes no longer fight with bones at banquet tables or hack at their adversaries by a bloody river as they had in the chansons de geste. Instead they go hunting or hawking, and they prove their skills as warriors in the ritual display of fighting ability, the tournament.

Heroines of the romances had no place in these mock battles. In the stories sung by jongleurs and trobairitz, they were admiring observers, giving favors to the victors and comfort to the defeated. In the imagined wars of the poets, rarely did they even act as surrogates for their men. Most often they were shown as innocent bystanders in need of protection.

To honor and protect women became a key part of the imagined hero's activities. In addition to birth, performance in the lists, and reputation in war, the knight of the twelfth- and thirteenth-century romances was distinguished from others by his behavior toward women. This too became a measure of true masculine nobility that set him above other men. King Mark reminded Tristan at his knighting ceremony, "Honour and love all women."[75] Chrétien's Lancelot explained that when a knight found a women alone, "If he cared for his fair name, he would no more treat her with dishonour than he would cut his own throat. And if he assaulted her, he would be disgraced for ever in every court."[76]

These new kinds of warfare and these new ways of distinguishing one man from another portrayed in the romances had their counterparts in real life.[77] By the middle of the twelfth century, tournaments—massed mock battles—involved tens and then hundreds of knights, the goal being surrender and capture, not bloodshed. The prizes were horses, armor, and fines (just like the ransom of the battlefield). No one intended to kill. When Geoffrey, son of Eleanor of Aquitaine, died, trampled at a tournament in Paris, chroniclers described it as an accident and a great tragedy. In the thirteenth and fourteenth centuries, the tournament became the joust, a ritualized encounter between two mounted, armored warriors, where victory came through decisions, points awarded for movements when mounted and for fighting on the ground when unhorsed. In the fifteenth and sixteenth centuries, the jousting provided the entertainment for royal and noble festivals to celebrate marriages like that of Princess Catherine of France and Henry V of England. Kings like the Tudor Henry VIII and Francis I of France vied with each other in the splendor of their armor before fighting their joust on the Field of the Cloth of Gold near Calais in 1520. Just as in the romances, women played only an observer's role in these staged encounters. Froissart described a joust at Windsor Castle in 1344 with the queen, three hundred "ladies" and young girls in attendance.[78]

In these centuries war became a craft open to selected artisans. Kings encouraged this wish for exclusivity. Between the twelfth and thirteenth centuries, simultaneously in Italy, Germany, Spain, France, and England, families reached agreements with their kings that restricted access to arms and to training: Roger II of Sicily in 1140; Frederick Barbarossa in 1152; in Aragon in 1234; in Castile in 1260; Louis IX's law courts from 1226–1270. By the middle of the thirteenth century the right to knighthood and thus to membership in the warrior elite became hereditary, just like the holding of land. Then in the fourteenth and fifteenth centuries, kings formalized both the highest ranks of this hereditary nobility and the models of proper behavior by creating the orders of chivalry.[79]

Wives and daughters of the twelfth to the fifteenth centuries benefited from these changed attitudes. In real life as in the romances, women became a man's means of proving himself. In the leisured gatherings of the great halls, the key aspect of "courtoisie" and a man's reputation as a "gentilhomme" came from the way in which he treated women. Outside the great hall, the same would be true—a man could be judged by his behavior toward women. In particular, when women were in danger from war or misadventure, the knight was supposed to protect them. According to the ideals of the newly created codes of chivalry, to war on women reflected badly on the warrior.[80] During the Hundred Years War when John, Duke of Bourbon, organized a group of sixteen knights in 1415, he made them swear that

> each and every one of us shall undertake to do everything possible to maintain the honour of all ladies and women of good birth; and if we find ourselves in company where evil or wicked things are said of ladies, then shall we be held to maintain the honour of womanhood, as we would our own, and do this with our bodies, if need be.[81]

According to the chroniclers, this duke escorted the Duchess of Brittany to the nearest castle defended by her people, rather than keep her his prisoner, explaining, "No, we do not war on ladies."[82] Two of Edward III's nobles rescued two maidens from rape by the archers. The English King set knights to patrol in captured towns to protect "women of rank" from the soldiers.[83] The fifteenth-century French noble, Marshal Boucicaut, on being told that the two women he was paying homage to were prostitutes, is supposed to have replied with exemplary chivalry, "Oh, I would rather have paid my salutations to ten harlots than have omitted them to one respectable woman."[84]

The noblewomen of the eleventh and twelfth centuries enriched the literary life of their households. They inspired and gave their patronage to male entertainers. They wrote and sang in their own voices with emotional force, with skill, with originality, and with humor. They held power in the literary world of courtly love through the passion they aroused and shared. In the world outside the great halls, they came to hold another kind of authority. Searching for ways to distinguish themselves from other men, their husbands, fathers, and sons defined the criteria of nobility: birth and prowess not only in the arts of war but also in their conduct toward women. All came to be enshrined in the fourteenth- and fifteenth-century codes that created the orders of chivalry all across Europe. All survived into the centuries to come as "chivalrous" behavior, long after the castles, the ladies, and the knights had disappeared.

Wives and Daughters in the World of Centralized Monarchies: The Thirteenth to the Seventeenth Centuries

Between the thirteenth and the seventeenth centuries, the political and economic circumstances of the noblewomen of Europe's castles and manors changed again. As in previous centuries, these changes determined the powers allowed privileged women and the ways in which they remained vulnerable in what was still a culture intermittently at war. In these centuries their men lost most of their feudal obligations as the royal dynasties and royal bureaucracies established hegemony and determined the conditions of peace and war. By the fifteenth century a family could hold land with no feudal responsibilities at all. The nobleman's role as the feudal warrior and manorial lord diminished. Instead, first in France and England and then elsewhere in Europe, royal authority competed with and then superseded older customs. By the fourteenth century, royal justice and statutes used the king's, not the noble's authority, and royal not feudal courts to enforce the law and claim fines.

The romances told of young male adventurers challenging each other to individual combat on a forest path. In reality, the very character of warfare changed throughout Europe, with a new emphasis on order and the efforts to enforce the king's peace. First the Church and then, when they could, the dynastic rulers determined when and where and why men could fight. The Church declared "truces of God," days of prescribed peace. As early as 1130 the Church outlawed massed and individual combat that led to death. By the end of the thirteenth century, the French and English kings of the Plantagenet and Valois dynasties had forbidden trial by combat even as a means of adjudication. Honor, bloodshed, and death no longer described how men fought or what they fought for. Now men battled in "just" wars, and kings decided what was "just" and what was not.[85] By the fifteenth century kings had stopped all wars but those in their own names.

War had become a profession. As a romance of the late fifteenth century explained, "War . . . is a proper and useful career for young men."[86] Warriors still fought seasonally, but the old fixed obligation of forty days of service no longer suited the needs of commanders. All noblemen did not have to fight. With the new strategies of siege warfare and massed attacks by foot soldiers, bowmen, and artillery, the knight's duty to fight, like the peasant's duty to plow, became a money payment, "an aid" paid in cash so that others could fulfill the old military obligation. The knights in Marie de France's lais hired themselves out for pay. In real life those vassals who wished to be warriors still took oaths of allegiance, but fought not so much out of loyalty as for wealth, for plunder, and for the ransoms and accouterments of those captured. Loyalty now had to be insured with "gifts," "benefits," and pensions.

The last semblance of the old ways of fighting disappeared in the final years of the Hundred Years War at the beginning of the fifteenth century. Thereafter war became artillery sieges of walled towns, or battles fought in open fields by pikemen and bowmen. Sovereigns and great families made war to gain control of kingdoms and the royal bureaucracy that administered them. Noblemen who chose to fight became the privileged officers in these wars. They gathered companies of soldiers and hired themselves and their men to opposing factions in conflicts like the Wars of the Roses, the Fronde, the French Religious Wars, and the English Civil War. In keeping with this new kind of warfare, Isabella of Castile kept one fifth of the booty for the crown in the wars against the Arabs of southern Spain. The winning commanders divided the rest.[87] She and her husband used the "aids" from the feudal levy on lands and towns to employ 50,000 foot soldiers and 10,000 mounted warriors, a standing army that continued to serve the queen in peacetime.[88]

In this new political environment, the noblewoman and her family's relationship to the land changed as well. If her husband chose to fight himself, money was essential to buy the warrior's horse, the armorer's skill, to pay the feudal "aids" and levies. Only money could free the warrior captured and held for ransom. In the fourteenth century, a French duke explained his wife's reluctance to have him fight. The couple just "had to pay so much money for the ransom that she would have begged me to stay."[89] The money could only come from the lands. By the beginning of the thirteenth century, elite families used their lands not only to provision themselves but also to generate revenue. Agriculture changed from subsistence farming to commercial production for profit. By then commerce and capital were replacing barter and the local exchange of goods and services in much of Europe. More efficient use of land, improvements in technology and power generation had led to increased production, had allowed specialized cultivation and the sale of surplus goods at fairs and markets.[90]

The lady, the châtelaine, the wife of the nobleman and lord adjusted to and participated in the evolution of the new mercantile economy. She learned to transform the goods and services owed from peasant families on the manors into the money needed by the family to finance the warrior, to pay feudal obligations, and to support itself in peacetime. Fees and rents gradually replaced days once owed for hauling and plowing on the lord's lands. The noblewoman's duties became increasingly entrepreneurial. She supervised the lands more closely to generate more revenue. She might travel herself to the fairs to buy and to oversee the sale of produce. In 1265 the Countess of Leicester made such journeys, probably about fifteen miles a day in midwinter over muddied tracks, the horses pulling her in a rough enclosed cart. The lady of the manor might supervise the bailiff in the leasing of their own fields, the domain (demesne.) Fees to use wine presses, grinding and fulling mills, and bake ovens were all ways in which additional revenue could be generated from the land and its peasantry. When, in the fifteenth and sixteenth centuries, the prices of finished products multiplied three and five times, as "land lords," she and her husband raised their rents and tried to enforce old manorial obligations for the fees they needed.[91]

These altered economic and political circumstances changed the lives of privileged wives and daughters in practical and tangible ways. As early as the fourteenth century in the eras of the king's peace, fighting turned from competition for castles and lands to warring over the rich prizes, the towns. More often than not the noblewoman, the châtelaine, could tend to her household without fear of attack, and her castle could become more of a dwelling and less of a fortress. Increased trade meant more comforts and a wider, richer range of possessions for her. These changes in the way noblewomen lived came gradually from the end of the twelfth century as their simple stone dwellings changed to the elaborate châteaus and palaces of the sixteenth and seventeenth centuries.[92]

By the sixteenth century everything about the way noblewomen and their families lived was different. They no longer traveled so often from one dwelling to another. In England, the lady left the glass in the windows.[93] When the minor English landholders, the Pastons of Norfolk, decided to improve their manor, they moved the fire to the wall, creating a fireplace with a chimney to replace the open hearth. This arrangement retained and reflected heat more efficiently, and with stone-walled fireplaces and chimneys, areas could be partitioned and still heated. Seventeenth-century Woburn Abbey, the home of the Duchess of Bedford, had ninety separate rooms, eighty-two of which had hearths.[94] Great halls came to have two-story sections at each end of the large space, with antechambers, a bedroom, and access by staircases to this private area for the lady and her lord.

By the fifteenth century, the noble family did not have to compromise so much of the design for defense; now comfort and ease came into play. There might be cold running water instead of a wooden cistern and stone ledges with open spaces down to the moat for discarding waste water. There would still be little privacy, since rooms were connected, and much of the daily activity was still performed in common, but separate areas were demarcated and partitioned for cooking, storage, and sleeping. Families constructed the great turreted castles like Vincennes, Saumur, and Windsor.

By the fifteenth century, interiors showed care in decoration, and resources were used to make space not only comfortable but pleasurable to look at. Mahaut, the fourteenth-century Countess of Burgundy, had different motifs in the rooms of her castle: roses, shields, and birds decorated the hangings and the furnishings. Women used leather to cover the walls, put colored and brightly striped hangings about the bed, hired workmen to carve wood paneling. Tables, beds, and cupboards stayed in place and did not need to be taken from one castle to another. Furniture was carved and elaborately decorated, like the queen's privy at Hampton Court at the end of the sixteenth century; the royal "close stool" was covered in red velvet and lace with carved gilt decoration.

In their marriage contracts, noblewomen listed the prized possessions that they brought with them, especially the fabrics: hangings, bedding, and coverlets. The twelfth-century French Countess Adele's bedroom was hung with embroidered depictions of Paradise, the Flood, Troy, and the battle of Hastings.[95] The thirteenth-century Countess of Leicester prized her massive bed with rope and leather lacings, linen hangings, and a gray fur and scarlet wool coverlet.[96] Princesses had more extensive possessions. In 1348 Edward III of England sent his daughter Joan, age thirteen, to marry the King of Castile with the equipment for a kitchen, pantry, buttery, spicery complete with spices, and the furnishings for a chapel.[97] Their possessions might be as sumptuous as those described in the romances. At the end of the seventeenth century Agrafena Mixailov, the daughter of a Russian prince, listed her dowry over two pages: towels, silk pillowcases, bedding, blankets, copper kitchen utensils, clothes of sable and ermine lined with silk, silver- and gold-brocaded coats, bright red and azure satins, and jewelry of "ruby chips and emeralds" set in gold, pearls, diamonds, sapphires; all valued at 2,000 rubles.[98]

That the Russian princess' dowry consisted almost entirely of goods for her new household and her wardrobe of clothing and jewelry reflected yet another aspect of the changed circumstances of elite women's lives: political and economic changes made it less common for daughters to be given land as their portion of the family's wealth. From the eleventh to the thirteenth centuries, their right of inheritance stood intact, but increasingly that inheritance came not as land but in movables and money payments. When a daughter did acquire land, it usually came to her not from her father but from her mother or another female relative.[99]

The evolution of the mercantile economy (with the accompanying changes in warfare and the increased costs of equipping and paying for the knight's service owed to the feudal overlord) explained the transformation of her inheritance. In the new economic and political circumstances, noble families had to keep the bulk of their revenue-producing lands together. As early as the twelfth and thirteenth centuries in France and England, families had begun to pass on this block of land, the patrimony, intact to one son and then on to his son, insuring adequate revenues and thus the survival of the family's authority from one generation to the next. As a fixed pattern this came to be called "entail" in England; when the eldest son had first right of inheritance, it was called "primogeniture." These practices meant that other sons inherited differently; families usually assigned them collateral or peripheral lands, money, or movables like their sisters. The increasing power and assertiveness of kings in the thirteenth to fifteenth centuries also discouraged the designation of lands as inheritance for daughters. Any sign of vulnerability, such as a female landholder, could lead to seizure of family properties and their consolidation into the ever-expanding royal domain. If, as was done in twelfth-century Mâconnais, a father passed land to his daughter, he gave it on the occasion of her marriage while he was still alive to ensure her claim.[100]

Examples of these ways of guaranteeing the survival of family power and prestige have been found as early as the ninth century. Local custom continued to be honored in that the daughter's share might even be equal in value to that of her brothers but—as in northern Spain, where at the beginning of the twelfth century the Visigothic law of equal inheritance was followed—she did not receive that share in land. Instead the son of the Vivas family provided for his sisters on the occasion of their marriages with clothing, household goods, and payments in gold.[101] Similarly, in 1270 the French king Louis IX made his will before going off on Crusade and specified that each daughter receive money, not lands, should he die.[102] In the seventeenth century the English families, the Verneys and the Fells, gave collateral lands and money to their daughters.[103] Oliver Cromwell, the Lord Protector of England, kept the farm lands and gave one daughter a salt works at Yarmouth. Some fathers required oaths to insure the son's exclusive right to the lands. At the end of the fourteenth century, for example, the daughters of the Counts of Armagnac renounced their claims to the family's land in return for 20,000 livres.[104]

In the shifting economic and political circumstances, a noble daughter lost equal access to the family's wealth in another respect. From the eleventh and twelfth centuries on, her payment came not at the death of the parents but when she married. Thus her dowry and her inheritance came to be combined in one payment, and this one gift tended to be smaller than custom would have dictated when the two gifts had been separated as marriage gift and daughter's portion.[105]

Though perhaps smaller, the amount of money and goods was significant enough to be the subject of negotiations between families as it had always been. Contracts of the fifteenth and sixteenth centuries from all over Europe show the variety of arrangements and how businesslike the proceedings had become. In December of 1463 Jehan D'Argenteau, Seigneur d'Ascency, met with a notary, a priest, and the father of Marie de Spontin on the road between their two villages where in the presence of witnesses they specified her dowry (part of the proceeds of two estates), and her widow's inheritance. If she produced children, she was to have half of his property for life.[106] A Russian contract of 1529, between guardians of Orina and the father of the groom, specified her inheritance and even a penalty of 200 rubles if the contract was broken.[107]

English fathers beginning in the fifteenth century found ways to ease the burden of payment. In sixteenth-century Cambridgeshire, they took advantage of new commercial practices to provide for their daughters. Members of the gentry, the lesser property holders, took out mortgages to generate money rather than sell land or give up savings.[108] They did not always pay all of the dowry at once. "Marriage portions," as they were called, appear as debts on their estates. The English phrase, "to pay court" as synonymous with finding a wife, dates from the 1590s and aptly describes the process. The fifteenth-century letters of the Paston and Stoner families show parents surveying young women on the basis of their value, and negotiations ended for lack of sufficient dowry.

These inheritance practices became part of European culture and prevailed into the nineteenth century. Even so, as with so much else that determined noblewomen's lives, the new customs could be circumvented and the principle of female inheritance reasserted when it suited noble and royal families. In particular, rather than lose the lands in the absence of sons, families fought to give the patrimony to their daughters. From the thirteenth to the sixteenth centuries, as in the era of feudalism, when a daughter was the only child to survive to adulthood, the concentration of entailed lands sometimes worked to her extraordinary advantage. She then became a potentially powerful heiress, the means by which the family retained its land and revenues. The commentaries on the law of the thirteenth-century English jurist Bracton reflected the existence of these exceptional women. He asserted their right to hold land though unmarried, to sue in court without a guardian, and to will their property. Battle Manor of Sussex in the south of England listed thirty-six women as burgesses (property holders) in 1240.[109] In the thirteenth century, Robert, Earl of Leicester, had no male heirs and so divided his lands between his two sisters.[110] In France from the fourteenth century on, the property might pass to a daughter with the father requiring that her husband take her name and arms, thus saving his lineage.[111] In the fifteenth and sixteenth centuries, such marriages might be arranged with males from the collateral branches of the family to neutralize men who might assert contrary claims. In the sixteenth century, the lands of the Baron du Vigean were inherited through the female line for three generations, from eldest daughter to eldest daughter and then to a son who resumed the title.[112]

The interests of the royal dynasty might also be served by female inheritance. English kings asserted claims through women to the French throne during the Hundred Years War of the fourteenth and fifteenth centuries. Exceptional circumstances in Spain in the fifteenth century led powerful men to favor the rights of daughters and made it possible for a woman to assert claims to the throne in her own name, and on her own behalf. Isabella of Castile (1451–1504), by her political skills and just such circumventions of custom and usage, became Queen Isabella of Spain.

Born in 1451, she was the eldest child of the second marriage of John II, King of Castile. After the accession of her half-brother Henry IV to the throne, Isabella and her younger brother Alfonso grew up with their mother away from the court. When Alfonso died in 1468, only Isabella and Henry IV's daughter, thought to be illegitimate, stood as possible heirs. Civil wars marked Henry IV's reign. At his death in 1474 the twenty-three-year-old Isabella seized the Treasury, proclaimed herself Queen, and won the allegiance of the great noble families with her promise to restore order. A third claimant, Ferdinand of Aragon, had become her husband in 1468. With Alfonso's death the establishment of an alliance between her and Ferdinand's families made the couple the most powerful force in Spain. By uniting Castile and Aragon, Isabella and Ferdinand created a dynasty with enough resources to impose peace and strong enough to pass the united kingdom on—again through the female line—to their daughter's son, the future Charles I of Spain (Charles V of the Holy Roman Empire).[113]

Contemporaries acknowledged Isabella for her piety and her pride in being the model wife. She prayed the canonical hours like a member of a religious order and sewed Ferdinand's shirts herself. Her portrait shows a round, smooth-skinned young face with a thin, white coif fitted to her head. The mouth is small; the eyes gaze straight forward. The stillness of the image belies the energy, independence, and singleminded determination that characterized her rule. Isabella intended to be both a queen in her own right and the founder of a new dynasty. Even in the marriage settlement with Ferdinand of Aragon, she retained full royal authority over her own kingdom of Castile, and never shared her power with her husband. She maintained her own treasury, promulgated all decrees, and she alone received the feudal homage of Castilian nobles. In contrast, her husband allowed her to rule jointly with him in his kingdom.[114]

With Ferdinand and on her own, she warred on Portugal, put down revolts in towns like Segovia and Seville, and supervised the "crusade" against Granada, the Arabs' last stronghold in Europe. She delayed a military campaign only once, for the birth of her second daughter, Catherine, a future wife of Henry VIII of England. Together, she and Ferdinand broke the power of the nobles' cortes and ruled through their councils and a newly created royal bureaucracy. Isabella and her confessor, the Franciscan Ximenes de Cisneros, took control of the Spanish Church. She made her own appointments, refused appeals to Rome against her decisions, and imposed rigid adherence to Christian doctrine and practice in her kingdom. In her will she called for a codification of the laws and thus perpetuated her influence even beyond her lifetime.[115]

As in other ages, changes in warfare, economy, and government affected how privileged women lived and how they participated in the warrior and landed culture. The changes from the thirteenth to the seventeenth centuries reduced noblewomen's vulnerability in war and eased their lives. These changes also limited elite women's access to land and thus to power. For a few extraordinary women like Isabella of Castile the changes made their place in the lineage more significant. In the absence of sons, the constraints of gender no longer applied. Daughters became the means of maintaining the noble or royal family's claims to property and authority. These women became liege lords and reigning monarchs in their own right and acquired more power than most noblemen.

Widows and Mothers: The Ninth to the Seventeenth Centuries

Rights of Widows Abused

Isabella of Castile never lived as a widow. Had she survived her husband, Ferdinand of Aragon, she would in all probability have continued to be exceptional, able to use the lands she held and the institutions they had created together to control the great landed families of fifteenth-century Spain. For most noblewomen from the ninth to the seventeenth centuries, however, becoming a widow, a woman "sole and unmarried," as the English called it, meant becoming vulnerable. In theory her husband's family had assumed responsibility for her on her marriage, the obligation to protect her person and to provide for her. The gift of her dowry at the time of the betrothal negotiations had been balanced by rights to part of her husband's wealth should he die before her.

In fact, however, a widow could easily become the pawn of the interests of her new family, or those of a stronger warrior, even the king. Lords and overlords considered widows fair game. Widows could lose title to all property, even their dowry. They might be forced to remarry. When enough property was involved, they might lose the right to care for their own children; others would want control of the heirs' lives and future holdings. A tenth-century Anglo-Saxon widow found that though her husband had had no difficulty claiming lands and holding them illegitimately, on his death his kinsmen reclaimed them and left her and the children propertyless. In the eleventh century in Lorraine in northeastern France, a widow sought refuge with the local monastery when her neighbor tried to force her to pay "quit-rent" and to swear fealty.

Custom and royal statute law placed widows under the protection of the king. For example, Anglo-Saxon and Viking laws of the late tenth and eleventh centuries made them royal wards and guaranteed that a woman "may decide as she herself pleases" about remarriage.[116] When it suited their needs, however, kings ignored their own laws. Even a widow's royal blood gave her no protection. In the twelfth century Joanna, widow of William II, King of Sicily, was confined and refused her dower (her widow's portion) by Tancred, her husband's successor. Such abuses continued across Europe into the seventeenth century. Sir Richard Verney's sister Cary bore a female child after her husband's death. His family resisted giving over the income-producing property, her jointure, even though all had been part of the arrangements made at the time of the marriage.[117] In 1669 the widowed Marica Usakova petitioned the czar against her nephews, who had assaulted her and her serfs, perhaps hoping to intimidate her into parting with her lands.

These widows were denied their rights because they had so much property. Other problems came with too little. Twelfth-century English feudal records tell of women like Alice de Bidun with lands worth only £10 a year and four children to provide for, Alice de Beaufow with lands worth just over £5 a year, livestock, and rents of only 36 shillings from her tenants.[118] In the thirteenth century the widowed wife of a knight from Metz in France sold lands to cover her husband's debts of 33 livres. She could never have covered them from her income of 19 livres a year, or from the sale of her movables valued at 30 livres.[119]

If she were young enough, a widow expected to remarry. A well-born woman might trust to her family to chose a new husband and so to ensure her continued comfort and safety. In the sixteenth century, Charlotte d'Arbaliste, a twenty-six-year-old French widow, wrote to her own and her dead husband's relations to tell them of Philippe du Plessis' proposal "for I would go no further in the matter without their permission."[120]

As with her first marriage, the noblewoman's property and the lineage she embodied influenced the choices and negotiations. From the ninth to the seventeenth centuries the more prosperous she was, the more illustrious the lineage, the more important her remarriage would be, and the more likely the occasions for abuse. The Viking conqueror Canute cemented his hold on the English throne in the eleventh century by marrying the widow of the man he defeated, just as his Germanic predecessors had done in the centuries before 800. In the fifteenth century, Anne, Duchess of Brittany, showed how valuable a widow holding property in her own name could be. Her first marriage to Maximilian of Austria was annulled when he failed to protect her duchy. Her second husband, King Charles VIII of France, died, but her lands were so valuable to the French dynasty that Louis XII, his successor, divorced his own wife so that he could marry Anne.

The twelfth- and thirteenth-century English kings Henry II, Richard, and John shamelessly abused widows. Planning to use them as a source of revenue, in 1185 Henry II had a list made of the "ladies" "in the gift of the king," the well-born widows over whom he held guardianship.[121] His sons Richard and John took widows' money and forced their remarriages. Richard arranged for Hawisa, the Countess of Aumale, to marry one of his fleet commanders on her first widowhood. When she was widowed again, John, Richard's successor, made her pay 5,000 marks not to be remarried.[122] In 1216 John married off the widowed daughter of one of his chamberlains to his mercenary Faukes de Breauté so as not to lose the lands of this heiress to a stranger. In honor of the match, Matthew Paris' chronicle reports the following verses:

> Law joined them, love and the concord of the bed.
> But what sort of law? What sort of love? What sort of concord?
> Law which was no law. Love which was hate. Concord which was discord.[123]

John's fifteenth-century successor, Edward IV, used acts of parliament to disinherit widows, and thus to accrue land to his own family. Noblemen protested such treatment of their women. In 1215 Article 8 of Magna Carta forbade the forced marriage of widows. Philip Augustus' Norman subjects made him pledge that he would not "force" the women under his protection. The Church always opposed the practice because of its belief in marital consent and because often the match involved consanguinity (marriage within prohibited degrees of kinship). But in the end, families continued to give way to superior force, and the Church accepted that its rights would lie not in prohibition but only in adjudicating disputes and in giving dispensations.

The same sort of practical considerations affected the widow's relationship with her children. From the ninth to the seventeenth centuries, in order to gain control of property, overlords and kings altered their own customs and broke their own laws. By tradition the widowed mother held the guardianship, or wardship, of the children of the marriage, usually under the protection of male family members. This meant that authority over the inheritance, the revenues from their lands, and arrangements for their marriages lay in her domain. From as early as the ninth century, the widow's inability to fight and thus to fulfill the basic feudal obligation of the land she ruled in her child's name caused men to question the efficacy of her authority, and to find ways to deny her power. In England, from the twelfth to the fourteenth centuries, kings and royal courts found ways to insure the knight service and access to the revenues to support it. The supervision of the heir, the wardship, came to be divided into supervision of the body of the child, including choice of marriage partner, usually allowed the widow, and supervision of the lands and access to its revenues, awarded to another, a man.[124]

Particularly unscrupulous kings forced changes of custody, awarded the children like prizes and even sold them. In the late twelfth century King Richard of England and the abbot of Bury St. Edmunds fought over a three-month-old baby girl. The abbot sold his claim to the archbishop of Canterbury for £100, who in turn sold it for £333.[125] King John's abuse of this right of wardship was another cause of the rebellion against him in 1215. His son, Henry III, commanded Mabel, the widow of Roger Torpell, to deliver her daughter Acelota to the bishop of Chicester or risk confiscation of her goods and lands.[126]

The sagas and the chansons de geste from the ninth to the twelfth centuries show widows of the warrior and landed elite alone and in need of protection. They also portray them as mothers and the stalwart supporters of their sons. In *Njal's Saga* a widow offers her daughter and some property in return for protection from a young thane. In *Raoul de Cambrai,* Aalais, abandoned by her lover, bears their son and when he is full grown encourages him to regain his lands. In addition to the poet's tales, the writing of a noblewoman, the mother of a warrior son, has survived. Dhouda, wife of Bernard of Septimania (in southern France), lived like a widow when her husband was away at war or in exile. Between November of 841 and 843 she composed a manual for William, her eldest son, then about sixteen. (Dhouda's writings are the only extant work by a ninth-century European woman.) In her guide for the young man she cited the Church Fathers and modeled her phrasing on the didactic literature of the era. Yet her love and her concern for her son can be discerned amidst the traditional forms and references.

Dhouda, though "far away," is "anxious and longing to aid" him. She sends these words to him assured that "although I cannot talk with you face to face, when you read these pages, you will know what you ought to do."[127] She warns him:

> We live in evil times, my son; I fear lest disaster come to you and to your soldiers as they fight. You will meet often men ambitious for their own ends, greedy, arrogant, disobedient, pleasing the world rather than the Lord of all.[128]

Dhouda also wrote of her own frailty and her fears for her own safety. She died that same year, her husband was executed in the next, and neither William nor her other son survived into old age: William was executed for rebellion in 850 and Bernard had to flee and died in 872 at the age of thirty-one.

Rights of Widows Asserted

Dhouda's sense of vulnerability as a propertied woman alone was typical. An exceptional woman, however, in favorable circumstances could use even her widowhood like women in the centuries before 800 to acquire independence and power. The seventeenth-century Englishwoman, Alice Thornton, called it "the third honorable estate" and within the customs of the warrior and landed culture of the ninth to the seventeenth centuries lay the means to power for an enterprising noblewoman.[129] An elite widow's traditional right to sustenance until her remarriage or until her death had in theory been guaranteed. Most often such provision for her took the form of moneys or property negotiated at the time of the betrothal. What began as a custom came to have the force of law, the designation of a fixed portion of the husband's estate for her own maintenance, from one third to one half of its value given in land, in movables, in money, or in some combination thereof. Commonly it was not hers to own or to bequeath, only hers for use during her lifetime or until her remarriage. Thus she was provided for, but the property passed intact back to the family who had given it, not on to the children of another lineage. Laws from the twelfth and thirteenth centuries set down these provisions for Norway, for Ireland, for Russia, for Denmark, for Switzerland, for Spain, for England and France. Danish and Spanish law also granted her a portion of what had been acquired during the marriage. Changes in English and French laws show the influences of the economic transformation of Europe. In France, the widow's portion came to be taken from the movables, not the lands, and in England it was turned into an annuity called the jointure.[130]

Brave women learned to use the laws and courts to protest abuses for themselves. Thirteenth- and fourteenth-century English court records tell of many such widows. Isabella, Countess of Arundel (d. 1282), confronted the king himself when he claimed the wardship of one of her vassals. Juliana, widow of Robert Underburgh, successfully kept her dower against three separate suits in 1329.

English widows used the courts to protect their claims to their children as well. The records show women protesting the arrangements made by overlords, taking their children back, "deforcing" the heir. In 1303 Johanna, widow of Adam de Lorymer, was awarded custody of her son Walter. The jury agreed that the guardian had no right to him and that she had not by taking him "ravished the heir."[131] In other cases, accused of deforcing the heir, the mother was allowed the "right of nurture" as long as she made no claim to any rights of wardship, to authority over the lands or its revenues. If children were in the award and grant of the king's court, an enterprising woman, like a man, might buy their guardianship. Margary, widow of Henry La Pringe, early in Henry III's reign, in the thirteenth century, bought the wardship of her husband's heir in this way.

When a strong woman came into her money or her property with relative ease, it might embolden her to assert her independence. At the least, a confident noblewoman would insist on choosing her own second husband. Eleanor, widow of the Earl of Pembroke, married Simon de Montfort secretly in 1228 against the wishes of her brother, King Henry III of England. She litigated and won the properties he tried to take from her. Other women used their money to purchase the right of choice. In Edward I's reign at the end of the thirteenth century widows purchased such rights for themselves and for their sons. It cost Gena, widow of Roger de Caleston, £40 for the two permissions needed.[132]

Other privileged English women from the fifteenth to the seventeenth centuries were able to enjoy the wealth that came to them as widows. Agnes Paston kept control of the property even after her eldest son came of age. Elizabeth Talbot (b. 1518), known to contemporaries as Bess of Hardwick, survived four husbands, and ended life as a countess with a jointure giving her an annual income of £60,000. She did very well with her properties, including forests and mines, and her business ventures, trading in timber, lead, and coal.[133] Margaret Poultney, Sir Ralph Verney's aunt, spoke for them all when in 1639 she insisted on not remarrying even though it meant a break with her family. "A widow is free," she explained.[134]

For other women, to be "free" meant more than property; they enjoyed the full exercise of their rights as feudal landholders. In twelfth-century Saxony, Sophia, widow of Berthold of Zähringen, used her knights to help her brother Henry the Proud besiege a rival's castle. She remained in charge when he had to go to the aid of his father-in-law, King Lothar. Gertrude, sister of Ekbert, the Margrave of Neisse, survived three husbands to be called by a chronicler "Saxony's almighty widow."[135] She used her vassals to conspire against Emperor Henry V's son-in-law, leading them herself.

Some royal widows also acted independently. They used their own lands and their skills to protect or advance the interests of their sons. In the thirteenth century Blanche of Castile (1187–1252) fulfilled her first duty as the wife of a king by producing twelve children; five survived. She taught her eldest son, the future Louis IX, to read, and acted as regent when he was a child from 1226 to 1252. She successfully put down conflicts with the nobility, repulsed an invasion by Henry III of England, and won a dispute with the French bishops. A contemporary chronicler described her as "the wisest of all women of her time, and all good things came to the realm of France while she was alive."[136]

In the fifteenth century Lady Margaret Beaufort (1444–1509), mother of the future Henry VII of England, acquired property for him while he was in exile and helped foment a rebellion against the king to gain him the throne. Once he had become king, Henry rewarded her with her own lands. Parliament granted her the right to rule like a man: "as anie other sole person not covert of anie husband."[137]

In fourteenth-century Scandinavia a royal widow, Margrethe of Denmark (1353–1412), achieved more than the gift of lands and the praise of contemporaries. Acting on her own behalf, she maneuvered and negotiated within the special circumstances of her era to rise from widow to regent to queen in her own name. She turned the principles of inheritance and her contemporaries' desires for peace and order to her own advantage. Queen Margrethe asserted the right to unify and rule in her own name not one but three kingdoms, those of her father, her husband, and her son.

Inheritance and the idea of the patrimony had already been established in fourteenth-century Denmark when King Valdmar IV was left with only two surviving daughters. He betrothed Margrethe at seven to King Haakon of Norway as part of an alliance between their two countries, with the idea that their child would be heir to both kingdoms. From the age of ten Margrethe lived in Oslo as Haakon's wife. (Marta Ulfsdotter, daughter of the Swedish noblewoman, Saint Bridget, was her governess.) She bore Haakon a son at seventeen in 1370, but saw little of her husband. She lived in a separate household and had to ask to borrow money from the Master of the Mint in order to maintain herself and her son. When her father died in 1375, there were four possible successors to the Danish throne: she and her son, her sister Ingeborg, and her nephew Albert. When the tenants-in-chief, the most powerful lords of Denmark, met at Odense in May of 1376, they chose Margrethe as regent for her boy Olaf, seeing this as the way to unification of the thrones of Denmark and Norway. The powerful Hanseatic League, eager at the prospect of increased trading privileges, gave its approval as well.

From the beginning of her regency, Margrethe forced her contemporaries to recognize her as a decisive, capable ruler. She allowed neither the death of her husband, nor that of her son in 1387 to threaten her power. She, not her son, is remembered. His burial stone reads, "Here rests Olaf, son of Queen Margrethe, whom she bore to Norway's King Haakon."[138] One after another, the noble councils chose her to reign. A week after Olaf's death in August of 1387, the Danes elected her "mistress and ruler with full authority as guardian of the realm."[139] In February of 1388 the Norwegians decreed she would rule for life, "the kingdom of Norway's mighty Lady and rightful Master."[140] She herself took the title of Queen of Sweden, and in 1389 defeated Albert, the rival claimant. She succeeded because she signified order, undisputed title to the throne, and unification of two and later three kingdoms. She created what contemporaries called the Union of Kalmar, a glorious age of relative peace for Scandinavia that lasted from 1397 until 1523.

The union survived because during Margrethe's reign she created the kind of centralized monarchy that Henry II had begun in England and Philip Augustus in France two centuries before, the kind of monarchy that Isabella and Ferdinand would create in Spain in the fifteenth century. She balanced the power of the nobility and that of the Hanseatic League. She appointed chancellors and then never allowed them too much authority. She refused to remarry. She named her successor, her grandnephew Erik of Pomerania, but never gave him any of her power. The Swedish nobility at her prodding gave up its tax exemptions and the right to build new castles without the monarch's consent. She took back the crown lands lost when Albert, her nephew, had claimed authority. She did the same in Denmark, reclaiming lands lost during her father's reign. Norway became a Danish fief. She died at fifty-nine in 1412 in the midst of a campaign to reclaim Holstein for the crown. Throughout her life she had used inheritance customs and the circumstances of her time to work to a widow's advantage.

Women like Eleanor of Aquitaine, like Isabella of Castile, like Margrethe of Denmark used the laws and the opportunities given to them to wield power, but this remained a man's power, used for the ends valued by men. Nothing of these women's actions, goals, or successes altered the institutions and the attitudes that left women of the nobility, however well-born, vulnerable to loss of privilege and abuse. Nothing of their lives secured a better position for women, a position that would protect and empower them regardless of circumstances. The traditional patterns of life and the traditional views of women had been circumvented, not challenged. For centuries to come even the most privileged European women would continue to live in a world in which they acquired power through accident of circumstance, not by equal right or design, where they held that power only because it suited the needs of the male warrior and landed elite.

4

THE NEW FLOWERING OF ANCIENT TRADITIONS

FROM THE NINTH to the seventeenth centuries, women exercising power and authority remained aberrations, their position made possible by unusual combinations of circumstances. In the eleventh and twelfth centuries of contending feudal lords and again in the fifteenth and sixteenth centuries of aggressive royal dynasties, some daughters, wives, and widows had inherited and held property in their own names. These noble and royal women fulfilled roles usually reserved to men. Their success, however, meant neither a permanent change in European women's status within the family nor in their political and economic authority. The laws of these centuries confirmed that noblewomen lived within a family defined by their relationship to men. The ideal behavior presented to young girls continued to echo the attitudes of the cultures before 800: the ideal woman was dependent, protected, and provided for, the dutiful, chaste, obedient daughter, wife, or mother.

In Literature

The literature of these centuries—the epics, lyrics, chansons de geste, and romances—presented the same divergent images as in previous ages and had the same overall effect. There were, on the one hand, the positive portrayals of the beautiful lady and the glorification of love. On the other, there were the negative portrayals of women as mere objects, as agents of forces bringing pain and even death to men. Even in the great halls of powerful countesses like Marie de Champagne and queens like Eleanor of Aquitaine, the lyrics and songs of courtly love, the romances with their tales of ladies and their knights subtly kept alive not only traditional images of female beauty and of the pleasures of love but also equally vivid negative images of the female nature and of the traditional fears of women's power over men. In the imagined world of literature as in the real world of war, property, and lineage, Europe's women could not escape traditions inherited from the Greek, Roman, Hebrew, and Early Christian cultures.[1]

The Positive Legacy

In describing the beloved, jongleurs and troubadours pictured an ideal of female beauty echoing that first expressed by Greek poets. The woman's "voice is pure as gold and as the zither," her body "beautiful, slender and fine," her eyes "bright and hazel." Her skin is white "as a Mayflower," her hair "fair and radiant."[2] The poets sang of love inspiring exemplary devotion. Siegfried in the epic *Nibelungenlied* told Kriemhild's brother that "she is to me as my soul and as my body, and I will do anything to deserve that she become my wife."[3]

The women and men portrayed in the lyrics and especially in the romances enjoyed all the sensual aspects of their love. Chrétien de Troyes' Queen Guinevere holds out her arms and draws Lancelot "into the bed beside her and showing him every possible satisfaction: her love and her heart go out to him. . . . Their sport is so agreeable and sweet, as they kiss and fondle each other, that in truth such a marvelous joy comes over them as was never heard or known."[4] Gottfried von Strassburg in *Tristan and Isolde* conveys the couple's delight in each other's bodies and in the lovemaking. Sent to nurse Tristan, Isolde "scanned his body and his whole appearance with uncommon interest. . . . She looked him up and down; and whatever a maid may survey in a man all pleased her very well. . . ."[5] When Isolde's husband King Mark found them, they slept:

> tightly enlaced in each other's arms, her cheek against his cheek, her mouth on his mouth. . . . Their arms and hands, their shoulders and breasts—were so closely locked together that, had they been a piece cast in bronze or in gold, it could not have been joined more perfectly.[6]

The troubadours even described union with the beloved as bringing spiritual bliss. Raimbaut d'Orange wrote that "holding her [he] feels he's holding God."[7] Gottfried von Strassburg also added this dimension to Isolde and Tristan's meetings. Theirs was more than a physical joining, it was similar to a mystical union with God:

> They shared a single heart.
> Her anguish was his pain; his pain her anguish.
> The two were one both in joy and sorrow.[8]

Negative Traditions

Yet for most of the jongleurs and troubadours, as for the poets of the ancient cultures, this ideal image of the beloved and the glorification of physical union also had their negative aspects. The singers of the lyrics, especially the twelfth- and thirteenth-century minnesingers of Bavaria, idealized the women they loved much as the Greek and Roman authors had. In their verses, however, the beloved became so stylized that the individual was lost amidst the emotions she inspired. The woman became the object of love and that alone.[9] Walther von der Vogelweide, a famous minnesinger, traveled from castle to castle, singing not of individual women or the variety of their lives, but of the variety of his feelings and the effects of love on him. Women were also portrayed as the objects of love in the epics, chansons de geste, and romances. Love, not the woman herself, caused the hero to act, to test and prove himself. For example, love, not the ladies, made the knights of Chrétien de Troyes' *Lancelot* "rich, powerful and bold for any enterprise."[10]

Even more damaging to women, the love and desire that they cause bring not only pleasure and happiness but also pain, misadventure, and sometimes death. The twelfth- and thirteenth-century heroines can be as harmful to the heroes as Circe, Dido, and Eve. To suggest the power of women, some lyric poets used the imagery of a battle, with man as the victim. Bernart de Ventadorn said the woman made war on him. Giraut de Bornelh described himself besieged with engines toppling towers, catapults, and attacking warriors. He cries for mercy.[11] Heinrich von Morungen hates the woman who "has wounded me / in my innermost soul."[12] Women of the romances acquire power and bring harm not by laying siege to their lovers but by using tricks and stratagems. They "obstruct" men's vision; they "steal" men's love. So Tristan was seduced. With her "smiles and laughter, her talk and prattle, her blandishments and coquetry" the woman "at last set him on fire."[13] Both Guinevere and Isolde outwit their husbands and thus hide their adultery. Isolde used traditional devices to still King Mark's suspicions.

> At night, when the Queen went to bed with her lord, she took him in her arms and, kissing and embracing him and pressing him close to her soft smooth breasts, resumed her verbal stalking by means of questions and answers.[14]

In some tales of the male singers and poets love of a women can even destroy the hero. In *Beowulf* a maiden's potion bewitches the warrior Hedin and turns him into a murderer.[15] In one version of the epic *Guillaume d'Orange* love torments and makes the young lover "reckless and a fool" and drives the old husband mad.[16] Love makes the heroes of the romances forget their codes of chivalry. Love deflected Chrétien de Troyes' Erec from his proper role as knight for he "devoted all his heart and mind to fondling and kissing [Enide], and sought no delight in other pastimes."[17] Lancelot is so enraptured with love of Guinevere that "he totally forgets himself."[18] Love keeps him from his goals; he lusts, succumbs to his passion, and then can no longer seek the Holy Grail. In *Tristan and Isolde* the love destroys both heroine and hero; death is the only resolution for their illicit passion.

In addition to these negative images of women and of the ways in which love affected men's lives, the literature of these centuries also presents examples of misogyny. Even the most complimentary of male poets can echo the traditional fears and condemnations of women. Some portrayed the same flawed woman as in past ages. In Gottfried von Strassburg's *Tristan and Isolde,* the traditional prejudices flow from the mouth of a rejected suitor. Women are

> ". . . all so constituted in body, nature, and feelings that you must think the bad good and the good bad. . . . You love that which hates you, you hate that which loves you. . . . Of all the games one can play on the board you are the most bewildering. The man who risks his life for a woman without security is out of his senses."[19]

The sagas and the epics gave their listeners traditional villainesses: Hallgerda in *Njal's Saga* takes her revenge on Gunnar by refusing to help him fight off an attack on the fortress. Wives of warriors initiate blood feuds. In the *Nibelungenlied,* it is the rivalry between Kriemhild and her sister-in-law Brunhild that leads to warring. As the poet explained: "From the quarreling of two women many a hero lost his life."[20] It is Siegfried's wife who reveals the one spot of vulnerability on her husband's body and thus makes possible his murder.

For some male poets, Eve, the first betrayer, the first seductress, continued to embody all women. Like Eve going against God's prohibition, "women do many things, just because they are forbidden," wrote Gottfried von Strassburg.[21] Like Eve, women bring men's destruction. By repeatedly committing adultery with Isolde, Tristan "did just as Adam did he took the fruit which his Eve offered him and with her ate his death."[22] Using the style of the graceful love lyric, a Celtic poet attributed all of the world's horrors to this first woman. He imagines her describing life had she not "picked the apple from the bough":

> There would be no ice glazing ground;
> There would be no glistening windswept winter;
> There would be no hell; there would be no sorrow;
> There would be no fear, were it not for me.[23]

The jongleurs and troubadours inherited the traditional beliefs about the female and her power over men. In the great halls and courtly gatherings they recast the ancient heroines and presented the ancient fears in new settings. The effect, however, was the same as in the literature of the older cultures: a contradictory and ambiguous image of women and a warning to beware of their powers.

In Law and Practice

From the ninth to the seventeenth centuries unusual circumstances gave some privileged women access to vast holdings of land and thus the means to independent authority. A noble family or a dynasty might give power to a daughter like Eleanor of Aquitaine or Queen Isabella of Spain as the only surviving heir. A widow like Queen Margrethe of Denmark might be allowed hegemony to avoid civil war. However, these royal and noble women exercising power in their own names remained exceptions. When circumstances altered, when order returned, the traditional attitudes about women and the traditional patterns of family relationships were reaffirmed. The ancient beliefs from the cultures before 800 in women's inferiority had not died. Rather, amidst the alternative courtly images of women and men as lovers, they had been muted. The old desire to ensure women's subordinate position had not disappeared but had merely been ignored for a time to satisfy political needs. With order came law and practice making the wives and daughters of warriors once again dependent on their male protectors and providers. As circumstances changed, kings and lords regularized access to property, made uniform the restrictions on who might hold it and how the power might be exercised over it. Women, considered as vulnerable as children, found themselves again without opportunity and without recourse, legally incapacitated as they had been in previous cultures.

The reassertion of the ancient subordinating attitudes and patterns can be seen in the changing role of the Queen Consorts of France. From the tenth to the fourteenth centuries these royal women gradually lost powers and authority they had once been granted and had exercised as the wives and royal helpmates of kings. In the tenth and eleventh centuries the Queens of France had supervised the royal household, acting much like a royal treasurer as there was no separation between the revenues of the king as monarch and as head of a family. By the middle of the fourteenth century all had changed—inheritance could not even pass through the female line.

The fourteenth-century French royal wife, Jeanne de Bourbon (1338–1378) exemplified the diminution of the consort's power in this age of order and of revitalized traditions of women's incapacity. She was the last Queen of France to be crowned in a separate ceremony. Although she sat to her husband Charles V's right on official occasions and attended his councils, she never helped him rule. Instead she was remembered for the elegance of her court, her piety, and her acts of charity and kindness to petitioners. Her chronicler, the courtier and writer Christine de Pizan, described her as the ideal wife whose husband "was often in her company" and who brought him nothing but "honor and reverence."[24]

The queens and princesses of Froissart's chronicle of the Hundred Years War are shown as exemplary wives, never as rulers. Philippa of Hainaut (1314?–1369), wife of Edward III, was the near perfect consort. She produced twelve "heirs of the body." In England she was never made regent; instead, the King's cousin, the Earl of Kent, cared for her. On the Continent she waited anxiously for news of the battles her husband waged and was praised for "she had suffered great anxiety."[25] When she did assist with royal duties, it was planning celebrations and giving gifts commemorating marriages. If she did assert herself, it was with appropriate humility. If she had power, it was indirect and subject to her husband's will. On one occasion she pleaded with the King to have mercy on the leaders of a town recently surrendered, "in the name of the Son of the Blessed Mary and by the love you bear me."[26] He acquiesed "although I do this against my will" for the most chivalric of reasons. Seeing "his gentle wife" as she knelt pregnant in tears before him:

> his heart was softened, for he would not willingly have distressed her in the state she was in, and at last he said: "My lady, I could wish you were anywhere else but here. Your appeal has so touched me that I cannot refuse it."[27]

As laws were collected and codified, as royal statutes and edicts were issued, the same kind of devolution of opportunity and authority happened to all noblewomen, not just to royal wives. Possibilities for privileged women's independent action through control of their own lands disappeared. Laws and courts designed to protect women reinforced the inherited traditions from the centuries before 800: the principles of female legal incompetence, female subordination, and the establishment of male guardianship. As part of their acceptance of these traditions, women gave up claims to their own lineage as well. French women adopted their husband's name as early as the fourteenth century; by the seventeenth century this had also become the practice in England.[28]

The compilations of local customs and the commentaries written on the laws from the thirteenth to the seventeenth centuries enunciated the premises and the reasoning by which women lost the right to make any decisions about the lands they inherited or acquired. It was presumed that, since a woman could not fight, she could not in her own person fulfill the obligations of the property. Then it followed that she was by the very nature of her being incapacitated. Therefore in law as in life, she could have no power and the man must act on her behalf.[29] By the sixteenth century, French custumals (formal collections of local or regional customs) described wives as having "legal disabilities." The contemporaneous revivals of Roman law, the glosses of Justinian's Digest, also worked to her further disadvantage. An early sixteenth-century humanist scholar's description resounds with echoes of traditional attitudes from the centuries before 800: "The woman is in the power of her husband, and is no longer in the power of her father . . . and cannot make valid contracts . . . nor administer common goods, nor her own, without the authority and consent of the said husband."[30] Seventeenth-century English jurists took this principle one step further: Coke's *Institutes* defined husband and wife as one person, one male person. *The Law's Resolutions of Women's Rights,* written in 1642, made explicit that women existed as married or to be married and were subject to their husbands. They "have no rights in Parliament, they make no laws, they consent to none, they abrogate none."[31]

Russia in the fifteenth century showed how far such restrictions could go. Although the formal and legal definitions of rights gave a wife authority to own property and to exercise independent authority over it, even to appearing in court, in fact, she rarely used these powers. Instead between the thirteenth and seventeenth centuries (perhaps in response to the continuous invasions by outside peoples), she became totally subject to her husband's authority: taking his name, sequestered in separate sections of their quarters like a Muslim wife, and even sold into slavery if need be to cover his debts. His exile meant her exile, her exile justified his disowning her. In the seventeenth century Peter the Great reaffirmed her separate authority, but the fact of reissuing the edicts suggests that they were not obeyed any better in the later period than they had been before.[32]

Other kinds of laws reinforced the traditional subordination even for these privileged noblewomen. Well into the thirteenth century, customs and statutes affirmed a husband's right to discipline his wife. In thirteenth-century Beauvaisis he could beat her "reasonably."[33] Thus in every way he was her guardian. Over and over again the statements come, from as early as the eleventh century in the Mâconnais in France, the twelfth century in Ireland, the thirteenth-century Spanish laws of Cuenca and Sepulveda, from German law codes for thirteenth-century Saxony and Mühlhausen—that though the property, even the movables, the linens and clothing she brought to the marriage, might be held by her in her own right, she was (as the thirteenth-century English courts described it) "under the rod" of her husband, unable even to question what he did.[34]

In the cultures before 800 women had been subject to their fathers and husbands. The reaffirmations of this subordination in the European era were more constraining. By the seventeenth century, uniformity of law and consolidation of jurisdiction in royal courts ended variations in customs and weakened institutions that had offered noblewomen countervailing authority to call upon. First the feudal nobility and then the Church bowed to the royal dynasties. Nowhere is this loss of alternative authorities and the unintentional independence they had provided for women more evident than in the changes in law and practice about marriage, its dissolution, and the legal responses to extramarital liaisons.

From the twelfth century on, annulment and divorce became matters not so much of personality as of property; similarly a woman's infidelity in this warrior and landed culture became an issue of lineage. Both situations could threaten the survival of the landholder's family, for a woman leaving the marriage reclaimed her dowry and an unfaithful woman could mean an illegitimate son. Where before in these situations a wife could appeal to her family, to the Church, or to local custom for negotiation or a chance at vindication when wronged, between the twelfth and the sixteenth centuries, she found herself increasingly left to the power of her husband. The Church turned over much authority in these matters to the monarchy and its courts allied with the husband.

In the centuries before 800, the Roman, Hebrew, Germanic, and Celtic customs had made the dissolution of marriage among the elite a relatively easy matter, especially when initiated by the husband. In front of witnesses, the couple declared their intent and followed customary rules about the division of their goods. The early Church had opposed any severing of the marriage vow, allowing separation for adultery, but prohibiting union with another. The Emperor Charlemagne tried to reconcile the customs of the Germanic peoples with the attitudes of the Church. He and his successor Louis the Pious followed the policy enunciated in the Church's name by Hincmar, bishop of Rheims, in 860: a couple might request separation or annulment of the Church (not of their families or the secular ruler). Separation might be allowed for any number of reasons, but only with annulment—literally erasing the fact of the union—could the couple begin anew with different partners.[35]

Annulment could be granted for only a few causes, all having to do with the couple's failure to fulfill the purpose of the union. The Church defined marriage as the means for the production of legitimate children, thus the union could be dissolved for impotence or for incest (marriage of partners related in the seventh degree and called consanguinity). This remained the official attitude of the Church into the seventeenth century when the Council of Trent reaffirmed these policies and, by making marriage a sacrament, gave even more significance and permanence to the couple's vows.[36]

Such rigidity, however, did not fit the needs or inclinations of kings and feudal lords. A barren queen produced no sons for the dynasty. A richer heiress could bring more power to a noble lineage. Marriage to a cousin within the prohibited degree, like that between Eleanor of Aquitaine and Louis VII, might create an alliance useful to both ambitious branches of a family. All of these considerations meant that secular rulers ignored or placed themselves in opposition to the Church's policies. Despite all his edicts, Charlemagne simply put aside his first wife to take another. Bishop Hincmar opposed Lothar II of Lotharingia's efforts to be rid of his queen, Tetberga, so that he might wed the mistress, Waldrada, who had produced heirs for him. Lothar imprisoned Tetberga and bribed papal legates, but only won from the Pope the right to a separation. In the eleventh century Bishop Ivo of Chartres condemned Philip I of France for abandoning Queen Bertha and their marriage, for a possibly more fecund Bertrade. Philip Augustus found himself excommunicated at the beginning of the thirteenth century for his treatment of Ingeborg of Denmark, but still kept her imprisoned to gain her consent to the dissolution of their marriage.[37]

Compromise came at the beginning of the thirteenth century as part of the accommodation between Church and state associated with Pope Innocent III.[38] The Lateran Council of 1215 modified the definition of incest from the seventh to the fourth degree, legitimating more of the marriages favored by the nobility and their kings. The canons also defined the procedures by which marriage might be annulled, although only a woman of a powerful family could make this system work to her advantage. The Church had acquiesed to the political realities. From 1215 on, the religious and secular authority acted as one in principle and in practice. Kings and powerful landholders maintained their right to be rid of their wives, but the Church insisted that it must arbitrate and consent to such aberrations, must give its dispensation. Thus marriage bonds could be dissolved but only by consent of the Pope or his representative. Increasingly the woman was no more than the object of negotiation.[39]

Despite the official condemnation of the Church, from the ninth to the seventeenth centuries, kings, dukes, knights, men of property maintained another time-honored male right; they commonly had sexual partners other than their wives. The early Church had condemned infidelity by either husband or wife as illicit sex, intercourse for illegitimate ends—for pleasure rather than procreation.

The only arguments among theologians were on the relative gravity of the sin for a woman and for a man. At the end of the eleventh century, Ivo of Chartres believed that by marrying a fornicating woman, the man gave her back her chastity. The Lateran Council of 1215 allowed marriage between adulterers after the death of one marital partner much as a ninth-century Carolingian capitulary (decree) had, as long as there was no evidence of murder in the death of the sinned-against husband or wife.[40] These pronouncements had little if any effect on the behavior of men of the warrior and landed elite. In Spain, Arab influence led some men to sequester their wives and to emulate the eastern institution, the harem. Frederick II, the thirteenth-century Holy Roman Emperor, followed this practice at his court in Sicily. The fourteenth-century English Duke of Lancaster married his concubine, Catherine Swynford. Philip van Artevelde of Ghent avoided a surprise attack from the French because, according to Froissart, his young woman companion, unable to sleep, "went out at about midnight to look at the sky and the weather and find out the time" and saw the preparations.[41]

"Adultery" in both religious and secular law came to signify only the action of a married woman with no similar popular meaning for the married man.[42] Though she and her lover would in theory be equally culpable, she would have the harder test of innocence and earn the greater punishment if guilty. In the tenth- to thirteenth-century laws of Cuenca, she needed twelve female witnesses to exonerate her. The harshest punishments were reserved for the woman who had intercourse with a man of another race: exile, burning, or immediate execution if she was caught in the act.[43] Appeal to the Church could mean "scourging" or whipping for her; after the Council of Trent in 1545 she would be sent into exile, excommunicated.[44] From the thirteenth to the seventeenth centuries, the Eastern Church in Russia maintained a policy of equal punishment, but in fact it was the wife who was sent to the monastery and the husband who had the right to resume relations with her or to abandon her.[45]

A married woman like those eulogized as heroines—Susannah of the Old Testament and Lucretia of the Roman legends—might even have to defend herself when she had, in fact, been raped. Froissart recounts the story of the wife of Sir Jean de Carrouges, left at home while her husband went on campaign "to help him in his advancement."[46] A neighbor, Jacques Le Gris, a companion in arms of her husband, came to visit, and asked to view the keep alone with her. Having closed the door, he "put his arms round her and said: 'Lady, I swear to you that I love you better than my life, but I must have my will of you.' " When she tried to cry out, he stuffed a glove in her mouth "and pushed her down to the floor. He raped her, having his desire of her against her will."[47] Leaving her, he threatened to dishonor her should she ever speak of what he had done. Though distraught she did not tell the servants, but confessed to her husband on his return. On the advice of friends, he appealed to his overlord, but the Count of Alençon believed Le Gris, who insisted on his innocence, and wondered "why the lady hated him."[48] It was only the husband's persistence that saved her in the end; he appealed to the king, whose court set a duel to the death to determine innocence or guilt. Le Gris, if he lost, would die in battle or by hanging. If he won, the wife would be burned to death. She was spared such a fate by her husband's defeat of Le Gris.[49]

In the end what happened between a married woman and her cuckolded husband depended upon him. In some areas of Europe, the husband always had the right to kill the couple caught in the act. In 1000 Count Fulk Nerra of Anjou burned his wife Elizabeth to death. Philip of Beaumanoir, a thirteenth-century French legal commentator, explained that if the husband warned the man and then found them together, he could kill the lover as long as he did it on the spot.[50]

Among the nobility the dissolution of marriage and a wife's infidelity, if used as grounds for separation, became more than a question of honor. Annulment and divorce meant loss or acquisition of property. Secular and religious courts then became the places for negotiation. When the Church supervised the separation and there had been no wrongdoing by either partner, as described in the thirteenth-century *Decretals* of Gregory IX, the wife was to have what she had brought to the marriage returned. Thirteenth- and fourteenth-century English ecclesiastical court records suggest that the Church supported return of the dowry, the movables at least, and the payment of support for the children.[51] In seventeenth-century France, the wife of the Marquis de Saint Germain-Beaupré went to court and was granted a separation and 8,000 livres pension from her husband, who had beaten her and expected unusual sexual practices.[52] Margaret Verney married in 1646. Soon she reported that she was unhappy and that her husband had struck her. In the 1650s she refused a reconciliation. In 1658 a separation settlement was agreed to, £160 per year for her "separate maintenance as if [she] were sole and unmarried."[53]

A wife's proved infidelity changed the bargaining significantly. Germanic customs, like that of the Viking Canute in the eleventh century, expected some form of compensation for the wronged husband and forfeiture of her property. Unfaithful wives, according to laws codified in thirteenth-century England, France, and Spain, lost their right to inheritance on their husband's death (the dower) and even the inheritance from their fathers. With important women holding much property, accusations of adultery might be very profitable and have secondary benefits. The fourteenth-century King Philip IV of France had two of his daughters-in-law arrested: Margaret of Burgundy, and Mahaut's daughter Blanche. Margaret died in prison, Blanche survived her time in Château Gaillard to die in a convent.[54] Then he claimed their lands. Frances Villiers, daughter of the seventeenth-century Chief Justice Coke of England, fled to France with her lover, escaping a forced marriage. She was charged at the Court of High Commission for adultery and lived as an outlaw and exile, forfeiting all of her property to her husband.[55]

A woman like Frances Villiers broke religious and secular law and personified what men had always feared about the female nature. She had disobeyed her father, been unchaste, and willfully defied both her husband and the secular authority. The newly allied institutions of Church and monarchy uniformly condemned her behavior and withdrew their protection. Such a woman lost all claims to property and to family. According to the attitudes of her era as of those of centuries past not only did Frances Villiers break the law and go against custom, but by her insubordination she also rejected every ideal of female behavior. To the privileged men of the seventeenth century she was truly dangerous. Disinheritance and exile kept her away and showed other women the consequences of defying the traditional patterns of behavior. She had stepped outside the protective relationship; she was no longer the dutiful, chaste, obedient wife and daughter.

The Ideal Woman

From the ninth to the seventeenth centuries, circumstances gave some royal and noble women unusual political power in a world that valued war, property, and feudal and dynastic authority. Some noblewomen played men's equals in love. The literature of the great halls idealized women for their beauty and for love's power to inspire men to great deeds. Even so the male poets, the learned men of these centuries, and royal and noble parents did not abandon the traditional views of women and of their subordinate relationship to men. The traditional descriptions of the women's proper roles and of the female's nature survived and found new expression in these later centuries.

The troubadours and jongleurs of the great halls gave their listeners beautiful, deceitful villainesses, but, in addition, they also portrayed the older positive ideals familiar to the cultures before 800: the modest, dutiful virgin daughter and the loving, obedient, chaste wife. Walther von de Vogelweide described the proper response of a maiden to his gift of a flower garland:

> She took what I offered her, even as a child that is honoured; her cheeks flushed red, as a rose in a bed of lilies; then her bright eyes were ashamed, yet she sweetly bowed in greeting to me.[56]

In the thirteenth-century story of the Count of Pontieus' exemplary daughter, she, like her Roman predecessor Virginia, offers to kill herself when she is raped by robbers on the pilgrimage route to St. James of Compostela, and thus ease her father's disgrace.

The sagas and chansons de geste emphasized a wife's duty to her husband and glorified her sacrifices on his behalf. Begthora dies with Njal as part of her duty to him. "I was given to Njal in marriage when young, and I have promised him that we would share the same fate," she explains.[57] Alde in the *Song of Roland* and Doña Jimena, the wife of the Cid, present clear models of loyalty: the one dying when she hears of the death of her betrothed, Roland, the other spending her hours praying for her husband, accepting his leaving for yet another campaign, kissing his hand like a suppliant.

The romances honor wives for their beauty and also when they prove "worthy, constant, [and] chaste."[58] Like the Church Fathers, Gottfried von Strassburg offered the ultimate reward for such a woman. She who retains her modesty, who accepts and does not flaunt prohibitions, "grows in virtue despite her inherited instincts and gladly keeps her honour, reputation, and person intact, she is only a woman in name, but in spirit she is a man!"[59] Chrétien de Troyes' heroine, Enid, has the perfect qualities: "gentle and honourable, of wise speech and affable, of pleasing kindly mien. No one could ever be so watchful as to detect in her any folly or sign of evil or villainy."[60] Despite her fine qualities, however, she wins the love and trust of her lord only through a series of desperate trials. He forbids her to speak. Repeatedly she does, usually to protect him. With Erec in theory dead, she becomes the prisoner of the local count, who abuses her as he chooses. Yet, she refuses intercourse with him and is reclaimed by her no-longer-doubting husband. He announces, "Sweet sister mine, my proof of you has been complete."[61]

Even the characterizations of the strong heroines in the chansons de geste and romances by inference sometimes reveal the more ancient negative views of the female nature and teach the same lessons of behavior. Even these otherwise capable women predictably betray innate female weakness and must be chastized and brought to obedience. Guiborc, the wife of Guillaume of Orange, takes on the task of reconstructing the castle, declaring, "I am the one who will take charge of it."[62] She dons a shirt of mail, helmet, and sword to protect their stronghold, but thinking her husband dead, faints and, comforted by the priest, cries. Raoul de Cambrai's mother, though she had protected, encouraged, and advised him always, at one point suffers a more traditional fate; she finds herself turned away, scorned, mocked by her son: "Let that knight be accursed and held for a coward, who takes counsel of a woman." He gives her instructions which echo Homer:

> "Go to your apartments, lady, and take your ease; drink pleasant draughts to fatten your body. Think out for your household what they shall eat and drink, and meddle not with other things."[63]

In the last resort, even the strongest heroine could be forced to accept a subordinate place in relation to a man. Brunhild, one of the heroines of the *Nibelungenlied,* was praised for her beauty and her prowess. She defeats knights at their own games of skill and strength. When she repelled and humiliated her husband on their wedding night, Siegfried decides to rape her. He crushes her on the bed, holding her "in such a way that her limbs and her whole body cracked," and she gives in. He treated her this way, he explained, as an example; "after this many a woman may not be so insolent to her husband."[64]

The lords and ladies of the castles and manors heard these songs and stories. They believed these tenets about appropriate behavior. They too made their contribution to the survival of the ancient attitudes and ideals. From the twelfth to the fourteenth centuries, noblemen and their learned chaplains wrote treatises presenting the traditional premises about the female nature and advocating the ideal roles and functions of older ages. Thus the traditional Greek, Roman, Hebrew, and Christian attitudes and relationships thrived, restated in the phrases and images of the new age. With the invention of the printing press in the mid-fifteenth century, these didactic works became the equivalent of best-sellers. The male authors in their righteous concern explained the task to their readers. Just as fathers and husbands had assumed responsibility for women's person and property, so now men must accept responsibility for women's behavior. They believed that just as Isolde "much improved herself" and "grew to be courteous, serene, and charming" under Tristan's tutelage, so under the guidance of a father or husband all women could be transformed.[65] A good teacher could turn the potentially destructive wanton into an exemplary maiden and a willing, chaste wife.

These male writers described the defects of the female nature and offered remedies in their plans of education.[66] Philip of Novara, the thirteenth-century Lombard Crusader, in his treatise on the *Four Ages of Man (Quatre tenz d'aage d'ome)* portrayed women as weak and easily led astray. Louis IX, the thirteenth-century king of France, believed that since goodness did not come naturally to the female, it would have to be inculcated by rote and reflex. He expected his daughters to fill their idle hours with reading and prayer and thus avoid temptation. At bedtime they would come with their brothers to his bedchamber to hear exemplary tales of those they should emulate or avoid. In his letter to his daughter after she married, he emphasized piety, good works, and obedience to her husband, "by which one can judge that you are a wise young lady and well instructed."[67] Vincent of Beauvais in the sermon he wrote for Queen Marguerite, Louis IX's wife, devoted one fifth of it to the education of their girls. In *De eruditione filiorum nobilium* he explained the different approach to their education:

> You have sons? Train them and care for them from babyhood. You have daughters? Guard their bodies and do not show a joyful face to them.[68]

Given his views, much that he recommended was meant to keep their sexuality in check, their virginity intact. He borrowed remedies from the Church Fathers Jerome, Ambrose, and Augustine. Warm baths, fasting, singing psalms, prayer in the middle of the night, strict control of dress, reading and handwork to keep them busy—all would train the young daughters to chastity and teach them that corporal beauty was, by definition, evil. He advocated punishing any sexual transgression with violence. As in the centuries before 800, so in these later eras, the young woman must always remember that it was *her* actions, *her* dress, *her* demeanor that excited others. Robert of Blois, a poet of the mid-thirteenth century, gave explicit instructions in his *Châtoiement des dames (Chastisement of Women)* on how to walk to church, how to look at men, about who could and could not touch her breast or kiss her mouth.[69]

The French chevalier from Anjou, Geoffrey de la Tour-Landry, wrote the most popular and influential of these didactic books. His series of lessons for his daughters, *L'Enseignement de ses filles (The Education of His Daughters),* composed in his old age in about 1371, came to be translated by Caxton in 1483 and to be printed in all the great capitals of Europe. By 1538 it had run to eight editions in three languages.[70] Page after page gives the traditional view of women's natural weaknesses and the traditional remedies and models of behavior. With priests and clerks to help him find exemplary tales, he used all kinds of female examples to prove the evil nature of women: Venus, Helen of Troy, Lot's daughters, and his contemporary, the Queen of Naples. He believed every maiden a potential wanton whose "young tender flesh, when it is excited . . . is easy to be tempted."[71] He enjoined his daughters to pray and to fast three days a week until their marriage, so as "to hold low your flesh, to keep you chaste and clean, in God's service."[72]

The old knight viewed his wife, Jeanne de Rouge, daughter of the king's chamberlain, as an ideal: "both fair and good, with all knowledge of all honor, all good and fair maintenance."[73] He recommended her example to his daughters as well as models from the Bible like Ruth and Deborah and from history like the thirteenth-century Queen Blanche, included for her acts of charity. He saw one life for women—in marriage responsible for the maintenance and harmony of the household. The ideal gentlewoman thanked God for her children and educated them properly. She expected nothing from her husband and was even prepared to balance out his evil with her goodness, ready to please him and bring him "out of his frenzy."[74]

Landry's view of the female nature was such that even education, admonition, and example would not be enough. He praised Esther as the perfect wife but believed that she acted rightly not only "out of her love for him [her husband]," but also out of "her dread of him and her fear of shame."[75] So the last lesson to be learned by his young daughters was fear—fear of the inevitable consequences of inappropriate behavior. His tales gave punishments for all kinds of improper actions on the part of wives: a broken nose for jealousy; mockery for vanity; madness for laxity in religious observance.

The ideal he offered was the woman who obeys without question. Not surprisingly he used Eve as the classic example of the insubordinate, willful female. Eve, he explains, "lost all worship, richness, ease, and bliss, and the obeisance of all things for the sin of disobedience."[76] For it was "through her look she fell into the foul and horrible sin to break God's commandment, through the which all the world and her offspring was dead, and lost, and damned."[77]

Though learned and noble men wrote the treatises and sermons, in reality mothers, not fathers, oversaw the day-to-day education of their noble daughters. This made no difference in the lessons. Just as Eleanor of Aquitaine, Margrethe of Denmark, and Isabella of Castile did not question the institutions of their time, however constraining they might be, so these noblewomen with property and education rarely challenged the images or the premises about the female nature, function, and role. Instead mothers accepted the traditional premises and attitudes and presented the ancient models of good behavior. For only by obedient, chaste behavior might that always vulnerable daughter be provided for and have a place of safety and sustenance.

Mothers instilled piety. Blanche of Castile, the thirteenth-century queen of France, ordered the psalter (a book of psalms) for her children's first reader. Traditionally mothers commissioned a Book of Hours for their daughters.[78] They saw only one future role and function for the young girl. When Margaret Paston was a little girl, her mother took her to St. Paul's and St. Saviour's Abbey "to pray that she may have a good husband."[79] Sir Edmund Molineux in 1551 sent his daughter Katheryn to a cousin's family to be educated in "vertue, good manners and learning to playe the gentlewoman and good huswyffes, to dress meat and oversee their households."[80]

Seventeenth-century French and English wives delighted in their household functions, in bringing order to their husband's affairs. Suzanne de la Porte, wife of Henri de Richelieu and mother of the future cardinal and adviser to the king of France, kept her "good temper" though "burdened with a confusion of affairs which she used to manage with so much order and prudence."[81] The wife of the seventeenth-century Comte de Souvigny handled all of his correspondence, giving him only the good news and dealing with everything else herself. These women offered their absolute obedience. Mary Verney accepted without question that her husband had sold off part of her inheritance, though she had hoped to bequeath some of the acreage to her younger children for "thou mayest as freely dispose of that as of myself."[82]

When noble and royal women wrote their own didactic works, few questioned the views of men or advocated other than the most traditional female virtues and functions. Christine de Pizan, the early fifteenth-century French widow, by her life as a writer and a successful courtier, belied the traditional image of the dependent female in need of protection and provision. Even so, her treatise, *The Book of the Three Virtues* or *The Treasure of the City of Ladies (Livre des trois vertus* or *Le trésor de cité des dames)* offers the same advice first proffered in the centuries before 800.[83] Pizan wrote the treatise, she told her readers, so that all women from the "great queens, ladies and princesses" to the peasant wife would know how to be virtuous. In the service of inculcating virtue, a mother's tutelage of her daughter fell into the traditional patterns. Pizan emphasized behavior, not intellectual attainments. These young "virgins" were to be "clean," "humble" in manner, "prudent," quiet, demure, obedient to their parents and in all ways "moderate and chaste."[84] A daughter's reading had to be carefully selected, nothing with "any vain things, follies, dissipation," could be given to the young innocents.[85] The goal of the lessons and admonitions remained that of the past. Mothers were to teach the young girls "to govern themselves," to hold in check the supposed unchaste and immoderate tendencies of the female nature that could bring dishonor to them and to their families.[86]

As an adult, a noblewoman was to cultivate the equally traditional qualities of humility, patience, steadfastness, piety, modesty, and most of all "discretion."[87] Pizan assumed that most well-born women would marry. As a wife, the noblewoman would be judged like the young daughter by her "good manners and behavior"; as in the cultures before 800, her most important asset was her "good reputation."[88] The model wife gave her principal care and attention to pleasing her husband.[89]

Even a less conventional education did not necessarily change propertied women's views. Lucy Apsley Hutchinson, the daughter of a seventeenth-century English lieutenant of the Tower, was unique in her opportunities. She received a classical education equivalent to that given a son, including Latin, Greek, and Hebrew. She read by age four, and though she had teachers for the more traditional womanly arts like dancing, music, and needlework, she remembered that "my genius was quite averse from all but my book."[90] Her learning "very well pleased" her father, but troubled her mother who believed that she was "so wholly addicted . . . as to neglect my other qualities," qualities that would make it possible for her to marry.[91] Her mother's worries proved unjustified, however, for Lucy Apsley was no rebel. She became the wife of Colonel John Hutchinson, a Parliamentary officer in the English Civil War. She presented him with their first child at sixteen and seven more in the course of the marriage. When he died in prison on the Restoration in 1660, she turned her considerable intellectual energies to a memoir of his life. In the best tradition of the model wife, she transformed this undistinguished gentleman into a paragon. She admired his attention to their estates, described his life as "nothing else but a progress from one degree of virtue to another." Even his body was exceptional: "a handsome and well-furnished lodging prepared for the reception of that prince."[92]

She had gladly subordinated herself to him, explaining that he had managed their lives "with such prudence and affection that she [who] would not delight in such an honourable and advantageable subjection, must have wanted a reasonable soul."[93] It would have been irrational for Lucy Hutchinson to act in any other way, for she accepted all of the traditional attitudes about women's character. The old premises flowed from her pen as easily as if written by a man. Females, she knew, were by nature "inferior" because of "ignorance and weakness of judgment." Most serious of all their deficiencies they are "apt to entertain fancies and be pertinacious in them." She warns: "We ought to watch over ourselves . . . and embrace nothing rashly."[94]

"Watch over ourselves," "embrace nothing rashly," wrote Lucy Apsley, for even a woman of her intelligence and education would rarely find support for any other view. Then, too, as her words demonstrate, there is more than acquiesence by women of this landed elite. The maidens strolling in the May sunshine, the ladies presiding at their high tables, Queen Isabella sewing her husband's shirts, Lucy Apsley eulogizing her Colonel Hutchinson fostered the stereotypes, the subtle denigration and the acceptance of subordination.

As Christine de Pizan had explained at the beginning of the fifteenth century, "a prudent woman" knew how to behave with her husband. In an unfortunate marriage she was to show "patience," use "dissimulation," "steadfastly" endure.[95] Christine de Pizan believed in women's vulnerability in the warrior and landed culture, and like her female and male contemporaries, she feared the consequences to a woman of any other response:

> If you speak to him harshly you will gain nothing, and if he leads you a bad life you will be kicking against the spur; he will perhaps leave you, and people will mock you all the more and believe shame and dishonor, and it may be still worse for you.

She concluded: "You must live and die with him whatever he is like."[96] From the ninth to the seventeenth centuries, the vast majority of well-born and noble women could see no other alternative for themselves or their daughters.

V

WOMEN OF THE WALLED TOWNS

•

PROVIDERS AND PARTNERS

1

THE TOWNSWOMAN'S DAILY LIFE: THE TWELFTH TO THE SEVENTEENTH CENTURIES

Alleyways, Streets, and Squares

The European towns of the twelfth to the seventeenth centuries bore little relation to modern urban metropolises, in appearance, size of population, and the activities of the inhabitants. There was less contrast with the countryside; the poor worked the fields just outside the walls, the wealthy had orchards and household gardens around their houses. Though the towns were crowded, the numbers of inhabitants never reached the millions of the eighteenth, nineteenth, and twentieth centuries. Cologne, fifteenth-century Germany's largest town, had a population of about 20,000.[1] Fourteenth-century Florence began to incorporate surrounding communes into its orbit, so that by the fifteenth century 40,000 women and men lived inside the walls or just outside. By the beginning of the 1600s, Genoa, Milan, Venice, Naples, Seville, and Paris had populations from 60,000 to 100,000. A very large town like London created new parishes, forty-seven inside the walls in 1695 and fifty-four outside to accommodate its over 400,000 inhabitants.[2]

Even the most populous urban centers provided smaller, more rural, and more controlled environments than would the sprawling neighborhoods and masses of humanity that filled the modern cities. Daily tasks familiar to a countrywoman, small-scale manufacturing and retailing, guild protections and restrictions characterized the occupations of the townspeople. A late fifteenth-century engraving of Cologne shows the roads coming from the mountains and along the River Rhine. The tiled rooftops and the spires of the churches rise just above the crenelated walls and towers. The country noblewoman in her covered cart, the young peasant girl and her mother from a nearby manor would have waited just outside the gates, with other women and men come to buy and sell goods or to find a few hours labor.

All of this scene was familiar to the townswoman who lived inside the walls. With cockcrow, with first light, the bells of the town cathedral rang, and the gates opened to herald the beginning of another market day. The streets, alleyways, and squares filled with wagons, animals, and people. Contemporaries described the carts pouring into fifteenth-century Vienna, carrying flour, eggs, bread, meat, crayfish, and poultry. At the time of the harvest 1,200 horses a day brought in the casks of new wine.[3] Gondolas carrying goods filled the Grand Canal of Venice in the early sixteenth century; there were twenty-five sailboats bringing nothing but melons.[4] Venice's bridge, the Rialto, was like a market unto itself, a bustling thoroughfare lined with shops and warehouses.

The biggest towns, like London, had many markets, each specializing in one kind of produce or another: fish at Billingsgate, leather near Leaden Hall.[5] The townswomen pushed by each other with baskets full of cabbages, of oranges, of secondhand clothing mended and patched to be sold again: the merchant's wife on her way to early mass; the laundress and the brick carrier on their way to the day's work; the herbalist returned from gathering her medicinal plants at sunrise; the silk spinner's apprentice on an errand for her mistress; street sellers like the baker's wife with her meat and cheese pies calling out the goods and the prices from the stalls set all around and across the square.

Away from the busy market squares, the town was a crowded, noisy, dark world in comparison to the light and space of the fields and countryside around it. Though townspeople represented only about 10 percent of the overall population of Europe, from the twelfth to the eighteenth centuries, in the oldest sections of towns like London and Paris as many as three to five hundred would be crowded together in one hectare (2.74 acres).[6] Because of the towns, sometimes whole regions took on an urban character. In Tuscany in northern Italy 26.3 percent of the local population had moved to ten of its towns by the beginning of the fourteenth century.[7] So many had come to Florence and its environs in the 1300s that the population density was greater than in the twentieth century (85 per square kilometer).[8]

The poorest women lived in the oldest, most crowded neighborhoods. By the end of the fourteenth century, the alleyways were cobbled and so less muddy, but no less dirty. Waste and refuge went out the door or the window of the wooden dwellings that lined the narrow streets. Into the fifteenth century in towns like Nuremberg, pigs and other animals scavenged freely just as they did in the manor forests in the fall.

In London in the fifteenth and sixteenth centuries, a pieceworker or a day laborer and her family lived in a single rented room, perhaps six feet by ten feet in size, fronting on the alleyway. In 1580 new building had been prohibited by the town council to try to ease the crowding, but the poorest women and men continued to piece shelters together out of whatever materials they could find.[9] In the campaign of 1381–82 Froissart, the chronicler of the Hundred Years War, described one of these dwellings in Bruges in Belgium, where the Count of Flanders had fled to hide: a "poor grimy hovel, blackened by the smoke of the peat fire." "An old sheet of smoke-stained cloth" shielded the fire; "a ladder with seven rungs" led to a "cramped little loft" overhead where the children slept. The woman nursed her baby by the open fire.[10] In Madrid the women day laborers lived outside the walls near the orchards and fields they worked, in one-story dwellings they and their families had built out of mud.[11]

When towns counted their inhabitants, these women would be too poor to appear on any tallies of household hearths. Not so with all women, however. Those who had become associated with the crafts and their guilds, either by marriage or in their own right, appear in the historical record. For they would be listed among the town's elite, members of those families that paid taxes: the 135 grocers, 150 goldsmiths, 170 skinners on the tax rolls of London in the last quarter of the fourteenth century; the 169 bakers, 75 fishsellers, 110 barber-surgeons of Antwerp listed in the second half of the sixteenth century.[12]

Craftswomen's houses were in newer sections of the town, substantial two- and three-story buildings with ten to twelve feet fronting on the street, twenty feet deep with a household garden at the back. Most of their houses would be constructed of wood; in Rouen most were made of half-timbered stucco, in Amsterdam, brick. The second story was commonly built to overhang the street by a few feet. As this was the principal living space for the family, the extension gave more room for sleeping, cooking, and eating. Curtains or walls divided off areas inside. Here the artisan or craftsman's wife performed her household duties, making the meals over the central stone hearth. In German and Dutch towns an earthenware stove, in southern France braziers, heated the rooms in the winter. A third floor in a prosperous artisan's family meant private space for the mistress and master, who could then send the apprentices and servants to the top floor. The workshop took up the whole of the ground floor. Shutters on the front of the building opened; the top served as an awning, the bottom as a counter from which the wife could sell the products made in the household.[13]

The richest townswomen, members of the families of the governing elite of merchant and banker guilds, lived in a still more prosperous neighborhood and enjoyed differences of scale and comfort way beyond women in other parts of the town. In fourteenth-century Genoa, the family of the merchant, Andrea Doria, lived together with their kin in the section known as San Matteo. It was a compound with shared houses grouped about their own square with a family-endowed church to service their religious needs. Initially, in the twelfth and thirteenth centuries in towns like Genoa, the houses were fortified with towers and thick outer walls, similar to the castles and manors in the countryside. Families warred for power and built their houses as protected refuges. When the warring ceased, in Venice in the fourteenth century, then in Florence in the fifteenth century, and gradually across Europe in Bourges, Nuremberg, London, Paris, Madrid, the winning families built the grand private houses still associated with their names.

Order and prosperity meant the luxury of houses with business and living space on different floors, with interior rooms giving privacy for different branches of the family and the public and private activities of the wife and her husband. In Prato, Italy, when the wool merchant, Francesco Datini, built a new home for his wife and himself late in the fourteenth century, he put the warehouse at the end of the garden, separating it from their living space. Palazzos of fourteenth- and fifteenth-century Florence, like that of Cosimo de' Medici, were on an even grander scale. The ground floor still consisted of offices and storerooms, but rooms above were designed for beauty and comfort, with high ceilings and frescoed walls. The Medici Palazzo had a row of large glass windows facing the garden.[14]

Elite merchant-banker families in other towns made the same changes. The Runtinger family of Regensburg, Germany, in the late 1370s took three adjacent buildings on Keplerstrasse (its twentieth-century name) and remodeled them so that the office and kitchen were on the ground floor, the public rooms on the second, the family area and special rooms for the women on the third. Instead of warehouses at one end of a courtyard, the fourth and fifth floors were used for storage, much as they were in the merchant houses of Amsterdam, where the overhang for the pulleys used to raise goods from the street still exists. The house of banker Jacques Coeur in Bourges, France, built in the 1440s, perhaps in deference to the old styles had one side much like a fortress, but the other was of soft stone carefully carved and decorated. Designed around a central courtyard to insure light for the rooms that did not front on the street, the interior design combined the public and private with fourteen large rooms, each with a series of small rooms clustered around it for dressing and for antechambers. By the beginning of the seventeenth century, the well-to-do of Madrid in Spain had given up all pretense of defense and had balconies along the third story so that the family could sit out and watch the public events staged in the Plaza Mayor.[15]

These walled towns created a vital new world for women. Young country girls, the daughters of peasants, immigrated in large numbers. Young men came to the town from far away, day laborers looking for any work, journeymen seeking to finish out their training before becoming masters of their craft. Young women tended to come from closer by, from the neighboring villages. A study of seventeenth-century Weissenburg in Bavaria, Germany, showed that two thirds of the female immigrants came from within a forty-mile radius.[16]

Perhaps they had first come to the town as little girls on their fathers' cart, fulfilling one of the tasks of his manorial contract. Once they came, the women stayed. They did not leave even when widowed in the last decades of their lives. In the towns women could find work, they could acquire possessions, even access to property, perhaps membership in a guild. Some could hope to make their own way without being dependent on a husband or father.

The tax records of towns like Florence in the early fifteenth century show that women could become almost independent, "masterless" in the language of German town ordinances. In Florence in 1427, 15.6 percent of all households were headed by women.[17] Records of parishes for London, in the early fifteenth and late seventeenth centuries, show the same possibilities for single and widowed women, women of substance listed as heads of households, as freeholders (owners of property in their own right).[18] More women came, more stayed, and created a pattern unique to European history: in the towns of the fourteenth, fifteenth, sixteenth, and seventeenth centuries, overall, there were from 20–30 percent more women than men. For example, in Bologna in 1395 the ratio of females to males was 100:95.6 and in Nuremberg in 1449 100:84. The Florentine figures for the same century show that the imbalances occurred especially in different age groups. Women outnumbered men from thirteen to seventeen, suggesting the greater immigration of young women, and again from the late forties to late fifties. This latter statistic also suggests earlier death for men and longer life for women once they had passed their childbearing years.[19]

The walled towns of Europe offered unique opportunities to the young woman brave enough to take the risk. As early as the thirteenth century in England, leaving the manor and running away to the nearby town could break the tie of serfdom imposed by her birth. If she could survive a year in a royal, chartered borough, she would become a free woman in the eyes of the law. In 1303 Matilda Siggeword and Alice White lived in Lincolnshire on the manor of Ingoldmalls. The lord's records tell of their escape but not of their return.[20] Like other young peasant women they probably went to the nearest municipality, hoping for a more prosperous life than was offered in the countryside. There they would work, marry, and remain, providing the labor and the skills that fostered the economic transformation of Europe from a predominantly rural to a predominantly urban environment.

The Poor

Young women in the towns, like the two who ran away from Ingoldmalls, usually lived as part of a family, even if they were unmarried. More than half of the young women working for wages in the towns hired themselves out as domestic servants in the houses of the prosperous. By the end of the fifteenth century, this percentage had risen to 60 percent, and domestic service remained the largest field of European urban women's employment until the fourth decade of the twentieth century, long past the transformation of the walled town into the modern city.[21] As early as the fourteenth century in Paris and London, enterprising women and men recruited and placed those with references as servants. A 1351 English statute fixed fees for the "recommanderess," as she was called in French, at 1 shilling 6 pence for placing a chambermaid and 2 shillings for a nurse.[22] In the fifteenth and sixteenth centuries, in Nuremberg, Munich, and Strasbourg young women still used such agents.[23]

Placement was not difficult. In the towns of the thirteenth to the seventeenth century just as in the countryside, once a family could afford to have a servant, it was a young woman they hired. Even in households with more employees, the vast majority were women. In Romans in sixteenth-century France, one in seven families had a *chambrière,* "the maid of all work."[24] In Aix-en-Provence in southern France, at the turn of the seventeenth century, the town records show that 56 percent of the women listed as servants served by themselves. (When the family hired more, such as in professional or noble households, the percentage of female servants was well over 60 percent: 94 percent in professional families; 67 percent for *noblesse d'épée* households—members of the older military nobility.)[25]

The young woman servant usually worked alone, from the first light until the family retired. On the one hand, the mistress considered her part of the household and felt a sense of responsibility for her, but on the other the mistress saw no limits to her authority over the servant.[26] In the fourteenth century, a Franco-Flemish phrase book helped a mistress give orders to her servant, the ignorant country girl who spoke a different language. "Janet" must straighten furniture, pillows, and curtains, tend fires, set tables, begin the dinner, wash and scour dishes and pots, buy wine.[27] In the absence of a male apprentice, a family often found a young female to do the work. In other households, such as in seventeenth-century London, the retailer's or the craftsman's wife helped in the trade, leaving her other duties to the woman servant.

In southern France, Italy, Spain, and Portugal from the fourteenth to sixteenth centuries, the young servant might find that she worked side by side

Midwives in attendance at the birth of the Virgin Mary to St. Anne. By the German artist,
e Master of the Life of St. Mary. Cologne, fifteenth century

2. Engraving of a beggar woman outside the town walls. By Jacques Callot, seventeenth century.

3. Townswomen making linen cloth. From an Italian manuscript, fourteenth century.

4. Dutch townswoman at her linen chest with her servant. Painted by Pieter de Hooch c. 1663.

5. The Virgin Mary giving her protection to townswomen and men. Sculpted by Michel Erhart. Germany, c. 1480.

6. Jewish women preparing the house for Passover. Seventeenth century.

7. Fishsellers. From an Italian manuscript, fourteenth century.

8. Women attendants bathing a patron. Fror a medieval manuscript.

9. The Italian painter Artemisia Gentileschi. Self-portrait, seventeenth century.

with a girl bought and enslaved. Records of enslavement are clear but scanty. A Florentine register of 1363 lists 119 females all under eighteen. The town's *catasto* of 1427 (an early tax survey) lists 301 slaves, almost all of them female.[28] In fourteenth-century Florence, one enslaved girl came as part of a bride's dowry. A merchant like Francesco Datini of Prato made money trading the enslaved, and he bought young girls for his own household because he believed them cheaper over the long term than free women. He paid 87 ducats for one young woman, three times the cost of a riding horse.[29] A typical letter of 1392 asked that they be twelve to fifteen years old, and none younger than ten. The buyer continued, "I don't care if they are pretty or ugly, as long as they are healthy and able to do hard work."[30]

As the Church condemned enslavement of Christians, the young women were usually of a different race. Far from their own lands, with no refuge or protections, they had little alternative but to obey the mistress and master. A fourteenth-century bill of sale from northern Italy guaranteed the young slave to be "healthy and whole in all her members, both visible and invisible" and then gave the owner power to do "in perpetuity, whatsoever may please him and his heirs."[31] Punishment for rebellion came swiftly and harshly. In fourteenth-century Florence, they used hot pincers to pull the flesh off a young woman who had poisoned her master.

Whether enslaved or free, young female servants were vulnerable to physical and sexual abuse. The seventeenth-century English Admiralty secretary Samuel Pepys had his wife whip their maid and lock her in the cellar as punishment. All employers believed they had a responsibility to ensure the proper character of the young woman, her obedience, modesty, and chastity. This belief made it all the more ironic when she fell prey to the lusts of the master or his sons. Even if the young woman had freely chosen to have intercourse—perhaps in hopes of gaining favor with the master, perhaps because of a genuine affection for a young apprentice or another servant who had promised marriage—if she became pregnant, she would be disgraced and vulnerable to punishment and ostracism.[32] Assuming that the employer turned her out, the parish and the town officials feared that she and her child would become their burden. Ordinances tried to ensure identification of the father and minimum support from him. Even with provision, the child would always live at a disadvantage. Almost every law and custom in Europe defined the infant as "illegitimate": without family, with no guaranteed rights of inheritance, and not even a family name.[33]

The best outcome for the woman and the child would be to remain with her employers. Records of families in fourteenth- and fifteenth-century Yugoslavia and Italy show some masters who accepted responsibility for their abuse of the young servant. Dubrovnik merchants dowered illegitimate daughters,

Florentine fathers paid into the town's dowry fund for them. Cosimo and Piero de' Medici of the great banking family had their illegitimate offspring raised with their other children. Cosimo's son Carlo became a priest; Piero arranged a good marriage for his daughter Maria. The artist Leonardo da Vinci's mother Caterina was seduced by the son of the family she worked for. They allowed her to remain with them after she became pregnant, and the father, a notary, arranged the boy's apprenticeship to an artist at fourteen.[34]

Most often, the young "maid of all work" married a fellow servant or perhaps a laborer or apprentice she had met in the course of her household duties. They would have saved a little from their wages and, both in their late twenties, would begin their own family. Without special skills or training, however, they became part of the 50–80 percent of the town's population with no privileges, living a marginal existence much like the fourteenth-century Belgian family who dwelt in the one room described by Froissart.[35] They lived in the poorest section of the town as a small family unit. There would be nothing extra to support sisters, brothers, or aging parents, just themselves and their children.

Like their counterparts, the peasant women and men who stayed behind to work the land, the couple would be partners: working at anything to bring in the money needed for shelter, clothing, and, most important, for food. The wives were particularly inventive. When one source of income closed to them, they found another. Boccaccio told a story of the bricklayer's wife who spun to supplement the family income and pawned their clothes when all else failed. Women did manual labor, seasonal work, sold house to house, hawked in the streets, did piecework of every variety, and still they and their families lived on the edge of subsistence. In Florence in the 1400s a bushel of wheat cost 15 soldi, yearly rent for the smallest house 375 soldi, when an unskilled female laborer made only 7 soldi a day.[36] A woman in sixteenth-century Seville in southern Spain made 1 real a day victualling the fleet when bread cost 5 reals.[37] Studies of wages and costs in sixteenth-century Speyer, Germany, show wages increasing for women and men by 2 and 3 times, but rye rose 15 times, wheat 13 times, meat 6 times, and even salt 6 times.[38]

Though it would be hard for both to make a living, women wage earners suffered special disadvantages. As in the centuries before 800, just like their counterparts in the countryside, women's wages were lower than men's, the jobs open to them were the least prestigious and the most vulnerable to economic fluctuations. These ancient traditions about women's work had survived into the centuries of the walled towns.[39] For example, as an unskilled worker in late fourteenth-century Florence, a wife made 7 soldi to her husband's 15 for a day's work.[40] In Bristol in the decades after the Black Death of 1348, a woman made 75 percent of the man's wages (before the plague she made 68 percent).[41]

Most familiar of the women's jobs would be those that arose out of the closeness of the countryside to the town. Much within the walls was like the manors the young peasant women had left. In the twelfth century, the richest sections of Paris resembled a collection of castles with gardens and orchards. Nuremberg, despite its narrow streets and four- and five-story houses, had open areas all through the town where lime trees grew and commons where animals grazed; as late as the sixteenth century, the most prosperous of Nuremberg citizens owned land just outside the walls to a radius of twenty-five square miles, including villages, fields, and woods.[42] The poorest of the townswomen worked tending livestock, weeding, helping with the harvest just as they had in their villages. A fifteenth-century Book of Hours shows the scene in October just outside the walls of a Burgundian town, the towers of the royal palace evident in the background. There the women work by the river while the fields are being harrowed and sown.

Women did every kind of day labor in the town, some familiar female tasks, others more traditionally male. In seventeenth- and eighteenth-century Aix-en-Provence, a woman made 8 sous per day working in the fields outside the town, but 12 sous as a washerwoman.[43] In Paris at the end of the fourteenth century, women who did special tasks on a daily basis negotiated their wages. Laundresses seem to have been particularly numerous and figure in the town records for Brussels and for London, where Beatrice le Wimplewasher and Massiota la Lavendere worked between 1488 and 1493.

Just as women in the countryside worked in the mines of Germany and Czechoslovakia, panning, hauling, and sorting the ore, so in the towns they did heavy labor at the construction sites. Women carried stones, brought water for the mortar mixers, bunched thatch for roofs, collected moss and bracken to cushion the roof tiles of houses. Women carried sand for the construction of the cathedral in Siena in 1336 and made about 2 soldi a day.[44] Women helped to build the college of Périgord in Toulouse, France, from 1365 to 1371, carrying stones and bricks, cleaning out latrines, and digging ditches. In July of 1562 in the towns of the Rhone Valley, women made from 1 sou 6 pence to 2 sous a day working on such projects.[45]

The energy and ingenuity of poor women led them to provide every kind of service they could think of. They would spin, card linen and wool, tat lace, do piecework for the textile retailers who lived in another section of the town.[46] They took in lodgers, made a room into an alehouse, ran an inn. They made food and peddled it in the street or from door to door if they were barred from the marketplace. Women made journeys outside the walls for secondhand clothing, tallow, wheat, beer, fish, anything that could be brought back and sold in the town. In fifteenth-century Brussels, women went back and forth between the town and the countryside selling goods to both groups and were fined because they infringed on the rights of the guild members. Florentine women also continued their trading into the sixteenth century, despite guild efforts to stop them. More than men, more than the citizens in their shops and stalls, these women serviced the retailing needs of the vast majority of the town populations.[47]

A woman might combine all of these labors and services—everything from servant to construction laborer to retailer—and become a camp follower of a mercenary army. As European armies of the fourteenth to the seventeenth centuries moved across the countryside from town to town, half were the soldiers, the other half were the women and men who looked after their everyday needs. At the siege of Neuss, 1474–1475, the Duke of Burgundy's army had one woman for every four men.[48] These wives, daughters, and companions set up the campsites, collected the fuel, cooked, washed, and cared for the men. They dug trenches before the battles and nursed the wounded and prepared the dead afterward. Often wages from services that they performed supported the family until the taking of another town provided spoils and pillage. A sixteenth-century German handbook designated a captain with two lieutenants to supervise the camp followers for a regiment, to ensure their proper behavior and the completion of their duties. To supplement their incomes, some might earn money in the camps as prostitutes. The sixteenth-century guide gave these women the task of cleaning the latrines.[49] In the sixteenth and seventeenth centuries, the women of Seville in Spain performed many of the same tasks for the fleets of the Atlantic trade.

The prostitute servicing the sexual needs of soldiers had her counterpart outside the army camps. In the walled towns, prostitution was a common option for the desperate wife and mother unable to find piecework or a day's labor, for the young servant turned out because of a sexual encounter. As in the centuries before 800, a woman of the towns could always sell the sexual use of her body for money. In fourteenth-century Florence, her neighbors offered Angela, the wife of Nofri de Francesco, a basket of bread a week to stop plying her trade. She told them that even so, even if they prosecuted her, she would continue to work. Could they give her 2 florins? she asked. Otherwise, she explained "she earned much more money" selling her body.[50] For this tradition continued as well, prostitution remained the one occupation in which women earned more than men.

Initially in the thirteenth and fourteenth centuries, women sold their favors working in a variety of ways, most often alone in a rented room, or together with other women in a brothel or a bath house. In Italian towns they walked behind the cathedral, in German towns outside the city walls to signify their willingness for sex. Even as independent contractors, women in prostitution tended to band together, finding an inn or a block or rooms fronting on the street with a man willing to rent to them. In 1427 in Florence an innkeeper rented rooms beneath a house he owned to prostitutes for 10 to 13 lira a month. This at a time when the women could charge only the small sum of 19 quattrini per client.[51] Some Florentine prostitutes were Spanish, Arab, French, German, and Greek. In the towns of seventeenth-century Somerset in England, the prostitutes worked out of an alehouse or the local inn. Ann Morgan of Wells had a sliding scale of charges and doubled her price for soldiers.[52]

Men of all ranks hired the women. Men of all ranks earned money from the prostitutes' work. As in the centuries before 800, some women came to prostitution through the actions of another. In fourteenth-century Florence, Niccolo di Giunta seduced women, then pimped them for other men and took their earnings. He also sold them to a brothel in Bologna. A Florentine court in 1417 recorded that some husbands did the same with their wives. In sixteenth-century Venice, the courts prosecuted older women who turned young girls to the trade with promises of clothes and finery. Still the practice continued. A French visitor to a May fair in seventeenth-century Venice saw young daughters displayed by their mothers, buyers inspecting the merchandise and paying 150 gold écus and room and board for a year for a young virgin. A young girl, sold for 10 to 20 ducats in sixteenth- and seventeenth-century Seville, was expected to make 4 to 5 ducats in a day's work and be active in the trade for as much as twenty years.[53]

For many poor women prostitution was but one of several jobs they performed. The fourteenth-century French poet, François Villon, identified prostitutes in his "Testament" by these other occupations: Blanche the Cobbler, Guillemette who also worked for a tapestry master, Jeanneton, a bonnet maker, and Catherine who sold purses. La Grosse Margot, his favorite, ran a brothel near Notre Dame. Like men in the earlier centuries, Villon wanted to believe that she loved him. Even so he shared her earnings and beat her when he chose. The Roman census of 1517 illustrated another tradition of such women's lives. They rarely lived alone, and almost always they were mothers with children to support.

François Villon took La Grosse Margot's money. Whatever the arrangements about her work, the prostitute always had to share her earnings with others. A contract between a woman and the keeper of a brothel in fourteenth-century Perugia guaranteed her only food and clothing.[54] As an independent agent in sixteenth- and seventeenth-century Venice, the prostitute gave over a fifth of her earnings to those who supplied the bed, linens, and food. Should she try for more, men were quick to punish. In Rome there was the supposed justice of the *trentuno* to punish prostitutes thought to have cheated or robbed a client: the threat of gang-rape by thirty-one men, or the *trentuno reale* (royal thirty-one), sixty-nine men.[55] It was rumored that some of the brothels of London were owned by the Lord Mayor. In 1309 the bishop of Strasbourg built one for the revenue. In 1390 Duke Albrecht IV of Austria owned the main brothel in Vienna. Men of the craft guilds and professional associations, in fourteenth- and fifteenth-century southern French towns, sought the right to operate the local brothel, by then a lucrative monopoly.[56] In the second half of the sixteenth century, Pope Pius V taxed the women's earnings even after he had officially banned them from the town, insisting that the tax would cut down the numbers of women in the trade. The tax went for streets and for bridges. In seventeenth-century Venice, prostitutes' taxes allegedly went to pay for the galleys.[57]

Initially areas of the town were designated for the prostitutes: outside the walls of thirteenth-century Toulouse; across the Thames in Southwark in thirteenth-century London; only on the Carriera Calida (Hot Street) given royal protection in Montpellier, France, in 1285.[58] As early as the twelfth century King Henry II of England took such regulation one step further. He licensed brothels, or "stewes" as they were called, and placed the bishop of Winchester in charge of seeing to it that certain regulations about the lives and work of the prostitutes were enforced. The English king set wages, rents, days when they could ply their trade, and required weekly health examinations.

In the fourteenth and fifteenth centuries, King Henry's idea of the legal, town-, or crown-supervised brothel became the pattern for all of Europe, the place where most prostitutes would have to work. By 1415 the Florentine town council had authorized three brothels, two built at town expense for 1,000 florins.[59] Lisbon built a town brothel both to protect other women from men's advances and to earn tax revenue. The *freie Tochter* (free daughters) of fourteenth- and fifteenth-century Nuremberg's brothel had preference before the independent contractors, but lived regulated lives with their clothing, their bathing, and even where they sat in church specified. They owed a cut of the profits to their patrons; one penny of each transaction went to the "abbot," the man who supervised the brothel. In Toulouse the right to run the "public house" was auctioned each year on the feast day of Saint Lucy. As in Nuremberg the women lived strictly supervised, almost like nuns. Though they protested in 1462 to the local royal court that they were "not free," the town continued to fine and whip any women who serviced men outside the authorized brothel.[60]

The appearance of syphilis in epidemic proportions, coupled with the religious enthusiasms of the sixteenth century, made it temporarily impossible for any women to work as prostitutes. As the virus spread from Italy to the west and north of Europe, townsmen identified women as the sole carriers. Secular and religious authorities tried to protect men by forbidding all contact. In 1520 Pope Leo X ordered even the courtesans out of Rome. In 1536 the Diet of the Holy Roman Empire forbade sex outside marriage. Italian, English, French, and German towns ordered both public and private brothels closed for reasons of health or morality. Toulouse's regulations punished prostitutes with dunking in an iron cage. Some German ordinances specified flogging, mutilation, and exile.[61]

By the end of the sixteenth century, however, women could work at the trade again. Town councils and kings had decided that they must allow prostitution. In 1537 the German town of Ulm closed its brothels, but in 1551 reopened them at the request of the guilds "to avoid worse disorders," explained the statute.[62] A sixteenth-century Spanish Catholic churchman described the function of the brothel:

> It is like the stable or latrine for the house. Because just as the city keeps itself clean by providing a separate place where filth and dung are gathered, etc., so . . . acts the brothel: where the filth and ugliness of the flesh are gathered like the garbage and dung of the city.[63]

There must be prostitution, townsmen reasoned, but only subject to even more regulation. To insure "public order," in addition to wages and places of work, the municipal authorities all across Europe determined who might be clients, who might be prostitutes, and when the service could be offered.[64] To protect themselves from syphilis, townsmen now required more forced inspections of women's bodies. In the second half of the sixteenth century, Philip II of Spain supervised this aspect of his kingdom as carefully as he did the preparations for the Armada. Guildlike regulations kept out women he deemed inappropriate: the married, those in debt. A "mother" and "father" supervised them, closed the brothel on holy days, and saw to their weekly health examinations.[65] If the women contracted syphilis, they could be sent to secular or religious hospitals. In Paris in 1654 the prison hospital of Salpetrière was built to house diseased prostitutes. The hospital regulations of 1684 specified daily prayers, coarse wool clothing, a bread and soup diet, and straw bedding. As for the women:

> They shall be made, at the discretion of the directors, to do the hardest work for as long as their strength allows. . . . Swearing, laziness, bad temper, and other faults shall be punished by making the offenders go without soup, by [putting them in] the iron collar, or by confinement in the *malaises* [a hole or cubicle in which the occupant could not sit, stand, or lie down] or by other means which the directors shall judge necessary.[66]

Townsmen showed other more contradictory attitudes toward the women who turned to prostitution to earn money. As early as the seventh century, when the Eastern Roman Emperor Justinian and the Empress Theodora had established the first convent for prostitutes, religious and secular authorities had assumed that given the opportunity women would leave this "life of sin" for marriage, or like the biblical Mary Magdalene, for a life of penitence and retirement. In 1227 Pope Gregory IX authorized the Order of St. Mary Magdalene; houses were founded in thirteenth- and fourteenth-century Paris, in Vienna, in Naples. Fifteenth-century Florence had its Convent of the Penitents. In 1520 Pope Leo X authorized the Augustinian rule of the Convertite in Rome, supported in part by an inheritance tax on courtesans.[67]

The religious enthusiasms of the sixteenth century intensified men's efforts. In the first half of the century, Roman friars became famous for their efforts to persuade courtesans to repent. In 1555 one of the town's most successful, Angela Greca, became a nun, to the applause of Church leaders. Saint Ignatius Loyola and his Jesuit order made the prostitutes their special concern. With money from his noblewomen followers, he supported the House of St. Martha for prostitutes. First authorized by the Pope in 1543 it had about eighty members; soon such groups existed in Florence, Bologna, Modena, Trapani, Messina, Palermo.[68] In Italy and in France townsmen sought to prevent young women from turning to the trade altogether. Seventeenth-century Aix-en-Provence dealt with the problem in three ways. It established the Filles du Bon Pasteur for penitents, a refuge for women arrested, and one for girls disowned by their parents and thus considered likely candidates for the trade.[69]

In fact none of these measures improved the basic situation of the women or altered the circumstances that made them turn to a trade inviting men's contempt and abuse. In 1577 it was a courtesan, Veronica Franco, a woman at the top of the profession, who spoke out and asked for more realistic measures to help prostitutes. She petitioned the senate of Venice for a refuge, but for one where women could bring their children. She knew that the women in the trade numbered in the thousands, that they had children, that they worked because they had to, and that the few public brothels, public hospitals, refuges, and convents could not begin to meet their needs. She was in her own way trying to explain what prostitutes have always known, that they are not some other species of female to be raped and punished or kept chaste and saved depending on the needs and whims of the men around them. Instead they, like other townswomen, worked, had families, and survived as best they could.

Poor women committed other kinds of illegal acts. In the seventeenth century, English townswomen stole to survive, like Sarah Young in Hertford in 1683 accused of pickpocketing. Though punishments for robbery remained harsh on paper, in fact the courts appear to have been lenient toward women. Sarah Young was acquitted. The English made a heroine of the thief Mary

Frith (1584–1659). An anonymous life of her, published in the seventeenth century, made her the daughter of a shoemaker who ran away from her servant's job and became a criminal. Her house became a center for thieves. She wore male dress and supposedly bought her way out of Newgate prison for £2,000.[70] In addition, the laws of Italian and English towns tended to assume that somehow the woman thief was not responsible and must have been influenced by a man. Roman law suggested she was neither strong enough nor intelligent enough to steal.[71] Seventeenth-century English law condemned the woman when she acted alone, but otherwise, as had been the case in the customary law of the Germanic peoples, repeated the traditional belief that she could not be considered guilty when acting with her husband or a man "insomuch as she is forbidden by the law of God to betray him."[72]

Considering their circumstances, it is remarkable that more townswomen did not become criminals—unlicensed prostitutes or petty thieves—for the vast majority were poor. They came to the walled towns as young girls to work as domestic servants. They married, created a partnership with a man, and lived from day to day, working at every kind of paid labor. Like peasant women in the countryside, their strength, persistence, and ingenuity made possible their survival and that of their children.

Guildswomen

The young woman who had come from the countryside and found a place in a prosperous household as a servant, or the widow left to her own resources who found employment with others, might rise within the staff of the wealthy family. A 1410 London grocer's household listed Margery as the principal servant, probably in charge of others. The Parisian Ménagier at the end of the fourteenth century recommended Dame Agnes, the widow, to his young wife. Dame Agnes hired the staff and instructed them in their duties. In the 1560s Benvenuto Cellini, the artist, wrote of his housekeeper, Mona Fiore da Castel del Rio, "a very notable manager and no less warmhearted."[73] She even cared for him when he was ill with fever.

The dream of every *chambrière,* however, was to leave service and to marry—not just another servant, but a journeyman, a young man soon to become a member of a craft guild. Married to such a young man, an artisan, she would join the town's elite; she would become the wife of a member of a guild.

What had begun in the twelfth and thirteenth centuries as informal associations among women and men, engaged in the same crafts and living together in the same section of town, evolved into the formal organizations known as guilds in the course of the fourteenth and fifteenth centuries. The early groups had organized processions for the feast day of the craft's patron saint and collected alms in times of emergency. The later guilds did this and much more. Artisans, retailers, and merchants contracted together to regulate the training and admission of members, hours, and conditions of work, standards of quality, prices, profits, and access to market stalls. In times of emergency, the guild guaranteed the welfare of its members; in hard times, it collected money for sickness, for a daughter's dowry, for a burial. The guildsmen came to control even the affairs of the towns. By their professional regulations and their town ordinances, the members protected themselves against competition and loss of livelihood, defined the rights of citizenship, and became the acknowledged elite.

To become the wife of such a man, the young servant would have to save enough from her wages for a "dowry"—the money and goods she could bring to the union—in order to be a worthy partner of a craftsman. But a female servant made less than her male counterpart, and neither made very much, so collecting the money for her marriage took determination. In fourteenth-century Prato in Italy, her goal would be 14 or 15 florins. The merchant family, the Datinis of Prato, paid their maid 10 florins a year.[74] In the 1500s young women came from the countryside to Lyons and earned 2 to 5 livres a year and *bouche et couche* (bed and board) working in a family. The young men they aspired to marry, the journeymen, made 8 sous a day in the 1540s (20 sous to the livre) and thus would seem prizes worth labor and sacrifice to win.[75]

Records of young French servant women's wages and marriage documents, from the beginning of the eighteenth century, show how they managed to accumulate a dowry. In 1715 Angélique Simon worked in the doctor's household in Aix-en-Provence. Her wages were set at 21 livres for the year, and her goal was a dowry of 200 to 300 livres in goods and cash. Simon needed roughly ten times her yearly salary, and she could not save it all. For example, in April she needed an advance of 1 livre 8 sous for shoes; in July she had to have them repaired; in August she borrowed 6 sous for dressmaking goods to make clothing for herself. In the end she received only 2 livres 17 sous.[76] To solve this problem, servant women took other jobs. Anne Auvray in Bayeux worked as a seamstress in addition to fulfilling her duties in the household.[77]

By her marriage to a guildsman, a woman became partner in his craft, mother to their children, and supervisor of their household. Guild members and their wives kept their families small. The *catasto* of 1427 for Florence shows that family size largely depended on prosperity. Those families paying the lowest taxes were the smallest, and the wealthy had the largest households. Records for fifteenth-century Germany, for France in the sixteenth century, and for England in the seventeenth century show this same pattern: the mother, father, and two or three children making up the average family.[78]

Most guild families lived modestly, simply but comfortably. Rushes covered the floors of the shop on the ground floor and the living space above. Walls would be whitewashed; only the wealthiest in the craft had wood paneling or glass windows. Dutch families gave the smaller, original rooms of their houses to boarders, once they had begun to prosper and could add more space.[79]

They had all the essential possessions but few luxuries. At the end of the thirteenth century, a butcher's family in Colchester, England, listed as their valuables two beds, a trestle table, tablecloths, towels, two silver spoons and a silver cup, some iron and brass pots for cooking, two barrels, and three pounds of wool.[80] Pictures from contemporary manuscripts show rooms with wooden benches, draped wooden beds, chests, straight wooden chairs and tables, metal and wooden bowls and spoons, and woven baskets. As her marriage gift a typical Florentine artisan woman received ribbons and fabric to use as trimmings on her clothes. In fourteenth-century Paris the Ménagier, writing for the benefit of his young bride, assumed that clothes would remain in the family from one generation to another and recorded remedies for removing stains and for protecting furs from moths.

A fortunate woman might become a member of a craft guild in her own right. These crafts, called métiers in France, had evolved gradually from the twelfth and thirteenth centuries. An enterprising woman or man separated out a product or a process, incorporated with others involved in the same manufacture, and by a charter limited access to the right to practice the trade. The majority of the charters specified only male membership; some, however, allowed women to join, and others—especially in textile and clothing-related crafts—were exclusively female.[81]

Some lists of craft members show only a few women. Historians assume they were widows who inherited the right to guild privileges on the death of their husbands. Some guilds allowed a widow equal rights with male members. In thirteenth-century Lübeck she could even pass membership on to a new husband.[82] In the late fourteenth century in Cologne, Germany, individual women carried on their husband's crafts as shoemakers, tanners and saddlemakers, pewterers, copper beaters, blacksmiths, makers of felt hats, dyers, and weavers of the cotton-linen blend called fustian. Other guilds explicitly or implicitly excluded wives and widows from crafts like harness making, boot making, roofing, bronze casting, rope and sword making, working in timber and stone.[83] Widows in sixteenth-century Strasbourg and seventeenth-century London inherited both the presses and the role of publisher from their husbands. Margarethe Prüss, widowed twice, ran a Strasbourg printing shop and used it to publish radical Anabaptist tracts. From 1553–1640 almost 10 percent of London's publishers were women.[84] Other guilds treated guild membership like the widow's portion, the land inherited by a woman in the countryside. The widow enjoyed the privileges only until she remarried.[85]

During their lifetime together, the wives of artisans participated in their husbands' crafts and thus gained indirect access to the privileges of the guild by helping in the production of goods and by supervising the apprentices, servants, and journeymen. A weaver would need the efforts of his whole family, since his wife and daughters spun the thread for his loom. Wives took responsibility for selling the goods. Commonly the lower story of the house had shuttered windows that opened on the street so that she could sell right from the shop. Or she might sell in the market, door to door or in the streets. In sixteenth-century Nuremberg the market opened for three hours in the morning, in theory to make it easier for women to sell around and about their other work. It was there that the artist Albrecht Dürer's wife, Agnes, went to sell his prints. Buying and selling goods gave many women their own income and sometimes access to a guild in their own right.[86]

The earliest records of women's activities in the towns—from Paris in the thirteenth and fourteenth centuries, from Arles in the fifteenth and on into the seventeenth centuries, from London and numerous other English towns in the sixteenth and seventeenth centuries—show them participating in the retailing guilds. They paid for licenses to sell, acquired the right to use the choice market sites, and bought the privilege to be first to buy from the wholesalers. They belonged to these guilds at every phase in their lives, unmarried, married, and widowed. Their names appear on the tax roles attesting to the profits they made. Women sold fish in Les Halles in Paris and spices in the market in Ghent. Women bakers in Arles contracted with country families for grain and then sold their wares in town, three grades of bread: for the householder, for the laborers, and for the master's dogs. In the town of Reading, Steven Foorde's wife purchased the monopoly for selling lindsey-woolsey (a coarse fabric made of linen and wool). Wives not husbands held the right to the stalls in the cloth market of Ghent. Throughout Europe women marketed products that had no connection to their husband's occupation. Many women in Lyons, France, sold clothing accessories, Pernette Moilier, wife of a goldsmith, specialized in wimples. In Cologne the wife of the carpenter, Gotschalk von Weinsberg, kept selling fish, butter, and cheese after their marriage.[87]

In addition to acquiring guild privileges, women found other ways to eliminate or mitigate competition. Food sellers in seventeenth-century Manchester went from house to house on market days, offering their products at less than the men selling in the official stalls. Records for seventeenth-century Dorcester and London show that the women continued to act as retailers even outside the guilds, even when they could not acquire the proper licenses and had not paid the necessary fees. Elizabeth Beaseley sold tobacco illegally, the wife of Richard Morton cuffs and handkerchiefs.[88]

The easiest way for a woman to become a member of a guild was the same as for a man—to be apprenticed as a young child. Just as propertied and privileged families in the countryside sent their daughters to another's household to be trained, so retailing and manufacturing families in the towns sent theirs to learn a craft. Girls and boys in Paris from the twelfth to the fourteenth centuries began their apprenticeships at eight to ten years of age. Thirteenth-century girls sent to learn embroidery for headdresses served for six years.[89] At the end of the fourteenth century, in the exclusively female silk and thread makers' guilds of Paris, the girl had to be eleven and served a four-year apprenticeship.[90] In some crafts the parents, or the master and mistress, had to pay a fee for the apprentice. In fifteenth-century Mittelrhein in Germany, a young seamstress cost the master a gift of three pounds of wax for the church. English silkwomen in the fifteenth and sixteenth centuries required a fee of £5 and trained the girls from seven to fifteen.[91] Parents of apprentices to the lace makers in seventeenth-century Caen paid the first two months' room and board for their daughter (in theory the time it would take for her to know enough to earn her keep).[92]

Usually young girls stayed with the mistress until they married, in their mid-to-late twenties. During her apprenticeship the young girl owed her mentors obedience; they in turn gave her clothes, shoes, bed, and board. They could beat her when they thought it necessary. Some young silkworkers argued with their mistresses over money, others, according to wills of their mistresses, became prized employees. In 1456, for instance, Isabel Fremely left a green silk girdle and a pair of shoes to her dutiful apprentice.[93]

Having served her apprenticeship and having met any other requirements (such as making her "master piece"), like her male counterpart, the young woman became a member of the craft guild. From the thirteenth to the seventeenth centuries, records of towns like Paris and Florence show women both admitted to some predominantly men's trade associations and creating their own exclusively female craft guilds. They worked in a wide range of trades. The Parisian taille (list for tax purposes) of artisans of 1300 listed, out of a total of 200 crafts, eighty to ninety with both female and male members and twelve exclusively female.[94] By 1321 eight women belonged to the minstrels' guild.[95] They had also organized themselves as embroiderers, pin makers, and hat makers. By the fifteenth century women had more or less taken over the production of women's clothing and table linen, grouped together as *lingères,* or seamstresses. In 1397 the town of Cologne listed its forty chartered guilds; three involved in thread and silk manufacturing had only female members; twenty-eight, including metal-, wood-, and leather-working, needle making, and smithing, were open to women either in their own right or as widows. In 1500 a woman goldsmith, Drutgin van Caster, became an artisan for the Holy Roman Emperor Maximilian.[96] Records for other towns like Brussels, Ghent, London, Southampton, and Bristol show even more variety of occupation; women worked as saddle and spur makers and tailors, wagoners, brewers, candlemakers, tanners, and coopers. Though craftswomen's names appear in a wide range of occupations, they always predominated in those dealing with textile production and decoration. These were usually the guilds that admitted no male members. By the seventeenth century in Florence, women had all but monopolized certain aspects of textile manufacturing. They dyed, sorted and packed the wool, wove, and embroidered it.

Of all the textiles, the production of silk (everything short of the weaving of the finest measures of cloth) and thus its guilds became almost exclusively female. Women brought care, patience, and diligence to the processes involved in unwinding and preparing the delicate threads that first appeared in Sicily and northern Italy in the twelfth century. Soon women in Florence, Paris, and London had created guilds of silk workers with their own regulations and corporate status from the towns. The *Book of Crafts* compiled by Boileau, the provost of Paris, about 1270 listed seven silk crafts. The spinners, ribbon, and silk coif makers admitted only women; the makers of silk and metallic threads (used to decorate hats, harnesses, armor, pillow coverings, and ecclesiastical draperies) allowed both women and men.[97] In London by the middle of the fourteenth century, the silkwomen had organized themselves sufficiently to petition the lord mayor against the granting of a monopoly for the sale of raw silk to the Lombard merchants of Italy. In the second half of the fifteenth century, they petitioned Parliament twice, in 1455 and 1482, against foreign competition. By this time London silkworkers spun the thread, wove ribbons and bands, made tassels and laces, and sold the final product. Given the complaint in the petition, many probably acted with husbands in the importation of the raw silk and were involved with more than one aspect of the trade.[98]

Under the protection of the guilds, townswomen from the thirteenth to the seventeenth centuries knew unique opportunities. Through marriage to a guildsman, they gained in status and won some measure of economic security. They participated in their husband's craft. They might practice one of their own. Women in guilds with only female members enjoyed a kind of independence. They had their own income and their own protected craft. The image of their relative success brought young women to the towns and kept them from returning to the more predictable, but less prosperous life of the countryside.

Merchants' Wives

Daughters of successful guild members, whether from the family of a specialized craftsman or a retailer, could aspire to become part of the most privileged group of all, wives to men in the merchant and banking guilds that controlled the trade, the finances, and usually the political life of the town. These families made up the powerful merchant corporations and financial partnerships, with representatives living and dealing in towns all over Europe. These were the great entrepreneurial families of the new commercial capitalism that evolved from the thirteenth to the fifteenth centuries. Gradually they had taken control of the towns from the old feudal aristocracy and closed their ranks, defining their status and limiting access to their privileges. When total populations numbered from 40,000 to 60,000, the merchant elite numbered about 1,100 in Ragusa (Dubrovnik) in Yugoslavia in 1442. When the patricians of Nuremberg closed their ranks in 1521, they listed forty-three families at the top. In sixteenth-century London with a much larger population of 400,000, the male elite is estimated at 1,200.[99]

As wives of these wealthy and powerful husbands, the women of the merchant elite had many comforts and many responsibilities. More wealth and the more sumptuous life meant more possessions, more staff, and greater expectations placed on the wife's abilities to supervise and provide a pleasant refuge and suitable show place for her husband. In the houses of such families in Florence, Regensburg, Paris, Bourges, or London, there was light, vivid colors, warmth, space and numbers of rooms, with water readily accessible. This environment was much more comfortable and luxurious than the castle or manor of even the most powerful noblewomen and their lords. Increased light came from windows of white glass, or of translucent oiled parchment, which also acted as additional insulation; in daytime wooden shutters would be pulled open to let in the daylight. In 1321 a London father gave his daughter the property adjoining his own with the reservation that she must leave ten feet between their houses so as not to block his light.[100]

Braziers, tiled stoves, fireplaces with flues and chimneys to take the smoke away warmed the rooms, now separated and subdivided, with business conducted on the ground floor. The family lived and slept on the floors above. The signora, or wife, of a north Italian banking family had as many as twelve private rooms and a large public space on the middle floor of her palazzo (a large town house). In Florence, in Paris, in Amsterdam, throughout Europe, tapestries now hung on the walls, not just around the master bed. Families commissioned painters to decorate their rooms; they paneled and white-washed the walls, covered ceilings with bright cloth, laid floors with parquet, and added ceramic tile for decoration. Bright colors filled the family rooms. Like her counterparts in the country, the patrician's wife strewed the floors with herbs and flowers in the summer, with straw in the winter. By the sixteenth century, the family's activities could continue into the dark hours as the wealthy wife had tallow candles, oil lamps in holders on the walls, and metal chandeliers that hung from ceilings. The well in the courtyard gave the family water, which would be scented and heated for bathing.[101]

Throughout the rooms there would be much more furniture and many more possessions than in a castle or manor house. Unlike the landed noble-woman, who had limited space and had to move from dwelling to dwelling, the signora, the burger's wife, lived year round in one place. She no longer had to fold and store the furniture. The large rectangular table remained standing in front of the fireplace. Chests, stools, chairs, and benches were arranged along the walls. The master of the house had his carved chair with cushions. By the sixteenth century, chests had evolved into cabinets for storing linens and clothing. Eating utensils now included plates, cups or goblets, forks and spoons. In fifteenth-century Italy, glass and ceramic be-came cheaper than metal, and different kinds of tableware began to multiply.

With the wealth of these families, wives could have many kinds of clothes. Margherita Datini, wife of the fourteenth-century merchant of Prato, was frugal in all of her household management. When she spent money, it was on her clothing. She had linen shifts, linen under socks and long hose, linen nightcaps for sleeping, two woolen dresses for household use and a few long-sleeved ones for public occasions. Much was interchangeable: linings could be added to one cloak or another, sets of sleeves could be added to a dress. Her prize possessions were her eleven surcotes, one in blue damask with ermine trim, another in soft gray with miniver trim on the hem. In 1392 her husband ordered an overgown for her from Romania, a "roba" in silk embossed with velvet. She had cloth headdresses to match some of her dresses, two straw hats, and a beaver hat. She also owned two gilt buckles and two purses. Her jewelry consisted of a few necklaces, buckles, and rings with gem stones, all valued at 133 florins and kept by her husband in a chest in his room.[102] In the 1456 inventory of her dresses Lucrezia de' Medici, wife of Piero de' Medici, listed even more sumptuous clothing: "crimson and gold brocade damask faced with ermine," "black damask faced with red taffeta," a blue taffeta silver brocaded cloak and sets of gold brocaded sleeves.[103]

Perhaps most indicative of the more sumptuous life of the wealthy townswoman were the elaborate meals she provided for her household. Margherita Datini probably supervised the making of two meals in the day: the first, a broth served with bread, cheese, and fruit and nuts like dates and almonds. Near midday, the family would have "dinner," with as many as three courses. The Datinis began with a broth or pasta, then stuffed meats, fish, poultry, served roasted, boiled, or in pies with beans, peas, onions—all covered with sauces. The third course would be sweets, sometimes made in a pie with meat.[104] The fourteenth-century Ménagier of Paris included recipes and dinners in the treatise he wrote for his young wife. He rejected some dishes like stuffed peacocks as unsuitable for "a citizen's cook" and others like a "mock hedgehog" made out of mutton tripe as "too much to do."[105] For a banquet, when he expected her to provide for as many as forty guests, he left nothing to chance, told her what to order from whom, gave the menu, and listed the servants she would need to produce and serve the meal. The matron of a more modest household in the fifteenth century would typically serve a potage, a thick souplike mixture of whatever was available. All was cooked in the same pot, with different foods wrapped separately and put in to boil for different lengths of time. The staples of their diet would be fish, eggs, beans, and leeks.[106] With the spread of printing in the sixteenth century, household "cookery" books appeared in England. By the seventeenth century, a London housewife could study many aspects of the art in herbals and almanacs, most written by men.

For much of the food, the merchant wife depended on the gardens and orchards that surrounded her town house. The wife of the Ménagier had cherry and plum trees, raspberry and currant bushes and grape vines. She grew spinach greens, lettuce, leeks, cabbages, parsley, beans, peas, turnips, radishes, parsnips, and pumpkins. Her herb garden had everything from lavendar, borage, hyssop, and savory to sage, mint, basil, and marjoram. In the market she bought fruits like dates, figs, raisins, apples, pears, peaches, pomegranates, and oranges. Almonds, a delicacy, were a staple of the kind of cooking she supervised. She or her husband purchased other cooking and medicinal herbs like camomile, pennyroyal, fennel, saffron, coriander, tansy, and rue. Francesco Datini bought the spices his wife required: pepper, cinnamon, and nutmeg. For meat, poultry, fish, eggs, and dairy products, a wealthy woman had both what she had preserved and what she could buy. Merchant households made sausages and salted meats and fish. In sixteenth-century Nuremberg, the market was filled with fish and game: hares, pheasants, wild ducks, pike, and trout. Poultry, eggs, and milk came from the country. In some English towns the farmer or his wife brought their cow to town and sold fresh milk at the door.[107]

A fifteenth-century merchant's wife wore keys on her belt, the keys to the chests, doors, and outbuildings, for her supervisory responsibilities went far beyond providing meals. The Ménagier of Paris in the fourteenth century instructed his new wife on the care of their possessions. Through the housekeeper, Dame Agnes, she was to be sure that pillows and bedclothes were shaken, floors swept, to see to the hourglasses (which she was to fill with marble dust) and to the preparation of rose water for the wash basins. She also had a maître d'hotel, Master Jehan le Dispensier, to help with the supervision of the gardens and orchards. Dispensier saw to the storage of grain and wine. She was to supervise the shepherd, cowherd, dairy maid, and the farmer and his wife who cared for the livestock. The Ménagier gave pages and pages of instructions, and expected her to see to the household garden, even to planting, cutting, and grafting.

Other wives of these elite families had additional duties beyond the house and the town garden and orchards. As early as the thirteenth and fourteenth centuries in Italy and Germany, it was customary for wealthy townsmen to acquire lands outside the town walls: farm lands, vineyards, meadows, olive groves, and untilled fields. All these were left to the wife to supervise when the husband traveled to fairs and markets or worked in his offices. At the end of the fourteenth century, Francesco Datini lived most of his life in Florence or Pisa and wrote once or twice a week to Margherita in Prato. His letters from 1407 give minute instructions on how, when, and what to do for the 300 acres that he had acquired and expected her to administer.[108] In addition to seeing to horses' shoes, watering orange trees, milling grain, Margherita Datini also had to provision her husband wherever he was. Dutifully, she arranged for everything from fresh produce, to preserves, to wine, oil, meat, eggs, and even bread to be carefully wrapped, packaged, and sent off to him. In sixteenth-century Florence, Isabella Guicciardini (married to the younger brother of the Renaissance historian) supervised her husband's estates while he traveled for as long as six months at a time as a provincial governor. She oversaw all aspects of the planting, the harvest, storage, and marketing. She maintained the canals and buildings, kept the accounts, and paid his debts.

Sixteenth- and seventeenth-century merchant and bankers' wives, in the towns of Italy, Germany, France, England, and the Netherlands, learned to read and write and to do accounts, because only then could they keep track of all the transactions they supervised. In the seventeenth century, Samuel Pepys taught his wife basic mathematics in Arabic numerals. Before then, townswomen learned simple accounting but with Roman numerals, adding up accounts with an "exchequer cloth"—material divided into marked squares. They placed copper or iron counters on the appropriate squares, counted the piles, and then entered the totals in the account books.[109]

Many patrician women became the trusted business surrogates for their husbands. In fourteenth-century Ghent, Clare van der Ponten's husband gave her his legal proxy so that she could sell wool and buy land for him. In the middle of the fifteenth century, Contessina, the wife of Cosimo de' Medici (founder of the Florentine dynasty), had a scribe to help her keep up with her voluminous correspondence. In addition to writing to sons and daughters-in-law with advice, to her staff supervising the palazzo in Florence and the summer villa, she wrote to her husband. At one point in their marriage she relayed information from his business mail, passing on suggestions from others and giving out his instructions. Sixteenth-century German women did the same for their merchant husbands. Most of the entries in Matthaus Runtinger's account books were made by his second wife, Margarete. Other townswomen took even more responsibility. In 1622 an Englishman, James Howell, wrote admiringly of Dutch women:

> In Holland the Wives are so well versed in Bargaining, Cyphering & Writing, that in the Absence of their Husbands in long sea voyages they beat the trade at home and their Words will pass in equal Credit.[110]

The wives of the great merchant entrepreneurs, the manufacturers, and the bankers lived well in the towns. More wealth, however, brought more responsibility. Their households, their families' lands, and sometimes their husbands' businesses consumed their time and their energies. They rarely acted for themselves, however, or on their own behalf. They accepted the role of willing helpmate that the male writers of their centuries popularized.

2

DANGERS AND REMEDIES

Natural Disasters and War

Only the most privileged townswomen had any hope of insulating themselves and their families from the effects of the natural disasters that came to the urban setting just as they did to the countryside. All were vulnerable to extremes of cold and rainfall. In the coldest winters, rivers froze, water mills went stiff, pumps failed, and the already tenuous water supply became even more uncertain. A fifteenth-century Parisian noted in his diary the cold of the 1427 winter with frost and hailstorms into the month of May. In December of 1439 wolves came into the streets and killed four housewives. Even the mild December of 1421 had meant heavy rains, frost, ice, and flooding of the Seine.[1] In the Netherlands in the thirteenth century, a dike failed, and 50,000 died.[2] Spring and summer rains could bring flooding as well. In 1333 the Arno of Florence overflowed its banks and took the buildings along the river.

Harsh winters often meant food shortages. Towns, especially those of northern Italy, tried to regulate prices and to provide against maldistribution, but still-uneven production, natural catastrophes, failed crops, and problems of transportation meant that poor women and their families might starve to death. Between 1375 and 1791 the *signoria* (the town council) of Florence recorded 111 families who died in this way.[3] In Paris in the winter of 1438 the poor had no bread and took to eating "turnips and cabbage stumps." The fifteenth-century diarist wrote: "All day and all night young children and women and men were crying out, 'I am dying! Alas, dear God, I am dying of hunger and cold!' "[4] Even in the best of times, the environs of Florence could only produce enough grain for five months of the town's needs.[5] Shortages, whatever the cause, made the price of grain, the staple of townswomen's and their families' diets, rise dramatically. In 1316 in Strasbourg women paid 30 shillings for a quarter of wheat, then 40 shillings. That same winter in Louvain women bought a measure of wheat for £5 at the beginning of November, £7 by the end of the month, £10 at Christmas, and £12 by Easter.[6]

Fire, so essential for much of life in the town—for heat, for light, for cooking, even for the practice of many crafts—was one of the townswomen's principal enemies. Open fires were everywhere in the crowded wooden-housed towns of the twelfth to the seventeenth centuries: in the houses, the shops, in the yards of tanners and dyers. The big fires like the one in London in 1666 destroyed whole sections of the towns. Fire, in fact, caused the townspeople to rebuild in stone and brick and to abandon the wood and thatch of earlier centuries.

In times of war, fire, shortages, and famine came to the townswomen just as they did to the women of the countryside. The peasant woman ran with her husband and children to the forest to escape the bands of soldiers. The noblewoman, or châtelaine, supervised the defense of her manor or castle. Townswomen and men could play no such direct role and had no refuge. In the changing world of commerce and capital, towns became rich prizes for noble and royal commanders, and the siege became the accepted method of winning victory. From as early as the eleventh century, town councils authorized money for walls, towers, gates, and drawbridges, all designed to make the town impregnable to assault, like the castle with its fortifications. Vienna in 1458 also had a moat, flooded with waters from the Danube. Townspeople rarely fought themselves; instead the councils of guildsmen, the burgermeisters, employed professional soldiers—as early as the twelfth century in the smallest Italian hill towns. By the fourteenth and fifteenth centuries, women in Nuremberg and Cologne were protected by the town's own soldiers and own artillery. The Hansa towns of northern Europe had their own navy. Mercenaries did the fighting, commanders and councils made the truces and peaces. Townswomen might help to fill the need for unskilled labor; they dug trenches in London in the Civil War of the seventeenth century just as they had in the 1500s when the town had been threatened by invasion from Spain. Women filled buckets to prepare against the inevitable fires; they might make torches to light with oil; they carried and heaved chains across streets. They may have taken a turn at the watch.

Most often in time of war townswomen, whether laborers, craftswomen, or merchant wives, looked first to the survival of their families, especially by finding enough food. At the end of the fourteenth century, Margherita Datini was instructed by her husband to bring as much as she could from their farm outside the town of Prato where they lived. Peasants from their lands came with her, believing that they would be safer inside the walls. Refugee women straggled through the gates of Paris in June of 1419, fainting with the heat and "worn out with carrying their children," explained a contemporary diarist.[7] Supplies of basic goods and prices fluctuated wildly during the five sieges of Paris by the English from 1410 to 1431. In times of fighting meat, cheese, eggs, flour, wine, fish, charcoal, firewood, even garlic became prohibitively expensive, however rich the townswoman might be. In times of truce the prices fell just as dramatically. Desperate in May of 1418, some women ventured outside the walls of Paris to find provisions in the countryside; the enemy troops raped them.

Most often when the town fell to the attackers, women and children became the chief victims. The conquerors expected to steal, to run riot, to loot, and to take their revenge on the townspeople. In the Hundred Years War, England's Black Prince took Limoges in 1370 and, according to the chronicler Froissart, cut the throats of an estimated 3,000 women, men, and children. Froissart described the townswomen as "too unimportant to have committed treason," yet they were victims nonetheless. An Italian silk merchant caught in Mons in 1383 when the Duke of Burgundy's forces took the town could not believe the "truly horrifying sight" of a woman

> sitting with a two-year-old child in her arms, a three year-old clinging to her shoulders and a five year-old holding her hand, by the door of a furiously burning house. She was pulled up and moved some distance away to prevent herself and the children coming to harm, but, as soon as she was let go, rushed back in the door of the house, despite the great flames which were billowing from it, and was finally burnt inside with her three children.[8]

The French and German religious wars of the sixteenth century proved just as brutal. The Humanist Olympia Morata fled Schweinfurt with her husband. He was taken on the road, and she remembered, "I had nothing with which to ransom him." Those around her took her clothing:

> I was left with nothing but an undershirt. I lost my head covering and my shoes . . . the first day. I said to myself, "This is all I can take."[9]

At one point she hoped just to lie down and die. Weakened by the ordeal, she contracted malaria, but before her subsequent death, the intrepid woman gained her husband's release and a position for him at the university in Heidelberg.

In the Thirty Years War of the seventeenth century, when mercenaries and companies of soldiers in the tens of thousands fought across central Europe, the hardship and devastation reached new levels. Contemporaries recorded 33 percent of the towns destroyed or deserted, Hameln, Göttingen, and Magdeburg in rubble, others with their populations decimated.[10] In the

1630s, townspeople of every region had witnessed horrors: women pushed out of second-story windows, mothers dead and their children hiding in cellars with the houses burned to ashes above them, refugees driven from Strasbourg in the tens of thousands one winter. A contemporary pictured a scene in the Polish town of Calw in the winter of 1634: a dead horse lying in the street with ravens, a dog, and a starving woman gnawing at the carcass.[11] Until the end of the seventeenth century, towns remained rich prizes, and townswomen and men vulnerable to looting, physical abuse, and starvation.

Disease and Childbearing

At worst war meant victimization and death for a townswoman and her family. At the least it meant shortages. War also brought disease. The soldiers of the Thirty Years War spread typhus through the streets and alleyways of the towns they conquered. Even in times of peace, crowding and rudimentary sanitary conditions made everyone from the poorest to the wealthiest vulnerable to infectious disease. In York in the sixteenth century only 11 percent of the population lived to be forty, and this was typical of town life in these centuries.[12] The vast majority of the town's population lived crowded together and used water from a common well in the square at the end of a rutted street. Refuse was an ever-present problem. Towns attempted to provide clean water and to control sanitation. In Venice, from the fifteenth to the eighteenth centuries, water was delivered daily to locations around the town. Towns passed regulations against throwing waste and garbage into rivers and canals. Both London and Nuremberg had public privies where women were expected to bring the family's sewage. The rich had their own cesspits.[13] In Madrid town officials decreed that sewage should not be thrown from windows but rather set out to be collected at ten o'clock on winter evenings and at eleven in the summer.

In the wealthiest families, servants brought water into the house and heated it so that the mistress and master could bathe. Everyone else in the town used public bath houses. Female and male bath house keepers appear on the list of guilds for the 1292 Parisian tax rolls.[14] In the first half of the fifteenth century, when Joan of Arc stayed in Bourges, she and Marguerite La Touroulde, wife of a royal official, bathed together at one of the public baths. With the spread of syphilis in the sixteenth century, town councils closed the baths down. From thirty-nine in Frankfurt-am-Main in 1387, the number dropped to nine in 1530. Protestant townsmen likened them to brothels and thus disapproved of them even more.[15]

In these centuries it was no idle opening in a letter to ask, "How are you?" Contessina, Lorenzo de' Medici's mother, wrote to him often of many matters, but almost always asked after his health. She worried particularly about "the plague." Well into the seventeenth century, the bubonic plague, the Black Death (*Pest Jungfrau,* Maiden's Disease, the Viennese called it) came decade after decade, claiming the lives of townswomen and men. One third to one half of the town populations died in the first outbreak in the summer and fall of 1349, when the disease spread from the Italian ports of Messina, Genoa, and Venice along the trade routes from market town to market town, to Pisa, to Florence, to Marseilles, and then all across northern and eastern Europe. Contemporaries estimated 2,000 dead in seventy-two days in Frankfurt-am-Main, 5,000 to 6,000 in Vienna. Figures for the sixteenth and seventeenth centuries are no less dramatic. The historian Henry Kamen has estimated that a minimum of one eighth of the town's population died with each appearance of the plague.[16] Forty thousand died in the 1575–1578 epidemics in Messina in Sicily, and one quarter to one third of the population of Venice. In the small French town of Romans, 51 percent of the population died. In Amsterdam, in the towns of Bavaria in southern Germany, similar figures emerged in the seventeenth century. Paris had six outbreaks and London five. Englishmen tried to keep a tally for the years 1593–1665; they noted 156,463 deaths.[17]

With the plague, as with other diseases like influenza, typhus, dysentery, measles, and smallpox, poor women and men and their children suffered and died in the greatest numbers. Town councils tried to institute and enforce quarantine measures, employed physicians, established procedures for destroying contaminated clothing, and collecting and disposing of the dead, but sheer numbers defeated them. Boccaccio described 1348 in Florence in the *Decameron:* the poor stuck in their houses, with only a rotting corpse to alert the neighbors to death, with bodies picked up by strangers and taken to trenches in the churchyards, and with no women there to attend the dying, to mourn the dead.[18] One fall in Paris in 1418, a contemporary reported that 50,000 had died of an unnamed fever over a five-week period:

> They had to dig great pits in the cemeteries of Paris and lay thirty or forty in at once, in rows like sides of bacon, and then a bit of earth scattered over them.[19]

"The contagion only ever hits the poor people . . . God by his grace will have it so," wrote a citizen of Toulouse in his journal in 1561. "The rich protect themselves against it," he explained.[20] But in the major epidemics, everyone gave themselves over to panic. In 1450 Contessina de' Medici urged Lorenzo and his brother Giovanni to "protect themselves" by fleeing. She had heard that the plague had struck Rome, "even among better class people."[21] At the end of the seventeenth century, Glückel, a Jewish merchant's wife, left Hamburg with her family for Hameln, hoping to escape contagion, but the panic had already spread. When they stopped for the night, to break the journey and to celebrate the holy days before Yom Kippur, women of the house learned that Glückel's four-year-old daughter Zepporah had a sore and swelling under her arm. No one listened to the mother, who tried to convince them that the child must be healthy as she still "plays in the grass and eats a buttered roll as nicely as you please."[22] Instead Glückel stayed with her two younger children while Zepporah was sent to another village with only a servant to care for her.

The women of the towns died for other reasons as well. The 1427 *catasto* of Florence reveals different patterns of mortality for women and men. In two age categories, males outnumbered females: from 0–four years old (124.57 males: 100 females) and from twenty-three to twenty-seven (117.61 males: 100 females).[23] The statistics suggest that infant girls died in greater numbers than their brothers, perhaps because they were fed less.[24] As adults in their late twenties, women's higher mortality reflects their increased vulnerability because of childbearing. Even with an easy delivery and no complications from the birth, for rich as well as poor women, infections (like pelvic inflammatory disease) could bring lingering debility; frequent births could lead to malnutrition or to tuberculosis.

The childbearing years of the women of the wealthier families have been best recorded. These women birthed children roughly every two years. Families with seven, ten, and fourteen live births appear to have been common. Sometimes this was the labor of a single wife, sometimes of a series of wives in turn. As one died, another replaced her. Gregorio Dati, a silk merchant in fourteenth-century Florence, noted births and deaths in his account books. He begins with his betrothal to Berta, his second wife, in March of 1393; they were married in June, and by July 1402, nine years later, she had produced five sons and three daughters, the first one named for his first wife. Dead three months after the last boy's birth, she was soon replaced by Ginevra, who bore four daughters and three sons. Ginevra died in July 1420 with the birth of a baby girl. Dati's fourth wife bore at twelve-, thirteen-, and fifteen-month intervals like her predecessors, five children in ten years.[25] This was an unusually large family, but other wealthy women in fifteenth-century merchant and professional families in English and French towns produced averages of seven, eight, and eleven children.[26]

To generate so many offspring these women of the elite started bearing young and, if they survived, continued to bear into their forties.[27] Glückel, the merchant's wife from Hameln, arrived in Hamburg to assist in her daughter's confinement, only to give birth eight days later herself. A merchant wife of Toledo in early sixteenth-century Spain, Doña Beatriz de Ahumada, mother of Saint Teresa of Avila, was a second wife, married at the age of fourteen. She bore ten children, three girls and seven boys, until she died at thirty-three. Saint Teresa remembered her mother as always sick.[28] Other women's experiences were similar: Margherita and Francesco Datini's closest friend, Ser Lapo, saw his wife birth fourteen children, but she lived as a semi-invalid.[29]

Just as in the castles and manors and in the rough cottages of the countryside, women of all parts of the town died in the attempt to bear. Women of the great Italian merchant and banking families like Clarice Orsini (Lorenzo de' Medici's wife), Lucrezia Borgia, and Beatrice D'Este all died after childbirth. Venetian women of property in the fourteenth and sixteenth centuries typically made their wills with their first pregnancy. The silk merchant Dati lost his first wife Bandecca in 1390 from illness nine months after a miscarriage; his second wife Ginevra died "after lengthy suffering." He remarked on her "strength and patience" in her last days.[30] An artisan's wife from seventeenth-century France wrote of her sixteen-year-old daughter's death in her arms: "Would that God had put me in the coffin instead of her. I will never forget her to my last breath."[31] In 1388 a friend wrote to Francesco Datini:

> Your maid has been about to give birth since Tuesday evening and it is the most pitiful sight ever seen. . . . She must be held down, or she would kill herself and there are six women who care for her in turn. This morning they think the child is dead in her belly.[32]

At the birth, whether in a poor or a rich household, there would be a midwife, family, and friends. If it appeared likely that the infant had died—as in the case of Datini's servant—a barber surgeon was called in to extract the fetus. The tearing and forcing of the surgeon's instruments might cause hemorrhaging or infection. A long labor brought death from dehydration and exhaustion, from internal bleeding if the uterus ruptured.

Although wealth did not ensure survival for the mother, it did affect the chances of the infant. The poorer the woman, the less likely a large family. In Italian, French, and Spanish towns in the fifteenth, sixteenth, and seventeenth centuries (the earliest for which records are available) a newborn baby had a 50 percent chance of surviving to adulthood. In York from 1538 to 1812 mothers most frequently bore in April and October, with the highest infant mortality in September and April, just before new harvests would reach the towns. Even for the rich, infant mortality was high. For the second wife of fourteenth-century silk merchant Gregorio Dati, three of her eight children died of plague, one of a respiratory infection; of Dati's third family, only two survived. Only seven of the cloth manufacturer Buonaccorso Pitti's eleven children lived to adulthood.[33] Although the fact of higher mortality among the poorer members of the towns and communes holds consistently, otherwise statistics show fluctuations and suggest no clear explanation for the changes in rates of infant mortality.[34]

Faith in God, Repentance, and the Saints

Like the women of the countryside, women of the towns explained these disasters, this presence of death and uncertainty with their religion. Grain shortages in Lyons in 1504 and 1529 brought women and men out into the streets dressed in white penitential robes, to march in processions and to pray for forgiveness. The view of Margherita Datini is typical; she wrote to her husband in November of 1398 that the plague "is like unto the Judgment in the Gospels, whereof we know not whether it will come upon us by day or by night."[35] At the end of the seventeenth century, Glückel of Hameln described her merchant husband's death and the loss of her former comforts: "My sins found me out, and I was not worthy of it."[36] The financial disasters of her second marriage she viewed as God's test of her and a lesson in patience; she was grateful her circumstances were no worse. Glückel of Hameln was Jewish, Margherita Datini Christian.

All in the towns, women and men shared this sense of the sinfulness of humanity, the fact of God's judgment, and the imminence of punishment in life. Christian women feared punishment after death. Death had particular significance for all women, for they, like their sisters in the countryside, prepared the bodies of their loved ones for burial, washing and wrapping them in plain white shrouds knotted at top and bottom. Often the women took on the obligation of prayers. Christian families paid to have masses said; the wealthy might have nuns or monks to prepare their bodies and to pray for them in specially endowed chantries, or chapels. Matilda Tody, fourteenth-century widow of a London wool merchant, made a pilgrimage to Rome to pray for her husband and to prepare for her own death.

Preparation for death, prayers for the dead became more significant in the late thirteenth century when townswomen saw images of Heaven and Hell all around them. Over the great doorways of the cathedrals at Amiens, Chartres, Bourges, Bordeaux, Rouen, Poitiers, Paris, townswomen saw the reliefs of Jesus as Christ the Judge enthroned—Heaven on one side, Hell on the other. At Rheims horned devils hold the damned with chains, ready to place them in a cauldron; in contrast, Abraham gathers the saved into his lap, and they smile the happiness of those in paradise. Bourges Cathedral showed townswomen the ultimate Christian reward, the day of resurrection, when pure nude adolescent bodies rise from the dead; a young woman prays in wonder, a young man stands and lifts away the lid of a coffin.

What women might miss in the reliefs of the cathedral fronts, or in the stained glass of its windows, they would see acted in the mystery plays that made up the entertainment for many of the town's festival days. Plays portrayed Eve and Adam, Noah and the Great Flood; they showed the devil costumed in black or green, "harrowing" the guilty into the mouth of a black cavernous Hell, or Lucifer as a giant head and body emerging from flames holding snakes. Great wagons pulled to the market squares functioned as stage and set; some were two stories high, with doors, balconies, and towers.[37]

How to expiate one's sins, remedy wrongdoing? How to escape the consequences of Eve and Adam's transgression, the dreadful punishments meted out in life and after death? Jewish women, like Glückel of Hameln, found solace and refuge in the traditional celebrations of the Sabbath, in keeping their home and themselves ritually clean and pure, in the holy days of Passover and Rosh Hashanah. For Christian townswomen answers and comfort came as they knelt in the cathedral chapels or stood together with others in the town square listening to the traveling preachers. The thirteenth century brought the Franciscan and Dominican friars to the towns, fired with religious zeal and eager to bring salvation. They told of the saints. Townswomen, even the wives of artisans, might also be able to read about the saints in *The Golden Legend,* a popular, widely circulated collection of stories made by the Dominican preacher Jacobus de Voragine. A woman might model her life on one of these holy figures, or she could pray to her for help in gaining mercy and forgiveness. The thirteenth-century servant woman of Lucca, Saint Zita, became famous for her selflessness in the service of the family she had been sent to at twelve, a life of giving charity and caring for the sick. Parisians into the seventeenth century joined processions in honor of their patron saint, Geneviève. In the fifth century she was believed to have staved off invaders, in the later centuries to stop the rains.

Women prayed to the saints for guidance, for strength in special situations, for forgiveness. To Christian believers, Mary Magdalen gave particular solace and proved that repentance and salvation came to all who had faith, however sinfully they had lived. Portrayed as a prostitute, as an adulteress, she was according to the gospels not only forgiven but honored by Jesus. *The Golden Legend* of the thirteenth century told townswomen that Jesus appeared to her first after his resurrection just "to show that He had died for sinners."[38]

Mary Magdalen first appeared in popular festivals in the Latin tropes (short scenes or dialogues) performed as part of the Easter liturgy. In thirteenth-century Orléans a typical scene showed the moment of her meeting with the resurrected Jesus when she mistook him for the gardener. Their encounter became part of the pageants of the resurrection enacted outside the churches in the market squares all across Europe. In addition, the Magdalen appeared in many other scenes of Jesus' life, transformed into a composite character, all showing how faith, humility, and penitence gained the highest favor. She became the woman anointing and drying Jesus' feet with her hair, the sister of Martha who sat at his feet to learn and whose brother, Lazarus, he raised from the dead. She was Mary of Magdala freed of seven demons and the adulteress saved by Jesus. At the foot of the cross she comforted the Virgin and then first witnessed Jesus risen from the dead.[39] Not even the rational inquiry of the Catholic Humanists or the skepticism of the Protestant Reformation discouraged the combination of "Marys" or curtailed her significance. The Humanist Jacques Lefèvre d'Étaples insisted she was many different women, but in the sixteenth and seventeenth centuries, Protestants and Catholics embraced a new popular reverence for this perfect model of the sorrowing, repentant sinner. The stories of her life went beyond anything mentioned in the gospels—preaching in France, marriage to Saint John—and she assumed special significance for Catholics; only she, of all the saints, had the Apostles Creed recited on her day. In honoring the Magdalen in this way, the Church followed popular belief.[40] Theologians and commentators gave her symbolic value comparing her to the Church itself. The thirteenth-century Dominicans and Franciscans rejoiced in her tears, her connection to Jesus, her vision, and saw her as the means of their own way to God. Women mystics took comfort in her example.[41]

All across Europe Christian townswomen found solace in worship of saints like Mary Magdalen. In other ways their religious beliefs and the rituals of their churches defined and directed their lives. The churches determined the calendar and the festivals. Religious ceremonies marked the significant events of townswomens' lives: marriages, births, and deaths. In addition, the Catholic Church gave them another powerful female image: Mary, the virgin mother of Jesus. This Mary was unique, at one and the same time the Queen of Heaven and a woman with a child like themselves.

Prayers to the Virgin Mary

By the fourteenth century, towns like Nuremberg had a mechanical clock chiming the hours from the market or guildhall square. Even so, for the majority of women and men, the churches continued to mark the passage of time. The Ménagier of fourteenth-century Paris used a psalm, the Miserere, for timing the boiling of nuts in one of the recipes he gave his young wife. The bells of the cathedral tolled the canonical hours (intervals designated for prayers). The ecclesiastical calendar marked the week, the months, the days of rest, the days of work. As late as 1600 in Paris, townswomen celebrated fifty-nine feast days.[42]

Well into the seventeenth century, these holidays became the ritual times of religious observance and celebration, of penitence and revelry. Guilds, social and professional confraternities, associations of women and men organized processions, dances, and plays all across Europe. In sixteenth-century Madrid during Easter week, guilds, each with its own saint and procession, competed in the richness of their displays and the numbers of their penitents. Celebrations for Corpus Christi (the holiday in honor of the sacrament of the Eucharist) meant masquerades and floats, tournaments in the Plaza Mayor and bullfights and bonfires.[43]

These festivities celebrating Jesus as Savior, as well as the popular stories of the Magdalen, came with an apparent shift in emphasis in the Church's attitudes toward sinning women and men. The Church came to focus not on the exemplary individual who lived without sin but on the rest of humanity, subject to error and in need of ways to mitigate the consequences of ill-considered actions. Purgatory evolved—a place described in the twelfth century by the Cistercian Bernard of Clairvaux as half flame and half frost, where the dead might do penance and eventually be freed to rise to heaven. In the twelfth, thirteenth, and fourteenth centuries, the townswomen saw Jesus portrayed less often as the enthroned King of Heaven. The crucifix showed him as a man suffering and dying on the cross. From the judge, he had become the redeemer. At the same time, his mother, the Virgin Mary, appeared not only in her representation as the Queen of Heaven but also as the humane, accessible mother ready to intercede with her actual son on behalf of her spiritual daughters and sons.

Mary as Mother

For the theologians of these centuries, Jesus' mother Mary became the symbol of this aspect of the Church's role—intercessor for the frail sinner—and the embodiment of the ideal Christian believer. The French mystery play of Adam described Mary as the fulfillment of a holy prophecy: a woman would be the serpent's greatest enemy for

> She'll be avenged of thee yet! . . .
> She yet shall bring thy head full low;
> There yet shall spring from her a Root
> That all thy cunning shall confute.[44]

Humanity damned by one woman was to be redeemed by means of another. Everything about Mary's life demonstrated her special qualities to Catholic writers. The disobedience of Eve found its counterweight in Mary's unquestioning acceptance, her absolute obedience to God's will. The Ménagier of Paris recommended the Virgin Mary to his wife as the model of humility. For the twelfth-century mystic and reformer Bernard of Clairvaux, Mary was the holiest of mortals, the person allowed to carry within her body the redeemer of humanity. According to *The Golden Legend:*

> She was truly full of grace, for of her fulness captives have received redemption, the sick their cure, the sorrowful their comfort, and sinners their pardon; the just have received grace, the angels joy, and the Blessed Trinity glory and honor, and the Son of Man the substance of human flesh.[45]

Even though the gospels make few mentions of Jesus' mother beyond the events surrounding his conception and birth, popular and theological imaginations filled out the details of her life. Mary was portrayed as a woman like other women who knew joy and pain, and found salvation.[46] Every event of her life became a potential feast day, and her worship came to rival that of her son. From the fifth century on, believers made images of the Annunciation, the moment Mary learned she was to bear a child. The angel Gabriel came upon the young virgin in a variety of scenes; sometimes she was reading, other times spinning; she was alternately humble, surprised, frightened. Every possible combination of situation and response was portrayed in reliefs on cathedral doors, in the paintings of every major artist of the fifteenth and sixteenth centuries. The pregnancy and the birth of the child also became legendary. The Franciscans popularized the idea of the Nativity, in a shed with animals for companions; townswomen and men believed that two midwives had been called to attend to her. When one questioned Mary's virginity, her hand withered and was restored only by her adoration of the holy infant.

A troubled townswoman could take comfort from the image of the gentle mother birthing and suckling her child. Even more powerful was the image of Mary at the scene of her son's death. Townswomen and men raised chapels to this image of her, the Mater Dolorosa. In the fifteenth century men established brotherhoods and feast days to the Lady of Sorrows. By the seventeenth century in Seville her statue in a golden crown with diamonds for tears and a dagger in her breast was the key figure of the Holy Week processions, not the image of her crucified son.[47] From the sorrowing Mother of God, Mary gradually became the sympathetic mother to all, the Mater della Misericorde (Mother of Mercy). Because of her suffering, townswomen and men visualized her as the one who understood the plight of all daughters and sons. Because of her special tie to the deity, they believed that she could speak most effectively on their behalf. The Virgin Mary became the intercessor of first choice to whom women and men prayed for forgiveness; believers and theologians called her the "mediator of grace."[48]

With the return of the Crusaders and pilgrims from the Holy Land in the twelfth century, popular worship of the Virgin Mary in the towns took on new dimensions. Relics appeared in chapels established in her name, special prayers and festivals glorified her imagined roles and marked the events of her life. Eastern chapels boasted bits of her clothing. Now in the West women and men believed in the power of her supposed wedding ring, in trimmings of her hair and nails, even the appearance in Loreto in 1296 of the house they assumed she had lived in. By the tenth and eleventh century, the monasteries had prayers to be recited throughout the day called the Offices of the Virgin. By the thirteenth century, Books of Hours with her prayers were decorated by the great religious communities and were among a wealthy woman's prized possessions. From the learned and the illiterate, holy and profane, the devotions multiplied. Kings like the thirteenth-century Alfonso V of Castile even wrote hundreds of love lyrics to her.

By the sixteenth century all but a few of the festivals and rituals for the separate veneration of the Virgin Mary had evolved. In a sixteenth-century Spanish town, women and men celebrated her Purification on February 2, the Annunciation on March 25, the Visitation on July 2, and the Presentation on November 21. A townswoman might make her own prayer to the Virgin, but more often she repeated the Ave Maria, Mary's special prayer:

> Hail Mary, full of grace, the Lord is with thee: blessed art thou among women, and blessed is the fruit of thy womb, Jesus. Holy Mary, Mother of God, pray for us sinners, now and at the hour of our death.

The Dominicans popularized the prayer's repetition as a form of penance. In the late fourteenth century Margherita Datini had three treasured rosaries in coral, gilt, and black bone. Moving her fingers along the beads, she counted the prayers to the Virgin set as her penance.

Everyone believed in the miraculous effectiveness of the prayers. Churchmen collected and recounted the stories of the Virgin's miracles.[49] Townswomen of every rank—a young apprentice, an oyster seller, a damask weaver, a carpenter's wife—heard the stories in sermons delivered in the town squares, saw them enacted in mystery plays. The Virgin always bested the devil in his plots to steal souls. She frightened away tempter demons. Her relics put out flames. Always she saved those who had been loyal to her in their prayers. In particular, townswomen looked to the Virgin Mary in all that pertained to their shared motherhood, in pregnancy, birthing, and protecting their offspring once born. In a favorite story of northern France the Virgin saved the life of a pregnant woman washed overboard in a storm, and then made possible the birth of her child without pain. Parisians of the fourteenth century believed that putting the books of *Miracles of Our Lady* on the abdomen speeded delivery. Well into the sixteenth century, even a queen like France's Anne of Austria prayed to the Virgin for fertility in her marriage. Contemporaries considered her pregnancy after twenty-two years of marriage another miracle of the Virgin.

Mary as Queen of Heaven

Historians still debate the significance for European women of devotion to this humane image of Mary whether in the towns or the countryside. Some suggest that women could find their traditional roles and functions glorified by the Virgin's portrayal as suffering mother and sympathetic intercessor. The older, more distant image of Mary did not disappear, however. In the name of the Virgin Mary as Queen of Heaven churchmen, nobles, women, and men of the towns raised the great cathedrals and lady chapels all across Europe—from 1150 to 1250 eighty were dedicated to this incarnation of the Virgin in France alone.[50]

As stories were fabricated about Mary the Mother, so traditions evolved around the Virgin as the majestic, allegorical Queen. Where the stories of Mary's motherhood seemed to bind her to other women, the traditions surrounding her special status in Heaven separated her from other women and from all other mortals. Most significantly, as the mother of Jesus the Savior, Mary was believed to have been granted the privilege all other Christians had been promised, the resurrection of her body. From the ninth century the Western Church celebrated her resurrection and her rising bodily into Heaven on August 15, the feast of the Assumption. In Antwerp in the sixteenth century the artist Dürer watched the processions celebrating the Assumption: all the guilds, even laborers, lords, town officials, religious orders, and widows, grouped and walking together. All dressed in white to signify their mourning for her death.[51]

Throughout the centuries women and men responded to both of these images of the Virgin. Popular worship and the Church's official veneration of this female figure gave comfort and promised salvation. Her glorification, however, never brought any actual changes in European women's position in society, nor did it signify any change in either the religious or the secular view of the female. Even as mother, Mary always stood apart from other women: virginal, powerful, another kind of being.[52]

3

THE WORLD OF COMMERCIAL CAPITALISM: THE THIRTEENTH TO THE SEVENTEENTH CENTURIES

Protected and Provided For

The towns of the thirteenth to the seventeenth centuries were significant far beyond the numbers of their inhabitants and the amount of land they occupied. Into the eighteenth century, only about 10 percent of Europe's female and male population lived in this busy, walled world.[1] The expansion and the significance of the walled towns came with the transformation of Europe's economy from subsistence and barter to specialization, commerce, and the use of money. The new system of economic relationships, capitalism, occasioned and necessitated major commercial centers.

As capitalism grew, towns grew. As early as the last decades of the eleventh century, exchanges in the town marketplace of local produce and the occasional foodstuff or bolt of cloth from another region had given way to the long-distance trade and transportation characteristic of Europe's thirteenth-century mercantile centers.[2] By the thirteenth century, towns had grown to become production and processing centers for local and regional needs. Merchants, goods, and capital traversed the Mediterranean and the Alps, going from fair to fair so that women like the Countess of Leicester and Margherita Datini might have their spices, and men like the Count and the merchant their armor and their profits. The enterprising in regions like northern Italy, parts of Belgium, and northern France specialized in single aspects of manufacture, formed corporations and financial partnerships, financed the ventures of others, and sent representatives to live and deal in other towns.

The changes in Europe's economy meant increased opportunity for all, both enterprising women and men. From as early as the fifteenth century, however, the commercial, capitalist towns became a world in which women also experienced new kinds of disability and vulnerability. At first some townswomen enjoyed access to craft associations, participation in new professions and new commercial enterprises. Gradually, however, the new opportunities offered by the changing and expanding economy were closed to them. Gradually, they found their lives restricted and constrained once again. In addition, women had in the new economic circumstances been allowed more legal autonomy. This too was taken from them. For example, after the "democratic" revolutions of the 1300s, when craft and merchant guilds seized control of town governments from the old patchwork of feudal authority, new definitions of citizenship excluded some men (previously included) and all women. Some women had actually held citizenship under the old governments, though they were never chosen for office.[3]

Privileged women of the merchant and banking elite lost legal privileges in other ways. In the changed economic circumstances, a woman's expected marriage became part of a family's business considerations; marriage might mean a financial alliance, access to a craft, the dowry and widow's portion a gain or loss in the account books. As royal jurisdiction superseded the customary negotiated charters of towns with their lords, political circumstances changed as well. Capital and property became the subject of royal law, and even wealthy women, whether married or widowed, lost their ability to act independently except in the most circumscribed situations.

The same kind of codification harmed guild women more than men. To protect their livelihoods, craftsmen limited access to membership and its privileges. Wives and daughters previously welcomed to the work found the conditions of their participation restricted and preference given to male relatives. In many of the women's crafts, especially in textile manufacturing, entrepreneurs avoided their guilds and craft associations altogether, subdivided the tasks, and used the cheaper labor of the countryside.[4] Even in the new professions that evolved, women found themselves gradually relegated to the least skilled, least lucrative, least prestigious aspects of the work. Poor and laboring women suffered in other ways. The new secular authorities took over what had once been the concern of the churches. Towns became the principal source of charity, and in their regulated acts of giving took a punitive attitude toward those destitute and unfortunate women who lived outside the protection and provision of a family.

In this mercantile world of the towns, men defined new functions and roles for themselves that had adverse consequences for women. No longer could men pose as warriors, as "protectors" in the traditional sense of the word. In time of siege, townsmen and townswomen suffered deprivation and helplessness together. Both suffered when the walls were breached and taken; no husband could protect his wife, no father his daughters from the victorious mercenaries as they rampaged through the streets. As circumstances altered, so did men's sense of their role. From "protectors" they transformed themselves into "providers." In this new role, townsmen still saw life as a war, a competition, but now the battle to be fought and won was played out in the marketplace rather than on a muddied field or in the lists of the tournament.

At the end of the fourteenth century the Parisian Ménagier differentiated himself from his counterpart in the countryside in this way: The landed lord "seeks to win prize and praise in the world," while he found honor as "the citizen [who] seeks to win wealth for himself and his children."[5] Stories and tales portrayed this new hero: Tristan disguising himself as a merchant and victorious because of his cleverness, not just his ability to fight. "Being a very astute and capable man" one of Boccaccio's heros became a king even though he had once made a living lending money.[6] A popular tale (or fabliau) "The Widow" praised such a man: "One thing about my husband, he was a good provider; he knew how to rake in the money and how to save it."[7] By the mid-fifteenth century in England, the word "husband" was used as a verb meaning to keep or to save. By the seventeenth century, a woman honored her husband for these virtues. Glückel, the wife of a late seventeenth-century German-Jewish merchant, presented him as a model to her sons who in future must "diligently go about your business, for providing your wife and children a decent livelihood is likewise a mitzvah [blessing]—the command of God and the duty of man."[8]

The creation of this new male heroic image had profound effects on the lives of women, first in the towns and by the nineteenth century all across Europe. To function as the provider gave the man of the family new responsibilities. To meet those responsibilities, men of the towns espoused attitudes and supported laws to protect their means of livelihood. Protection in most instances meant limitations and regulations that affected women far more than men. All of the ancient traditions governing women's economic and political capacities and right to participation intensified. Deprived of access to citizenship, to separate legal authority outside the family, to specialized training, women fell prey to the age-old assumption of female incompetence and to its traditional corollary, the need for male guardianship. In the world of commercial capitalism, women of the towns would in law and practice be protected and provided for by men, but at an awesome price, the loss of their potential and actual autonomy.[9]

The Business of Marriage and Charity

In the world of the towns, just as in the countryside, young women of all occupations and classes expected that they would marry. The sixteenth, seventeenth, or eighteenth-century servant who worked in the house of a craftsman's wife saved from her wages to have goods and money to share with her marriage partner. As a working woman with her own dowry, as an apprentice learning her own craft, her marriage arrangements signified the beginnings of a partnership. Though the couple might speak of love, they would also negotiate a marriage contract before a notary. They were usually in their late twenties, two competent workers joining to create a household.[10] Sometimes the parents of the young woman became involved in the arrangements. Genoese contracts from the mid-thirteenth century note the consent of the parents. More typically as in fifteenth-century Avignon, documents identify the woman by her own name with no mention of her family.[11]

Daughters of craftsmen might not have to earn their own dowries; part or all of it could come from their families. The same custom had evolved in the towns as in the countryside: The parents gave a daughter her inheritance on the occasion of her marriage, not when they died. A twelfth-century Genoese blacksmith and his wife gave half their property, valued at 12 Genoese pounds, as dowry. In return the daughter and her husband lived with them and cared for them (much as a peasant daughter or son would after inheriting access to their parents' house and land).[12] Contracts from French towns of the fifteenth and sixteenth centuries show dowries of every variety. Jacommette, the future wife of a blacksmith in fifteenth-century Avignon, brought cooking pots and her clothes; a contemporary, Perrenette Pastoris, negotiated to marry the weaver Rostan Bengier and brought a dowry of 24 livres, a mattress, bedding, and sixteen sheets, table linens and three new dresses.[13] A marriage contract from Bayeux at the end of the seventeenth century suggests country connections. The bride's dowry consisted of furniture, linens, dresses, and assorted livestock: a cow, a year-old heifer, and more than a dozen sheep.[14]

Just as lords and princes used marriage to advance family interests, so did rich mercantile and manufacturing families. As early as the twelfth century in the towns of northern Italy, marriages created connections for trade and manufacturing, neutralized competitors, cemented political alliances and tied mercantile families to patrician ones. Beginning in the thirteenth century, the richest families of Italian towns crisscrossed the landscape with their marriage alliances: between the Orsini of Rome, the Visconti of Milan, the Estes of Ferrara, and the Medici of Florence. The Estes aspired to royalty and in 1234 arranged the marriage of their daughter Beatrice to Andrew, King of Hungary. Fathers in Ragusa (Dubrovnik) in the same century used marriages to create ties to Venice and up and down the Dalmatian coast.[15]

The arrangements of Buonaccorso Pitti, a Florentine cloth manufacturer, show the considerations of an astute businessman. He kept the records of the arrangements for his marriage in his account books of 1391. He asked a member of one of the town's leading families to pick a bride for him from "among his relatives," reasoning that "a connection of his could obtain his goodwill." Pitti hoped that the influential alliance would help in a dispute he was having with another family. The daughter of a cousin suited his needs perfectly, and he noted in between the sums and debits of his accounts that he "sent back word that I would be very happy and honored and so forth." The betrothal and marriage took four and one half months to bring to fruition.[16]

The Medici family in the late fifteenth and early sixteenth centuries used marriage to protect and advance their interests. In the last quarter of the fifteenth century Lorenzo de' Medici brokered a match for the Salviati and Valori families, prevented a marriage with his enemies the Pazzis, and with his daughter Clarice's marriage in 1508 to Filippo Strozzi restored the Medici to their former prominence after the rebellions of the 1490s. In the sixteenth century the Medici intermarried with the royal houses of France and Spain.

In English and French towns, marriage became the means of keeping political authority in the hands of a few families. A list of London's court of aldermen for 1580 shows fathers, sons, sons-in-law, brothers, brothers-in-law, and widowers all sitting together governing the town.[17] In Vraiville, France, in the seventeenth and eighteenth centuries, the mayors for ten generations consistently married their daughters among the thirteen elite families.[18]

Women of the landed nobility brought access to property as marriage partners, so did townswomen. Among this mercantile, patrician caste, the exchange of money and goods was as important as the exchange of persons and the ties of family. Just as in the countryside the emphasis in the marriage arrangements had shifted from the Germanic and Celtic custom of gifts from the groom's family to gifts from the bride's parents. The young woman brought a dowry to the match. This dowry had become her sole inheritance, her portion of the family wealth.[19] These customs arose in the towns in the twelfth and thirteenth centuries in response to the same economic and political changes that had influenced noble, landed fathers. In the new commercial economy, the privileged men of the towns consolidated holdings so as to pass a central portion of capital and possessions to only a few offspring. They wanted to avoid having too much property and capital go to females, who would marry and take family wealth to others. No mercantile or manufacturing dynasty believed it would long survive following the older customs of equal or more equitable inheritance. As a Venetian lawyer argued in the sixteenth century: "Everyone knows it is men who continue families."[20] Fourteenth-century Florentines entailed lands held outside the walls just as counts and dukes entailed their property. In Paris by the sixteenth century the town's wealthy men had created patrimonies, and only peripheral holdings went to daughters.

Even if only peripheral assets, a bride's dowry in the world of commercial capitalism could be the means of improving the economic fortunes of the groom and his family. As the historian Susan Mosher Stuard has pointed out, dowries meant the opportunity to accumulate capital, a way to redistribute funds when inheritance laws had restricted those who might inherit. Younger sons, barred from access to a father's wealth, could turn to that of their new wives. The entrepreneur could afford to expand his trading operations, the banker to finance some new enterprise. Gregorio Dati, the late fourteenth-century silk merchant, acknowledged the significance of dowries in keeping his business going. He married four times. Each death and remarriage made possible new infusions of capital: restoring stolen assets in one instance, removing debts in another, making reinvestment and expansion possible in a third.[21]

Northern European mothers and fathers similarly negotiated for capital and for economic alliances. In some towns brokers arranged the betrothals, working for a commission on the dowry. The Cely family, members of the English merchant guild of Calais that controlled the wool trade between England and the Low Countries, wrote back and forth in the 1480s about marriage prospects, as they did about other aspects of their business.[22] The letters of the seventeenth-century Verney family of English merchants show the daughter Susan actively involved with her parents in identifying appropriate matches. In a letter to her older brother in November of 1644, she described a widower with an income of £500 a year as a man with "good commendations."[23] In Avignon in the fifteenth century, women like Madeleine Cocorde and Hélène Cambi brought money, letters of credit, even a house and grain to their marriages.[24] In Hameln, Germany, in the seventeenth century, the Jewish merchant's widow, Glückel, made a match for her daughter Zepporah with one of the principal Jewish financial men in Amsterdam by paying out a small fortune, 2,200 Reichstalers, as her dowry.[25]

To generate capital like this taxed the resources of even the wealthiest families. At the end of the fourteenth century, one of Gregorio Dati's fathers-in-law arranged to pay him over a year. As Dati's own daughters came of age, he needed to raise a minimum of 500 florins for each of their dowries.[26] To help with the financing, fourteenth-century English craft guilds established emergency funds that fathers could draw upon for daughters ready to marry or to enter a nunnery (which required a dowry as well). In Florence and Venice, mothers, grandfathers, uncles, and aunts often contributed. From 1425 until 1525 Florence's mercantile elite created a dowry trust, the *monte delle doti,* which fathers invested in on their daughters' births and then withdrew from on their marriages. Financially it remained a losing venture for the town, but the elite continued it, subsidizing it with tax revenues because it performed such a valuable service for them.[27] In Venice from 1551 to 1620 the lay brotherhood of the Scuole Grande di San Rocco created a dowry trust for impoverished members to draw on. It was so popular that by the seventeenth century it had to be run like a lottery.[28]

Townsmen tried other ways to control expenditures. Northern Italian communes regulated the cost of the wedding. Beginning in 1299 and continuing into the fifteenth century, Venice limited the value of the gifts, the number of attendants, the cost of the bride's clothes. A law from Prato also limited the types and number of courses and dishes served. Just as Spanish townsmen of the tenth to the thirteenth centuries had limited the amounts husbands had to pay for their wives, so Italians now regulated the size of dowries. In 1235 in Ragusa (Dubrovnik), a maximum dowry was established. In 1420 the Venetian senate established fines for those paying dowries of over 1,600 ducats. Such regulations also protected families from having to diminish the patrimony of the favored son. Even so, when it suited fathers, they ignored the regulations. They wanted their daughters to marry well, and when in the thirteenth century or at the beginning of the sixteenth century inflation and a relative shortage of young men made the price of dowries soar, parents raised the money despite the laws.[29]

Traditionally the Church had tried to make betrothal and marriage subject to its authority. The joining of two families, however, with all of its social and financial implications, was too important to leave to the Church, whether Catholic or Protestant. Instead the wealthy, politically powerful townsmen, like the landed nobility, subsumed both betrothal and marriage under the secular authority. In particular, they ignored the Church's insistence on the consent of the couple and legislated the parents' primary right to choose their childrens', especially their daughters', marriage partners. Florentine law made sixteen the age of consent for a daughter and twenty for a son. French law of 1639 set it at twenty-five for the young woman and thirty for the young man, but still forbade independent choice. Royal and civic ordinances stipulated loss of goods, even disinheritance, for a daughter or son who married without consent.

Protestant townsmen made the same regulations. Geneva in the sixteenth century set similar ages and established the need for consent of the mother, a guardian, and some other family member should the girl's father die before arrangements for her betrothal had been made. The Genevan Consistory excommunicated a young woman and her aunt for arranging a marriage without the uncle's consent. Fearing secret marriages in the seventeenth century, Catholic and Protestant townsmen required repeated public announcement of betrothals (the posting of the banns) and public exchanges of vows.[30]

In addition to the laws, other practices evolved that ensured parental approval. Wealthy fathers established guardianships in their wills, or more commonly negotiated the betrothal while their daughters were still very young. In fifteenth-century Florence, Avignon, Pisa, Venice, Toulouse, and London the fathers arranged the choice of husband and the terms of the dowry while the daughter was fourteen or fifteen, rarely would she be older than seventeen or eighteen. Special circumstances of wealth or position might occasion betrothal at nine or eleven, with the wedding ceremony to take place after she had reached puberty. At the end of the seventeenth century, Glückel of Hameln betrothed her daughter Zepporah at twelve, her daughter Hannah at eleven. In this case her husband's death prompted her to see to early provision for her daughters.

From a mother's and father's perspective, marriage among the town elite was not just alliances and dowries. The choice of a husband had to be well considered, for he would be the means of a daughter's future happiness and security. As a result, elite parents of the town looked for the prosperous man, seven to ten, or even more years her senior. Such a groom had, in theory, established himself in business and thus could provide for the young bride properly. In fifteenth-century Florence, daughters married at eighteen or less to men in their early thirties. The silk manufacturer Dati worked his way through a number of wives, the last was thirty years his junior. Protestantism changed nothing of this pattern. The young women tended to be in their twenties, but still the husband was six or seven years their senior.[31]

Marriages became the occasions among the mercantile and manufacturing elite for families to show their wealth and status. The exchange of vows took place in a simple ceremony before a priest or a minister at the door of the church or at the altar. Every other aspect of the occasion gave parents the opportunity to celebrate and display their position in the town. In 1466 in Florence the groom Bernardo Rucillai paid almost 6,000 florins for the festivities surrounding his marriage to Giovanna, granddaughter of Contessina and Cosimo de' Medici. She in turn brought 2,500 florins and a trousseau of fourteen dresses, embroidered with gold, pearls, and gemstones. Headdresses, a diamond necklace, rubies, pearls, and twenty rings completed her dowry.[32] Other grooms gave the wedding dress. Marco Parente, when he married Caterina Strozzi in 1458, presented her with jewelry valued at 165 florins.[33]

The wedding of Clarice de Jacopo Orsini and Lorenzo de' Medici in June 1469 was a magnificent display. His family had picked Clarice, fifteen or sixteen in the spring of 1467 when Lorenzo was twenty. His mother Lucrezia de' Medici went to Rome to inspect the young woman, viewing her on the way to mass at St. Peter's and then on a visit to the Orsini house. She wrote to her husband Piero that the girl was of a "good height," with long, delicate fingers, abundant reddish hair, gentle manners, and a nice complexion. As she explained, her face "did not displease me."[34] She concluded: "There is no handsomer girl at present unmarried in Florence."[35] Perhaps most important, Clarice's modest ways suggested that she would easily adapt to their household and learn their customs. Lorenzo saw her himself once at mass, but had assured his mother that he would accept whomever his parents chose. The negotiations lasted a year. The prospective bride brought 6,000 florins as dowry and an alliance with a great family of Rome.

After a wedding by proxy in Rome, Clarice journeyed to Florence for the ceremony and celebrations organized and paid for by her husband's family. This was a joining of merchant princes as significant as any among the landed nobility. The festivities lasted throughout the week, beginning with a procession on horseback with trumpeters and fifers on Sunday, the fourth of June, through the streets to the Medici palazzo. Young people danced in front of the house. The Medici letters describe the rainy Monday weather that "wet the beautiful dresses" and marvel at the "robe of white and gold brocade and a magnificent hood" that Clarice wore on Tuesday.[36] In all there were five banquets, serving 200 each, inside the house, outside in the garden courtyard, each to entertain a different group of citizens. Guests danced on a stage erected in the courtyard, watched jousting and a mock tournament, attended masses and the ceremony itself.[37]

According to traditions as old as the laws and customs of the Roman, Hebrew, Celtic, and Germanic peoples, by her marriage a young woman passed from the guardianship of one male to the guardianship of another. Although a laborer and her husband could not afford the luxury of such a relationship, at the other end of the economic spectrum, the wealthiest townswomen found themselves legally subordinated to their spouses. An artisan's wife might function like a partner or have her own craft. A merchant's wife might attend to the business in her husband's absence. Even so, gradually from the thirteenth to the seventeenth centuries, the laws of the towns and of the new royal dynasties gave more and more authority to the husband—authority over any money or property she might have in her own right or that the couple might have acquired together (called the community of acquests). A Parisian tradeswoman, according to the 1270 edict of Louis IX, could do business but only with her husband's consent, and he had to appear for her in civil suits. In thirteenth-century Ragusa (Dubrovnik), the husband controlled any capital brought to the marriage; he left supervision of the movables and lands to her. Venetian law of the fourteenth and fifteenth centuries similarly gave control of the money portion of the dowry to the husband. With the shift of dowries from property to capital, these statutes effectively limited a woman's control of her wealth. German towns like Lübeck and Munich in the thirteenth century gave women rights to possess property, to buy and sell goods, but by the 1500s the final say in these matters had been granted to the husbands. In sixteenth- and seventeenth-century France, the ideal of the king and his subjects appeared mirrored in the image of the husband and his family. The term in French law *femme couvert* described the wife's situation literally as well as figuratively: a woman "covered" protectively by the rights and obligations of her guardian.[38]

This increasing legal emphasis on a woman's lack of independent authority made it even more difficult for her to separate or divorce, to move out of the provision and protection of the marriage bond. The Catholic Church condoned separation but only under specific circumstances—a wife's adultery, a husband's impotence—and made remarriage problematic and a subject for papal dispensation. Protestant town ordinances and royal laws took much the same attitude. Sixteenth-century Nuremberg allowed divorce for infidelity or desertion. While Zurich courts awarded money damages and allowed remarriage in cases of adultery, Calvinist Geneva favored reconciliation. Seventeenth-century Anglican statutes cited consanguinity and impotence as causes for separation, but did not condone remarriage.[39]

Practically, women and men did break up their households, but as among the landed nobility, the dissolution was the subject of negotiation. Records for thirteenth-century Ragusa (Dubrovnik) show the wife given lands and goods, the husband retaining the capital she brought to the union. In fifteenth-century Avignon, couples went before a notary. The resulting contracts show some money and perhaps jewelry, linens, and movables given to the wife, much like the portion she would have received as a widow. Most secure of all alternatives would be that shown in Italian family records, where the estranged wife returned to her parents.[40]

For to live outside the family, even amidst the opportunities of Europe's towns, remained difficult for women. More often than not women in this situation, particularly those without special skills and with children, would barely be able to survive. Throughout these centuries women figured disproportionately in the numbers of urban poor. In a 1561 accounting for Segovia, Spain, 60 percent of the adult poor were women. The 1570 census of the indigent poor for Norwich, England, showed 832 women and 461 men. A Paris hospital's records for the years 1612–1661 showed that all but eight of those cared for were women of various ages.[41]

In the countryside indigent women and men turned to the Church and to individuals for charity. In these centuries the wealthy of the towns made individual charitable gestures similar to those made by ladies and their lords. For example, in fifteenth-century Florence, Contissina, wife of Cosimo de' Medici, created a special endowment for the Convent of St. Lucia. Merchant families in Florence and Venice commonly provided dowries for impoverished young women and widows.[42] In 1401 citizens' wives in Nuremberg arranged feeding, clothing, and shelter for lepers during Holy Week. Like the landed elite, as early as the twelfth century, townspeople endowed hospitals, refuges for the old and sick, and foundling homes for infants abandoned in the streets by parents unable to care for them. London had its hospital of St. Giles, Paris its Hôtel Dieu, which in the fifteenth century had five wards with fourteen nuns, twelve priests, six clerks, and two chaplains to attend to the inhabitants.[43] By the seventeenth century, orders like the Poor Clares and the Sisters of Charity made a specialty of such service. As early as the fifteenth century, support for such institutions had become a matter of business. A firm like the Banchi, the silk manufacturers of Florence, deducted from guild members' wages not only to make contributions to childbirth costs and for the dowries of artisan families, but also for the town's foundling home.

The arrangements of individuals, lay societies, and business came to be considered inadequate in the sixteenth and seventeenth centuries. Gradually councilmen, aldermen, and burgermeisters took charge. Private donations and Church-administered charity gave way to public distributions and public institutions. The townsmen provided, but not in the same spirit. These successful men had little compassion for the poor and their circumstances. Images of lazy, disorderly, sinful women and men replaced the traditional sympathetic portrayals of the poor as needy and unfortunate. Theologians embraced and encouraged the change in attitude. In 1526 the Spanish churchman Vives wrote: "Let no one among the poor . . . be idle." He wanted to punish "imposters" and set everyone to work.[44] In the seventeenth century, Saint Vincent de Paul thought the poor must have sinned and that the hospitals they took refuge in were a fitting penance.

The elite of the towns blamed and feared the poor, and their actions reflected their attitudes. In their administration of charitable institutions, in their legal definition of those who could and could not receive the town's charity, townsmen created punitive rather than protective measures.[45] Poor laws enacted in Venice in 1529 and 1545 required licenses of those asking for alms and declared that "Poor men and women beggars who are at present wasting their time must be informed that they shall not have alms, but must do some work."[46] The same suspicion and impatience characterized care of the poor as it became regularized and institutionalized throughout the towns of Europe. Lyons was typical. Beginning in the 1520s and 1530s, the designated poor were counted, required to fulfill residency requirements, categorized, assigned to appropriate shelters, and given predetermined amounts of money based on accountant-like definitions of their need. If young, they were to be trained. In Lyons poor girls learned cotton spinning, sewing, and how to unwind silk cocoons from women hired by the town.[47] Such arrangements satisfied the goals of the well-to-do. The haphazard almsgiving and charity of the past disappeared. As an Aix-en-Provence pamphlet of the late 1600s explained, these new regulations kept people out of doorways, eliminated the malingerers and cheaters, identified the "truly poor," and efficiently administered aid in one place at one time.[48]

Poor townswomen survived as best they could, asking help when all else failed. In 1356 the Florentines gave alms to Monna Francesca, a mother with four children, who could not work because of a broken arm, and to three other women because they were pregnant and disabled or alone: to "old and sick" Nexetta, and to Monna Lippa, "a poor and sick market peddler."[49] The town records of seventeenth-century Aix-en-Provence show women in similar circumstances—those who lived marginally, forced into penury by unforeseen circumstances. Laboring wives could not support the family alone when their husbands deserted. The old, the sick, and the very young could not make enough to live on.[50]

Town records show that in desperate times women and men unable to feed their children abandoned them to the foundling homes established by the religious and secular authorities. Records from Florence tell of a baby girl brought naked to San Gallo in January of 1445:

> a newborn child all full of blood on the head and on the rest of her back. The umbilical cord had not been tied, and we could neither see nor know who had brought her. . . .

The friars tied the cord, washed her in white wine, and tried to warm her. They baptized her Smeralda the next morning.[51] The records also tell mothers' stories. One young woman gave up her baby so that she could work as a servant, another so that she could use her breast milk to become a wet nurse. Those were the only wages she could earn and "she was dying of hunger."[52] The majority of such infants died.[53] Older children were put out to service or worked in the foundling home. This was probably the fate of the little girls ages six, five, and four brought by their father the weaver Antonio di Tommaso de Jacopo to San Gallo after his wife had died. Frightened, destitute women also killed their newborn infants. Records of Nuremberg from 1533–1599 report fifty-five such deaths: one baby stabbed, one strangled, another left to bleed to death—its umbilical cord untied. Records of the 1570 quarter sessions for Essex towns show one desperate woman who slit her infant's throat, another left hers in a ditch, another drowned hers. A servant left hers in a chest hoping it would die; the infant's father had advised burying it in the dung heap.[54]

The vast majority of women left to deal with an illegitimate baby came originally from the countryside. Most had been put out to domestic service in the town or perhaps apprenticed. Their sexual partner was a lodger, another servant or apprentice, a soldier or a student, the master or his son. More often than not, court and notarial statements describe it as a quick coupling with little time for affectionate words or gestures. With a social equal, promises of marriage and the discussions of a future together probably stilled any fears the young woman might have had about pregnancy. The parish court in seventeenth-century Somerset recorded that the young woman Hannah and her lover had been interrupted in the act of intercourse. Her accusers heard her refuse "for fear she would be with child," and then give in when John promised to "marry her the next morning."[55] With men above their station, the testimony from young mothers indicates that they had similar expectations. One had been promised work, another marriage on the death of the wife, another that she would be set up in a trade. Women explained that when nothing came of the promises, they continued to give their sexual favors hoping for a change of heart on the part of the man.[56]

Before the sixteenth and seventeenth centuries, judgments of such women had been left to the ecclesiastical and manor courts. The churchmen forgave intercourse and the birth of a child without marriage. At worst these and the manor officials had fined and sometimes flogged both partners.[57] Customary and Church law had also taken a lenient view of the mother's part in the death of her illegitimate infant. Definition of infanticide as murder occurred only in isolated instances.[58] In the villages of the countryside, local and family pressures usually prevailed and ensured that the man married the young women when she became pregnant. In addition, in the popular mind, cohabitation, or intercourse, was thought similar to a private betrothal and so was considered binding in the rural communities. Local custom in some regions even encouraged pregnancy before marriage, to guarantee the young woman's ability to bear.[59]

The same pressures and assumptions did not operate in the relative anonymity of the walled towns. The couple might have truly meant to marry, but changes in economic circumstances made it impossible. They had not saved enough so they could not afford to set up their own shop or household. In such a situation the young journeyman or the young male servant could always move on to another town without the prospective bride.[60]

Without a certain means of livelihood, without family to appeal to, and with an illegitimate child, women in this situation would be dependent on the more fortunate for survival. Even though townsmen came to see care for such women and their offspring as a civic duty, the men's attitudes became increasingly more disapproving. At the end of the fourteenth century, when asked his advice in such a situation, the Prato merchant Francesco Datini recommended sending "the creature" (a young slave woman made pregnant) to the hospital.[61] Other fathers suggested abortions or took the women elsewhere to bear the child. She herself might flee to another town to find someone to help her. English parliamentary laws sought to punish her. In 1609 the statute specified a year in the House of Correction for birthing a child out of wedlock.

Catholic and Protestant theologians connected poverty and sinfulness. The young pregnant woman without a husband personified the sinful. In the eyes of male religious and secular authorities, if she killed her infant, she compounded her transgression, she was the "fornicator" trying to avoid the consequences of her wrongdoing. The laws and the provisions of town councils reflected this contempt and distaste. Ordinances passed in Germany, France, Switzerland, and England, in the course of the sixteenth century, gave the minimum of assistance at the time of the birth, did not try to force marriage, and focused instead on ensuring the survival of the infant. The Carolina, the 1532 codification of German law ordered by Charles V, the Holy Roman Emperor, showed the new attitude toward infanticide. Killing an infant became murder carrying the death penalty.[62]

The 1586 edict of Henry II of France assumed that the unmarried woman would try to kill the infant. The presence of a midwife and witnesses was supposed to prevent this. In addition, the midwife had another responsibility. This French royal edict, like many others issued by the townsmen or their kings, hoped to spare the town monetary responsibility for the mother and her new infant. During labor, when the woman was thought more willing to cooperate, the midwife was instructed to collect information for the *déclaration de grossesse* (declaration of pregnancy)—specifically, the name of the father and the mother's place of birth. A father could be made to pay for the child if he could be found; if not, the mother could be returned to her own village. Either way, reasoned the authors of the statute, the townsmen would be spared expense.

English laws followed the same reasoning and established the same procedures. Rarely, however, could the father be made to take responsibility for the child. The experience of Joan Sage, a seventeenth-century woman from Somerset, is typical of what happened. She made her way to a strange parish and, while in labor, confessed to the ministering women. Having given the name of the parish where the child had been conceived, she was sent away. She "remained for a very short time, and sithence is wandering around the country begging."[63] Nothing improved in the secular attitudes or statutes into the nineteenth century.

Laboring and poor women knew the harshest life—at best, companions to their husbands in the quest for the means to survival, at worst alone, condemned and abandoned when circumstances denied them employment. From the thirteenth to the seventeenth centuries, wealthy townswomen were expected to live protected and provided for within a family. Fathers and husbands made the key decisions about their lives and acted as guardians over their property and capital. Only among craftswomen and men was there some measure of equality and a relationship within the family much like a partnership.

Old Crafts

Laboring women and pieceworkers lived often on the edge of subsistence. In contrast, those women who acquired the skills of a craft and became members of a guild knew relative prosperity and security. Yet even in these protected associations of artisans, women were more disadvantaged and more vulnerable than their male counterparts. In the European towns of the thirteenth to the seventeenth centuries, fewer women were admitted to guild membership, and once admitted fewer women gained the full status and privileges of master. Some historians, most well known among them Alice Clark, have speculated that there was a "golden age" for guild women and that only with the evolution of commercial capitalism did they find themselves at a disadvantage.

In fact, study of the crafts so far completed shows that the traditional patterns of female employment were never seriously challenged. As early as the 1300s, in the first century of the heyday of guild authority, and in the 1500s and 1600s, when records appear again in numbers, women's circumstances are the same. Their access to training and to protected status was less than men's.[64] Lists of guilds and tax records for Paris and Oxford in the late thirteenth and late fourteenth century show the unequal situation. Of twenty occupations in the Paris records, only nine have ten or more women members. The embroiderers have the largest number of females employed, but only eighty in contrast to the hundreds of men in hat making, weaving, in working with fur and in what could be considered female crafts like the making of silk coifs and women's clothes.[65] The Oxford poll tax return of 1380 lists ninety-three trades in all, of which only twelve include women; they worked as washerwomen, peddlers, shoemakers, tailors, innkeepers, brewers, net makers, chandlers, and more typically, as spinners and wool combers.[66] When records reemerge again for late sixteenth-century London, where men had routinely moved from apprentice to journeyman to master, only about 10–15 percent of women had attained the status of master. Overall estimates of women's membership in guilds set the numbers much lower, at 2 to 3 percent, and most of those were widows of members. The majority of women following the trades did not belong at all. Instead they paid a fee to the guild, like the Florentine women tailors of 1578, or ignored the regulations altogether and worked at the craft illegally.

Commonly in the predominantly male guilds, women had been allowed access only because they were the wives or daughters of members. Like their counterparts among the landed nobility, privileges came because of their relationship to a man, not in their own right. In thirteenth- and fourteenth-century Paris and Oxford, those crafts with fewer than ten women listed represented wives, daughters, and widows assisting their husbands or fathers. They did not have full master status in the guild. While a husband could more or less automatically assume his wife's rights and privileges on her death, the same was not true for the widow of a master. For example, the Provisioners guild in Cologne, Germany, at the end of the fourteenth century allowed the widow full rights, but placed restrictions on her if she remarried. The new husband, the craftsmen reasoned, should provide for her from his own resources.

In times of economic prosperity, regulations went unenforced, and women willing to work for lower wages found acceptance. In times of economic difficulty, however, craftsmen took measures to protect themselves and their privileges. All placed skilled women at a disadvantage. The guildsmen bolstered and enforced old regulations or created new ones to minimize women's access to membership and to markets, or excluded them explicitly. In Bristol in the late fifteenth century, when many in the textile industries found themselves without work, an ordinance forbade the employment or membership of women on the assumption that they were taking jobs from men. The silk confraternities of late sixteenth-century Lyons allowed only two apprentices per master and specified that, if they were females, they had to be daughters or sisters of members.[67]

The same motives governed responses to new trades in the seventeenth century. The London Stationers Company and the watchmakers guild of Geneva gradually barred women from their craft associations. From 1553 to 1640 almost 10 percent of the publishers in the London company were female, women printing, binding, and selling books.[68] First the guild forbade any but male apprentices, then in 1676 it closed the trade to all but legitimate sons of members. The Geneva watchmakers wrote their regulations first in 1601. By the end of the century members who taught the craft to their wives or daughters were fined.

Most significantly, when artisan women had their own guilds, they still functioned at a disadvantage. Exclusively female trades were limited to only a few areas of manufacturing and retailing—overwhelmingly those involved with textiles, clothing, and accessories. Within these areas the traditions about women's wage work prevailed. As in the centuries before 800, women found themselves in the least skilled and prestigious, least lucrative aspects of production, in those aspects most vulnerable to economic fluctuations. The new world of commercial capitalism did not alter the patterns, rather it occasioned their re-creation and intensified them.

The shift from localized, subsistence production to long-distance trade and manufacturing centers caused a variety of economic pressures. This new world of the entrepreneur, of increasing specialization of production and marketing, of new technologies undermined the guilds, their economic privileges, and their special status. Each change and each protective response from townsmen worked to the disadvantage of craftswomen and harmed them more than it did their male counterparts.[69]

The experience of women in the textile trades dramatically illustrates their circumstances and the effects of this economic transformation of Europe. From the fourteenth to the seventeenth centuries, the overwhelming majority of women in guilds in the towns of France, Germany, Italy, England, Belgium, and the Netherlands worked in the textile and clothing trade. In Paris, London, and Florence women exercised a virtual monopoly on all of the processes involved in the production of silk cloth. Yet even in these protected and privileged crafts, the traditions of women's work operated inexorably both before and after the advent of commercial capitalism. Even in the silk industries in London, the silkwomen never had the right to make the most lucrative product, "whole cloth." Male weavers took over that process. Most silkwomen worked not for themselves or for each other, but for a male entrepreneur for whom they did "piecework." Paris guild statutes specified male overseers, and like other textile crafts silkwomen worked by a "putting out system" where merchants brought the raw silk and paid the craftswoman by the piece. Piecework could be picked up and put down; pieceworkers were paid by the item, not by the fixed hours of employment; they were artisans increasingly difficult to organize, doing work particularly vulnerable to the demands of a free market where the more you made, the less you were paid.[70]

In these circumstances there would be little that even a guild could do to protect its members. In 1285 in Douai, Flanders, twenty women—spinners, weavers, carders, and stretchers of raw wool—brought complaints against the estate of the entrepreneur Jehan Boinbroke. He had thought of a multitude of ways to cheat the women who did piecework for him: charging high interest on debts incurred for materials, seizing collateral before the debt fell due, making payment in kind with overvalued or rotten wool, giving lower valuation than the market price for finished work, cheating on the weight or the quality of the materials. When asked why they had accepted such treatment, the women explained that their protests had been ignored or ridiculed. Some had been threatened with fines or trouble with the town authorities. Boinbroke could always take the work elsewhere. The women only dared to present their case after his death.[71]

As early as the fourteenth century in the towns of northern Italy, the new technology and specialization characteristic of commercial capitalism worked to the disadvantage of women in textile manufacturing. Fifteenth- and sixteenth-century English guild regulations prohibited women from working certain fabrics, from being trained in the use of the new machines. For example in 1511 the Norwich worsted weavers insisted that the new looms that could weave pieces twenty to forty yards long and more than a yard wide required more strength than women could manage. "Whole cloth" increasingly became the province of men, leaving only the bands and ribbons of the smaller looms for women to work. Similarly, from the end of the thirteenth century, the weaving of what the Parisian guilds identified as "oriental tapestry" came to be considered too arduous for women. Other technology worked against women, taking even those tasks left to them in earlier decades. In 1598 the knitting frame was invented, in 1604 the ribbon frame that created twelve bands at once. Women protested the ribbon loom but its use spread throughout the 1600s.

Florentine women maintained their access to weaving but found themselves relegated to the least valued aspect of the manufacturing, even in the silk industry, an area of their specialization. The account books of the silk merchant and entrepreneur Andrea Banchi for the years 1458–62 show how, despite their guild affiliation, the women did not do as well as the men. Banchi hired clerks and shop boys for his central office. He employed winders, throwsters, boilers, dyers, and enough weavers—both women and men—to keep about thirty looms active. Even with their salaries set by statute, in every category women weavers always made less than the guild regulation required; men made the specified rate or more.[72] Though Banchi manufactured and sold everything from satins and taffetas to the most elaborate velvet brocades, women wove only the inferior cloths. Monna Antonia, a widow and typical silk weaver, made his crimson taffetas, taking four to five weeks to produce fifty braccia (units of length, between 15 and 39 inches each) and earning about 32 florins a year. A man wove the heavy brocaded velvet, taking six months to produce 50 braccia, but earned as much as 171 florins for the year, as much in fact as a manager of a Medici bank.[73]

Always more vulnerable to economic dislocation, women suffered when the competition of long-distance trade affected the firm's profits. From 1458 to 1462 Banchi cut the wages for weavers; women's pay fell 25 percent while men's fell less.[74] The same adverse consequences of economic change affecting skilled women in Banchi's employ affected women in textile manufacturing throughout the towns. By the middle of the seventeenth century, men had taken over the more lucrative luxury fabrics, leaving women to work in wool, the coarser satins, and taffetas, all fabrics bringing lower wages.[75] The same phenomenon in the sixteenth and seventeenth centuries left women with the linen weaving in Leyden and led them into seamstressing in Geneva.[76]

Gradually specialization and technology in the textile trades, even in the manufacture of silk, consigned women to only one aspect of production. Overwhelmingly they became the spinners, working by the piece for entrepreneurs, especially vulnerable to any fluctuations in the market, to arbitrary practices of their suppliers and employers. With the improvements in the spinning wheel, the fly wheel in the 1400s, the pedal in the 1500s, technology reinforced traditional patterns. Women could spin more and more, supplying more weavers, usually men.

Entrepreneurs soon increased their profits at the expense of these artisan women, now relegated to spinning. As early as the fourteenth century the woolen merchant Francesco Datini had found that the women of the countryside would spin for less. It took him longer to convert the raw English wool into cloth, six months, but the money saved by avoiding the guild regulations made it worthwhile. Other merchants found they could make a similar savings at the expense of townswomen. In England in the seventeenth century, the town council of Reading required that employers hire a set number of women a week, but such regulations were ignored.[77]

These adverse consequences, the hardships caused by change, did bring protests from artisan women as they did from men. From as early as the 1300s the records show women both skilled and unskilled rioting and participating in protests against the merchant oligarchies, against adverse economic conditions in general, and against food shortages. Silkwomen petitioned their town governments in the fifteenth century. In 1456 in Cologne, Germany, they gained restrictions against unlicensed spinners (Beguines, pious women living and working together). Women joined the grain riot in Lyons in 1529; in 1649 they protested a salt tax in Naples by taking over the house of the tax collector and burning his goods; in early May of 1652 protests and a week's rioting and pillaging grew around a woman who walked through the streets of Córdoba carrying the body of her son who had starved to death.[78] The English Civil War of the 1640s and 1650s, with its questioning of traditional secular and religious authorities and attitudes, emboldened many townswomen of London to go beyond protests. They insisted that they held the same rights as male citizens.

Having joined in political demonstrations since 1641, in February of

1642, 400 women led by Anne Stagg, a brewer's wife, presented the women's own petitions: one to the members of the House of Commons and two to the House of Lords. These women were not political radicals. Instead they complained of the decline in trade and the loss of business with the queen gone from London. They explained that they were unable to "pay rents, support ourselves or our families."[79] In August 1643, another group, wearing white silk ribbons to show their loyalty to the King, petitioned the Commons again about economic conditions. They appealed to "the Physicians . . . that can restore this languishing Nation," and bring about the "renovation of trade." The next day 2,000 to 3,000 women assembled crying out for "Peace and our King," hurled abuse and assaulted the republican members of Parliament, and blocked the exit to the House of Commons. Male demonstrators went unmolested while troops dispersed the women, leaving slashed faces and three or four female demonstrators dead.[80]

From 1647 to 1649, women followers of the more radical sects and leaders—like John Lilburne of the Levellers—presented petitions. They too protested economic conditions but also made political demands. Rebuffed by members of Parliament, they insisted on their right to petition and thus on their right to political liberties equal to those of men. Ten thousand women signed the April 1649 petition to the House of Commons, in which they described the "multitudes ready to starve and perish for want of work, employment, necessaries and subsistence." They explained their unorthodox behavior:

> We are so over-prest, so overwhelmed in affliction, that we are not able to keep in our compass, to be bounded in the custom of our Sex . . . considering that we have an equal share and interest with men in the Commonwealth. . . .[81]

The sergeant at arms of the Commons grudgingly accepted their petition but told them that women could not "understand" such affairs, that the answer had gone appropriately to the men of their households and that "you are desired to goe home, and looke after your owne businesse, an meddle with your huswifery."[82]

When in May 1649, Lilburne's women followers in protesting martial law insisted: "Have we not an equal interest with the men of this Nation in those liberties and securities contained in the Petition of Right . . . ? Are any of our lives, limbs, liberties or goods to be taken from us more than from men, but by due processe of Law. . . ." they were ignored.[83] With the order imposed by Cromwell's Protectorate, the craftswomen's protests against economic hardship and their assertions of the rights of citizens ended. Women would not make such demands again until the end of the eighteenth century.

Acquisition of a skill and the protection of a guild afforded some women relative prosperity, but neither of these circumstances nor the advent of commercial capitalism freed them from the traditional disadvantaged patterns of female employment; neither gained them lasting equal status with men. Into the twentieth century the vast majority of European urban women continued to function without the full rights of citizens and to earn their livings in the least prestigious, least paid, most vulnerable and unskilled occupations.

New Professions: Art and Medicine

Women of the older crafts and modes of production, the elite with access to membership in guilds, had not been able to alter the traditional patterns of female employment. They had suffered further diminution of privilege and opportunity with the emergence of commercial capitalism. Unfortunately, women in the new trades and professions of the sixteenth and seventeenth centuries fared no better. Art and medicine were fields suited to women's traditional talents, and neither violated traditional images of women's proper spheres of endeavor. Both could be seen as extensions of traditional tasks. Yet in neither field were women able to maintain expertise and status equal to that granted to male artisans and practitioners. Instead they were gradually denied access to training and to membership in professional associations, and forbidden the most lucrative aspects of the new crafts.

ART

Before the sixteenth century, women functioned as artists much like men, within the Church or through participation in the family craft. Art in the thirteenth, fourteenth, and fifteenth centuries was principally illuminated manuscripts. The 1292 list of occupations for Paris included eight women illuminators.[84] Records from Belgian, German, and Italian towns give the names of other women, for example, Saint Catherine of Bologna, a fifteenth-century miniaturist and calligrapher, and Elizabeth Scipens, a member of the sixteenth-century Bruges guild of scribes in her own right. In 1521 Albrecht Dürer, the artist, met Susanna Horenbout (1503–1545), taught illumination by her father. Though she was just eighteen, he bought her picture of Jesus and marveled "that a woman can do so much."[85] Antonia (1456–1491), the artist Ucello's daughter, painted for her Carmelite house. A very few women assisted as sculptors and appear on the Paris tax records as early as the late 1200s. Sabina von Steinbach is supposed to have finished her father's work on the facade of Strasbourg Cathedral in the thirteenth century. Contemporaries in the sixteenth century like Vasari noted the relief of the sculptor

Properzia de' Rossi (c. 1500–1530) for the church of St. Peter in Bologna.

By the fifteenth century, the town elite employed artists and sculptors to decorate their houses and their churches and to immortalize their families in paint and stone. Women were active at all levels of the new crafts, from printing and engraving to painting and sculpture. Yet even in the favorable circumstances of the towns, in Italy in particular, when men like Leonardo da Vinci, Michelangelo, Caravaggio, and Titian vied with each other for commissions, only one woman, Artemisia Gentileschi (1593–1652?), emerged their equivalent in genius and reputation. Only one woman acquired the opportunity to paint on a grand scale and the recognition her talent deserved.

Like most women artisans Artemisia Gentileschi received her first training and encouragement from her father. Orazio Gentileschi, a painter himself (a follower of Caravaggio), already had a reputation as a teacher and probably gave his daughter her first lessons when she was still a little girl. Her first surviving canvas is thought to have been done when she was seventeen. Everything about this painting of the biblical heroine Susannah and the Elders suggests her unusual talent. She chose an unorthodox posing of the figures—Susannah nude seated on a bench turning away in fear with the two old men leaning over the wall behind her, leering and conferring. The scene has particular force, depicting the young woman's vulnerability and the ugliness of the men's plans, as the painting mirrors an event in the artist's own life.[86]

In 1610 Artemisia Gentileschi's father had sent her to study perspective with another master, a family friend, Agostino Tassi. Tassi was attentive from the first. Then one day in 1611, when they were alone in the workshop, he succeeded in raping her though she tried to fight him off. It was Orazio, her father, who brought suit when, no longer believing Tassi's promises of a wedding, she spoke out. Throughout five months of trial, though tortured with thumb screws, Artemisia Gentileschi insisted that she had struggled, that Tassi had given her a ring and promised to marry her if she kept silent. He spent eight months in prison, but then was acquitted, his defense being that she had already had other lovers, he was but one of many. The month after his release, in November 1612, she married and went to Florence with another artist of the circle, Pietro Antonio de Vincenzo Stiattesi.[87]

A painter of lesser determination might have given up. The early sixteenth-century sculptor Properzia de' Rossi had been persecuted and labeled a courtesan. Though she had won her commission for the reliefs for Bologna's Church of St. Peter in a competition, she died four years later at thirty, a pauper in the commune hospital. In contrast, under the protective mantle of marriage, in a different town, Artemisia Gentileschi persevered and prospered. By 1629 she had gained admission to the Academy in Florence, the equivalent to the artist's guild. Commissions for large canvases came—biblical scenes and heroines, allegorical figures, portraits, from patrons in Rome, Venice, and Naples as well as Florence. In 1638 she went to England to help her father finish the ceiling of the Queen's House at Greenwich.

The few sources for her life show how aware Gentileschi was of all that being a woman and an artist entailed. Letters to a patron, Don Antonio Ruffo in Sicily, in 1649 show her arguing over fees—100 scudi per figure—proud of her talent, of "what a woman can do," and aware of how different her vision could be.[88] She consciously varied traditional poses and the presentation of subjects to show women strong and capable of their heroic actions. She portrayed Judith, the young Hebrew widow, as handsome and robust. She is cutting off Holofernes' head with the heavy, man's sword in her hand while her maid holds him down. Even Gentileschi's self-portrait shows the practical realities of the working painter: She stands strong, tall, with the sleeves of her dress pushed up, strands of brown hair loose about her face, all of her energy and concentration directed toward the brush, the palette, and the image she is creating on the canvas. She wrote to Ruffo in November of 1649, "You will find the spirit of Caesar in the soul of this woman."[89]

Other women of the sixteenth and seventeenth centuries in Italy and then in Belgium and the Netherlands painted and worked, acquiring the necessary training and finding commissions. Yet it is only because of their ingenuity and their persistence that so many managed to attain any kind of professional status and recognition by contemporaries. For by the seventeenth century women artists had already been forced to the periphery of the new profession. Always these women worked at a disadvantage. The price of training was to be an anomaly, the prodigy (usually of an artist father), set apart and denigrated by the novelty of the accomplishment. To marry meant needed respectability but often less opportunity to paint and thus fewer works. Some women died in childbirth, others left more children than paintings. Rarely given commissions for the large canvases and the monumental works favored by Europe's patrons, they painted subjects considered decorative but not significant. Women specialized as miniaturists, as the painters of still lifes and as botanical illustrators, or as the creators of "genre" paintings, popular scenes like those done by the seventeenth-century Dutch artist, Judith Leyster. Some consciously copied their father's styles, remained in the family workshop, and thus never received any credit for works completed.[90]

Vasari, the sixteenth-century Renaissance biographer of artists, chose to write about the sculptor Properzia de' Rossi, not so much because of her artistry as because of her virtuosity: She was "skilled not only in household matters . . . but in infinite fields of knowledge. . . . She was beautiful and played and sang better than any woman in the city."[91] Other Italian women artists of the sixteenth and seventeenth century achieved recognition for similar reasons. Men considered them unusual. Critics praised the religious and classical pieces done by Lavinia Fontanta "because they are after all by a woman who has left the usual path and all that which is suitable to their hands and fingers."[92] Men loved self-portraits of the young women. As one wrote to Sofonisba Anguissola's father in 1558: "There is nothing I desire more, than the image of the artist herself, so that in a single work I can exhibit two marvels, one the work, the other the artist."[93]

Lavinia Fontanta (1552–1614) managed to move beyond the designation of prodigy, but another obstacle awaited her. At the height of her career in her forties, her commissions included biblical scenes, family portraits, an altarpiece for the chapel of Philip II of Spain at the Escorial. She left more extant works than any woman artist before 1700. Even so, the contrast with a contemporary male artist's production is still extreme. Only a few more than fifty of her paintings have been identified, although she must have completed twice that number. Her male peers painted in the hundreds, and their works have survived to enhance their reputations.

Of all the seventeenth-century women artists, the Dutch painter Judith Leyster (1609–1660) led a life most similar to that of a male artist. The daughter of a Haarlem brewer, she came to know Frans Hals and used aspects of his and Caravaggio's style. She gained admission to the painter's guild. Like Gentileschi her fame as an artist lies in the unique way she chose to portray the traditional. She did scenes of women at home, clean, trim, intent on their household tasks and best exemplified by her painting "The Proposition." The lecherous man soliciting the young housewife was a common theme, but Leyster's portrayal clearly focuses attention on the woman's refusal even to acknowledge the man's suggestion. She holds herself quiet, bent over her sewing. He looks disreputable both in his appearance and in his gesture of secretive solicitation. For Leyster, however, marriage in 1636 and the birth of three children harmed her career. She painted less and turned from the scenes that had won her recognition to still lifes.

Still lifes and portraits lent themselves to women's restricted circumstances. Denied training in anatomy, they could study these subjects from life. Denied important and prestigious commissions, functioning on the periphery of the profession, by painting these subjects they could still earn a living. In sixteenth- and seventeenth-century Italy and England, patrons commissioned women to do portraits and especially miniatures, a new art form. Few have many extant works. Joan Carlisle, a seventeenth-century portraitist and probably the first professional woman artist in England, was in her day compared favorably with the acknowledged male master, Sir Peter

Lely. Giovanna Garzoni (1600–1670) remained unmarried and died a wealthy woman from the miniatures she had done for all of the great families of Venice, Naples, and Rome.

Garzoni was also known for her still lifes; she used water colors on vellum to do careful representations of natural phenomena. She and other women artists created many of the conventions of this popular seventeenth century genre. Clara Peters (1594–1657?) of Antwerp, with twenty-four known works, at first painted traditional scenes with fruits, flowers, and objects mixed in the composition. Later she chose everyday foods, and favored juxtaposing the natural and the man-made, using metals and glass to show reflections, including miniature self-portraits. Rachel Ruysch (1664–1750), the daughter of a botany and anatomy professor in Amsterdam, gained guild admission in the Hague and the patronage of court figures. She commanded the highest fees for her works and achieved an international reputation for the accuracy and originality of her scenes. (She added animals to her compositions: a lizard eating a bird's egg, a snake or a grasshopper posed beside traditional objects.) She kept painting though married and mother to ten surviving children. She rivaled a male artist in the numbers of her compositions, 800, but as with so much of European women's formal artistic efforts only ninety-five remain.[94]

In most ways, Ruysch remained a fortunate exception. The majority of female still-life painters gained little recognition. Their works became illustrations for botanical books with limited circulation, or textile and porcelain designs. Few gained access to guilds, few were lucky enough to find their way to the attention of wealthy urban or royal patrons. Unable to make a living at their craft, they passed into the history of art as gifted amateurs.

Medicine

Unlike sculpture, painting, and the other skilled occupations where men predominated as masters, Europe's women had always taken primary responsibility for medical care of their families and had the monopoly in matters of childbirth. In the thirteenth century, when practitioners first began to be salaried and organized into guilds, townswomen were active in all branches and at all levels of medicine, as university-trained physicians, as herbalists, as (barber) surgeons, and as midwives.

Into the fifteenth century, the Italian university medical faculties allowed women to take the degree and become "physicians." Dorotea Bocchi earned her degree from the University of Bologna in 1390 and succeeded her father, lecturing in medicine and moral philosophy. In 1423 Costanza Calenda, daughter of the dean of the Faculty of Medicine in Salerno, lectured on medicine at the university in Naples.[95] To be a physician, to receive the degree of Doctor of Physic placed a woman well above most health practitioners. The very rich called them in to diagnose, to ascribe a cause, and to suggest a cure. Rarely did physicians touch the patient or prescribe medicines; instead they spoke from their theoretical knowledge of the body. They spoke of humors and put forward hypotheses based on logic; for example, cold food should cure a fever.

Given the physician's theoretical and often incorrect orientation, the richest and the best educated townspeople preferred to call on others when they were sick: women who, because of their practical experience, their knowledge of plants, and their common sense, relieved pain and cured illness more effectively than those with university degrees. In some towns these herbalists, or medical empiricists, joined in guilds. The apothecaries were such an association. Often these women acquired a practice by reputation and belonged to no professional group. Empiricists used everything from bloodletting, purges, poultices, and salves to prayer, astrological readings, and alchemy.

From the thirteenth to the seventeenth century in the towns of France, Italy, England, Germany, and Switzerland women also gained recognition and membership in guilds of "surgeons": those specializing in bloodletting, in cutting the body for tumors and amputations, and in removing the fetus when it had died in the course of labor and could not be expelled naturally. As early as 1292 women appear on the Paris roll as "barber surgeons," and until 1694 French corporations automatically allowed widows to continue if this had been their husband's craft.

Among the thirty-seven English women surgeons of London, licensed as a result of an act of Parliament in 1511, there were women who practiced surgery in combination with other aspects of medicine.[96] From the thirteenth to the seventeenth century, this was often the case; women practiced more than one medical speciality. Contemporaries called such individuals "doctors." In sixteenth-century Bern, Marie Colinet had been trained as a surgeon by her husband; she also set bones, used magnets to remove fragments of metal from eyes, and acted as a midwife. Her skills gained her the rights of citizenship in Paris, a rare privilege for a woman and a testament to the respect in which her fellow townsmen held her. Tax rolls and other town records in France and Germany name other women doctors, eight on the Paris list in 1292.[97] Some towns favored women with this kind of broad expertise, especially for the care of female patients. A fourteenth-century Italian license for a woman doctor explained that "due regard being had to purity of morals, women are better suited for the treatment of women's diseases."[98]

Women with a wide range of medical specialization served monarchs.

Hersend (c. 1250) accompanied Louis IX of France to the Holy Land as his physician; her husband was the king's apothecary. Queen Philippa, wife of King Edward III of England, employed Cecilia of Oxford as her surgeon. In the seventeenth century Lady Anne Halkett acted as surgeon, midwife, and doctor to the army of the King of Scotland.[99]

Yet with medicine, as with the other skilled trades and occupations, the traditional patterns that governed women's employment gradually changed the nature of their participation in this new profession. University graduates and craftsmen established regulations and licensing and defined specialties. Women found themselves excluded from needed training and from professional organizations. Women shared their knowledge. Men took it as their own. In their books male writers trivialized women's practical expertise and discredited those who used it in treating patients. By the eighteenth century, university-trained male physicians had established themselves as the experts and the preferred practitioners in all aspects of medicine, even those having to do with childbirth. Men controlled all except midwifery and nursing. Even in these fields women came to care only for the very poor or those in the countryside far from male specialists. When the patients were more prosperous, women ministered only under the supervision of a male physician.

Beginning in the thirteenth century, Italian towns licensed doctors. German and French towns followed suit in the fourteenth century, English and Spanish towns in the sixteenth. By the beginning of the seventeenth century, to be a doctor, a surgeon, an apothecary, or a midwife required some sort of examining and licensing. The case in August 1322 against Jacoba Felicie de Almania brought by the masters of medicine, the university-trained elite, shows how the new attitudes affected women doctors in Paris. The masters argued that Jacoba practiced without training and thus endangered lives. In her defense, Jacoba hoped to prove that despite her lack of classical education she was an effective doctor. Three of her patients attested to her honesty, her skill, and her dedication. Each made it clear that she had been paid only after her cure had worked. Jean Faber had a "sickness in his head and ears"; Odo de Cormessico's illness could not be cured by the masters; the third, Jeanne, had been given up for dead by the university-trained physicians. Jacoba treated them with medicines which she mixed herself, purges, herbal drinks, special bandages, and inhalations. Jeanne's recovery was credited to Jacoba's having "so laboured over her that by the grace of God she was cured of the said illness"; for unlike the learned Doctors of Physic who merely prescribed, Jacoba stayed, watched, and nursed the patients herself. She favored the regulation setting qualifications but insisted that this regulation was not meant to curtail her practice, but was directed "against ignorant women and inexperienced fools." She argued that women must be permitted to practice medicine for "a woman would allow herself to die before she would reveal the secrets of her illness to a man, because of the virtue of the female sex and because of the shame which she would endure by revealing them." The verdict came in November: without the proper university degree Jacoba was barred from medicine and, as this was a Church court, given its harshest punishment, excommunication.[100] Other Parisian trials from this period condemned women for the other new "crime," acting outside of their licensed specialty—the surgeon who prescribed like an apothecary, the midwife who treated like a doctor.

The experiences of women practitioners in England exemplify the way in which women were gradually excluded from the medical professions throughout Europe. By the beginning of the seventeenth century, only the newly organized apothecary's guild admitted female members. Men first denied women access to training and then prohibited their membership in the professional associations. In 1390 an examination was required for masters of medicine in London. University training was necessary for success, and women were barred from the universities. In 1421 the masters petitioned parliament to act against members of the barber-surgeon's guild who did medicine as well as surgery. They specifically condemned women:

> who possessing neither natural ability nor professional knowledge, make the gravest possible mistakes (thanks to their stupidity) and very oft kill their patients; for they work without wisdom and from no certain foundations, but in a casual fashion, nor are they thoroughly acquainted with the causes or even the names of the maladies which they claim that they are competent to cure.[101]

When in 1540 King Henry VIII authorized a separate London surgeon's guild, women were forbidden membership. The same happened in Salisbury in 1614. Surgery became the exclusive practice of "free brothers"; the guild forbade outsiders: "no such woman, or any other, shall take or meddle with any cure of chirurgery."[102] Bit by bit men had cut women off from first one and then another aspect of medicine.

Most surprising, women lost expertise and status even in the supervision of childbirth, an area of medicine exclusively theirs by tradition. Both university physicians and surgeons' guilds joined together to control the midwives' qualifications, to circumscribe their function, and eventually to discredit them altogether.

Fifteenth- and sixteenth-century engravings and paintings of childbirth scenes show the midwife as a matronly figure—a thoughtful, competent professional. Her sleeves are rolled up. She wears a white apron and headdress with the edge of her skirt tucked into her belt. Although there are other women in the pictures, it is the midwife who looks busy and tired. In one engraving the mother is sitting up in bed, the midwife resting on a stool beside her. She is about to bathe the baby, about to test the temperature of the water in the wooden tub beside the bed with her bare foot.[103] For the right to attend the birthing woman, the midwife took an oath before the local bishop so that she could baptize infants likely to die. By the fifteenth and sixteenth centuries in towns like Lille, Paris, Regensburg, and Strasbourg, either because she was the official midwife paid by the town council or because of physicians' and surgeons' interest in her activities, the midwife had to appear before the secular authorities as well. A doctor and practicing midwives usually questioned her and issued a license. In Lille in 1460 Catherine Lemersne, wife of a baker, gained certification as a midwife from a doctor and then a license from the town magistrate. In 1472 Agnes LeClerc received her license on the basis of the testimony of other women in the craft.[104]

By the late fifteenth and early sixteenth centuries, the types of examinations show that the concern of those in authority was not so much with the midwife's expertise in childbirth as with her ability to recognize when she should call in the male professionals—the physician and the surgeon. They also wanted to insure that she practiced no other branch of medicine. In Nuremberg, Regensburg, and Frankfurt-am-Main, the town regulations specified that the midwife must call in the doctor and the surgeon if the fetus appeared to be dead and the mother too weak to deliver. Paris regulations of 1560 included this and went on to insist that the midwife not treat diseases and that she report anyone working without a license.

By the last decades of the sixteenth century, physicians, surgeons, town officials, and midwives themselves wanted more than licensing. They wanted to require formal training of some sort, if for no other reason than to give women practitioners access to the new knowledge of female anatomy. Though by tradition the first clinic for teaching midwives was established in Padua in the sixteenth century, the Paris regulations of 1560 and the course of study designed by the surgeons of the Hôtel Dieu became the model followed by towns throughout France and then all across Europe. They formulated comprehensive requirements for apprenticeship, for study and examination by surgeons, doctors, and licensed midwives. A sample set of questions for a midwife's examination in Amsterdam in 1700 gives an idea of the level of expertise required. Questions tested their knowledge of anatomy, stages of labor, normal and abnormal positioning and developments, and measures to be taken when bleeding occurred before the onset of labor.[105]

Qualified midwives, often the women at the top of the profession in the capitals of Europe, acknowledged the need for regulation and for training and did their best to participate in the process, to make their knowledge and experience available to other women. The male elite of the profession sabotaged their efforts. In 1634 the College of Physicians in London refused the petition of the town's most important practitioners, Elizabeth Cellier, Jane Sharp, and Hester Shaw (who made up to £1,000 per delivery) to incorporate as a guild in order to ensure education and to enforce standards.[106] In 1635 the Parisian Faculty of Medicine refused an offer by the leading midwife of the town to teach.

Denied the right to give formal training or to set up the means of self-regulation, women shared their knowledge by writing medical guides. The first by Trotula, a physician from Salerno (Europe's first medical school), appeared in the thirteenth century. *De passionibus mulierum (Concerning the Disorders of Women)* was widely used both in manuscript and in a printed version into the sixteenth century.[107] The sixty-three chapters are a wonderful combination of the theoretical explanations of classical authors like Galen and Hippocrates and the practical conclusions from her own observations. There are discourses on the hot and dry nature of the male soothed and complemented by the cold and wet nature of the female, on the female's greater propensity to illness because of her "weaker quality," "especially around the organs involved in the work of nature."[108] There are the practical recommendations for worms, lice, deafness, toothache, abscesses, hemorrhoids, and stones.

A significant portion of the book is devoted to female ailments and to childbirth. Again Trotula combines the theoretical nonsense of the classical authors and the commonsense recommendations of the experienced practitioner. She believed bloodletting would bring on menstruation and recommended wearing weasel testicles as a contraceptive, but also gave very sound advice to women once pregnant, during labor, and after birth. She suggested baths and light foods near the due date, had drinks to ease the pain of labor, gave instructions for gentle massage (much like the *effleurage* of the twentieth century). She described repositioning the fetus to ease passage through the birth canal and sewing tears in the vagina with silk thread. She had poultices for breast abscesses, methods of manual replacement of a prolapsed uterus, salves, tampons for bleeding and for infections of the womb.

In 1609 her work came to be superseded by the collected expertise of another female practitioner, Louise Bourgeois (1563–1636), midwife to Marie de' Medici, Queen of France, and Europe's acknowledged authority on the art of midwifery. Born outside Paris, she married Martin Boursier, a barber surgeon and an assistant to Ambroise Paré, head surgeon of the Paris hospital for the poor, the Hôtel Dieu.[109] By the age of twenty-four she had borne three children. She then gave up lace making to apprentice herself to a midwife. She was practicing in Paris by 1593 and in 1601 became midwife to the Queen.

In 1609 her book appeared: *Observations diverses sur la stérilité, perte de fruict, fécondité, accouchements et maladies des femmes et enfants nouveaux naiz (Several Observations on Sterility, Miscarriage, Fertility, Childbirth and Illnesses of Women and Newborn Infants).* Translated into German and Dutch, plagiarized extensively by English writers, it became the basic text for anyone interested in childbirth. With her knowledge of anatomy from the surgeons, her practical training and supervision of almost 2,000 births, she combined in her fifty chapters of explanations and observations the best that the seventeenth century had to offer its childbearing women.[110] She described the symptoms of pregnancy. She gave illustrations and explained the causes of miscarriage and premature birth. Like twentieth-century obstetricians, she recommended rest to stop hemorrhaging and induced labor for placenta previa (premature separation of the placenta from the uterine wall). She told how and when to intervene in delivery and gave twelve possible presentations of the child. She also had recommendations on the care of the mother after the birth and on the choice of a wet nurse. Bourgeois meant the book to be an aid to midwives, especially if they were able to observe dissections. She was careful never to advocate that they practice any other sort of medicine.

Midwives to other royal families, or who had similar extensive experience, also wrote texts. Justina Dietrich (c. 1645–?), who attended the wife of Frederick I of Prussia, published her book in 1689, over 250 pages of information from her cases with forty plates of illustrations.[111] In England, Jane Sharp's *The Midwives Book on the Whole Art of Midwifery,* published in 1671, drew on her thirty years of cases and included anatomy and information from the texts of other midwives.

Both Bourgeois and Sharp spoke against a change in childbirth practice occurring especially among the wealthy and privileged, the use of what the seventeenth century called "male midwives." Bourgeois wrote of their misdiagnoses, their impatience, and the damage they did to women, especially by male surgeons brought in when the labor had become critically difficult.[112] Sharp thought little of their university training, their lectures, or their apprenticeships. She cited the Bible and women's vaster experience to justify their exclusive right to the profession. (Still subject to the prejudices of her day, however, she gave as another justification that midwifery was not worthy of the male mind.)

But having written such excellent texts, midwives had given over their last advantage. Once women shared their experience and their knowledge, the fact that men had been forbidden the birthing room except for the most unnatural labors ceased to have significance. Already in the sixteenth century in Worms, in Zurich, in Venice, men had written books posing as childbirth experts ready to enlighten the ignorant midwife. One, by the south German apothecary, Eucharius Rösslin (*Der S[ch]wangern Frauwen und Hebammen Rosegarten [The Pregnant Women's and Midwives' Rose Garden]*, 1513, called the *Byrth of Mankynde* in English) went into multiple editions in the sixteenth and seventeenth centuries though he had no practical experience at all.

Seventeenth-century surgeons in France, in the Netherlands, and England possessing the new knowledge of anatomy also wrote texts incorporating both the work of Vesalius and Fallopio and the observations from the books of the skilled midwives. Some (like Hendrik Van Deventer, François Mauriceau, and Nicolas Culpepper) saw their works as aids to women practitioners. Other surgeons expected to replace them. Ambroise Paré, the sixteenth-century barber surgeon of the Hôtel Dieu in Paris, wrote two treatises published in 1551 and 1573. Most important for the future of childbirth practices, he saw himself supervising the labor from the beginning, positioning the woman, deciding on the stages and probable difficulty, leaving almost no role for the midwife beyond comfort and companionship. Generations of medical historians have credited Paré with the creation of the medical speciality of gynecology and with the rediscovery of information and obstetrical techniques thought lost since antiquity. While Paré probably knew more than anyone of his day about anatomy—from his service to the French army and then at the Hôtel Dieu—it does not necessarily follow that women did not know of key obstetrical maneuvers. For example, they too had been repositioning the fetus in the womb, a procedure he is credited with reinventing (called "podalic version" in modern medical texts). In the thirteenth century Trotula had written:

> If the child does not come forth in the order in which it should, that is if the legs or arms should come out first, let the midwife with her small and gentle hand moistened with a decoction of flaxseed and chick peas, put the child back in its place in the proper position.[113]

Subsequent writers deprecated female practitioners. The English scientist and physician, William Harvey, in his 1651 pamphlet on reproduction, blasted midwives, who by their ignorance, their misguided attentions "in fact, run great risks of life."[114] In some parts of Europe superstitions about the wisewomen of the village in league with the devil became part of the townspeoples' images of midwives. To gain her license in London in 1558, Margett Parry vowed to the bishop not to use "sorcery."[115]

The Englishman James McMath's book, *The Expert Mid-wife*, published in the seventeenth century, reflected all of these attitudes. The text purported to instruct, but omitted anatomical drawings as unnecessary and potentially dangerous—an inspiration to "debauchery." McMath described the much reduced role for the midwife that increasingly characterized her work among townswomen. She was to "sit and attend Nature's pace and progress . . . and perform some other things of smaller moment, which Physicians gave Midwifes to do."[116]

New professions initially open to women like medicine and art gradually became subject to the traditional patterns of women's employment. Without the specialized training available to men, women could not excel at all levels in the new fields. Men controlled the guilds, the academies, the university degrees, patronage, and licensing. They systematically excluded women from all but the least prestigious areas of activity. Only in the late nineteenth century would women artists again become acknowledged professionals. Only in the late nineteenth century would women once again be admitted to the practice of medicine.

Capitalist Entrepreneurs

The guild women found themselves excluded, the artists denied training and commissions, the women medical practitioners regulated and discredited. But one other area offered opportunities for elite women of the walled towns. Like men of the wealthy manufacturing and merchant families, they had access to capital and to markets. Like men they could become agents of the economic transformation of Europe: entrepreneurs, merchants, bankers.

Circumstances favored their equal participation. The initial context of this economic revolution, of commercial capitalism, had been the woman's world, the family. The word "company" came from the Latin phrase, *cum pane*—those who broke bread together. In the towns, wives though still subordinate, were significant. With the joining of families in marriage, a wife brought alliances and capital. When merchant husbands traveled, when entrepreneurs wrote to their agents, they trusted their brothers, their sons, their wives, and their mothers. Wealthy townswomen commonly were literate and learned simple accounting, the only measurable skills in these new kinds of economic enterprise. Success in this new capitalist world came because of qualities of mind and personality, because of circumstances and opportunities taken. None bore any relation to gender.

As in the centuries before 800, widows of the thirteenth to the seventeenth centuries had the most opportunity to excel and to exercise power. Contemporaries expected them to act with some measure of independence. As the matriarch, a widow had status within the family. Most significantly she had access to capital. As townsmen made their ordinances about the rights of widows, they followed the traditions of Roman, Germanic, and

Celtic customs, guaranteeing the widow access to capital or income from property to ensure her provision. From the twelfth century in northern Italy, Germany, the Netherlands, and Belgium widows had rights to anywhere from one third to one half the amount of their dowries on the death of their husbands.[117] In some regions a widow also had claims on part of the property acquired during the marriage through means other than inheritance, the "acquests" (called "community property" in the twentieth century). By the fourteenth century this was the practice in Toulouse in France, by the sixteenth century in Paris. Negotiations at the time of the betrothal might grant widows a larger portion than that dictated by custom or town charter. Widows in fifteenth-century London had by their marriage contracts rights to the jointure, a guaranteed income from property equivalent to the cash amount of their dowry. Sixteenth-century Parisian contracts gave widows one third the value of the dowry as an annuity and use of one half the acquests and one half the husbands' property during their lifetime. Marriage negotiations in Lyons in 1500 promised them amounts equal to their dowry and even in some cases an additional sum "for the good and agreeable services which she had done for him during their marriage."[118]

By custom in the towns, widows held guardianship over their children. Conflicts did not arise as among the landed over the child's person and property. In contrast in their wills and in the decisions of the courts, townsmen often gave their wives authority over all the rest of their holdings, even appointing them executors instead of male administrators, especially their sons. A thirteenth-century German merchant wanted the mother to run the property, for fear that the son would waste it. Courts in Exeter and Bristol in the fourteenth century awarded widows the right to administer the community property, and then named them the executors of the rest.[119]

In these centuries the custom evolved that the townswoman's portion came to her in cash. This and the right to administer her husband's property on her own or on behalf of the heirs made a widow a valued individual in the world of commercial capitalism. Townsmen realized the long-term significance of her position and her assets. Money in her hands might not serve the interests of the family. Raising the money to pay her might force debts on the newly created patrimony and draw off capital needed to sustain the new mercantile dynasty. This happened to families in sixteenth-century Italy, England, France, and Belgium. Widows might, as women in thirteenth-century Siena did, bequeath the money not to sons but to daughters and granddaughters.[120] Worst of all if the woman remarried, a new husband would control the capital, and children born of the new union might threaten the inheritance rights of the offspring of the first marriage. All meant loss of family assets and their absorption into the hands of others.

To prevent such diminution, townsmen from the end of the thirteenth century ringed their widows with restrictions, first in their wills, then in their ordinances, and finally in the edicts of the kings with whom they allied. Thirteenth-century Venetian husbands' wills prohibited remarriage to protect the children's inheritance. In 1348 Fetto Ubertini of Florence stipulated that his wife Pia would lose the house and her yearly income if she remarried, or if she took her dowry out of the estate. Sienese wills of the 1290s made the same stipulations to keep the capital within the family. Town ordinances limited the amount of her dowry a widow could bequeath and made contracts invalid which did not give equally to all the children (thus insuring that male children could not be disinherited by a mother choosing to give her wealth only to her daughters). French town ordinances, the customs of the Paris parlement, and royal edicts, all limited the widow's control of property if she remarried and had children. They specified how much she might give a new husband and, probably in an effort to discourage fortune hunters, punished widows who married men of lower status.[121]

When a woman did not remarry, however, when the capital came to her to use in her lifetime, the records of the towns offer tantalizing glimpses of the enterprise and energy of some widows. They proved worthy competitors with male manufacturers and merchants. The marriage pattern among the elite of older man and younger woman meant that there were many relatively youthful widows, and this favored women becoming heads of households and thus more independent. The overwhelming majority of women heading households were widows.[122]

English, German, and Swiss widows from the fifteenth to the seventeenth centuries administered and expanded their husbands' businesses or developed their own retailing enterprises. In the midfifteenth century, Margery Haynes of Castle Combe, Wiltshire, took over the three mills owned by her husband and expanded the holdings to include another house and a shop. Joan Dent, a seventeenth-century weaver's widow, amassed £9,000 before her death at eighty-four, selling mercery (cloth), hose, and "haberdashery" door to door. In 1630 Edith Doddington had a license to trade wheat, butter, and cheese throughout four counties in southern England. Mary Hall, Barbara Riddle and Barbara Milburne owned or administered collieries (coal mines).[123] About 1560, when Barbara Uttmann's husband began to lose money with his mines in Annenberg, she started commissioning women to make lace. After his death she organized almost 900 women and took over the supervision of the mines.[124] The most successful local entrepreneur of her day was Elizabeth Baulacre (1613–1693) of Geneva. Widowed at twenty-eight after only four years of marriage, she had managed to save a third of her dowry and inherited the inventory of her husband's dry goods store. She used it to create a new retail business, the sale of gilded decoration, *dorure.* On her death she had the second largest individual fortune in the town.[125]

Town records also show women engaging in the most lucrative and competitive aspect of the commercial revolution, long-distance trade. As early as the thirteenth century, Mabilia Lecavella of Genoa became the wine seller to the King of France, rich enough to buy, lease, and sell lands outside the town with her profits. Regulations from 1294 for Hansa towns like Lübeck show women merchants of northern Europe with rights to sell and borrow on their property. Fourteenth- and fifteenth-century widows in Coventry, in London, in Bristol, in Exeter, and in Dartmouth appear on the customs rolls as merchants. They contested claims for payment, traded in armor and in wool with Spain, in wine and dyes from Bordeaux and Flanders. In 1472 the widow Alice Chester paid £41 for the use of the crane at the port to unload her Spanish iron. With profits from this and cloth from Portugal and Flanders she built herself a new house. Margaret Greenaway fulfilled her husband's contract for biscuits with the East India Company; Jane Langton sold silk to the royal family.[126] In the fifteenth and sixteenth centuries, women of northern Europe participated in the great mercantile associations. They belonged to the Handelsgesellschaft (Merchant's Society) in Ravensberg; thirty-nine of the almost 300 members were women.[127] They stayed on in Calais as members of the prestigious English trading company, the Merchant of the Staple, exercising their husbands' rights to trade in wool, wine, herring, and skins.[128]

Yet with all these skills and assets, in these favorable economic circumstances, privileged townswomen went no further. They did not capitalize on the opportunities presented to them. They never became the leaders or the founders of great mercantile dynasties in their own right. Instead, like women before 800, like the feudal châtelaines and dynastic queens presented with lands and kingdoms to administer, the townswomen preferred to act as caretakers. Even the most astute never questioned the all-pervasive fact of male dominance. They made no significant challenge to the laws, institutions, and attitudes that fostered the subordination of women. Even the most successful made no effort to ensure other women—even their own daughters—access to the power they had enjoyed.

Instead, women of the great Italian and German merchant families demonstrated their intelligence, their skill, their ingenuity in a wide range of commercial enterprises, but they planned, worked, amassed, and protected a family fortune only to hand it over to a new husband or their sons. Women like Alessandra Strozzi, Lucrezia de' Medici, Barbara Fugger, and Glückel of Hameln, all successful capitalist entrepreneurs, remained dutiful wives and mothers.

In 1422 Alessandra de Filippo Macinghi (1406–?), age fifteen, married

Matteo di Simone Strozzi. She came from a new merchant family with ties on her mother's side to the Albertis (the family of the Humanist writer and architect). Her husband's family had moved from finishing woolen cloth to insurance underwriting, to investments in real estate and town funds. He became involved in the government, went on three diplomatic missions, but inspired the enmity of the Medicis and was forced into exile. There he died in 1436 leaving her, at twenty-nine, pregnant, with four other small children. With her dowry and income from his real estate properties to live off, Alessandra Strozzi returned to Florence and proceeded to provide for her three sons and two daughters.

She favored her sons, arranging merchant apprenticeships for them as each reached thirteen and had to go into exile as their father had done: first Filippo, then Lorenzo, then Matteo. She counted on them as she explained to the eldest, Filippo, to "raise up our house again."[129] To support their ventures she liquidated land holdings and sent them the proceeds. She literally shopped for their wives, calling the prospective brides *mercatanzia* (merchandise).[130] She saw them as sources of capital for the young men. To keep what was left of the boys' patrimony she cut the dowries of her daughters Caterina and Lesandra to 1,000 florins—artisan level, as she described it. She still could report with pride that Caterina left for her marriage with the dowry and "more than 400 florins on her back."[131] The sons-in-law she picked proved to be good choices. They too helped and advised her, so that the family fortune was kept intact and growing. After 1464 and the death of Cosimo de' Medici, she began negotiating and maneuvering to have the *signoria* (Florence's governing council) lift the ban of exile. Her efforts succeeded, and in 1466, having supported the Medici in a new political fracas, her family's honor stood restored.[132]

Lucrezia Tornabuoni de' Medici (1425–1482) and Barbara Basinger Fugger (d. 1497) played the same kind of role for their sons. As a widow, Lucrezia de' Medici held the family business affairs together and dispensed political favors and advice while her eldest son, Lorenzo, "Il Magnifico," followed his other interests. After the Pazzi Conspiracy of 1478, it was she who stayed in town with him. On her death in 1482 he wrote to Duke Ercole D'Este of his "terrible loss":

> I am more full of sorrow than I can say, as besides losing a mother . . . I have lost the counselor who took many a burden from off me.[133]

Barbara Fugger acted in the same way. She maintained the family textile business after the death of her husband and with the profits she generated made it possible for her tenth child, Jacob, to become a mining tycoon and the principal banker to the Hapsburgs, the rulers of the Austrian Empire.

At the end of the seventeenth century, Glückel, the merchant's widow of Hameln (1646–1724), wrote her memoirs, a remarkable endeavor in this era. Until her husband's death in 1689, she had lived the traditional life of a prosperous Jewish woman. Betrothed at twelve, at fourteen she married Chayim, "the perfect pattern of a pious Jew" as she described him, and bore him twelve children.[134] With Chayim's death Glückel knew the double vulnerability of being a widow and a Jew. (She had already been expelled from Hamburg with her family.) In 1689 only four of the children were married, eight remained to be raised, and her husband's business was in debt. Despite her grief Glückel took control and provided for herself and her young family.

Glückel had already demonstrated her abilities and strengths, taking responsibility for her sisters and brothers on the death of her own father, and early in her marriage advising her gentle husband on his business. He sold gold and dealt in seed pearls. When he became too weak to do the traveling his trade depended upon, Glückel made the contract establishing a partnership with the young man who traveled for him. When her husband died, she balanced the account books, arranged an auction and loans to raise capital, and thus paid his debts.

For ten years she continued the business, journeying to markets and fairs herself, active in the buying and selling. Partner to one son, she educated the others to the business—trade in precious gems. In 1700 she married Cerf Levy, the richest banker in Lorraine, son to the leader of the local Jewish community; Glückel could arrange such an excellent match because of her achievements as a merchant and the fortune she had amassed. By Jewish law she could have continued her own business, making contracts, buying and selling in her own name. She chose, however, to give it dutifully over to her new husband. And she had no reason to doubt his ability to manage it and keep it intact for her. She described him as "exceedingly able, and a great business man, and highly esteemed by Jew and Gentile." Bad times came, however, and creditors "pressed him so sorely that he crashed on the rocks and all was lost."[135] She saw to the repayment of her daughter and son, who had also given Levy their money, but all of her fortune and his was lost. Of his death in 1712 she wrote, "He went to eternal peace, and left me sitting with my cares and woes."[136] Able to claim less than a third of what she was owed as a widow, by 1714 she had no alternative but to move in with her daughter.

The letters of these women, the account books of their years administering the family affairs, their memoirs show that in the last analysis they thought of themselves as caretakers, defined and motivated by their relationship and their responsibilities to the men of their families. To act in any other way, to see themselves in any other role would have meant a negation of all that their laws, their religion, and their education had told them. The attitudes were too constraining, the constraints too strong even for these exceptional women. Having acquired the power that their male counterparts competed for, they denied its significance and gave it all away. To do otherwise meant rebellion, and for most women of the towns rebellion was unthinkable.

4

THE INVISIBLE AND VISIBLE BONDS OF MISOGYNY

Old Tales Retold

In the walled towns, in the world of money and trade, men had created a new heroic image of themselves. They were the tireless providers for their families. The new circumstances brought no equivalent change in the images of women. In a thousand different ways everything in the townswoman's culture recalled and repeated the most misogynistic traditions about her nature, about her inherent danger to men, and about her proper behavior within the family. Women heard the old tales retold in the fabliaux—the popular stories shared before the fireside—in the sermons delivered by parish priests from the pulpit or by traveling friars in the town square. Women saw the old images redrawn in the books they learned to read, and in the mystery plays—the entertainments staged by the guilds on feast days.

Men of the towns found new ways to voice old fears. Secular and religious spokesmen revived with renewed vigor and clarity the full panoply of ancient negative portrayals of women. They were shown as the instigators of vice, by nature dangerous and potentially out of control: ungrateful daughters, insatiable seductresses, adulterous wives, and shrews. Townsmen made laws meant to control and restrain women, with everything from fines to execution as punishment. They made women the butt of vicious jokes, enlisting ridicule to bolster their sense of mastery.

Townsmen also presented portraits of the ideal woman, also meant to restrain women's supposed natural instincts and thus their behavior. In the stories and sermons, in the educational treatises, the perfect woman appeared as the loving wife fulfilling the obligations owed her husband: bearing his children, minding his household, creating the perfect marriage.

In the world of the towns, one woman above all others symbolized all women. Eve was the first of the female sex and the first to transgress. In the mid-twelfth-century Parisian festival play *The Mystery of Adam,* Eve reveals the basic flaws of the female nature and the ways in which her inherent weakness and propensity to sin had first harmed man. The serpent easily tempts a creature so gullible and vain. Eve in turn beguiles Adam. He blames her for their sin of disobedience:

> How quickly to perdition brought'st thou me,
> When thou mad'st sense and reason both to flee! . . .
> Ne'er more canst thou bring man felicity.
> But aye opposed to reason thou wilt be.[1]

Though Eve acknowledges her sin, God's curse in the play echoes the biblical text, vividly damning the female:

> In pain throughout their life they'll fare;
> In sorrow they'll be born of thee,
> And end their days in misery.
> To such distress and direful need
> Thou'st brought thyself and all thy seed.[2]

What the mystery plays portrayed in story form, the sermons of the twelfth- and thirteenth-century friars made even more explicit. In their admonitions, explications, and exemplary tales, they condemned Eve and her female descendents. Just as Eve acted, so by their like nature would all women act. Believing this, the Franciscan Saint Bonaventure and his thirteenth-century contemporaries took all that they deemed undesirable traits of human character and attributed them to the female. The Augustinian Jacques de Vitry did this in his sermons. He made simony, hypocrisy, usury, knavery, the desire for rich and unnecessary clothing, the daughters not the sons of the devil. His and other preachers' exemplary tales tied foolishness, the tendency to weep, envy, malice, and laziness to women.[3] Even in the art of these centuries women became synonymous with all that sin implied. The thirteenth-century sculptor of Rheims cathedral showed Eve holding the serpent in her arms and stroking him. In German and Belgian cathedral sculpture, woman became *Frau Welt* (Madame World), a figure symbolizing lust: half beautiful, half rotten and decayed; half woman, half animal.

Just as in ages past there seemed no way for a woman to counter these negative images. In the eyes of male contemporaries denial could signify rebellion. A woman's rebellion in the towns of the twelfth to the seventeenth centuries necessitated control just as it always had. Proper female behavior—acting in a way that men deemed appropriate—meant acquiescence to the male-defined ideal, yet another way of bowing to male authority. Both responses led inevitably to the subordination that men required of women. Both led to the reaffirmation of this, the oldest of Europe's traditions about women. The belief in the subordination of women survived to pass into the modern era.

Sumptuary and Adultery Laws

All of the traditional prejudices about women as wives appeared in the sermons and tales of the towns. Saint Bonaventure, the thirteenth-century friar, preached that women were "an embarrassment to man . . . a continual worry . . . a never-ending trouble."[4] The popular satire, the *Fifteen Joys of Marriage,* written by a cleric (published in Lyons c. 1480), derided the institution of marriage and warned of the many ways wives rob their husbands: from the adulteress spending the patrimony while he is away to the young wife and son conspiring against the old man. The fabliaux, the popular tales of the inns, guildhalls, of merchant and artisan gatherings, show even more unscrupulous villainesses and their stratagems. In the Spanish story "The Reluctant Monk," when poison fails to kill him, the scheming wife persuades her husband to become a monk, thus acquiring jewels, house, and all his possessions.[5]

Townsmen believed that modest dress and respectable deportment of the women of their families indicated proper attitudes and behavior. The late fourteenth-century Ménagier of Paris gave careful instructions to his young wife about her clothes. She was to dress according to what they could afford, according to their "estate," and to carry herself "honestly" not like "certain drunken, foolish or ignorant women" who go out with "their hair straying out of their wimples and the collars of their shifts and robes one upon the other, and walk mannishly and bear themselves uncouthly."[6]

Although religious and secular authorities did legislate against cross dressing—women wearing men's clothing—this was not a major concern of the officials. In these centuries, lesbianism was considered both a sexual aberration and very unusual. It came under the jurisdiction of the Church. There are only isolated instances in sixteenth-century France, Spain, and Italy of women burned or hanged for sodomy, sexual acts between two women.[7] Instead, town councils worried about appropriate dress in the same terms as the Ménagier. They passed "sumptuary laws" in an attempt to restrict the social pretensions of women and men. Some even sought to impose propriety in styles of dress, specifying the décolletage of women's necklines and the length of men's short tunics.[8] In the ordinances concerning women's clothing there was, in addition, the suggestion of their innate prodigality and immodesty—of their willingness to waste their husbands' money and to attract improper attention by their dress. In Venice the council passed, repealed, and repassed such laws from 1229 to 1505 on the use of gems in hair ornaments, the cost of jewelry, the numbers of cloaks allowed, the length of trains, the number of silk dresses in a trousseau, and sleeve lengths. Given the expense of clothing, the senate in 1504 decried the constant changing of styles as "the most injurious to the substance of gentlemen and citizens."[9]

The cost of trousseaux and weddings occasioned Florentine regulations in 1355, 1373, 1430, and 1433. According to the *signoria*'s ordinance of 1433, men passed the law to protect themselves for they could not "restrain the barbarous and irrepressible bestiality of women, who, not considering the fragility of their nature, but rather with that reprobate and diabolical nature, they force their men, with their honeyed poison to submit to them."[10] The ordinance declared that having gained money, women dressed like their social betters. The 1433 law specified that only a knight's wife could wear ermine collars and that a servant could be whipped for wearing a headress or shoes of a quality above her class. As prosperity spread with the commercial transformation of Europe, so too did regulation of dress. In the fifteenth, sixteenth, and seventeenth centuries, English, German, and Spanish towns, councils, and kings dictated attire by rank.[11] Even when townsmen of the seventeenth century saw no willful disobedience or prodigality, their laws showed little faith in wives' abilities to control expenditure. A German ordinance of 1618 reflected the attitude becoming prevalent all over Europe: that women were "somewhat weak and easily taken advantage of."[12] As English statutes made clear, whether out of innocence or stupidity, a wife might go into debt with her extravagant ways and thus make her husband financially liable.

Townsmen believed a woman's clothing could reflect another aspect of the female nature. Just as their prodigality with clothing meant wasting a husband's money and reaching above one's station, so improper dress implied a readiness for sexual misconduct. In the fourteenth century the English Dominican John Bromyard and the Italian Franciscan Saint Bernardino spoke of young women by their adornment acting the whore, the "wretched incendiary of the Devil."[13] Even so, townswomen continued to dress in ways men thought suggestive and immodest. In response, some sumptuary laws had provisions to enable townsmen to identify the women who worked as prostitutes. Thirteenth, fourteenth, and fifteenth-century ordinances from Siena, Florence, Leipzig, Bern, Vienna, Augsburg, London, and Seville required these women to wear every variety of accessory to signify their trade: a yellow cloak with blue trim, or a yellow handkerchief, or a green coat and a red beret, or no hat but a short veil. A Florentine decree of 1388 wanted a bell worn "so that the token of their shame should enter into eye and ear."[14]

As in ages past, a young woman's proper dress assumed importance, because a daughter raised in the protected world of a prosperous family was supposed to be a sexual innocent. By popular tradition she was to know nothing of lust. Should she be tempted sexually, she was to remember that her virginity was her most precious possession. One of Boccaccio's heroines equated her maidenhead with her "honor"—her "sole remaining family heirloom, [which] I am determined to safeguard and preserve . . . as long as I live."[15] These expectations for daughters, however, changed nothing of the traditional views of the female's nature, of the female's dangerous sexuality. Everything from didactic literature to the tales of the inns and firesides explained that women remained virginal only out of ignorance. Once she had known sex, a woman's true nature would stand revealed; with her sexual instincts awakened, she became insatiable. In Boccaccio's *Decameron* of the late fourteenth-century, innocent heroines, once bedded, become caricatures of lust. The fourteen-year-old maiden seduced by the hermit Rustico becomes indefatigable and exhausts him to the point of impotence. For all of his efforts it "was rather like chucking a bean into the mouth of a lion."[16] This specter of the voracious mouth, the Hell of a woman's lusting body, was repeated endlessly, translated in one form or another into stories in English, French, and German. Chaucer in the fourteenth century and La Fontaine in the seventeenth retold the tale of the two clerics who steal from the miller, seduce his wife, and then take the daughter as well. She is too much for them. Explains one: "I've bent her back seven times tonight, but she still hasn't got her fill."[17]

Given the assumptions of woman's insatiable nature, townsmen assumed that the wives of their tales would tire even the most vigorous husband, and then turn to other men. The bitter humor of these tales lay in the gullibility of the husband and the cleverness of the wanton wife.[18]

Two tales portraying the lustfulness and cleverness of women and the dangers of female sexual power to men were retold throughout Europe in these centuries. In the most famous, Aristotle, the very personification of reason, was portrayed on all fours, ridden by a beautiful woman. According to the tale, Phyllis, a supposed courtesan of Alexander the Great, took her revenge on the disapproving, aged philosopher by enticing him from his study and humiliating him when he begged for her favors. The message was clear: the wisest of all men could be brought down by his senses and a woman's deceitful sexuality.[19] In the other, extant in both German and Italian versions, a blind, distrustful husband shackles his young wife to insure her chastity. One night he frees her to climb into a tree to pick its fruit. God and Saint Peter watch while she then has intercourse with a student. Taking pity on the old man, Saint Peter restores his sight. Even so the woman deceives her husband by convincing the old man that her adulterous actions have occasioned the miracle of his restored sight. In the Italian version, the young lover openly admits their sin and brings the husband back in to the garden. After repeating the act, they tell him it must be an enchanted tree for why would they ever have intercourse in front of him? Stupefied by their logic, the husband accepts what they say.

There was no humor, however, when women actually did deceive their husbands and commit adultery. The traditions of the ancient cultures before 800 prevailed and were enacted once again in the secular and religious laws men promulgated and in the actions they condoned outside the law. Roman and Germanic custom had decreed that a woman's adultery shamed her male kin and justified beating her, even killing her. The townsmen's laws maintained the harsh attitudes and condemned women more harshly than men; in cases of infidelity the woman was always punished, the male lover sometimes, the unfaithful husband rarely. This remained true while the Church authorities had jurisdiction and later, in the course of the sixteenth and seventeenth centuries, when the secular authorities took over prosecution and sentencing of all sexual offenses including adultery. In fifteenth-century French and Italian towns, the Church courts determined guilt or innocence, the secular officials carried out the punishment. They whipped and banished women and fined men. French secular law of the sixteenth century allowed the courts to pardon a husband who killed his adulterous wife.

The adoption of Protestantism in some sixteenth-century towns melded religious and secular authority, giving one court and one set of regulations power to judge and to punish offenders. In theory women and men were to be treated equally; in fact, as the sixteenth century progressed, the double standard of behavior and punishment for women and men prevailed. Initially in sixteenth-century Geneva, women were whipped, and their male companions banished. By 1604 the man only had to pay a fine while the woman was whipped and then banished if she committed adultery again. In Hesse in the same century, though on paper the punishments were the same, in fact the only examples of executions for such offenses are of women.[20]

Abuse of Women: Violence and Ridicule

Violence toward townswomen who acted in ways deemed inappropriate by men remained commonplace in these centuries. Prostitutes, like adulteresses, lived outside the customs that protected properly behaved wives and daughters. In Spanish laws of the tenth to the thirteenth centuries in Cuenca, a "shameless woman" could be beaten, raped, even killed. The acceptance of violence toward these townswomen made it more acceptable toward others. From the moment of her birth, in fact, a female was at greater risk than a male. For another inherited tradition survived in the walled towns. When women and men found they could not raise another child, whether out of poverty or the shame of an illegitimate liaison, they sacrificed the female baby more often than the male. Female infants tended to be in poorer condition and so died in greater numbers.[21] Records of Florence's families in the fifteenth century show how even wealthy parents favored infant boys over infant girls. In the tax census, the *catasto* of 1427, the ratio of women to men was 100:150, very skewed from the natural ratio of 100:105 at birth.[22] Florentine families waited until the daughter had reached two or even five years of age before contributing for her to the Dowry Fund. Those little girls whose families had given little tended to die, while those for whom large donations had been made survived to marry.[23]

Once older, the young townswoman risked being raped. Even within the care of her family, a girl might not be safe. In 1412 the Florentine Antonio di Tome took in his orphan niece Margherita. When she brought him food while he was cutting fuel in the forest, he raped her. When she protested, he threatened to kill her.[24] Records of towns, from the fourteenth to the seventeenth centuries in Italy, France, and England, tell of women raped by all kinds of men: their lodgers, the masters they served, by soldiers, a boatman hired to ferry them, students, gangs of journeymen. A Paris guild asked the city to alter its thirteenth- and fourteenth-century requirement that a wife report for her husband at curfew—twilight—when the man could not come for the night watch. The curfew hour, the guildmen said, was dangerous to a woman, alone or with a child "whether handsome or ugly, old or young, weak or strong."[25]

Though laws, customs, and town charters theoretically punished men with everything from fines in Venice to execution in Nuremberg, in fact, as in earlier centuries, the burden of proof lay with the woman who had been attacked. From the thirteenth century on, charges of rape by a young woman could be countered by accusations that she was a prostitute. In the sixteenth century, it had been the young Roman artist Artemisia Gentileschi's word against that of her teacher and his friends. Englishmen in the seventeenth century assumed that pregnancy proved consent on the woman's part. They reasoned, like European men of previous ages, that only with orgasm could conception occur, and thus the pregnant woman stood condemned. Husbands threatened and assaulted their wives. In the late fourteenth century, when the English religious zealot Margery Kempe wished to take a vow of chastity, her husband demanded settlement of his debts and easing of her religious observances. When she could not accept his terms, " 'Well,' said he, 'then I shall meddle with you again.' "[26]

In the same way that rape became the fault of the victim and was rarely punished as a crime, so physical abuse of a wife seemed a well-deserved chastisement and signified a man in control of his marriage. Only when the woman died did contemporaries take action. In 1366 a Florentine court condemned Donato Pellegrino who had killed his wife after a friend had convinced him she was dishonest and thus had a bad reputation.[27]

Townswomen suffered other kinds of random violence. Giovanna di Baldino, a widow, died in 1377 from a beating by a member of the Medici family. When she refused to give him her lands, he "seized her by the hair and with spurs that he had in his hands, beat her unmercifully."[28] Only rarely could women defend themselves. The English courts offer a few examples of their complaints. A seventeenth-century London master hung his female apprentice by her thumbs, stripped, and whipped her. She sued for assault. The English Archbishop Laud heard a case in 1639 brought by Sarah Westwood, a widow newly remarried with an infant child. Her second husband, the journeyman she had hired to help with her felt making business, had turned her out of her own house and threatened murder. She told the archbishop that she planned to poison the man.[29]

Townsmen acted in other ways that harmed and brought them control of women. As in earlier eras women, especially wives, became the butt of every kind of humor, and marriage the source of endless jokes. "If you find things going too well, take a wife," stated a sixteenth-century proverb.[30] An English fabliau tells of a man sent to Hell by Saint Peter because he had married three times though "set free" twice.[31]

For men of the towns marriage became a question of male mastery. A thirteenth-century Franciscan in his chronicle defined "the rule of women" as a form of servitude. The clerical author of the *Fifteen Joys of Marriage*, in the last decades of the fifteenth century, compared marriage to prison. Six of the unpleasant situations he satirized involved a wife asserting mastery over her husband: arguing, interfering, taking a lover, spending his wealth, outsmarting him. In marriage he explained "the game is to the best player" and a wife "always seeks to have as much authority and power in the house as her husband—or more if she can."[32] Wives even use their pregnancies to rule their husbands, he wrote. They vomit, demand special care, prolong their confinement, and always incur endless expense.

The street plays and public festivals on holidays portrayed the same theme. In the English mystery plays, Noah's wife disobeyed and became a stock comic character. She will not come on board the ark. Noah fears her anger and her scolding, and they hit each other (an early version of *Punch and Judy*). Hans Sachs's plays for sixteenth-century Nuremberg made a joke of the husband promising the wife who lost his money "two black eyes when we meet again."[33] The French popular spring festivals, the charivaris, of the fifteenth, sixteenth, and seventeenth centuries, commonly ridiculed the shrew, the domineering wife. In Lyons in 1560 the Mother of Folly (a man dressed as a woman) read out the sentence, mocking a tanner hit by his wife. He was to ride through the streets on an ass: "to repress the temerity and audacity of women who beat their husbands . . . and if husbands let themselves to be governed by their wives, they might as well be led out to pasture."[34]

In this context a few cuffs on the ear given by a husband became a sign that a wife was being kept in her proper place. French and English sermons told of disobedient wives drowned, their backs broken, dead from poison, all because they had done what their husbands had forbidden.[35] In a Boccaccio tale, the wife learns her lesson after being mauled by wolves.[36] The Ménagier of late fourteenth-century Paris endeavored to warn his young wife of the consequences of disobedience. In his book of instructions, he cited Eve and Lot's wife, and explained that by similar behavior she would lose her husband to "evil and dishonest women who obey them in all things and honour them."[37] Because "the stain of disobedience" could never be wholly washed away, she must always bow to her husband or be called "rebellious, arrogant and sly."[38]

He used all of his rhetorical skill to frighten her away from sexual misbehavior. This was the ultimate form of disobedience, for a wife's chastity was highly valued by townsmen just as it had been by European men of previous eras. Even a suggestion of wrongdoing, he warned, could never be undone. Like the didactic fathers and priests who wrote for the daughters of the landed nobility, the townsman instructed his young wife on when and where to go out (to church with a companion and speaking to no one else), on how to carry herself to avoid attracting attention (with "eyelids lowered and still and look straight before you about four rods ahead and upon the ground").[39] She must be distant with all males but her husband and her kin, read no books, only his letters, as anything else might lead her into temptation. Should she be tempted, he told a chilling tale of a young bride whose husband could give her no sex. The new bride confides her longings to her mother, who urges her to test his willingness to forgive. The wife does so: chopping down his favorite pear tree, killing his favorite greyhound, and overturning all the food at the Christmas banquet. By his silence the husband appears to accept each action. Then the day after the Christmas scene, he calls in the barber surgeon to rid her of the "bad blood" which he suggests has caused her strange behavior:

> Then he bade her right arm to be warmed at the fire and when it was warmed, he bade bleed her; and she bled until the thick red blood came forth . . . then bade her draw her other arm from out her dress. The lady began to cry mercy.

> Nought availed it, for he had the second arm warmed and bled, and they took so much that she swooned and lost all speech and became in hue as one dead. . . .[40]

The bleeding leaves her so weak that she has to be carried to bed. Suitably chastened, when she regains her strength and next sees her mother, she vows never to think of infidelity again.

The Ideal Marriage and the Perfect Wife

Most men did not need to resort to the harsh measures of the husband in the Ménagier's tale when they wanted to insure their wife's proper attitude and proper behavior. Instead, they looked to the traditional images and presented models for their women to emulate. Alessandra Strozzi, searching for a bride for her eldest son, Filippo, described an exemplary young woman in her letter of 1465: "We have information that she is affable and competent. She is responsible for a large family (there are twelve children, six boys and six girls), [for her] mother is always pregnant and isn't very competent."[41] "Affable and competent," able to manage a large household, these were the virtues extolled. The fifteenth-century Franciscan Saint Bernardino explained: "As the sun is the ornament of the heavens, so the wise and prudent wife is the ornament of the home."[42]

When wives achieved this ideal, their husbands praised them. Francesco Datini wrote to Margherita in 1397 "that you have ordered the house in a fashion to do you honour, pleases me."[43] When Elizabeth Withypoll died at the age of twenty-seven in 1537, her husband, a member of the Merchant Taylors Company of London, dedicated a plaque to her memory. The greatest tribute he could pay was to list her accomplishments: needlework, handwriting, singing, three languages including Latin, her virtues and her faith.[44] To bear and raise children was the wife's other duty. Alberti, the Florentine Humanist, in his 1434 treatise on the family suggested that a wife be chosen for her strength and physical size so that a husband could be assured of numerous, healthy offspring. Clarice Orsini, Lorenzo de' Medici's wife, reproduced dutifully and created a loving atmosphere. Eight-year-old Piero wrote to his father in 1479:

> We are all well and studying. Giovanni is beginning to spell. . . . Giuliano laughs and thinks of nothing else, Lucrezia sews, sings and reads, Maddalena knocks her head against the wall . . . Luisa begs to say a few little words. Contessina fills the house with her noise.[45]

Men described marriage with the ideal wife as more than the practical joining of two families to create a household and to produce offspring. In the fifteenth century, the Franciscan Saint Bernardino suggested in his sermons that marriage might be a "friendship."[46] The Ménagier quoted the phrase of a Flemish weaver to his young bride and spoke of "married friends." He described marriage as a bond with identical obligations for each, a state in which "all other loves be put afar off, destroyed and forgotten, save the love of each other."[47] Protestant leaders and writers in the sixteenth century like Bucer and Zwingli believed that all women and men must marry. They looked to a mutually companionable, loving relationship. The Englishman John Milton used the absence of such feelings as a justification for divorce.

Yet friendship and affection did not mean equality. Inherent in the learned and popular descriptions of marriage were all of the ancient premises about women's inferior nature and about man's obligation to "govern her."[48] The ideal wife accepted these traditional premises and acted the willing subordinate. As God explained to Eve in a Parisian mystery play:

> Love Adam, hold him dear as life . . .
> Ever to him submit thy heart
> And from his teaching ne'er depart.

The consequences of any other behavior were clear to the audience: "Unless she heed, to folly she will bend."[49] Margherita Datini's brother-in-law imagined himself with the "bridle in my hand."[50] The Ménagier likened the good wife to a pet dog for "even if his master whip him and throw stones at him, the dog followeth, wagging his tail and lying down before his master to appease him. . . . He always has his heart and eye upon his master."[51] A late fifteenth-century Italian Humanist, Giovanni Caldiera, saw a hierarchy of state and authority beginning with a husband in his family. Like a subject the wife was to obey, like a subject she should view her ruler as wise and all powerful.[52] The popular tales repeated the refrain. One of Boccaccio's characters explained: by "nature, custom, as well as in law" women have been placed below men, "by whose opinions their conduct and actions are bound to be governed. It therefore behooves any woman who seeks a calm, contented and untroubled life with her menfolk, to be humble, patient, and obedient. . . ."[53] The sixteenth-century German Hans Sachs expressed men's sentiments more simply in his play *The Old Game.* The husband explained to his wife:

> When you do as I bids you, like a humble, willing and obedient wife. Then I'm ready enough to share my last crust of sour bread with you, and to see you lack no clothing nor finery. Then it's a pleasure to look after you, and to give you good counsel.[54]

To the wife fell the responsibility for making the union companionable. In their didactic writings, townsmen told women how to behave with the demanding or difficult husband. Francesco Barbaro in his 1428 treatise told wives to happily and "quickly execute" a husband's wishes, to be "humble and timid" in the face of his rages, and when he was "troubled, inopportuned, bad tempered . . . with a sweet voice and modest voice try to make him happy, to comfort him, to console him."[55] The Ménagier expected wives to mitigate their husbands' faults, to protect them from the consequences of their own wrongdoing. "By good patience and gentle words you slay his proud cruelty," he explained. "Subtly, cautiously and gently" she could restrain him "from the follies and silly dealings" he might be drawn into.[56] Like the Chevalier of La Tour Landry who wrote for his noble daughters, so the Ménagier used the story of Jehanne la Quentine to instruct a young wife, a woman who "reclaim[ed] her husband by kindness."[57] Learning of his affair with a poor young woman, Jehanne never angers, rather she helps the lover care for him, and thus when her husband learns of her gesture, she wins him, and he never strays again. Such generosity had real life counterparts. Margherita Datini arranged for the dowry and marriage of one of her husband's lovers and raised the child of the union as her own.[58]

The fifteenth-century Florentine Humanist Leone Battista Alberti, in his description of the perfect family, also liked the prospect of educating his wife himself; this was an ideal situation, for then a man could be sure of her training, sure to inculcate the necessary virtues. Alberti even envisaged teaching his bride spinning, sewing, and how to train the servants for their duties. The Ménagier began writing instructions for his fifteen-year-old wife on their wedding night. In the first three pages he explained that he would be out working on her behalf, that she should so manage her "business" that he would "desire to return to his good home and to see his goodwife, and to be distant with others."[59] Always "gentle and amiable and peaceable" she was to create a veritable "paradise of rest." He assured her "no man can be better bewitched than by giving him what pleaseth him."[60] He painted a picture of the beleaguered provider and husband who had to "journey hither and thither," "in rain and wind, in snow and hail, now drenched, now dry, now sweating, now shivering, ill-fed, ill-lodged, ill-warmed and ill-bedded." He shared with her the image that would sustain him in his trials:

> the care which his wife will take of him on his return, and of the ease, the joys and the pleasures which she will do him, or cause to be done to him in her presence; to be unshod before a good fire, to have his feet washed and fresh shoes and hose, to be given good food and drink, to be well served and well looked after, well bedded in white sheets and nightcaps, well covered with good furs, and assuaged with other joys and desports, privities, loves and secrets whereof I am silent. And the next day fresh shirts and garments.[61]

Such care, he explained, would insure the longevity of her husband and thus her well-being.

Some townswomen balked at the premises and the requirements of these marriages. In the fourteenth century Margherita Datini protested her husband's lecturing, instructing, and shouting. Glückel of Hameln bemoaned the responsibilities, especially of her motherhood:

> Every two years I had a baby, I was tormented with worries as everyone is with a little house full of children, God be with them! and I thought myself more heavily burdened than anyone else in the world. . . .[62]

But most privileged women accepted the male ideal of the dutiful, obedient wife and mother and took pride in demonstrating the skills and qualities of character it necessitated.

In addition, given the principal responsibility for the education of their daughters, many townswomen passed on the traditional images, values, and admonitions about women to the next generation of wives and mothers. Religious and secular writers advised mothers that household skills and the arts of pleasing a husband took precedence over the academic. The fifteenth-century Franciscan Saint Bernardino went so far as to suggest that reading would only be an appropriate pursuit should the girl be intended for the Church. Beginning in the thirteenth century in most of the towns of Europe, the Church provided some sort of education for the more privileged boys and occasionally reading, simple arithmetic, and religious training for girls. In Paris, artisan and merchant daughters might attend a school run by the Beguines, or after the decrees of the sixteenth-century Council of Trent, parish schools run by the Ursulines, the Sisters of Charity, or the Sisters of Notre Dame de la Visitation. Sixteenth-century Protestant towns like Wittenberg in Germany and Geneva in Switzerland required primary education for their girls. At the end of the seventeenth century, in countries like France and Sweden, royal decrees stipulated education for girls and boys.[63] Thomas Mulcaster admitted girls to the Merchant Taylors' School he founded in London in 1561. In his educational treatise he designed a comparable course of study for girls and boys. Even so he saw girls becoming wives, wanted them to be good managers, and saw the principal responsibility for creating and maintaining a companionable marriage falling to the woman. Most mothers of wealthy families would have agreed with the Ménagier when he told his bride that her two most important activities in life would be "the salvation of your soul and the comfort of your husband."[64]

Over and over again, in many versions, in many languages young girls heard the story of the perfect wife. They embroidered its scenes in their needlework. "Poor Griselda" as she was called, exemplified every admirable quality: modest origins, beauty, humility, gentle strength, and unquestioning obedience. Boccaccio makes hers the last tale of the *Decameron.* The Ménagier included it in the instructions to his wife. The story begins when the Marquis of Saluzzo takes Griselda, a poor village maiden, to be his wife. Stripped of her peasant clothing, she is transformed into a beautiful creature worthy of her master's household. She performs all of her tasks and duties with grace and humility, bearing two children, even helping in the management of his estates. Griselda is so obedient she does not question her husband when he takes first one and then the other child, not even when he announces that she is to be cast out for another wife. Thus she wins his trust, the return of her children, and the right to remain his obedient wife.[65]

Contemporaries called this heroine "Poor" or "Patient" Griselda perhaps in acknowledgment of the harshness of the tests presented to her. Yet few criticized the roles she fulfilled or the qualities she exemplified. Instead most townswomen taught their daughters to be like her. They acknowledged that happiness and safety came with acceptance of the proper relationship to men. Regardless of physical, legal, or economic circumstances the vast majority of townswomen believed that they must function as willing subordinates within the traditional framework of the family and a male-dominated society.

Notes

PART I TRADITIONS INHERITED:
ATTITUDES ABOUT WOMEN FROM THE CENTURIES BEFORE 800 A.D.

PART I: 1. Buried Traditions: The Question of Origins

1. Dolní Věstonice information drawn from B. Klíma, *Dolní Věstonice* (Prague: Nakladatelství Ceskoslovenské Akademie Ved, 1963). The head is 2 inches (4.8 cm.) in length, and photographs of it are in J. Jelínek, *The Pictorial Encyclopedia of the Evolution of Man* (London: Hamlyn, 1975), pp. 361, 376.
2. N. K. Sanders, *Prehistoric Art in Europe* (Baltimore: Penguin, 1938), pp. 12–13.
3. Bachofen's *Mother Right* was first published in 1861, Morgan's *Ancient Society* in 1877, and Engels' *The Origin of the Family, Private Property and the State* in 1884.
4. Cited in Marielouise Janssen-Jurreit, *Sexism: The Male Monopoly on History and Thought,* translated by Verne Moberg (New York: Farrar Straus Giroux, 1982), p. 57.
5. Italics in original. Friedrich Engels, *The Origin of the Family, Private Property and the State,* edited by Eleanor Burke Leacock (New York: International Publishers, 1973), p. 120.
6. See especially, Frances Dahlberg, ed., *Woman the Gatherer* (New Haven: Yale University Press, 1981), Nancy Tanner, *On Becoming Human* (Cambridge: Cambridge University Press, 1981), Karen Sacks, *Sisters and Wives: The Past and Future of Sexual Equality* (Westport, Conn.: Greenwood, 1979), and Peggy Reeves Sanday, *Female Power and Male Dominance: On the origins of sexual inequality* (Cambridge: Cambridge University Press, 1981).
7. See Dahlberg, *Woman the Gatherer,* passim. and Jane Beckman Lancaster, *Primate Behavior and the Emergence of Human Culture* (New York: Holt, Rinehart and Winston, 1975), p. 80.
8. The statuette is 4 inches (11.5 cm.) high and is photographed in Jelínek, pp. 379, 403.
9. Map of these sites in Grahame Clark, *World Prehistory in New Perspective,* 3rd ed. (Cambridge: Cambridge University Press, 1977), p. 99.
10. The most important recent attempt is Marija Gimbutas, *The Goddesses and Gods of Old Europe: 6500–3500 B.C.* (London: Thames and Hudson, 1982). Lepensky Vir information from Gimbutas and Dragoslav Sréjović, *Europe's First Monumental Sculpture: Lepensky Vir* (New York: Stein and Day, 1972).
11. Alexander Marshak, *The Roots of Civilization: The Cognitive Beginnings of Man's First Art, Symbol and Notation* (New York: McGraw-Hill, 1972), p. 244ff.; Philip Rawson, *Primitive Erotic Art* (New York: G. P. Putnam's Sons, 1973), pp. 10–11, 20.
12. Nea Nikomedia information from Robert J. Rodden, "Excavations at the Early Neolithic Site at Nea Nikomedia, Greek Macedon (1961 season)," *Proceedings of the Prehistoric Society,* Vol. XII (NS, Vol. XXVIII) (1962), pp. 267–89, and Robert J. Rodden, "An Early Neolithic Village in Greece," *Scientific American,* vol. 212, no. 4 (April 1965), pp. 83–92.
13. These tablets are in Linear B, which has been deciphered. Jon-Christian Billigmeier and Judy A. Turner, "The Socio-Economic Roles of Women in Mycenean Greece: A Brief Survey from Evidence of Linear B Tablets," in Helene Foley, ed., *Reflections of Women in Antiquity* (London: Gordon and Breach Science Publishers, 1981).

14. Statuette is reproduced in color in Petros G. Themelis, *Mycenae Nauplion* (Athens: Delta, n.d.), p. 55, and in black and white (front and back) in H. A. Groenewegen-Frankfort and Bernard Ashmole, *Art of the Ancient World* (New York: Prentice-Hall and Harry N. Abrams, n.d.), p. 133.
15. For example, see George Thomson, *Studies in Ancient Greek Society: The Prehistoric Aegean* (London: Lawrence and Wishart, 1949), and Ruby Rohrlich-Leavitt, "Women in Transition: Crete and Sumer," in Renate Bridenthal and Claudia Koonz, eds., *Becoming Visible: Women in European History* (Boston: Houghton Mifflin, 1977).
16. For a discussion of the deficiencies of Cretan archaeology with regard to women, especially as practiced by Sir Arthur Evans, see Sarah B. Pomeroy, "Selected Bibliography on Women in Antiquity," in John Peradotto and J. P. Sullivan, eds., *Women in the Ancient World: The Arethusa Papers* (Albany, N.Y.: SUNY Press, 1984), pp. 347–53.
17. Cretan art depicted men with compressed waists and exaggeratedly curved chests, so the torsoes of these figures are similar. Cretans tended to paint women with white skin and men with red, but this convention was sometimes broken.
18. For Crete, see Manolis Andronicos, *Herakleion Museum and Archeological Sites of Crete* (Athens: Ekdotike Athenon S.A., 1977); the objects mentioned are reproduced in plates 9, 19, 20, 29, 39, 40, 43, and 44; also Costis Davaris, *Guide to Cretan Antiquities* (Park Ridge, N.J.: Noyes Press, 1976) and Sinclair Hood, *The Minoans: Crete in the Bronze Age* (London: Thames and Hudson, 1971).
19. Martin P. Nilsson, *The Minoan-Mycenean Religion and Its Survival in Greek Religion,* 2nd ed. rev. (Lund, Sweden: CWK Gleerup, 1950), p. 391; Peter J. Ucko, *Anthropomorphic Figurines of Predynastic Egypt and Neolithic Crete with Comparative Material from the Prehistoric Near East and Mainland Greece,* Royal Anthropological Institute, Occasional Paper No. 24 (London: Andrew Szmidla, 1968), Table 6.2, p. 316.
20. Sarah B. Pomeroy, *Goddesses, Wives, Whores, and Slaves: Women in Classical Antiquity* (New York: Schocken, 1975), p. 15
21. Sarah Blaffer Hrdy, *The Woman That Never Evolved* (Cambridge, Mass.: Harvard University Press, 1981), pp. 69, 189.
22. Marshall D. Sahlins, "The Origins of Society," *Scientific American,* vol. 203, no. 3 (September, 1960), p. 80.
23. Lancaster, p. 12; Hrdy, p. 187.
24. Psychological studies of intelligence tend not to separate out cultural training from "innate" ability. For a summary of twentieth-century studies to 1974, see Eleanor Emmons Maccoby and Carol Nagy Jacklin, *The Psychology of Sex Differences,* 2 vols. (Stanford, Calif.: Stanford University Press, 1974).
25. John Money and Anke A. Ehrhardt, *Man and Woman, Boy and Girl: The Differentiation and Dimorphism of Gender Identity from Conception to Maturity* (Baltimore: Johns Hopkins University Press, 1972), pp. 118–23; Nancy Chodorow, *The Reproduction of Mothering: Psychoanalysis and the Sociology of Gender* (Berkeley: University of California Press, 1978), Part 1.
26. Lancaster, p. 82.
27. Michael S. Teitelbaum, ed., *Sexual Differences: Social and Biological Perspectives* (Garden City, N.Y.: Doubleday, 1976).

28. Maccoby and Jacklin, p. 243.
29. Teitelbaum, pp. 65–66.
30. For Freud, see especially "Some Psychic Consequences of the Anatomical Distinction Between the Sexes" (1925) and "Femininity" (1933). These essays have been reprinted in the many editions of Freud's writings. "Femininity" is included in Rosemary Agonito, *History of Ideas on Women: A Source Book* (New York: G. P. Putnam, 1977). The concept of the male's psychosexual frailty is in Money and Ehrhardt, p. 117.
31. Chodorow, p. 183.
32. On this subject, see Michelle Zimbalist Rosaldo, "Women, Culture, and Society: A Theoretical Overview," in Michelle Zimbalist Rosaldo and Louise Lamphere, eds., *Women, Culture, and Society* (Stanford, Calif.: Stanford University Press, 1974), p. 28, and Bruno Bettelheim, *Symbolic Wounds: Puberty Rites and the Envious Male,* rev. ed. (New York: Collier, 1968).
33. For an early statement of this position, see Karen Horney, "The Dread of Women" (1932) in her *Feminine Psychology* (New York: Norton, 1973), p. 145.
34. William N. Stephens, "A Cross-Cultural Study of Menstruation Taboos," *Genetic Psychology Monographs,* vol. 64 (1961), pp. 392, 401.
35. Cited in Patricia Martin Doyle, "Women and Religion: Psychological and Cultural Implications," in Rosemary Radford Ruether, ed., *Religion and Sexism: Images of Women in the Jewish and Christian Traditions* (New York: Simon and Schuster, 1974), p. 27.
36. See Chodorow, p. 183; Horney, "The Dread of Women" and "The Distrust Between the Sexes" in *Feminine Psychology,* pp. 107–108; Clara M. Thompson, *On Women,* edited by Maurice R. Green (New York: Meridian, 1986), passim.
37. Chodorow, p. 183.
38. On this, see Horney, "The Distrust Between the Sexes," in *Feminine Psychology,* p. 115.
39. Sanday, p. 33; Alice Schlegel, ed., *Sexual Stratification: A Cross-Cultural View* (New York: Columbia University Press, 1977), p. 356.
40. Sanday, Chapter 4; Marvin Harris, *Culture, Man and Nature: An Introduction to General Anthropology* (New York: Thomas Y. Crowell, 1971), pp. 581–82.
41. Sanday, pp. 209–10.
42. See Marvin Harris, "Why Men Dominate Women," *New York Times Magazine,* November 13, 1977, p. 118 ff. and, for a discussion of Harris, Sanday, pp. 172–87.
43. Kathleen Gough, "The Origin of the Family," in Rayna R. Reiter, ed., *Toward an Anthropology of Women* (New York: Monthly Review Press, 1975), p. 69
44. Harris, "Why Men Dominate Women," p. 117.
45. Sanday, pp. 186–87, 210–11. For Sanday's analysis of this pattern, see Chapter 8, passim.
46. Richard A. Preston, Sydney F. Wise, and Herman O. Werner, *Men in Arms: A History of Warfare and Its Interrelationships with Western Society,* rev. ed. (New York: Frederick A. Praeger, 1962), p. 10.
47. Sanday, p. 210.
48. See Marylin Arthur, "Early Greece: The Origins of the Western Attitude Toward Women," in Peradotto and Sullivan, *Women in the Ancient World,* pp. 7–58;

Pomeroy, *Goddesses,* Chapter 2, and John H. Otwell, *And Sarah Laughed: The Status of Women in the Old Testament* (Philadelphia: Westminster Press, 1977).

49. For this argument, see Pomeroy, *Goddesses,* p. 8ff.
50. For the major works traditionally attributed to Homer, we have used the fine translations by Robert Fitzgerald: *The Iliad* (Garden City, N.Y.: Doubleday, 1974) and *The Odyssey* (Garden City, N.Y.: Doubleday, 1962). References are given in books and lines. Various translations of the Homeric Hymns have also been used.
51. *Iliad,* V, 892.
52. *Iliad,* XIV, 173–87.
53. *Iliad,* V, 348.
54. *Iliad,* XV, 14, 19–25.
55. *Iliad,* XIV, 217–21.
56. Cited in Arianna Stassinopulos and Roloff Berry, *The Gods of Greece* (New York: Harry N. Abrams, 1983), p. 168.
57. On the Vestal Virgins, who also had the power to pardon criminals, but themselves were liable to death if they broke their vow of virginity, see Mary R. Lefkowitz and Maureen B. Fant, *Women's Life in Greece and Rome: A Source Book in Translation* (Baltimore: Johns Hopkins University Press, 1982), pp. 174, 249–50, and J. P. V. D. Balsdon, *Roman Women: Their History and Habits* (New York: Harper & Row, 1962), pp. 234–43.
58. Exodus 15:20; Numbers 12. Biblical citations are from *The New Oxford Annotated Bible with the Apocrypha,* revised standard version (New York: Oxford University Press, 1977).
59. *Odyssey,* XXIV, 196–201.
60. *Odyssey,* VI, 182–84.
61. *Iliad,* VI, 440–46.
62. *Odyssey,* XXI, 350–53.
63. Genesis 2:23–24.
64. Genesis 12:11. Their names are Abram and Sar'ai at that time.
65. Genesis 24: 64, 67; Genesis 29:17, 20.
66. Genesis 3:20; Genesis 17 and 18; Genesis 30:1.
67. Genesis 30:23; Exodus 20:12.
68. *Iliad,* VII, 101; Nahum 3:13; Jeremiah 30:6.
69. Balsdon, p. 17; Lefkowitz and Fant, p. 173.
70. Lefkowitz and Fant, p. 175; Max Kaser, *Roman Private Law,* translated by Rolf Dannenbring (Durban, S.A.: Butterworth's, 1965), p. 68.
71. Numbers 30:1–16; Leviticus 27:1–7; Genesis 19:1–8.
72. Genesis 1:27; Genesis 3:16.
73. Genesis 3:17.
74. *Odyssey,* XXII, 462–64. Goddesses and nymphs do consort with both gods and men without dishonor, and there is mention of mortal women who bear illegitimate children (albeit to gods), who later marry with no loss of esteem. See *Iliad* XVI, 175–92 and III, 44–45.
75. Valerius Maximus in Lefkowitz and Fant, p. 176; for Romulan law, Lefkowitz and Fant, p. 173.

76. By later eras, Roman wives could divorce as well. Citations from Lefkowitz and Fant, pp. 173, 174, 202, n. 3.
77. Deuteronomy 24:1–4. Jewish women gained the right to divorce in the medieval period.
78. Numbers 5:11–31.
79. Deuteronomy 22:22–24; Deuteronomy 22:21, 28–29.
80. Genesis 16:1–5; Genesis 30:1–24.
81. Genesis 20:3; Karl Hermann Schelkle, *The Spirit and the Bride: Woman in the Bible* (Collegeville, Minn.: The Liturgical Press, 1979), pp. 25–26.
82. Lefkowitz and Fant, p. 174; Pomeroy, *Goddesses,* p. 78.
83. This penalty was not part of biblical law. Genesis 38:24; Leviticus 21:9.
84. Leviticus 15.
85. Leviticus 12.
86. Lefkowitz and Fant, p. 88, and Pliny the Elder, *Natural History* II, translated by H. Rackham (Cambridge, Mass.: Loeb Classical Library, Harvard University Press, 1962), p. 547.
87. For Greece and Rome, see Marylin Arthur, "Liberated Women: The Classical Era," in Bridenthal and Koonz, op. cit., passim., Pomeroy, *Goddesses,* pp. 48–52; for the Jews and Christians, Bernard P. Prusak, "Woman: Seductive Siren and Source of Sin? Pseudepigraphal Myth and Christian Origins," and Rosemary Radford Ruether, "Misogynism and Virginal Feminism in the Fathers of the Church," both in Ruether, ed., *Religion and Sexism;* for the argument in general, Katharine M. Rogers, *The Troublesome Helpmate: A History of Misogyny in Literature* (Seattle: University of Washington Press, 1966), Chapter 1.

PART I: **2. Inherited Traditions: The Principal Influences**

1. For this definition of Europe, see Marc Bloch, *Feudal Society,* translated by L. A. Manyon (Chicago: University of Chicago Press, 1966), Vol. I, p. xix, and Pierre Riché, *Daily Life in the World of Charlemagne,* translated by JoAnn McNamara (Philadelphia: University of Pennsylvania Press, 1978), p. ix.
2. Much Greek writing after Homer has been lost, and the poetic, dramatic, legal, and prosodic works which remain provide a slanted view of traditions pertaining to women. For instance, most of the Greek writings preserved come from Athens, and yet historians know that the role of women in Athens was the most restricted among the separate city-states of the fifth and fourth centuries B.C. For Greek writings, the Loeb Classical Library editions were used. Halfway through the writing of this book, Mary R. Lefkowitz and Maureen B. Fant's *Women's Life in Greece and Rome: A sourcebook in translation* (Baltimore: Johns Hopkins University Press, 1982) appeared. For the convenience of readers the Loeb references were changed to citations from this book when possible. The same was done for Roman writings: When the Loeb text was contained in Lefkowitz and Fant, the latter was cited. Roman law was not codified until late in the Imperial era, after the acceptance of Christianity, and was not adopted uniformly throughout the Empire, so Roman laws must be used with care.

   The same is true of Celtic and Germanic customary laws. They were selectively written down and then only for disputed points of law. By then, Celtic and Germanic cultures had been so influenced by Rome and Christianity that their original customs may never be completely recovered. Germanic customs and laws have been collected and studied; some have been translated. The Germanic peoples referred to in the text are the Alamanni, Anglo-Saxons, Burgundians, Franks, Lombards, Ripuarians, Visigoths, and Vikings. Epics of these peoples have also been used when possible.

Sources for the Hebrews include the Old Testament, the legal writings of the Babylonian and Palestinian Mishnahs (written down about 200 A.D.), and the teaching and commentaries which eventually came to comprise the Talmud. Contemporary Roman histories also contain information on the Hebrews, as well as the Celts and Germans.

Christian sources include the New Testament, the letters and tracts of Church leaders, decrees of councils and popes, commentaries, and books of penance. Only some have been preserved, some are occasionally contradictory, and most may reflect intention and not reality. All these primary sources must be used with extreme care.

PART I: **3. Traditions Subordinating Women**

1. The Greek is Xenophanes, cited in Jack Lindsay, *The Ancient World: Manners and Morals* (New York: G. P. Putnam's Sons, 1968), p. 41; papyri from M. I. Finley, *The World of Odysseus,* rev. ed. (New York: Viking, 1965), p. 11.
2. Katherine Fischer Drew, ed. and trans., *The Lombard Laws* (Philadelphia: University of Pennsylvania Press, 1973), p. 129.
3. From *Politics,* cited in Lefkowitz and Fant, pp. 63–64.
4. In the sixteenth century, the Council of Trent, by confirming Saint Thomas Aquinas' use of Aristotle, perpetuated his thought within the Roman Catholic Church into the twentieth century. On Aristotle and women, see Susan Moller Okin, *Women in Western Political Thought* (Princeton, N.J.: Princeton University Press, 1979), Part II, and G. E. R. Lloyd, *Science, Folklore and Ideology: Studies in the Life Sciences in Ancient Greece* (Cambridge: Cambridge University Press, 1983), Part II, Chapter 4.
5. Cited in Naphtali Lewis and Meyer Reinhold, *Roman Civilization: The Empire, A Sourcebook II* (New York: Harper & Row, 1966), pp. 543–44.
6. Donncha Ó'Corráin, "Women in Early Irish Society," in Margaret MacCurtain and Donncha Ó'Corráin, eds., *Women in Irish Society: The Historical Dimension* (Dublin: Arden House, 1978), p. 9.
7. Cited in Constance F. Parvey, "The Theology and Leadership of Women in the New Testament," in Ruether, ed., *Religion and Sexism,* p. 126.
8. Cited in Léonie J. Archer, "The Role of Jewish Women in the Religion, Ritual and Cult of Graeco-Roman Palestine," in Averil Cameron and Amélie Kuhrt, eds., *Images of Women in Antiquity* (Detroit: Wayne State Press, 1983), p. 284.
9. From the Babylonian Talmud, cited in Raphael Loewe, *The Position of Women in Judaism* (London: Society for the Promotion of Christian Knowledge and the Hillel Foundation, 1966), p. 22.
10. This prayer has been the subject of much debate. Its lack of symmetry points to womanhood being seen as an inferior state, but some Jewish apologists have argued that it only signifies women's freedom from religious obligations because of their household duties, which were highly valued in maintaining Jewish practice. For this

argument, see Judith Hauptman, "Images of Women in the Talmud," in Ruether, ed., *Religion and Sexism,* p. 196; also Loewe, p. 43, and Archer, in Cameron and Kuhrt, eds., p. 284.

11. Loewe, p. 18.
12. Thomas Kinsella, ed. and trans., *The Táin* (Philadelphia: University of Pennsylvania, 1970), pp. 250–51.
13. From *Diseases of Women,* Vol. I, cited in Lefkowitz and Fant, p. 90.
14. In his *History of Animals;* also see Lloyd, p. 171, note 198.
15. In *On Dreams,* cited in Lefkowitz and Fant, p. 89.
16. From *History of Animals,* cited in Lloyd, pp. 99–102.
17. Pliny the Elder, *Natural History,* Vol. II (Cambridge, Mass.: Loeb Classical Library, Harvard University Press, 1962), p. 547.
18. From *The Generation of Animals* (Cambridge, Mass.: Loeb Classical Library, Harvard University Press, 1963), p. 175.
19. From *On the Usefulness of Parts of the Body* in Lefkowitz and Fant, pp. 215–16. Soranus, the Roman physician of the first century A.D., is the exception in this regard; he based his work on empirical research rather than on accepted beliefs. For Soranus and women, see Lloyd, Part II, Chapter 5, and Lefkowitz and Fant, pp. 219–25.
20. From *Timaeus* in Lefkowitz and Fant, pp. 81–82.
21. For the Hippocratic Corpus, see Lloyd, p. 84, and Lefkowitz and Fant, pp. 91, 93; for Aretaeus, Lefkowitz and Fant, p. 225. For the persistence of this belief in European culture into the nineteenth century, see Ilza Veith, *Hysteria: The History of a Disease* (Chicago: University of Chicago Press, 1965).
22. Lloyd, p. 84; Lefkowitz and Fant, p. 87.
23. From *Timaeus,* in Lefkowitz and Fant, p. 81.
24. From *Generation of Animals,* in Lefkowitz and Fant, pp. 83–84.
25. From *On the Usefulness of Parts of the Body,* in Lefkowitz and Fant, p. 215.
26. Mark Golden, "Demography and the Exposure of Girls at Athens," cited in Sarah B. Pomeroy, "Infanticide in Hellenistic Greece," in Cameron and Kuhrt, eds., p. 217.
27. J. C. Russell, *Late Ancient and Medieval Population,* Transactions of the American Philosophic Society, NS-XLVIII (Philadelphia: The American Philosophical Society, 1958), p. 133.
28. Pomeroy, *Goddesses,* p. 91.
29. The history is Dio Cassius', cited in Lewis and Reinhold, eds., Vol. II, p. 500.
30. Cited in Lefkowitz and Fant, p. 111.
31. Cited in Julia O'Faolain and Lauro Martines, eds., *Not In God's Image: Women in History from the Greeks to the Victorians* (New York: Harper & Row, 1973), p. 6; also Pomeroy, *Goddesses,* pp. 30, 85.
32. Lewis and Reinhold, eds., p. 345.
33. This may have reflected their earlier age at marriage. Cited in Lefkowitz and Fant, pp. 243–44 as four fifths; Lewis and Reinhold, eds., Vol. II, p. 352, say girls received only three fifths of boy's money.
34. Pomeroy, *Goddesses,* p. 68.
35. David Herlihy, "Life Expectancies for Women in Medieval Society," in Rosemarie

Thee Morewedge, ed., *The Role of Women in the Middle Ages* (Albany, N.Y.: SUNY Press, 1975), pp. 8–9.

36. See Sophocles' "Hymn to Man" in *Antigone,* lines 338–70. For discussions of this point, see Arthur, in Bridenthal and Koonz, eds., pp. 65–66; K. J. Dover, "Classical Greek Attitudes to Sexual Behavior," in Peradotto and Sullivan, eds., pp. 148–49.
37. John H. Finley, Jr., trans., *The Complete Writings of Thucydides: The Peloponnesian War* (New York: Random House, 1951), p. 109.
38. Vergil and Juvenal used this argument as well. See Arthur in Bridenthal and Koonz, eds., pp. 82–84.
39. Thanks to David Berger for help with this point. Archer in Cameron and Kuhrt, eds., p. 286, n. 10.
40. Loewe, pp. 28–29, 24.
41. Cited in Judith Hauptman in Ruether, ed., *Religion and Sexism,* p. 192; also see Archer, in Cameron and Kuhrt, eds., pp. 282–84.
42. Drew, *Lombard Laws,* p. 127.
43. *Táin Bó,* p. 251. Her warrior image itself was overlaid by the contemporaneous *Cáin Adomnáin,* a Christian text which argued that the church had improved Ireland by forbidding women to fight in battle. Davies in Cameron and Kuhrt, eds., p. 159.
44. Lefkowitz and Fant, p. 104.
45. Drew, *Lombard Laws,* p. 92.
46. Livy, *History of Rome,* translated by B. O. Foster (Cambridge, Mass.: Loeb Classical Library, Harvard University Press, 1953), p. 62.
47. Homer, *Odyssey,* VI, 287–88.
48. Pomeroy, *Goddesses,* p. 57.
49. Roy Arthur Swanson, trans., *Odi et Amo: The Complete Poetry of Catullus* (Indianapolis: Bobbs Merrill, 1959), p. 62.
50. Deuteronomy 22:21. See Deuteronomy 22 for Hebrew law; relevant Roman law from the Theodosian Code is cited in O'Faolain and Martines, eds., pp. 49–50.
51. Deuteronomy 22:23–29; in Roman law, the girl raped in her own home lost her inheritance, but not her life. O'Faolain and Martines, eds., pp. 49–50.
52. Isaiah 3:16–17; for Rome, see Casper J. Kraemer, Jr., trans., *The Complete Works of Horace* (New York: Random House, 1936), pp. 230–32.
53. For the Lombards, see Drew, *Lombard Laws,* p. 93; for the Burgundians, Katherine Fischer Drew, ed. and trans., *The Burgundian Code* (Philadelphia: University of Pennsylvania Press, 1972), p. 46; for the Visigoths, Heath Dillard, "Women in Reconquest Castile: The Fueros of Sepulveda and Cuenca," in Susan Mosher Stuard, ed., *Women in Medieval Society* (Philadelphia: University of Pennsylvania Press, 1976), p. 92; for the Alamanni, Riché, *World of Charlemagne,* pp. 53, 119; for the Anglo-Saxons, Dorothy Whitelock, ed., *English Historical Documents,* Vol. I (New York: Oxford University Press, 1968), p. 377.
54. See for example, P. G. Foote and D. M. Wilson, *The Viking Achievement* (New York: Praeger, 1970), p. 112.
55. Cited in Lindsay, p. 91; see also Lefkowitz and Fant, p. 120. Surrendering the toys of girlhood in this way was also the custom in Rome. Balsdon, p. 182.
56. Mary Barnard, trans., *Sappho: A New Translation* (Berkeley: University of California

Press, 1958), p. 32; for a literal translation, see David A. Campbell, trans., *Greek Lyric I* (Cambridge, Mass.: Loeb Classical Library, Harvard University Press, 1982), p. 139.

57. Campbell, Vol. I, p. 133.
58. Cited in Myles Dillon and Nora K. Chadwick, *The Celtic Realms* (London: Weidenfeld and Nicolson, 1972), p. 241.
59. For the acceptance of marriage at menarche for girls: Pomeroy, *Goddesses,* p. 68, for Greece; Balsdon, p. 173, for Rome; Loewe, p. 22, for the Hebrews; Drew, *Lombard Laws,* for the Lombards.
60. Drew, *Lombard Laws,* p. 192.
61. Richard McKeon, ed., *The Basic Works of Aristotle* (New York: Random House, 1941), p. 1302.
62. This was in Aramaic. Loewe, p. 23.
63. Hauptman in Ruether, ed., *Religion and Sexism,* p. 185, for the Hebrews; Davies in Cameron and Kuhrt, eds., p. 153, for the Celts; Drew, *Lombard Laws,* p. 197, for the Lombards; Whitelock, Vol. I, p. 151, for the Anglo-Saxons. The influence of Christianity reinforced these concerns: laws of Canute in Whitelock, Vol. I, p. 429, and of Hakon in L. M. Larson, *The Earliest Norwegian Laws* (New York: Columbia University Press, 1935), p. 368.
64. Greek and Roman families provided a dowry, and this Mediterranean custom slowly extended to much of the rest of Europe as well. For a discussion of this complicated subject, see Diane Owen Hughes, "From Brideprice to Dowry in Mediterranean Europe," in Marion A. Kaplan, ed., *The Marriage Bargain: Women and Dowries in European History* (New York: Haworth Press, 1985).
65. For Greece, David Schaps, *The Economic Rights of Women in Ancient Greece* (Edinburgh: University Press, 1979) and Pomeroy, *Goddesses,* pp. 62–65; for Rome, Balsdon, pp. 186–89.
66. Hauptman in Ruether, ed., *Religion and Sexism,* pp. 186–87; Loewe, p. 34.
67. For the "widow's portion," see Hughes in Kaplan, ed., pp. 22–23; for the bride gift, variously called the *wittemon, meta,* or *dos,* see JoAnn McNamara and Suzanne Wemple, "The Power of Women Through the Family in Medieval Europe," in Mary Hartman and Lois W. Banner, eds., *Clio's Consciousness Raised: New Perspectives on the History of Women* (New York: Harper & Row, 1974), pp. 103–104; Jean Markale, *Women of the Celts,* translated by A. Mygind, C. Hauch, and P. Henry (London: Gordon Cremonesi, 1975), pp. 32–33.
68. McNamara and Wemple in Hartman and Banner, eds., pp. 106–107; see also Hughes in Kaplan, ed., on traditions regarding the *Morgengabe* and the bride's rights over property among the Burgundians, Franks, and Visigoths. For a review of Germanic law codes and their impact on women's lives, see also Parts II and IV of Anderson and Zinsser.
69. This would become the "trousseau." See Suzanne Fonay Wemple, *Women in Frankish Society: Marriage and the Cloister, 500 to 900* (Philadelphia: University of Pennsylvania Press, 1981), p. 47.
70. Cited in Richmond Lattimore, *Themes in Greek and Latin Epitaphs* (Urbana, Ill.: University of Illinois Press, 1962), p. 58.
71. Cited in Lewis and Reinhold, eds., p. 285.

72. Cited in Lefkowitz and Fant, p. 136.
73. Semonides, *On Women,* cited in Lefkowitz and Fant, pp. 15–16.
74. Proverbs 31:10–13, 25–29.
75. Cited in Richard Hamer, *A Choice of Anglo-Saxon Verse* (London: Faber and Faber, 1970), p. 99.
76. Pomeroy, *Goddesses,* p. 86; Dover in Peradotto and Sullivan, eds., p. 146.
77. Deuteronomy 22:22.
78. See O'Faolain and Martines, eds., pp. 51, 60.
79. For a summary of all Germanic laws on this subject, see J. R. Reinhard, "Burning at the Stake in Medieval Law and Literature," in *Speculum,* Vol. IV, no. 2 (April 1941), pp. 196–97; for specific tribes: Drew, *Burgundian Code,* p. 68; Drew, *Lombard Law,* p. 93; Whitelock, Vol. I, p. 376, for the Anglo-Saxons; and Katherine Scherman, *The Flowering of Ireland: Saints, Scholars and Kings* (Boston: Little, Brown, 1981), p. 32, for the Celts.
80. For Lucretia, see Livy, pp. 99ff.; the tale of Susannah was included in the book of Daniel as Chapter 13 for many centuries; today it is one of the books of the Apocrypha.
81. For the Hebrews, see Phyllis Bird, "Images of Women in the Old Testament," and Hauptman in Ruether, ed., *Religion and Sexism,* pp. 66, 198.
82. In *Against Naeara,* cited in Pomeroy, *Goddesses,* p. 8, and Sarah B. Pomeroy, *Women in Classical Antiquity: Four Curricular Models* (New York: Hunter College, 1983), p. 28.
83. For Greece, Pomeroy, *Goddesses,* p. 64; for Rome, Balsdon, p. 210, and Pomeroy, *Goddesses,* p. 158; for the Hebrews, Bird in Ruether, ed., *Religion and Sexism,* pp. 52, 62; for the Celts, MacCurtain and Ó'Corráin, pp. 5–6; among the Germanic peoples barrenness could occasion divorce, but it was more like dissolution of the marriage with no punitive aspects. The wife's property and bride price were returned to her. Divorce by mutual consent was also allowed. For specific examples, see Wemple, p. 42, and JoAnn McNamara and Suzanne F. Wemple, "Marriage and Divorce in the Frankish Kingdom," in Stuard, ed., pp. 99–100.
84. I Samuel 1:6, 8.
85. The number of cases from Athens is very small; only three are known so far where the wife initiated divorce. Pomeroy, *Goddesses,* p. 64. Later Greek marriage contracts from first century B.C. Egypt gave the wife the right to divorce if her husband was unfaithful. For an example, see Sarah B. Pomeroy, *Women in Hellenistic Egypt: From Alexander to Cleopatra* (New York: Schocken, 1984), p. 88ff.
86. On this subject, see Judith P. Hallett, *Fathers and Daughters in Roman Society* (Princeton, N.J.: Princeton University Press, 1984).
87. Elizabeth A. Fisher, "Theodora and Antonina in the Historia Arcana," in Peradotto and Sullivan, eds., p. 290.
88. Later laws attempted to help the wife in some cases. See Bird and Hauptman in Ruether, ed., *Religion and Sexism,* pp. 52–53, 189–90.
89. Pomeroy, *Goddesses,* p. 29; Claude Mossé, *The Ancient World at Work.* Translated by Janet Lloyd (New York: Norton, 1969), Chapter 1.
90. *Odyssey,* IV, 131–35.
91. *Iliad,* XII, 425–30.

92. Xenophon, *Memorabilia and Oeconomicus,* translated by E. C. Marchant (Cambridge, Mass.: Loeb Classical Library, Harvard University Press, 1923), p. 153.
93. Lattimore, p. 271.
94. Proverbs 31:19.
95. Mossé, p. 107.
96. Groenewegen-Frankfort and Ashmole, p. 141.
97. *Works and Days,* in Dorothea Wender, trans., *Hesiod and Theognis* (New York: Penguin, 1973), p. 173.
98. From *Politics* in McKeon, *Basic Works,* p. 1277.
99. Soranus, *Gynaecology,* Vol. I, cited in Lefkowitz and Fant, pp. 162–63. For Soranus' account of procedures for a normal birth, see Lefkowitz and Fant, pp. 224–25.
100. Drawn from Thomas Africa, *The Immense Majesty: A History of Rome and the Roman Empire* (Arlington Heights, Ill.: AHM Publishers, 1974), Genealogical Chart 3, p. 414.
101. P. Munz, *Life in the Age of Charlemagne* (New York: Capricorn, 1971); Pauline Stafford, *Queens, Concubines, and Dowagers: The King's Wife in the Early Middle Ages* (Athens, Ga.: University of Georgia Press, 1983), pp. 60–62; Wemple, *Frankish Society,* p. 90.
102. The contemporary sources for Agrippina's life are uniformly hostile. They are Suetonius, *The Twelve Caesars,* and Tacitus, *The Annals,* xii–xiv. For a modern account, see Balsdon, pp. 107–22.
103. Alfred John Church and William Jackson Brodribb, trans., *Complete Works of Tacitus,* edited by Moses Hadas (New York: Modern Library, 1942), p. 130.
104. For examples, see Grace Harriet Macurdy, *Hellenistic Queens: A Study of Woman-Power in Macedonia, Seleucid Syria, and Ptolemaic Egypt* (Baltimore: Johns Hopkins Press, 1932).
105. From Gregory of Tours, *The History of the Franks,* translated by Lewis Thorpe (Baltimore: Penguin, 1977), p. 613. For a fuller account, see Wemple, *Frankish Society.*
106. Pomeroy, *Hellenistic Egypt,* p. 27.
107. Hosea 2:2; Hosea 1:2.
108. Cited in Lefkowitz and Fant, p. 240.
109. Cited in Lindsay, *The Ancient World,* p. 52.
110. Cited in Otto Kiefer, *Sexual Life in Ancient Rome* (London: Abbey Library, 1934), p. 55.
111. Isaiah 1:21.
112. Kiefer, p. 63; Balsdon, p. 224.
113. Other remedies included magic amulets and various potions. For a more complete description, see Norman E. Himes, *Medical History of Contraception* (Baltimore: Williams and Wilkins, 1936), pp. 83–92.
114. Lucian, *Dialogues of the Hetairae* (Cambridge, Mass.: Loeb Classical Library, Harvard University Press, 1961), p. 53.
115. Pomeroy, *Goddesses,* p. 89.
116. Judith Herrin, "In Search of Byzantine Women: Three Avenues of Approach," in Cameron and Kuhrt, eds., p. 167.
117. Charles Diehl, *Byzantine Empresses,* translated by Harold Bell and Theresa de

Kerpely (New York: Alfred A. Knopf, 1963), p. 57; Joseph Dahmus, *Seven Medieval Queens: Vignettes of Seven Outstanding Women of the Middle Ages* (Garden City: Doubleday, 1972), p. 35.

118. Cited in Jennifer S. Uglow, ed., *The International Dictionary of Women's Biography* (New York: Continuum, 1982), p. 464.
119. Cited in Lindsay, pp. 294–95.
120. For this argument, see Fisher in Peradotto and Sullivan, eds., pp. 306–307.
121. For examples: Horace, *Satires,* in Kraemer, p. 10; Constance Carrier, trans., *The Poems of Propertius,* Vol. II (Bloomington, Ind.: Indiana University Press, 1963), p. 92; Marcus Argentarius in Willis Barnstone, trans. and ed., *Greek Lyric Poetry* (New York: Schocken, 1976), p. 214.
122. Horace, *Epodes* 8, in Kraemer, p. 103. For another translation and similar poems by other Roman authors, see Amy Richlin, *The Garden of Priapus: Sexuality and Aggression in Roman Humor* (New Haven: Yale University Press, 1983), Chapter V.
123. See especially Juvenal, Satire 6, *Against Women;* also poems by Martial, Catullus, and Horace.
124. *Theogony,* in Wender, p. 42.
125. *Works and Days,* in Wender, p. 62.
126. Included in the Apocrypha as Ecclesiasticus, or the Wisdom of Jesus the Son of Sirach, Chapter 25:24.
127. Ecclesiasticus 42:14; Ecclesiastes 7:26, 28.
128. For Lilith, see John A. Phillips, *Eve, The History of an Idea* (New York: Harper & Row, 1984), p. 39, and Herbert Spencer Robinson and John Knox Wilson, *Myths and Legends of All Nations* (New York: Bantam, 1961), p. 24.
129. Hesiod in Wender, p. 33.
130. See Olivia Coolidge, *Legends of the North* (Boston: Houghton Mifflin, 1951), passim.
131. Cited in Lefkowitz and Fant, pp. 14–16.
132. Hesiod, *Theogony,* in Wender, p. 43.
133. Eubulus, cited in Lefkowitz and Fant, p. 18.
134. From the *Ecclesiazousae,* cited in Jeffrey Henderson, *The Maculate Muse: Obscene Language in Attic Comedy* (New Haven: Yale University Press, 1975), p. 214.

PART I: 4. Traditions Empowering Women

1. For evidence of goddess worship among the early Hebrews, see Raphael Patai, *The Hebrew Goddess* (New York: Ktav Publishing House, 1967), and Merlin Stone, *When God Was a Woman* (New York: Harcourt Brace Jovanovich, 1976), Chapters, 8–10.
2. Material drawn from Robinson and Wilson, pp. 84, 215; Edith Hamilton, *Mythology* (New York: New American Library, 1942), pp. 43, 331.
3. Athenian women celebrated the Thesmophoria, three days in autumn dedicated to rituals to ensure fertility in the coming year. The Thesmophoria was connected to Demeter and her daughter Koré (Persephone), who were also the focus of the Eleusinian Mysteries, an important Greek cult about which little is known. On this topic, see Pomeroy, *Goddesses,* pp. 77–78. For the Bacchanalia, only fragments of

rituals remain, and the sources that describe them are so hostile as to be suspect. The *maenad,* or woman out of control, fits so well with views seeking to subordinate women to men's control that any correspondence to the real lives of women must be made with caution. For this subject, see Lefkowitz and Fant, pp. 113–14, 250–53; Ruth Padel, "Women: Model for Possession by Greek Daemons," in Cameron and Kuhrt, eds., pp. 3–19, Sheila McNally, "The Maenad in Early Greek Art," and Charles Segal, "The Menace of Dionysus: Sex Roles and Reversals in Euripedes' *Bacchae,*" in Peradotto and Sullivan, eds., pp. 107–42, 195–212.

4. For Diana of Ephesus, see Michael Grant, *History of Rome* (New York: Charles Scribner's Sons, 1978), pp. 32–33; for Ceres, Pomeroy, *Goddesses,* pp. 214–17.
5. R. E. Witt, *Isis in the Greco-Roman World* (Ithaca, N.Y.: Cornell University Press, 1971), Map I.
6. Cited in Arthur in Bridenthal and Koonz, eds., p. 77; Pomeroy, *Goddesses,* p. 218; see pp. 217–26 for a discussion of Isis.
7. Illustrated in John Boardman, *Greek Art,* rev. ed. (New York: Praeger, 1973), pp. 191, 193.
8. Sometimes associated with the "decline" of cities in the Greco-Roman world, priestess-magistrates actually appeared when these cities were at the height of their power. On this subject, and for a detailed description of the priestess Menodora, see Riet Van Bremen, "Women and Wealth," in Cameron and Kuhrt, eds., pp. 233–42.
9. T. Sulimirski, *The Sarmatians* (New York: Praeger, 1970); Mary R. Lefkowitz, "Influential Women," in Cameron and Kuhrt, eds., p. 62, n. 3.
10. On this subject: Abby Wettan Kleinbaum, *The War Against the Amazons* (New York: McGraw-Hill, 1983); Joan Bamberger, "The Myth of Matriarchy: Why Men Rule in Primitive Society," in Rosaldo and Lamphere, eds., pp. 263–80.
11. Cited in Bonnie S. Anderson, "Early English Periodicals for Women," unpublished, p. 11.
12. Judges 4; Judges 5:1–15.
13. Judith 13:6–10, 16:25 in the Apocrypha.
14. Kinsella, p. 208.
15. Graham Webster, *Boudica: The British Revolt Against Rome A.D. 60* (Totowa, N.J.: Rowman and Littlefield, 1978), pp. 87–89.
16. Kleinbaum, p. 33.
17. From his *Life of Mark Antony,* cited in Lefkowitz and Fant, pp. 150–51.
18. Cited in Lefkowitz and Fant, p. 150.
19. Material on Cleopatra drawn from Michael Grant, *Cleopatra* (New York: Simon and Schuster, 1972), and Pomeroy, *Hellenistic Egypt,* pp. 24–28.
20. For Galla Placidia and Pulcheria, see Steward Irvin Oost, *Galla Placidia Augusta: A Biographical Essay* (Chicago: University of Chicago Press, 1968), and Eleanor Shipley Duckett, *Medieval Portraits from East and West* (Ann Arbor, Mich.: University of Michigan Press, 1972).
21. For Athens, Pomeroy, *Goddesses,* p. 61; for the Hebrews, Hauptman in Ruether, ed., *Religion and Sexism,* pp. 194–95; for Rome, Balsdon, p. 222; for the Celts, Markale, p. 291, and David Herlihy, "The Structure of the Early Medieval Family: Evidence from Hagiography," unpublished manuscript, 1980; for Germanic cul-

tures, Wemple, *Frankish Society,* p. 46; Drew, *Lombard Laws,* pp. 145–46; Drew, *Burgundian Codes,* pp. 32–33, 67–69.

22. From *The History of Rome,* cited in Lefkowitz and Fant, p. 177.
23. For these developments, see John Crook, *Law and Life of Rome* (Ithaca, N.Y.: Cornell University Press, 1967), pp. 114–15; D. G. Cracknell, *Law Students' Companion No. 4: Roman Law* (London: Butterworths, 1964), p. 43.
24. Michele D'Avino, *The Women of Pompeii,* translated by Monica Hope Jones and Luigi Nusco (Napoli: Loffredo, 1967), pp. 21, 85; for Celtic and Germanic women's control over their own property, see Markale, p. 37, and MacCurtain and Ó'Corráin, for the Celts; Drew, *Lombard Laws,* pp. 32, 92; Drew, *Burgundian Codes,* p. 85, and David Herlihy, "Land, Family and Women in Continental Europe, 701–1200," in Stuard, ed., p. 32; for Vikings, Larson, pp. 118, 339, 368.
25. Cited in Lefkowitz and Fant, p. 159.
26. Macurdy, p. 20. For a more literal translation, which implies that Eurydice learned to read *because* she was the mother of growing sons, see Lefkowitz and Fant, p. 235.
27. See Lattimore, p. 276ff.; Lefkowitz and Fant, pp. 161–62.
28. Natalie Kampen, "Hellenistic Artists: Female," in *Archeologia Classica,* Vol. XXVII (1975), pp. 9–17.
29. Pliny the Elder, cited in Lefkowitz and Fant, p. 168.
30. Cited in Lefkowitz and Fant, p. 68.
31. For the contradictions in Plato's attitude toward women, see Dorothea Wender, "Plato: Misogynist, Paedophile, and Feminist," in Peradotto and Sullivan, eds., pp. 213–28.
32. For minor women philosophers, see Pomeroy, *Hellenistic Egypt,* pp. 65–68.
33. Cited in Lefkowitz and Fant, pp. 25, n. 7, and p. 23.
34. Cited in Sondra Henry and Emily Taitz, *Written out of History: Our Jewish Foremothers,* 2nd ed. rev. (Fresh Meadows, N.Y.: Biblio Press, 1983), p. 57; also Hauptman in Ruether, ed., *Religion and Sexism,* p. 203. Beruriah's horrible death shows the persistence of misogyny: Her husband sought to test her by having one of his students attempt to seduce her; dismayed by her attraction to him, Beruriah killed herself.
35. From W. R. Paton, ed., *Greek Anthology* (Cambridge, Mass.: Loeb Classical Library, Harvard University Press, 1916), p. 79. The writings of four women are included in the *Greek Anthology,* also called the *Palatine Anthology,* a collection of poems that reached its final form in the tenth century A.D.
36. From Paton, p. 87. Lefkowitz and Fant also include women's epitaphs composed by Anyte and Nossis, pp. 8–10.
37. Constance Carrier, trans., *The Poems of Tibullus* (Bloomington, Ind.: University of Indiana Press, 1968), p. 104. Sulpicia's poems are usually included in Book 3 of Tibullus. For a literal, prose translation of them, see Lefkowitz and Fant, pp. 131–32.
38. Cited in David A. Campbell, trans., *Greek Lyric, Vol. I* (Cambridge, Mass.: Loeb Classical Library, Harvard University Press, 1982), p. 49. Until this new, authoritative translation, work on Sappho in English was largely based on J. M. Edmond's edition of 1928, which added in a great deal of material now known to not be written by Sappho. The poetic translations of both Mary Barnard, *Sappho* (1958), and Willis

Barnstone, *Greek Lyric Poetry* (1972), are based on this early edition and must be used with care; many of the verses and fragments attributed to Sappho were not written by her. The following discussion relies on Campbell's literal translation, but lines have been divided to reflect Greek stanza form.

39. Campbell, p. 149; authors' line divisions. Biographical information from Campbell, pp. x–xii.
40. Cited in Campbell, p. 141; ix–x.
41. Campbell, p. 99.
42. Campbell, pp. 54–55.
43. Campbell, p. 131; authors' line division.
44. Campbell, p. 133.
45. Campbell, p. 131; authors' line division.
46. Campbell, p. 55; authors' line divisions, drawn in part from Lefkowitz and Fant, p. 4.
47. Campbell, pp. xi–xii.
48. Campbell, pp. 93, 101.
49. Campbell, pp. 80–81.
50. See Barnstone, p. 67, and Campbell, p. 67, authors' line division.
51. Campbell, p. 159.
52. Campbell, p. 39.

PART I: 5. The Effects of Christianity

1. John 4:57.
2. Matthew 5:5, 5:10; Luke 6:20–26; Mark 3:32–35; Luke 8:19–21. The four Gospels: Matthew, Mark, Luke, and John, are the best known sources for early Christian beliefs and practices. Recorded in the first century A.D. from the selective memory of Jesus' male followers, these and other books, such as the Acts and Letters of the apostles and Revelation, were the writings collected together as the books of the New Testament when Jerome created the Latin Vulgate at the end of the fourth century. This remained the basis of the authorized Catholic Bible into the seventeenth century. For interpretations of Jesus' ministry and the empowering character of early Christianity with respect to women, see Joan Morris, *The Lady Was a Bishop* (New York: Macmillan, 1973); Parvey, in Ruether, ed., *Religion and Sexism;* Rosemary Radford Ruether, *New Woman, New Earth: Sexist Ideologies and Human Liberation* (New York: The Seabury Press, 1975); Charles Caldwell Ryrie, *The Place of Women in the Church* (New York: Macmillan, 1958); George H. Tavard, *Women in the Christian Tradition* (Notre Dame, Ind.: Notre Dame Press, 1973).
3. Matthew 21:31.
4. Mark 12:41–44.
5. Matthew 13:31–33; also Luke 13:18–21.
6. Matthew 25:1–13.
7. John 16:20–22.
8. He also rebuked his disciples for criticizing her. The bleeding rendered her ritually "unclean" in contemporary Hebrew practice. Mark 5:25–26; also Matthew 9:20–22, Luke 8:43–48.
9. John 8:7.

10. John 11:25–26.
11. Luke 10:38–42.
12. Matthew 27:55; Mark 15:40; Luke 24:10; John 19:25.
13. Mark 8:2; Luke 8:2, 24:1–24.
14. This era is known as that of the Apostolic Church, the period of leadership by those who had been his male disciples and witnessed the resurrection.
15. See Ryrie, p. 35, and Luke 8:3.
16. Acts 9:36–42; also see Elisabeth Schüssler Fiorenza, "Word, Spirit and Power: Women in Early Christian Communities," in Rosemary Ruether and Eleanor McLaughlin, eds., *Women of Spirit: Female Leadership in the Jewish and Christian Traditions* (New York: Simon and Schuster, 1979), p. 35.
17. Galatians 3:28.
18. Fiorenza in Ruether and McLaughlin, eds., p. 36.
19. Philippians 4:3; Romans 16:1–2; Morris, pp. 119–20; Fiorenza in Ruether and McLaughlin, eds., pp. 35–36.
20. Romans 16:4; Acts 18:2–3, 18, 26.
21. Cited in Elaine Pagels, *The Gnostic Gospels* (New York: Random House, 1979), p. xvii. On Gnostic Christianity and women, see Pagels, especially Chapter III; Fiorenza in Ruether and McLaughlin, eds., pp. 44–51, and Parvey in Ruether, ed., *Religion and Sexism,* pp. 121–22.
22. Carolly Erickson, *The Medieval Vision* (New York: Oxford University Press, 1976), p. 186.
23. Cited in Fiorenza, in Ruether and McLaughlin, eds., pp. 41–42.
24. Marina Warner, *Alone of All Her Sex: the Myth and Cult of the Virgin* (New York: Pocket Books, 1978), p. 69. The emperors particularly known for the Edicts of Persecution are Antoninus Pius, Marcus Aurelius, Septimus Severus, and Diocletian.
25. Cited in Lefkowitz and Fant, p. 266.
26. In this era, Christians were often baptized close to the end of life, since it was believed that baptism washed away all previous sins.
27. Cited in Lefkowitz and Fant, pp. 269–70. Those who survived the arena left through the "Gate of Life."
28. From W. H. Shewring, trans., *The Passion of SS Perpetua and Felicity* (London: Sheed and Ward, 1931), pp. 38ff.
29. Scherman, p. 115.
30. See Balsdon, pp. 166–70. The historian Steven Runciman described her as one of the "most successful of the world's archeologists" for her efforts at the identification and sanctification of places in the Holy Land. See Steven Runciman, *A History of the Crusades,* Vol. I (Cambridge: Cambridge University Press, 1962), p. 39.
31. See Diehl, and Duckett, for the empresses; also Rosemary Ruether, "Mothers of the Church: Ascetic Women in the Late Patristic Age," in Ruether and McLaughlin, eds., pp. 71–98.
32. Gregory of Tours, pp. 141–42.
33. For tales of the first Christians in England, see Bede, *A History of the English Church and People,* translated by Leo Sherley-Price (Baltimore: Penguin, 1960), pp. 68–70, 113–20.
34. Morris, p. 12.

35. For the evolution of female religious orders: Mary Bateson, *Origin and Early History of Double Monasteries*, Royal Historical Society Transactions, New Series No. 13, (London, 1899); Lina Eckenstein, *Women Under Monasticism* (New York: Russell and Russell, 1963); Penelope Johnson, "Equal in the Order of Grace," unpublished article.
36. See Jean Daniélou, *The Ministry of Women in the Early Church* (London: Faith Press, 1961) and Ryrie.
37. For Paula and her circle, see Ruether in Ruether, ed., *Religion and Sexism*, pp. 175–80; quotation from p. 173.
38. Cited in Robert P. B. Moynihan, "Women's Spirituality in the Middle Ages," paper delivered at Yale University, April 4, 1981, p. 11. For the lives of Paula and Eustochium, also see JoAnn McNamara, "Cornelia's Daughters: Paula and Eustochium," unpublished manuscript, and Rosemary Ruether and Eleanor McLaughlin, "Women's Leadership in the Jewish and Christian Traditions: Continuity and Change," in Ruether and McLaughlin, eds., pp. 15–28.
39. Wemple, *Frankish Society*, p. 156.
40. For England, Eckenstein, p. 117; for the Franks, Wemple, *Frankish Society*, p. 156; for Saxony, Anne Lyon Haight, *Hrotswitha of Gandersheim: Her Life, Times and Works* (New York: The Hrotswitha Club, 1965), p. 4. JoAnn McNamara has argued that these endowments were a way to protect the wealth of a family from attack. The land went to the Church but could remain under indirect control of family members, since they often chose the rulers of the communities. JoAnn McNamara, "Female Monasteries and the Distribution of Wealth in the Early Middle Ages," Fordham University, March 21, 1986.
41. Cited in Susan Raven and Alison Weir, *Women of Achievement: Thirty-Five Centuries of History* (New York: Harmony Books, 1981), pp. 109–10.
42. Cited in Moynihan, p. 8.
43. Cited in Elizabeth Clark and Herbert Richardson, *Women and Religion: A Feminist Sourcebook of Christian Thought* (New York: Harper & Row, 1977), p. 61.
44. Cited in JoAnn McNamara, *A New Song: Celibate Women in the First Three Centuries of Christianity* (New York: Haworth Press, 1983), p. 130.
45. I Corinthians 14:34–35. This statement may be a later interpolation, not from Paul himself. Private communication from Rosemary Ruether.
46. I Corinthians 11:7–8.
47. I Timothy 2:11–13. This epistle is probably not by Paul, but by a later, first-century disciple.
48. I Peter 3:3–5. Again, the authorship of this epistle is subject to argument: it may have been written by a follower rather than by the apostle Peter himself.
49. I Peter 3:3–5.
50. I Timothy 3; Titus 1:5–9.
51. Cited in Prusak in Ruether, ed., *Religion and Sexism*, p. 103. For Origen, see Tavard, p. 68.
52. The *Secrets of Enoch*, the *Apocalypse of Moses* and the *Books of Adam and Eve* are Hebrew pseudepigrapha, which emphasize this. On this subject, see Prusak in Ruether, ed., *Religion and Sexism*, pp. 93–96, and Phillips, Chapters 3–5.

53. Cited in Prusak and Ruether in Ruether, ed., *Religion and Sexism,* pp. 105, 157. Translations combined.
54. Cited in Duckett, p. 40.
55. Saint Augustine, *The City of God,* edited by Vernon J. Bourke (Garden City, N.Y.: Doubleday, 1962), p. 278.
56. See Joan M. Ferrante, *Woman as Image in Medieval Literature from the Twelfth Century to Dante* (New York: Columbia University Press, 1975), p. 2.
57. Matthew 5:27–28.
58. Matthew 22:30, Luke 20:35, Matthew 19:21.
59. I Corinthians 7:1.
60. Cited in Ruether in Ruether, ed., *Religion and Sexism,* p. 16.
61. Cited in G. G. Coulton, *Life in the Middle Ages,* Vol. IV (New York: Macmillan, 1930), p. 11.
62. Cited in Morris, p. 111.
63. See Clara Maria Henning, "Canon Law and the Battle of the Sexes," in Ruether, ed., *Religion and Sexism,* p. 278, and "Theodore's Penitential" in John T. McNeill and Helena M. Gamer, eds., *Handbooks of Penance* (New York: Columbia University Press, 1938), p. 205.
64. For Isidore, Morris, p. 111; for Gregory the Great, Bede, pp. 78–79. See also Charles T. Wood, "The Doctors' Dilemma: Sin, Salvation and the Menstrual Cycle in Medieval Thought," *Speculum,* vol. 56 (1981), pp. 710–27.
65. Jewish law allowed women to enter the temple after seven and fourteen days respectively, but considered them "unclean" for the longer period. Gregory the Great's letter to the English Archbishop Augustine speaks to the Christian practice of forbidding women to enter the church for those specified times after childbirth. He wished them admitted. See Bede, pp. 76–77.
66. All of the major Gnostic beliefs and practices including those concerning women—the view of Eve as heroine, priestly power granted to female believers, and veneration of female aspects of the deity—would become buried traditions revived by European women and men in later eras. They resurfaced as the heresies of the twelfth century and after. On the Gnostics, see Pagels, and the summaries in Fiorenza in Ruether and McLaughlin, eds., and Parvey in Ruether, ed., *Religion and Sexism.* For descriptions of the way in which Gnostic Gospels either became designated as apocrypha or were rejected altogether in the second and third centuries, see Pagels, and P. R. Ackroyd and C. F. Evans, eds., *Cambridge History of the Bible,* Vol. I (New York: Cambridge University Press, 1976), pp. 297ff.
67. From the "Apostolic Church Order," cited in Fiorenza in Ruether and McLaughlin, eds., p. 56.
68. The "Teaching of the Apostles," cited in Daniélou, pp. 11, 19.
69. I Timothy 2:15.
70. Cited in Tavard, p. 107.
71. Cited in Tavard, pp. 89, 88, 89.
72. Cited in Wemple, *Frankish Society,* p. 22.
73. Matthew 19:5–6, 9. Divorce for infidelity on the part of the wife was allowed, but the husband was prohibited from remarrying. If he did, he too was considered an adulterer.

74. See Tavard, p. 60, and Herlihy, "Medieval Family," p. 3.
75. For examples: see "Theodore's Penitential," Chapters XII and XIV, and the "Judgement of Clement," in McNeill and Gamer, *Handbooks*; Anglo-Saxon laws in Whitelock, Vol. I, p. 362; Charlemagne's laws in Wemple, *Frankish Society,* p. 78; the Viking Codes in Larson, p. 249.
76. See the "Penitentials" in McNeill and Gamer, pp. 105, 184–85.
77. Cited in John T. Noonan, Jr., *Contraception: A History of Its Treatment by the Catholic Theologians and Canonists* (Cambridge, Mass.: The Belknap Press, Harvard University, 1965), p. 130.
78. See McNeill and Gamer, p. 166.
79. For a summary of the Church's teachings on these subjects, see Noonan, pp. 114–21; on abortion, also see Michel Riquet, "Christianity and Population," in Orest and Patricia Ranum, eds., *Popular Attitudes Toward Birth Control in Pre-Industrial France and England* (New York: Harper & Row, 1972), pp. 26–28.
80. Revelations 14:4–5.
81. Cited in John T. Noonan, Jr., "Power to Choose," in "Marriage in the Middle Ages," edited by John Leyerle, *Viator,* Vol. 4 (1973), p. 499. For this argument about female virginity, see JoAnn McNamara, "Sexual Equality and the Cult of Virginity in Early Christian Thought," *Feminist Studies,* Vol. III, nos. 3–4 (Spring-Summer 1976), pp. 153–54, and Ruether in Ruether, ed., *Religion and Sexism,* p. 160.
82. Cited in Ruether in Ruether, ed., *Religion and Sexism,* p. 159.
83. On this point see Ruether in Ruether, ed., *Religion and Sexism,* p. 165.

PART II WOMEN OF THE FIELDS: SUSTAINING THE GENERATIONS

PART II: 1. The Constants of the Peasant Woman's World: The Ninth to the Twentieth Centuries

1. Lynn White, Jr., *Medieval Technology and Social Change* (New York: Oxford University Press, 1973), p. 39; estimates vary from country to country; as low as 65 percent in 1750 England, as high as 76.5 percent in 1930 Yugoslavia. It is in the eighteenth century that the shift from rural to urban living begins to have significance for large numbers of European women.
2. Female illiteracy always outnumbered male; in the more remote rural areas of Greece, Portugal, Spain, and Yugoslavia, most countrywomen could not read into the 1960s. See Phyllis Stock, *Better Than Rubies: A History of Women's Education* (New York: G. P. Putnam's Sons, 1978), p. 225.
3. For this part the following books have proved the most useful sources:

   Ireland: Conrad Arensberg, *The Irish Countryman* (Garden City: The Natural History Press, 1968 ed.).

   England: Mary Chamberlain, *Fenwomen: A Portrait of Women in an English Village* (London: Virago Press Limited, 1977).
   Dorothy Hartley, *Lost Country Life* (New York: Pantheon Books, 1979).

George C. Homans, *English Villagers of the Thirteenth Century* (New York: W. W. Norton & Company, Inc., 1975).
G. R. Quaife, *Wanton Wenches and Wayward Wives: Peasants and Illicit Sex in Early Stuart England* (New Brunswick, N.J.: Rutgers University Press, 1979).

Denmark: Inga Dahlsgård, *Women in Denmark Yesterday and Today* (Copenhagen: Det Danske Selskab, 1980).

France: Marie Allauzen, *La Paysanne française aujourdhui* (Paris: Editions Gonthier, 1967).
Georges Duby, *Rural Economy and Country Life in the Medieval West,* translated by Cynthia Postan (Columbia, S.C.: University of South Carolina Press, 1968).
Pierre Jakez Hélias, *The Horse of Pride: Life in a Breton Village,* translated by June Guicharnaud (New Haven: Yale University Press, 1978).
Emmanuel LeRoy Ladurie, *Montaillou: The Promised Land of Error,* translated by Barbara Bray (New York: George Braziller Inc., 1978).
Martine Segalen, *Love and Power in the Peasant Family: Rural France in the Nineteenth Century,* translated by Sarah Matthews (Chicago: The University of Chicago Press, 1983).

Spain: William A. Christian, *Person and God in a Spanish Village* (New York: Seminar Press Inc., 1972).

Italy: Ann Cornelisen, *Women of the Shadows* (Boston: Little, Brown and Company, 1976).
Eliza Putnam Heaton, *By-Paths in Sicily* (New York: E. P. Dutton & Company, 1920).
Frank McArdle, *Altopascio: A Study in Tuscan Rural Society 1587–1784* (New York: Cambridge University Press, 1978).

Greece: Ernestine Friedl, *Vasilika: A Village in Modern Greece* (New York: Holt, Rinehart and Winston, 1962).

Yugoslavia: Vera St. Erlich, *Family in Transition: A Study of 300 Yugoslav Villages* (Princeton: Princeton University Press, 1966).

Soviet Union: Dorothy Atkinson, Alexander Dallin, and Gail Warshofsky Lapidus, eds., *Women in Russia* (Stanford, Calif.: Stanford University Press, 1977).
Mary Matossian, "The Peasant Way of Life," in Wayne S. Vucinich ed., *The Peasant in Nineteenth Century Russia* (Stanford, Calif.: Stanford University Press, 1968).
David L. Ransel, ed., *The Family in Imperial Russia* (Urbana: University of Illinois Press, 1978).

4. Cornelisen, *Shadows,* p. 123.
5. Cornelisen, *Shadows,* p. 154.
6. Joan Scott and Louise Tilly, "Women's Work and the Family in Nineteenth Century Europe," in Charles E. Rosenberg, ed., *The Family in History* (Philadelphia: University of Pennsylvania Press), 1975, p. 159, n. 44.
7. Arensberg, *Countryman,* p. 57.

8. William Langland, *Piers the Ploughman,* translated by J. F. Goodridge (Baltimore: Penguin Books, 1975 edition), p. 83.
9. Langland, Appendix B, p. 260; Hélias, p. 20; see Segalen, for a peasant woman's day in Tarn-et-Garonne, France, 1945—up with her daughter-in-law at four, to bed at midnight—pp. 101–102.
10. Antonina Martynova, "Life of the Pre-Revolutionary Village as Reflected in Popular Lullabies," in Ransel, ed., p. 176.
11. Langland, p. 125.
12. Cornelisen, *Shadows,* p. 122.
13. "Eve's Unequal Children," in Frances T. Magoun, Jr., and Alexander H. Krappe, trans., *Grimm's German Folk Tales* (Carbondale: Southern Illinois University Press, 1960), p. 585.
14. Martynova in Ransel, ed., p. 181.
15. See Magoun and Krappe, *Grimm's Folk Tales,* p. 565.
16. Barbara Evans Clements, "Working-Class and Peasant Women in the Russian Revolution, 1917–1923," *Signs,* Vol. VIII, no. 2 (Winter 1982), p. 220.
17. Chamberlain, *Fenwomen,* p. 28.
18. Chamberlain, *Fenwomen,* p. 28.
19. For example, see Fernand Braudel, *Capitalism and Material Life 1400–1800,* translated by Miriam Kochan (New York: Harper & Row, Publishers, 1973), p. 198.
20. For the continuity of design, see, for example, Maurice Beresford and John G. Hurst eds., *Deserted Medieval Villages* (New York: St. Martin's Press, 1971), pp. 104–105; G. G. Coulton, *Medieval Panorama: The English Scene from Conquest to Reformation* (Cambridge: Cambridge University Press, 1939), p. 69; Ladurie, *Montaillou,* p. 40; Mary Matossian in Vucinich, ed., p. 17.
21. The best descriptions of the evolution of the European landscape and relative continuity in the patterns of settlement (with population density the principal change through the centuries) come in two books by Georges Duby: *Rural Economy,* and *The Early Growth of the European Economy: Warriors and Peasants from the Seventh to the Twelfth Century,* translated by Howard B. Clarke (Ithaca: Cornell University Press, 1974). For specific examples cited: Beresford and Hurst, p. 117; Jean-Noël Biraben, "A Southern French Village: The Inhabitants of Montplaisant in 1644," in Peter Laslett and Richard Wall, eds., *Household and Family in Past Time* (New York: Cambridge University Press, 1974), p. 240; Friedl, p. 7. Anthropologists have distinguished variations in the broad patterns. For a useful summary see Conrad Arensberg, "The Old World Peoples: The Place of Old World Cultures in World Ethnography," *Anthropology Quarterly,* vol. 36 (1963), pp. 75–99.
22. Duby, *Rural Economy,* p. 375.
23. Hélias, p. 239.
24. Georges Duby in his two books on medieval rural economy describes the manner in which peasants increased the yield on their lands. In the ninth century more than 50 percent of the harvest would have been set aside for seeds for the next planting. By the thirteenth century the yield had increased to 4:1—four seeds harvested for every one planted—giving the peasant woman and her husband double the surplus. The increase came with changes over the three centuries: manuring the fields, three

field rotation, a metal blade plow that gave access to more diverse soils, that made deeper furrows, thus improving the drainage, a horse that, with an improved harness, could work twice as fast as the four-ox team. Still there could be fluctuations and famine. Only with the scientific and technological improvements of the eighteenth century would hazards be minimized and the yield increased again, to five times that of the earlier centuries. See Duby, *Rural Economy,* pp. 100–103. See also Fernand Braudel, *The Structures of Everyday Life: Civilization and Capitalism 15th–18th Century,* translated by Sîan Reynolds (New York: Harper & Row, Publishers, 1981). This is the revised version of his *Capitalism and Material Life 1400–1800.* Chapters 2 and 3 are concerned with cultivation practices, foods, and diet. Chapter 5 details changes in technology.

25. Olwen Hufton, *The Poor of Eighteenth-Century France 1750–1789* (Oxford: Clarendon Press, 1974), p. 44.
26. Hélias, p. 241.
27. Coulton, *Panorama,* p. 75.
28. Dahlsgård, p. 67.
29. Hélias, p. 245.
30. Homans, p. 358.
31. Duby, *Rural Economy,* pp. 29–30.
32. Doris Stenton, *The English Woman in History* (New York: The Macmillan Company, 1957), pp. 89–90.
33. See for example, Duby, *Rural Economy,* p. 365.
34. Segalen, p. 31.
35. Geoffrey Grigson, ed., *The Penguin Book of Ballads* (Baltimore: Penguin Books, Inc., 1975), pp. 46–47.
36. For this description of flax cultivation and preparation, see Jane Schneider, "Rumpelstiltskin Revisited: Some Affinities Between Folklore and the Merchant Capitalist Intensification of Linen Manufacture in Early Modern Europe" (New York: unpublished manuscript, 1985).
37. Jane Schneider, "Trousseau as Treasure: Some Contradictions of Late Nineteenth-Century Change in Sicily," Eric B. Ross, ed., *Beyond the Myths of Culture: Essays in Cultural Materialism* (New York: Academic Press Inc., 1980), p. 335.
38. White, p. 219.
39. Heaton, p. 339.
40. Stenton, p. 82.
41. Stenton, pp. 87, 90.
42. J. L. Flandrin, *Families in Former Times: Kinship, Household and Sexuality,* translated by Richard Southern (New York: Cambridge University Press, 1979), p. 102.
43. Matossian in Vucinich, ed., p. 34. See also G. G. Coulton, *The Medieval Village* (Cambridge: University Press, 1926), p. 359.
44. See for example, Segalen, p. 87.
45. Gladys Scott Thomson, *Life in a Noble Household* (Ann Arbor: University of Michigan Press, 1959), p. 137.
46. See Darrel W. Amundsen and Carol Jean Diers, "The Age of Menarche in Medieval Europe," *Human Biology,* vol. 45, no. 3 (September 1973) for discussion of previous suggestions that age of menarche varied.

47. See Darrel W. Amundsen and Carol Jean Diers, "The Age of Menopause in Medieval Europe," *Human Biology,* vol. 45, no. 4 (December 1973).
48. See E. A. Wrigley, *Population and History* (New York: McGraw-Hill Book Company, 1973), for the reasoning that leads to such exact estimates.
49. Granger Ryan and Helmut Ripperger, trans., *The Golden Legend of Jacobus de Voragine* (New York: Longmans, Green and Co., 1941), p. 323.
50. Coulton, *Village,* p. 182.
51. *Golden Legend,* p. 323.
52. Cornelisen, *Shadows,* p. 130.
53. Information about childbirth practices provided by the nurse-midwifery staff of the Maternity Center Associates, 48 E. 92nd Street, New York, N.Y.
54. Mary Chamberlain, *Old Wives' Tales: Their History, Remedies and Spells* (London: Virago Press Limited, 1981), p. 240.
55. Wrigley, p. 93.
56. See Roger Mols, "Population in Europe 1500–1700," in Carlo M. Cipolla, ed., *The Fontana Economic History of Europe: The Sixteenth and Seventeenth Centuries* (Glasgow: William Collins Sons & Co., 1976), p. 68; Quaife, *Wenches,* p. 78; Pierre Goubert, "Recent Theories and Research in French Population Between 1500 and 1700," in D. V. Glass and D. E. Eversley, eds., *Population in History* (London: Edward Arnold, 1965), p. 470.
57. David Hunt, *Parents and Children in History: The Psychology of Family Life in Early Modern France* (New York: Harper & Row, Publishers, 1972), p. 142.
58. See James Bruce Ross and Mary Martin McLaughlin, eds., *The Portable Medieval Reader* (New York: The Viking Press, 1966), p. 410.
59. Mary Collier, *The Poems of Mary Collier, The Washerwoman of Petersfield* (Petersfield, Hants.: W. Minchin, 1739), p. 5.
60. St. Erlich, p. 52.
61. St. Erlich, p. 279.
62. *Grimm's Folk Tales,* p. 425.
63. Kate Campbell Hurd-Mead, *A History of Women in Medicine from the Earliest Times to the Beginning of the Nineteenth Century* (Haddam, Conn.: The Haddam Press, 1938), p. 400.
64. See Chamberlain, *Remedies,* pp. 164–66.
65. Chamberlain, *Remedies,* p. 168.
66. Duby, *European Economy,* p. 29.
67. For discussion of child mortality: Goubert, "Recent Theories," in Glass and Eversley, eds.; Peter Laslett, *The World We Have Lost* (New York: Charles Scribner's Sons, 1965); J. C. Russell, "Effects of Pestilence and Plague 1315–1385," in *Comparative Studies in Society and History,* Vol. VIII, no. 4 (July 1966).
68. Grigson, ed., *Ballads,* p. 75.
69. *Grimm's Folk Tales,* p. 400.
70. Pierre Goubert, "Life and Death in the Peasant Village: A Regional Case Study of the Seventeenth Century Peasantry" in Isser Woloch, ed., *The Peasantry in the Old Regime: Conditions and Protests* (Huntington, N.Y.: Robert E. Krieger Publishing Company, 1977), p. 36, and Pierre Goubert, "The French Peasantry of the Seven-

teenth Century: A Regional Sample" in Trevor Aston, ed., *Crisis in Europe 1560–1660* (New York: Basic Books, Inc., Publishers, 1965), pp. 165–66.
71. See for example, Langland, p. 128.
72. Duby, *European Economy,* p. 8.
73. Studies show declining yields all across north, central and eastern Europe, anywhere from 8 percent to 55 percent from 1500 to 1750. See Aldo de Maddalena, "Rural Europe 1500–1750," translated by Muriel Grindrod, in Cipolla, ed., *Sixteenth Century,* pp. 337, 338, 341.
74. Laslett, *World Lost,* p. 115.
75. See Goubert in Aston, ed., p. 169. The potato blight from 1845 to 1849 in Ireland wiped out the poorest peasant class, the cotters, and the Russian famine of 1891 took 700,000 lives. Yves-Marie Bercé, "Peasant Rights and Rural Unrest," *History Today* (June 1982), p. 39.
76. St. Erlich, p. 51.
77. Laslett, *World Lost,* pp. 112–13.
78. Cited in Segalen, p. 63. The demographer J. C. Russell estimates half of the deaths in this age group came from this disease; see J. C. Russell, *Late Ancient and Medieval Population* (Philadelphia: The American Philosophical Society), 1958, p. 45.
79. Hélias, p. 245.
80. On background circumstances, see, for example, David Herlihy, "Population, Plague and Social Change in Rural Pistoia 1201–1430," *The Economic History Review,* 2nd series, Vol. XVIII, no. 2 (August 1965), p. 242.
81. William Woods, *England in the Age of Chaucer* (New York: Stein and Day Publishers, 1976), p. 48.
82. See Duby, *Rural Economy,* p. 308, Russell, *Medieval Population,* pp. 55–56 and J. C. Russell, "Population in Europe 500–1500," in Carlo M. Cipolla, ed., *The Fontana Economic History of Europe: The Middle Ages* (Glasgow: William Collins Sons and Co. Ltd., 1975), p. 41.
83. William H. McNeill, *Plagues and Peoples* (Garden City, N.Y.: Anchor Press, 1976), p. 149.
84. Woods, p. 50.
85. Duby, *Rural Economy,* p. 520.
86. Herlihy, "Plague," p. 71.
87. Gene Brucker, ed., *The Society of Renaissance Florence* (New York: Harper & Row Publishers, 1971), pp. 105–106. For seventeenth- and eighteenth-century examples, see Cissie Fairchilds, "Female Sexual Attitudes and the Rise of Illegitimacy: A Case Study," *The Journal of Interdisciplinary History,* Vol. VIII, no. 4 (Spring 1978), pp. 627–67.
88. Andreas Cappellanus, *The Art of Country Love,* translated by John Jay Parry (New York: W. W. Norton & Company, Inc., 1969), p. 150.
89. Cited in Fairchilds, "Female Sexual Attitudes," p. 646.
90. Cited in M. H. Keen, *The Laws of War in the Late Middle Ages* (Toronto: University of Toronto Press, 1965), p. 191.
91. James Bruce Ross and Mary Martin McLaughlin, eds., *The Portable Renaissance Reader* (New York: Penguin Books, 1980), pp. 214–15.

92. C. V. Wedgwood, *The Thirty Years War* (Garden City, N.Y.: Doubleday & Company, Inc., 1961), p. 199.
93. Wedgwood, p. 496.
94. Wedgwood, p. 401.
95. Cited in G. P. Gooch, *Maria Theresa and Other Studies* (New York: Longmans, Green and Co., 1952), p. 92.
96. See for example: Richard Johnson, "The Role of Women in the Russian Civil War," *Conflict*, vol. 2, no. 2 (1980), p. 206.
97. See Iris Origo, *War in Val D'Orcia* (Boston: David R. Godine Publisher Inc., 1984).
98. Cited in Jane Kramer, *Unsettling Europe* (New York: Random House, 1980), p. 52.
99. Cited in William M. Mandel, *Soviet Women* (Garden City, N.Y.: Doubleday & Company Inc., 1975), p. 159.

PART II: **2. Sustaining the Generations**

1. Marc Bloch, *Feudal Society*, translated by L. A. Manyon (Chicago: University of Chicago Press, 1966), Vol. I, p. 138.
2. Segalen, p. 24.
3. Cited in St. Erlich, p. 103.
4. Cited in Schneider, in Ross, ed., p. 341.
5. Cited in Arensberg, *Countryman*, p. 62.
6. Jean Scammell, "Freedom and Marriage in Medieval England," *Economic History Review*, 2nd Series, vol. 27, no. 4 (November 1974), p. 533.
7. See Duby, *Rural Economy*, p. 223, and Woods, *Age of Chaucer*, p. 28.
8. Thomas Kinsella, trans., *The Tain Bó* (London: Oxford University Press, 1969), p. 38.
9. Coulton, *Village*, p. 467.
10. Coulton, *Village*, p. 465.
11. Coulton, *Village*, pp. 467–68.
12. Homans, p. 141.
13. Duby, *Rural Economy*, pp. 283–84.
14. See David Levine, *Family Formation in an Age of Nascent Capitalism* (Toronto: Ontario Institute for Studies in Education, 1977), p. 29; Alan Macfarlane, *The Family Life of Ralph Josselin: A Seventeenth Century Clergyman, An Essay in Historical Anthropology* (Cambridge: Cambridge University Press, 1970), p. 93.
15. Schneider in Ross, ed., p. 232.
16. Hasia R. Diner, *Erin's Daughters in America: Irish Immigrant Women in the Nineteenth Century* (Baltimore: The Johns Hopkins University Press, 1983), pp. 11–12.
17. St. Erlich, p. 219.
18. Friedl, p. 66.
19. See Flandrin, p. 165; Natalie Zemon Davis, *Society and Culture in Early Modern France* (Stanford, Calif.: Stanford University Press, 1975), p. 105.
20. See Stenton, pp. 77, 80.
21. Homans, p. 197.
22. See Flandrin, pp. 78–79; Homans, pp. 140–42, 157; Emmanuel LeRoy Ladurie, "Family Structures and Inheritance Customs in Sixteenth-Century France," in Jack

Goody, Joan Thirsk, and E. P. Thompson, eds., *Family and Inheritance: Rural Society in Western Europe 1200–1800* (New York: Cambridge University Press, 1978), pp. 53–63.

23. Kinsella, *Tain Bó,* p. 54.
24. Katherine Fischer Drew, trans., *The Lombard Laws* (Philadelphia: University of Pennsylvania Press, 1973), p. 96.
25. Cited in Duby, *European Economy,* p. 45.
26. The records of the Frankish village of Thiasis in the Parisis show the gradual evolution of this status. The records at the beginning of the ninth century described 146 heads of households with eleven listed as slaves, 130 as *coloni,* and 19 as "protected" and owing chevage, a capitation tax. By the thirteenth century almost all were described as "servile." Bloch, *Feudal Society,* Vol. I, p. 262.
27. Duby, *Rural Economy,* p. 444.
28. Dahlsgård, p. 60.
29. See Peter Czap, Jr., "Marriage and the Peasant Joint Family in the Era of Serfdom" in Ransel, ed., p. 107; Max Beloff in Albert Goodwin, ed., *The European Nobility in the Eighteenth Century* (New York: Harper & Row, Publishers, 1967), p. 183. For a complete description of the varieties of serfdom in Russia in the eighteenth century see Isabel de Madariaga, *Russia in the Age of Catherine the Great* (New Haven: Yale University Press, 1981).
30. These dramatic shifts in the nature of agricultural production constituted an "agricultural revolution." In eras of relative peace and milder climate, the rural population increased yields from better use of fields, from fertilizing, from improved technology like the metal-bladed plow, and the use of horses instead of oxen. This combination of factors meant great periods of expansion with old fields extended, new fields created, new methods of trade distribution and specialized production on scales never imagined before.
31. Duby, *Rural Economy,* p. 238.
32. See Z. P. Pach, "Sixteenth-century Hungary: Commercial Activity and Market Production by the Nobles" in Peter Burke, ed., *Economy and Society in Early Modern Europe* (New York: Harper & Row, Publishers, 1972).
33. Homans, p. 73.
34. Sylvia Thrupp, "Medieval Industry 1000–1500" in Cipolla, ed., *Middle Ages,* p. 263.
35. William A. Christian, Jr., *Local Religion in Sixteenth-Century Spain* (Princeton: Princeton University Press, 1981), p. 13.
36. McArdle, p. 159.
37. W. H. Lewis, *The Splendid Century: Life in the France of Louis XIV* (Garden City, N.Y.: Doubleday & Company, Inc., 1957), p. 159.
38. Goubert in Aston, ed., pp. 165–66.
39. Eileen Power, *Medieval English Nunneries* (Cambridge: Cambridge University Press, 1922), p. 157.
40. Elda Zappi, " 'If Eight Hours Seem Few to You . . .' Women Workers' Strikes in Italian Rice Fields 1901–1906," paper delivered at The Berkshire Conference on Women's History, held at Vassar College, Poughkeepsie, N.Y., 1981, pp. 1–2.

41. Cited in Cornelisen, *Shadows,* p. 123.
42. Margaret Labarge, *A Baronial Household in the Thirteenth Century* (New York: Barnes and Noble, 1966), p. 114.
43. Alice Clark, *Working Life of Women in the Seventeenth Century* (New York: Harcourt, Brace & Howe, 1920), p. 63, note.
44. See Cissie Fairchilds, *Domestic Enemies: Servants & Their Masters in Old Regime France* (Baltimore: The Johns Hopkins University Press, 1984), p. 2.
45. Georges Duby, *The Chivalrous Society,* translated by Cynthia Postan (Berkeley: University of California Press, 1977), p. 208.
46. J. B. Ross, unpublished manuscript history of medieval women, 1979, p. 10.
47. See Quaife, p. 23.
48. Labarge, p. 68.
49. Isabel Ross, *Margaret Fell: Mother of Quakerism* (New York: Longmans, 1949), p. 263.
50. Ross, *Fell,* p. 264. Nineteenth-century French peasant families commonly hired a young girl to be a maid of all work until their own children were old enough to help with the daily and seasonal tasks. See Segalen, p. 63.
51. Heaton, p. 143.
52. Louise A. Tilly and Joan W. Scott, *Women, Work and Family* (New York: Holt, Rinehart and Winston, 1978), p. 45.
53. Tilly and Scott, p. 111.
54. Tilly and Scott, p. 45.
55. See Segalen, for a ninteenth-century French version of this arrangement, pp. 13–14.
56. The literature on this question is extensive; see the excellent summary by Ralph Houlbrooke, "The Pre-Industrial Family," *History Today,* vol. 36 (April 1986); for descriptions of the "nuclear family" pattern and examples, see his book *The English Family, 1450–1700* (New York: Longmans, 1984); Wrigley, p. 19; Laslett, *World Lost,* p. 91; and Segalen, pp. 57–59; Lutz K. Berkner, "The Stem Family and the Developmental Cycle of the Peasant Household: An Eighteenth-Century Austrian Example," *American Historical Review,* vol. 77, no. 2 (April 1972), p. 417. For the exceptions, see Michael Mitterauer and Reinhard Seder, *The European Family: Patriarchy to Partnership from the Middle Ages to the Present,* translated by Karla Oosterveen and Manfred Hörzinger (Oxford: Basil Blackwell, 1982), pp. 29–30. And St. Erlich, p. 58.
57. Wrigley, p. 93.
58. Russell in Cipolla, ed., *Middle Ages,* p. 57.
59. Herlihy, "Plague," Table II, p. 235.
60. Birth to death rate was 3.9 percent to 4.1 percent, see Russell, *Medieval Population,* p. 19.
61. See Eda Sagarra, *A Social History of Germany: 1648–1914* (New York: Holmes and Meier, 1977), p. 10; Hajo Halborn, *A History of Modern Germany 1648–1840* (New York: Alfred Knopf, 1969), p. 23. For a general discussion see Léopold Génicot, "On the Evidence of Growth of Population in the West from the Eleventh to the Thirteenth Century" in Sylvia L. Thrupp, ed., *Change in Medieval Society: Europe North of the Alps, 1050–1500* (New York: Appleton-Century-Crofts, 1964), p. 29.

62. Jan DeVries, *The Economy of Europe in an Age of Crisis 1600–1750* (New York: Cambridge University Press, 1976), p. 110.
63. The pattern for Eastern Europe is somewhat different after the sixteenth century. See "Women of the Cities," Part VIII in Volume II, for more information on the later centuries.
64. See Wrigley, pp. 17–19, 93.
65. Zappi, p. 7.
66. Russell, *Medieval Population,* p. 18.
67. See Laslett, *World Lost,* pp. 13–14.
68. E. A. Wrigley, "Family Limitation in Pre-Industrial England," in Orest and Patricia Ranum, eds., *Popular Attitudes Toward Birth Control in Pre-Industrial France and England* (New York: Harper & Row, Publishers, 1972), Table I, p. 60.
69. McArdle, pp. 57–58. See also the study of the Vallage region of Champagne in France from 1681 to 1735, where later marriage from twenty-four to twenty-six also coincides with a decline in population. See André Burguière, "From Malthus to Max Weber: Belated Marriage and the Spirit of Enterprise," in Robert Forster and Orest Ranum, eds., *Family and Society: Selections from the Annales,* translated by Elborg Forster and Patricia M. Ranum (Baltimore: The Johns Hopkins University Press, 1976), p. 244.
70. K. H. Connell, *Irish Peasant Society* (Oxford: Clarendon Press, 1968), pp. 117–18. Eastern European families once again proved exceptional, with marriage ages of thirteen in Russia, twenty in Hungarian villages; see Czap, in Ransel, ed., p. 111; J. Hajnal, "European Marriage Patterns in Perspective" in Glass and Eversley, p. 131, and Mitterauer and Seder, eds., p. 37.
71. Myron P. Gutman, *War and Rural Life in the Early Modern Low Countries* (Princeton: Princeton University Press, 1980), p. 185.
72. Flandrin, p. 59; for further examples of this phenomenon see also Laslett, *World Lost,* p. xi, and Russell, *Medieval Population,* pp. 58–59.
73. Cited in Quaife, p. 133.
74. See Norman E. Himes, *Medical History of Contraception* (New York: Schocken Books, 1970), pp. 137–73; John Noonan, *Contraception* (Cambridge: Cambridge University Press, 1965), pp. 255–56.
75. St. Erlich, p. 257.
76. St. Erlich, p. 295.
77. For discussion of this phenomenon see Emily Coleman, "Medieval Marriage Characteristics: A Neglected Factor in the History of Medieval Serfdom," *Journal of Interdisciplinary History,* vol. 2 (Autumn 1971), pp. 207, 211–12; Emily Coleman, "Infanticide in the Early Middle Ages," in Susan Mosher Stuard, ed., *Women in Medieval Society* (Philadelphia: The University of Pennsylvania Press, 1976), pp. 61, 64; Russell, *Medieval Population,* p. 22; Flandrin, p. 199.
78. See Russell, *Medieval Population,* pp. 14–15; David Herlihy suggests the exaggerated ratios may also come from under reporting of female births and deaths. See for example his article "Life Expectancies for Women in Medieval Society," in Rosemarie Thee Morewedge, ed., *The Role of Woman in the Middle Ages* (Albany, N.Y.: SUNY Press, 1975), pp. 21–22.
79. Russell, *Medieval Population,* p. 15.

80. Barbara A. Kellum, "Infanticide in England in the Later Middle Ages," *History of Childhood Quarterly,* Vol. I, no. 3 (Winter 1974), p. 368.
81. Quaife, p. 22.
82. See Russell, *Medieval Population,* p. 15.
83. See Wrigley, p. 88; Theodore H. Rabb and Robert I. Rotberg, *The Family in History* (New York: Octagon Books, 1971), pp. 208, 283, 39; Tilly and Scott, pp. 91–93.
84. Coleman in Stuard, ed., p. 60.
85. R. H. Helmholz, "Infanticide in the Province of Canterbury During the Fifteenth Century," *History of Childhood Quarterly,* vol. 2, no. 3 (Winter 1975), p. 381. Although some infants did perhaps die in this way, research in the twentieth century suggests that the infants probably died not of "overlaying" but of the syndrome known as Sudden Infant Death (crib death). All of the characteristics of the disease would hold true for the fifteenth century as they do for the twentieth. See Kathleen Stassen Berger, *The Developing Person* (New York: Worth Publishers, 1980), p. 194, and Todd L. Savitt, "Smothering and Overlaying of Virginia Slave Children: A Suggested Explanation," *Bulletin of History of Medicine,* vol. 49, no. 3 (Fall 1975), pp. 401–403.
86. Iris Origo, *The Merchant of Prato* (London: Jonathan Cape, 1957), p. 215. See also Christiane Klapisch-Zuber, *Women, Family, and Ritual in Renaissance Italy,* translated by Lydia Cochrane (Chicago: The University of Chicago Press, 1985), pp. 133–56.
87. Cited in Origo, *Merchant,* p. 216.
88. David L. Ransel, "Abandonment and Fosterage of Unwanted Children: The Women of the Foundling System," in Ransel, ed., p. 105.
89. See Hufton, and William L. Langer, "Infanticide: A Historical Survey," *History of Childhood Quarterly,* Vol. I, no. 3 (Winter 1974), for discussion of the interplay between circumstance and feeling.
90. Christiane Klapisch, "Household and Family in Tuscany in 1427," in Laslett and Wall, *Household and Family,* p. 273; Laslett, *World Lost,* p. 78.
91. See Homans, eds., p. 180; Jack Goody, "Inheritance, Property and Women: Some Comparative Considerations," in Goody, et al., *Family and Inheritance,* p. 31; Berkner, pp. 400–401, Atkinson in Atkinson et al., p. 32.
92. Cited in Homans, p. 181.
93. George Vernadsky, trans., *Medieval Russian Laws* (New York: W. W. Norton & Company, Inc., 1969), p. 77.
94. Homans, p. 183.
95. Coulton, *Village,* p. 453.
96. Cited in Duby, *Rural Economy,* pp. 485–86.
97. Stenton, p. 84.
98. Cited in Homans, p. 188.
99. Cited in Homans, p. 182.
100. Homans, pp. 152–53.
101. Cited in Homans, p. 145.
102. Rayna R. Reiter, "Men and Women in the South of France: Public and Private Domains," in Rayna R. Reiter, ed., *Toward an Anthropology of Women* (New York: Monthly Review Press, 1975), p. 262.

103. Lawrence Stone, *The Family, Sex and Marriage* (New York: Harper & Row, Publishers, 1977), p. 25. For the same custom in Central Europe in the eighteenth century, see for example Berkner, p. 401.
104. Klapisch, in Laslett and Wall, eds., *Household and Family,* p. 60.
105. Henry S. Lucas, "The Great European Famines of 1315, 1316 and 1317," *Speculum,* Vol. V (1930), p. 362.
106. Cited in Thomson, p. 362.
107. Cited in Laslett, *World Lost,* p. 31.
108. Hufton p. 165; see also p. 68, note, and Clark, p. 73.
109. Quaife, p. 134.
110. Homans, p. 185.
111. Barbara A. Hanawalt, "The Female Felon in Fifteenth-Century England," in Stuard, ed., p. 133.
112. Hannelore Sachs, *The Renaissance Woman,* translated by Marianne Hertzfeld (New York: McGraw-Hill Book Company, 1971), p. 48.
113. Davis, *Culture and Society,* p. 146.
114. Marion S. Miller, "Pane e Lavoro: Agrarian Strikes of Women Straw Workers in the Tuscan Contado 1896–97," paper delivered at The Berkshire Conference in Women's History, Vassar College, Poughkeepsie, N.Y., 1981, p. 1.
115. Miller, p. 5.
116. Miller, p. 6.
117. Cited in Flandrin, p. 113.
118. "The Girl Without Hands," in *Grimm's Folk Tales,* p. 118.
119. See "The Clever Peasant Lass," "The True Bride," "The Drummer Boy," and "The Iron Stove," in *Grimm's Folk Tales* for variations on the story of the young woman fulfilling tasks to capture the groom in contrast to the more familiar story of the prince tested to win the bride.
120. See Jo Ann Melanie Magdoff, "The Madonna and Work: An Analysis of Women's Identity in an Industrializing Town in Central Italy," paper delivered at The Berkshire Conference in Women's History, held at Mount Holyoke College, South Hadley, Massachusetts, August 1978.
121. Mémé Santerre, "A French Working Woman Recalls Her Childhood as a Linen Weaver," in Ena Olafson Hellerstein, Leslie Parker Hume and Karen M. Offen, eds., *Victorian Women* (Stanford, Calif.: Stanford University Press, 1981), pp. 39–44.
122. Hélias, pp. 35–37.
123. Jeanne Favret-Saada, *Deadly Words: Witchcraft in the Bocage,* translated by Catherine Cullen (New York: Cambridge University Press, 1980), p. 7.
124. Cited in Dorothy Atkinson, "Society and Sexes in the Russian Past" in Atkinson et al., eds., p. 24.
125. Christian, *Person and God,* p. 31.
126. Katherine Simms, "Women in Norman Ireland," Margaret MacCurtain and Donncha Ó'Corraín, eds., *Women in Irish Society, The Historical Dimension* (Dublin: Arlen House, 1978), p. 20, and Flandrin, p. 122.
127. See Segalen, pp. 43–46.
128. Flandrin, p. 122.
129. *Grimm's Folk Tales,* p. 554.

130. Cornelisen, *Shadows*, p. 71.
131. Cited in Kellum, p. 373.
132. This and subsequent cases from Hanawalt in Stuard, ed., pp. 130–31.
133. Hanawalt in Stuard, ed., n. p. 139.

PART II: 3. The Extraordinary

1. Curiously, with all of these sources the charges against Joan have been forgotten and others substituted. Popular tradition has altered the facts of history. Stories about Joan say that she was condemned as a witch, inspired by the devil. This was not the case; the persecution of the witches lay 150 years in the future. Joan was never declared a devil worshiper or a witch; rather she was found to be a "schismatic," one who "divided" the Church by not accepting its authority. In the end her inquisitors did not doubt the orthodoxy of her Christian faith, only her claim to divine inspiration and the unwomanly behavior and dress she believed this inspiration justified and necessitated. In the end the qualities of single-mindedness, courage, and independence that had given her success made possible her execution.
2. W. P. Barrett, trans., *The Trial of Jeanne d'Arc* (London: George Routledge & Sons, Ltd., 1931), p. 55.
3. Barrett, *Trial*, p. 100.
4. Régine Pernoud, *The Retrial of Joan of Arc*, translated by J. M. Cohen (New York: Harcourt, Brace and Company, 1955), p. 78.
5. Régine Pernoud, *Joan of Arc, by Herself and Her Witnesses*, translated by Edward Hyams (London: Macdonald, 1964), p. 36.
6. Pernoud, *Joan of Arc*, p. 35.
7. Pernoud, *Retrial*, p. 77.
8. Pernoud, *Retrial*, p. 101.
9. Pernoud, *Retrial*, p. 101. There are no descriptions of Joan that give specific details of her appearance, not even the color of her hair. The evidence of those who knew her suggests that she must have been attractive, healthy, and, given her activities on the battlefield, strong. The haircut comes from one of the charges at her trial, see Barrett, *Trial*, p. 230.
10. Pernoud, *Retrial*, p. 136.
11. Cited in John Holland Smith, *Joan of Arc* (London: Sidgwick & Jackson, 1973), p. 71.
12. Pernoud, *Joan of Arc*, p. 82.
13. Pernoud, *Joan of Arc*, p. 90.
14. Cited in Frances Gies, *Joan of Arc: The Legend and the Reality* (New York: Harper & Row, Publishers, 1981), p. 87.
15. Barrett, *Trial*, p. 147.
16. Pernoud, *Joan of Arc*, pp. 98–99.
17. Pernoud, *Joan of Arc*. p. 93.
18. Pernoud, *Retrial*, p. 153.
19. See Pernoud, *Retrial*, pp. 126–27, 142.
20. Pernoud, *Retrial*, p. 139.
21. Pernoud, *Joan of Arc*, p. 52.
22. Pernoud, *Joan of Arc*, p. 40.

23. Pernoud, *Retrial,* p. 154.
24. Pernoud, *Joan of Arc,* p. 128.
25. Pernoud, *Joan of Arc,* p. 137.
26. Pernoud, *Retrial,* p. 168, note.
27. Pernoud, *Joan of Arc.* p. 100.
28. For Cauchon's former relations with the English, see Gies, *Joan of Arc,* p. 14.
29. Pernoud, *Retrial,* p. 184.
30. Pernoud, *Joan of Arc,* p. 103.
31. Pernoud, *Joan of Arc,* p. 182.
32. Pernoud, *Joan of Arc,* p. 206.
33. Pernoud, *Joan of Arc,* p. 203.
34. Barrett, *Trial,* p. 246.
35. Barrett, *Trial,* p. 257.
36. Barrett, *Trial,* p. 255.
37. Barrett, *Trial,* p. 134. The canon law of the fifteenth century allowed male dress for women in extenuating circumstances, for protection, out of poverty. The final charges stated "you have continually used male dress . . . wearing your hair short . . . leaving nothing to show that you are a woman." See Smith, *Joan of Arc,* p. 159.
38. Pernoud, *Joan of Arc,* p. 207.
39. Cited in Smith, *Joan of Arc,* p. 164.
40. Cited in Smith, *Joan of Arc,* p. 165.
41. Barrett, *Trial,* p. 316.
42. Pernoud, *Joan of Arc,* p. 218.
43. Pernoud, *Retrial,* p. 241. Once the Earl of Warwick had changed the guards when she complained. During her imprisonment at Beaurevoir before the trial, she had jumped from a seventy-foot tower, hoping to escape when she believed that her new English guards planned to rape her.
44. Cited in Smith, *Joan of Arc,* p. 169.
45. Daniel Rankin and Claire Quintal, trans., *The First Biography of Joan of Arc* (Pittsburgh: University of Pittsburgh Press, 1964), p. 53.
46. Pernoud, *Retrial,* p. 216.
47. Keith Thomas, "A Working Girl," *The New York Review of Books,* June 25, 1981, p. 7.
48. Words from the Rehabilitation; also, for the refutations of the charges, see Gies, *Joan of Arc,* p. 233.
49. Gies, *Joan of Arc,* p. 241. See also excerpts in Charity Cannon Willard, *Christine de Pizan: Her Life and Works* (New York: Persea Books, 1984), pp. 206–207.
50. Richard Kieckhefer, *European Witch Trials: Their Foundations in Popular and Learned Culture, 1300–1500* (Berkeley: University of California Press, 1976), p. 28.
51. The principal modern historians of witchcraft—Gustave Henningsen for northern Spain, Christina Larner for Scotland, Alan Macfarlane for England, Erik Midelfort for southwest Germany, William Monter for Switzerland, western France, and northern Italy—have each noted coincidences that might explain the phenomenon in one particular region. For example, young adolescents in Spain made many of the 7,000 accusations; in Scotland, young women were persecuted for infanticide in this period, but the old for witchcraft; waves of hysteria occurred in border lands where

cultures cross, like the cantons of Switzerland, western France, and northern Spain; massive outbreaks of persecution occurred particularly in areas like southwestern Germany where the authority of the Church was traditionally weak and heresy a recurrent malady. Mary Martin McLaughlin helped particularly with the formulation of the key questions, and thus indirectly with the answers presented in this section.

52. Keith Thomas, *Religion and the Decline of Magic* (New York: Charles Scribner's Sons, 1971), p. 272.
53. *Grimm's Folk Tales,* p. 445.
54. Michael Olmert, "Points of Origin," *Smithsonian,* vol. 12, no. 7 (October 1981), p. 52.
55. Saint Teresa of Avila (1515–1582) recorded in her autobiography that when the Devil appeared to her she made him vanish by throwing holy water at him.
56. Alan C. Kors and Edward Peters, *Witchcraft in Europe 1100–1700: A Documentary History* (Philadelphia: University of Pennsylvania Press, 1972), p. 203.
57. Kors and Peters, p. 198.
58. The theologians Nicolas Eymerich (c. 1376), Jean Gerson (c. 1398), and Johannes Nider (c. 1435) tied heresy to sorcery, sorcery to pacts with the Devil, and thus created the new heresy of demonology.
59. Cited in Julio Caro-Baroja, *The World of the Witches,* translated by O. N. U. Glendinning (Chicago: University of Chicago Press, 1964), p. 94.
60. A Talmudic saying appropriated by the learned ecclesiastics, cited in Rosemary Radford Ruether, *New Woman, New Earth: Sexist Ideologies and Human Liberation* (New York: The Seabury Press, 1975), p. 89.
61. Kors and Peters, p. 120; see also pp. 121 and 124 for other traditional statements. Complete editions of the *Malleus Maleficarum* are in print in English. Kors and Peters have done a good abridgement and the quotations are taken from their version.
62. Kors and Peters, pp. 127, 121, 120.
63. See Atkinson in Atkinson et al., eds., p. 16, note.
64. Coulton, *Panorama,* pp. 115–16.
65. Kors and Peters, pp. 121, 127.
66. See Elizabeth L. Eisenstein, *The Printing Press as an Agent of Change: Communications and Cultural Transformations in Early-Modern Europe,* Vol. I (New York: Cambridge University Press, 1979), p. 439; Geoffrey Parker, "The European Witch-craze Revisited," *History Today,* vol. 30 (November 1980), p. 24.
67. For example: treatises by a Spanish friar and a German Calvinist, demonologies from French judges, Jean Bodin (adviser to the king of France), Humanists, and Jesuits, and James I, King of England. Bodin and the Flemish Humanist Binsfeld wrote the scenario for what came to be known as the Witches' Sabbath.
68. See, H. C. Erik Midelfort, *Witch Hunting in Southwestern Germany 1562–1684: The Social and Intellectual Foundations* (Stanford, Calif.: Stanford University Press, 1972), p. 59.
69. Kors and Peters, p. 215.
70. See Kieckhefer, p. 96, and E. William Monter, "The Historiography of Witchcraft:

Progress and Prospects," *Journal of Interdisciplinary History,* Vol. II, no. 4 (Spring 1972), p. 450.

71. See Midelfort, pp. 32, 72, Table 14.
72. For Lorraine, see Ruether, *New Woman,* p. 103, and Monter, "Historiography," p. 32.
73. See Parker, "Witchcraze," p. 24. For statistics on Belgium and Luxembourg, see E. William Monter, "French and Italian Witchcraft," *History Today,* vol. 30 (November 1980), p. 32.
74. See Thomas, *Decline of Magic,* pp. 451, 448, and Alan Macfarlane, "Witchcraft in Tudor and Stuart Essex" in Mary Douglas, ed., *Witchcraft Confessions and Accusations* (New York: Tavistock Publishers, 1970), p. 84.
75. See Christina Larner, *Enemies of God: The Witchhunt in Scotland* (Baltimore: The Johns Hopkins University Press, 1981), pp. 60–63; Parker, "Witchcraze," p. 24.
76. See for example, Gustav Henningsen, *The Witches' Advocate: Basque Witchcraft and the Spanish Inquisition 1609–1640* (Reno: University of Nevada, 1980), p. 216, and for Sweden, Caro-Baroja, p. 208.
77. Robert Muchembled, "The Witches of the Cambrésis," in James Obelkevich, ed., *Religion and the People 800–1700* (Chapel Hill, N.C.: The University of North Carolina Press, 1979), Table 4, p. 233.
78. Kors and Peters, p. 159.
79. Kors and Peters, p. 170.
80. Kors and Peters, p. 244.
81. Kors and Peters, p. 161.
82. See, for example, Thomas, *Decline of Magic,* p. 517.
83. See Kors and Peters, pp. 136–37, 144.
84. Those who have studied the phenomenon of witchcraft have sought scientific reasons for some of the women's visions and experiences. Did they in fact see and feel what they described? Historians and botanists speculate that a woman who knew about herbs, like the wisewomen, like the cunning folk, might have experimented with monkshood, nightshade, hemlock, henbane, toadskins, and the grain fungus ergot. All include chemicals that cause sensory paralysis, can depress the cardiovascular system, can give a sense of transformation into other shapes, cause delirium, hallucinations, fatigue.
85. E. William Monter, *Witchcraft in France and Switzerland: The Borderlands During the Reformation* (Ithaca, N.Y.: Cornell University Press, 1976), pp. 156–57.
86. Cited in Sachs, p. 50.
87. Viktor von Karwill, ed., *The Fugger Newsletters, Being a Selection of Unpublished Letters from the Correspondence of the House of Fugger During the Years 1568–1605* (New York: G. P. Putnam's Sons, 1925), pp. 106–13.
88. See for example, Larner, pp. 176–77.
89. For example, Johann Weyer, court physician to Wilhelm of Julich-Cleve-Berg, wrote in 1563 that they were just crazy old women, in need of a doctor's care, deluded by the Devil, but not in his power, not his worshipers.
90. Henningsen, p. 37.
91. Cited in Henningsen, p. 350.

92. Cited in Henningsen, p. 188.
93. Henningsen, p. 301.
94. Jeanne Favret-Saada's study *(Deadly Words)* of a northern French village gives the most complete picture of the continuing belief in the power of the supernatural and those who know how to use it into the 1970s.
95. Heaton, pp. 347–48.
96. Arensberg, *Countryman*, p. 171.
97. Friedl, p. 77.

PART II: 4. What Remains of the Peasant Woman's World

1. In 1980 women made up 60 percent of the rural population in Poland, 56 percent in Romania. See *Review and Evaluation of Progress Achieved in the Implementation of the World Plan of Action: Employment*, prepared for the World Conference of the United Nations Decade for Women: Equality, Development, and Peace, held at Copenhagen, July 14–30. 1980 A/CONF. 94/8/Rev. 1, p. 44.
2. For Soviet examples, see Susan Allott, "Soviet Rural Woman: Employment and Family Life," in Barbara Holland, ed., *Soviet Sisterhood* (Bloomington, Ind.: Indiana University Press, 1985), p. 203.
3. See *Review and Evaluation*, pp. 44–45.
4. Cited in Allott, in Holland, ed., p. 183.
5. Cited in Allott, in Holland, ed., p. 188.
6. Gail Warshofsky Lapidus, *Women in Soviet Society: Equality, Development, and Social Change* (Berkeley: University of California Press, 1976), pp. 176–77. See also Ethel Dunn, "Russian Rural Women" in Atkinson et al., eds., pp. 174–76.
7. See Cornelisen, *Shadows*, for examples for Italy.
8. Allauzen, pp. 15–16.
9. Allott, in Holland, ed., p. 191.
10. In 1852, a survey in the Orléans diocese of those who received communion on Easter showed that 88.9 percent of the adolescent girls and 31.4 percent of the women participated. The corresponding numbers for males were 28.6 percent of the boys and only 6.4 percent of the adult men. Cited in J. Michael Phayer, *Sexual Liberation and Religion in Nineteenth-Century Europe* (London: Croom Helm, 1977), p. 95.
11. Christian, *Person and God*, p. 138.
12. Maria Rosa Cutrufelli, *Des Siciliennes*, translated by Laura Revelli (Paris: des femmes, 1976), p. 251. On the general dismissal of women as of less value, dependent on and subordinate to men, see also Cornelisen, *Shadows*, pp. 15–30, Chamberlain, *Fenwomen*, pp. 12, 23, Lapidus, p. 176, Allott in Holland, ed., p. 203.
13. Cited in Cornelisen, *Shadows*, p. 150.
14. See St. Erlich, p. 347.
15. St. Erlich, p. 255.
16. St. Erlich, p. 255.
17. See Renate Bridenthal, "Something Old, Something New: Women Between the Two World Wars," in Renate Bridenthal and Claudia Koonz, eds., *Becoming Visible: Women in European History* (Hopewell, N.J.: Houghton Mifflin Company, 1977), p. 427.

18. Cited in Allott, in Holland, ed., pp. 185, 201; see Lapidus, pp. 179 and 180, on shifts in Soviet population from the countryside to the cities.
19. Allauzen, p. 187.
20. Reiter in Reiter, ed., p. 258.
21. Kramer, p. 50.

PART III. WOMEN OF THE CHURCHES: THE POWER OF THE FAITHFUL

PART III: 1. The Patterns of Power and Limitation: The Tenth to the Seventeenth Centuries

1. This section describes the activities of Christian women empowered by their faith. While significant, Judaism was not central to the beliefs and institutions of European society, and unlike Christianity, it did not offer its pious women a life outside the traditional female roles and functions. See Part V, "Women of the Walled Towns," for descriptions of a Jewish woman's life.

PART III: 2. Authority Within the Institutional Church

1. The term "double monasteries" is somewhat misleading, as it suggests equal numbers of women and men. Usually, however, these were establishments founded primarily for either women or men. If it was a community of nuns, monks (those ordained as priests) would have been resident to serve the women's religious needs. If it was a community of monks, with some women also following the rule, an abbess not an abbot might still preside. See Penny Schine Gold, *The Lady and the Virgin: Image, Attitude and Experience in Twelfth-Century France* (Chicago: University of Chicago Press, 1985), pp. 101–102.
2. Mary Bateson, *Origin and Early History of Double Monasteries,* Royal Historical Society Transactions, New Series 13 (London, 1899), p. 154.
3. Lina Eckenstein, *Woman Under Monasticism* (New York: Russell & Russell Inc., 1963), p. 191.
4. Pierre Riché, *Education and Culture in the Barbarian West, Sixth Through Eighth Centuries* (Columbia, S.C.: University of South Carolina Press, 1976), p. 457.
5. Cited in Eckenstein, pp. 136–37.
6. Cited in S. M. Gonsalva Wiegand, trans., *The Non-Dramatic Works of Hroswitha* (St. Meinrad, Ind.: The Abbey Press, 1936), p. 9.
7. See, for example, Katharina M. Wilson, "The Saxon Canoness: Hrotsvit of Gandersheim," in Katharina M. Wilson, ed., *Medieval Women Writers* (Athens, Ga.: The University of Georgia Press, 1984), pp. 31–32. Gandersheim had both canonesses and professed nuns following the Benedictine Rule. There is some question as to whether or not Hrotsvit had taken full vows. Wilson believes that the very nature of her activities and the subjects she chose to write on suggest that she remained a canoness. See Wilson, ed., p. xii, and Suzanne Fonay Wemple, *Women in Frankish Society: Marriage and the Cloister 500 to 900* (Philadelphia: University of Pennsylvania Press, 1981), p. 174.
8. The most complete manuscript organizes the works into three books: Book I is the

eight legends of the saints, Book II the six plays, Book III the history of Otto I and of the monastery in the form of epics and the short poem. Wilson in Wilson, ed., pp. 32–33.

9. Robert Herndon Fife, *Hroswitha of Gandersheim* (New York: Columbia University Press, 1947), p. 36.
10. Cited in Wiegand, p. 9.
11. Cited in Christopher St. John, trans., *The Plays of Roswitha* (New York: Benjamin Blom, 1966), pp. xxvi, xxvii.
12. St. John, *Roswitha,* p. 119.
13. St. John, *Roswitha,* pp. 85–87.
14. A fifteenth-century Humanist, Conrad Celtes, rediscovered her writings in the monastery at Ratisbon and published them, calling her "the German Sappho." Even so, in the nineteenth century an Austrian historian insisted that the Latin was so good that Celtes must have written the works himself. See Anne Lyon Haight, ed., *Hroswitha of Gandersheim, Her Life, Times, and Works, and a Comprehensive Bibliography* (New York: The Hroswitha Club, 1965), p. 3, and Fife, p. 29.
15. Anne Bagnall Yardley, " 'Ful weel she soong the service dyvyne': The Cloistered Musician in the Middle Ages," in Jane Bowers and Judith Tick, eds., *Women Making Music: The Western Art Tradition 1150–1950* (Chicago: University of Illinois Press, 1986), p. 19; Eckenstein, p. 243.
16. Cited in Eckenstein, p. 254; see also p. 243.
17. Muriel Joy Hughes, *Women Healers in Medieval Life and Literature* (New York: King's Crown Press, 1943), p. 121.
18. Cited in Eckenstein, p. 264.
19. Cited in Kent Kraft, "The German Visionary: Hildegard of Bingen," in Wilson, ed., p. 118. For a description of her visions and an analysis of their meaning in the context of mysticism in general, see Peter Dronke, *Women Writers of the Middle Ages: A Critical Study of Texts from Perpetua to Marguerite Porete* (New York: Cambridge University Press, 1985), pp. 144–201. Some of the illustrations from *Scivias* and *De operatione dei* have been reproduced in *Illuminations of Hildegard of Bingen* with commentary by Matthew Fox, O.P. (Santa Fe: Bear & Company Inc., 1985).
20. Carol Neuls-Bates, *Women in Music: An Anthology of Source Readings from the Middle Ages to the Present* (New York: Harper & Row, Publishers, 1982), p. 12.
21. See Kraft in Wilson, ed., p. 116, and Peter Dronke, *The Medieval Lyric* (New York: Cambridge University Press, 1977), pp. 75–76.
22. Cited in Joan Ferrante, "The Education of Women in the Middle Ages in Theory, Fact and Fantasy," in Patricia H. Labalme, ed., *Beyond Their Sex: Learned Women of the European Past* (New York: New York University Press, 1980), p. 24.
23. Cited in Kraft in Wilson, ed., p. 113.
24. See Caroline Walker Bynum, *Jesus as Mother: Studies in the Spirituality of the High Middle Ages* (Berkeley: University of California Press, 1982), p. 15, and Haye van der Meer, S. J., *Women Priests in the Catholic Church? A Theological-Historical Investigation* (Philadelphia: Temple University Press, 1973), p. 125.
25. Cited in van der Meer, p. xxiii.
26. Cited in Julia O'Faolain and Lauro Martines, eds., *Not in God's Image: Women in*

History from the Greeks to the Victorians (New York: Harper & Row, Publishers, 1973), p. 130.

27. Cited in Eileen Power, *Medieval English Nunneries* (Cambridge: Cambridge University Press, 1922), pp. 373, 346.
28. Cited in Power, *Nunneries,* p. 344.
29. *The Ancren Riwle,* a treatise advising recluses written about 1220, described this condition much as the Church Fathers had: a "ruddy face" grown "lean and green as grass," "dusky" eyes, pains, nausea, a giddy, aching brain and a belly "strut out like a water bag." Cited in Clarissa W. Atkinson, *Mystic and Pilgrim: The Book and the World of Margery Kempe* (Ithaca, N.Y.: Cornell University Press, 1983), p. 187.
30. Cited in van der Meer, p. 122.
31. Cited in Eckenstein, p. 317.
32. C. K. Scott Moncrieff, ed., *The Letters of Abelard and Heloise* (New York: Cooper Square Publishers, Inc., 1974), p. 81. The authenticity of Heloise's letters has long been a subject of historical debate. For a recent summary of the arguments, see Dronke, *Women Writers,* pp. 108–109, 140–43.
33. Scott Moncrieff, ed., *Letters,* p. 80.
34. Betty Radice, trans., *The Letters of Abelard and Heloise* (New York: Penguin Books, 1978), p. 150.
35. Scott Moncrieff, ed., *Letters,* p. 41.
36. Cited in Eleanor Shipley Duckett, *Medieval Portraits from East and West* (Ann Arbor: The University of Michigan Press, 1972), p. 251.
37. See Power, *Nunneries,* pp. 2–3.
38. For England, see Power, *Nunneries,* p. 3; for Italy see Suzanne Fonay Wemple, paper delivered at The Berkshire Conference on the History of Women, held at Vassar College, Poughkeepsie, New York, June 1981; for the Lowlands see Christopher Brooke, *The Twelfth Century Renaissance* (New York: Harcourt, Brace & World Inc., 1969), p. 8; for Saxony see Sybille Harksen, *Women in the Middle Ages,* translated by Marianne Herzfeld (New York: Abner Schram, 1975), p. 30; for Cistercian records, see R. W. Southern, *Western Society and the Church in the Middle Ages* (New York: Penguin Books, 1970), p. 317, note.
39. By 1200 there were sixty-six Cistercian foundations for women, with forty to a hundred members, and only fifteen for men; Mary McLaughlin "Hildegard of Bingen," paper delivered at Barnard College, New York, February 26, 1981; see also Gold, *Lady,* pp. 82–83.
40. For the Abbey of Fontevrault and the relationship between the nuns and the monks, see Gold, *Lady,* pp. 91–99, 112.
41. For this pattern from the early Church repeating in the twelfth and thirteenth centuries, see Gold, *Lady,* pp. 89–111.
42. See Rosemary Ruether and Eleanor McLaughlin, "Women's Leadership in the Jewish and Christian Traditions: Continuity and Change," in Rosemary Ruether and Eleanor McLaughlin, eds., *Women of Spirit* (New York: Simon and Schuster, 1979), p. 73.
43. In Germany between the twelfth and the seventeenth centuries, a number of communities of canonesses existed where young women lived according to the

Augustinian rule but never were enclosed and felt free to leave should they wish to marry. First in 1483 and then again in 1617, the Church insisted that they live more conventionally and follow the stricter Benedictine Rule. See Joan Morris, *The Lady Was a Bishop* (New York: Macmillan, 1973), pp. 66–67.

44. Johannes Jørgensen, *Saint Bridget of Sweden,* Vol. I, translated by Ingeborg Lund (New York: Longmans Green and Co., 1954), p. 99.
45. Cited in Sylvia L. Thrupp, *The Merchant Class of Medieval London* (Ann Arbor: University of Michigan Press, 1977), p. 189.
46. Gene Brucker, ed., *The Society of Renaissance Florence* (New York: Harper & Row, Publishers, 1971), p. 42.
47. Mary M. McLaughlin, "Women in Early Medieval Monasticism," comments delivered at The Berkshire Conference on the History of Women, held at Mount Holyoke College, South Hadley, Massachusetts, August 25, 1978.
48. Frances and Joseph Gies, *Women in the Middle Ages* (New York: Thomas Crowell Company, 1978), p. 90.
49. Power, *Nunneries,* p. 647.
50. See Bowers and Tick, eds., p. 4, and Jane Bowers, "The Emergence of Women Composers in Italy 1566–1700," in Bowers and Tick, eds.
51. Ferrante in Labalme, ed., p. 37, note.
52. Power, *Nunneries,* p. 241, note.
53. See Ferrante, in Labalme, ed., pp. 15–16, and Germaine Greer, *The Obstacle Race: The Fortunes of Women Painters and Their Work* (New York: Farrar, Straus & Giroux, 1979), pp. 159–64.
54. Power, *Nunneries,* pp. 138–39.
55. Power, *Nunneries,* p. 152.
56. Cited in Power, *Nunneries,* p. 382.
57. Cited in Derek Baker, " 'A Nursery of Saints': St. Margaret of Scotland Reconsidered" in Derek Baker, ed., *Medieval Women* (Oxford: Basil Blackwell, 1978), p. 123.
58. Power, *Nunneries,* p. 265.
59. For example, in the thirteenth century the archbishop of Rouen, Odo Rigaud, found that the convent Villarceaux was the most frequent offender against the injunction not to teach. In 1255, 1257, 1261, and 1268 he had to admonish Bondeville, another community, for this offense against the rule, *detecta* as the Church called these infractions. Visitation records for convents in thirteenth-century Normandy show the nuns running small schools, doing handwork, begging for alms, borrowing and even pawning the abbey's treasures. Penelope Johnson, "Poverty in Female and Male Norman Monasteries in the Mid-Thirteenth Century," paper delivered at Fordham University, New York, March 21, 1986.
60. Cited in Atkinson, *Kempe,* pp. 5, 7.
61. The historian Mary McLaughlin suggests that they created a new kind of literature, perhaps the forebear of the novel. Instead of the rhetoric and logic of the Classical Latin of the Scholastics, this was personal writing in the vernacular, stream of consciousness, filled with psychological insights, claiming no other authority than the private vision of the individual imagination.
62. C. H. Talbot, ed. and trans., *The Life of Christina of Markyate: A Twelfth Century Recluse* (Oxford: Clarendon Press, 1959), p. 67.

63. Talbot, ed., *Christina,* p. 105.
64. Long Text, Catherine Jones, "The English Mystic: Julian of Norwich," in Wilson, ed., p. 229.
65. Long Text, Jones in Wilson, ed., p. 286.
66. Long Text, Jones in Wilson, ed., p. 280.
67. See Bynum, *Jesus,* for descriptions of groups in the Low Countries.
68. John Howard, "The German Mystic: Mechthild of Magdeburg," in Wilson, ed., pp. 153, 154.
69. Cited in Alice Kemp-Welch, *Of Six Medieval Women* (London: Macmillan and Co., Ltd., 1913), p. 74.
70. See Kemp-Welch, pp. 80, 79.
71. Saint Teresa of Avila, *The Book of Her Life,* vol. 1., *The Collected Works,* translated by Kieran Kavanaugh and Otilio Rodriquez (Washington D.C.: ICS Publications, 1976), p. 40.
72. St. Teresa, *Life,* p. 214.
73. St. Teresa, *Life,* p. 74.
74. St. Teresa, *Life,* p. 188.
75. St. Teresa, *Life,* pp. 65, 43.
76. They learned from the male visionaries' methods of prayer and meditation, took the image of God as the Bridegroom, and, like them, described the mystical union in sexual terms.
77. Long Text, Jones in Wilson, ed., p. 278.
78. Cited in Howard in Wilson, ed., p. 167.
79. St. Teresa, *Life,* p. 66.
80. Cited in Howard in Wilson, ed., p. 170.
81. Cited in Bynum, *Jesus,* p. 153.
82. See Caroline Walker Bynum, *Holy Feast and Holy Fast: The Religious Significance of Food to Medieval Women* (Berkeley: University of California Press, 1987); see also Saint Teresa of Avila, *Spiritual Testimonies,* Vol. I, *The Collected Works,* p. 331.
83. Long Text, Jones in Wilson, ed., p. 286.
84. Long Text, Jones in Wilson, ed., p. 288.
85. Long Text, Jones in Wilson, ed., p. 289.
86. Talbot, ed., *Christina,* p. 119.
87. Cited in Kemp-Welch, p. 65.
88. Cited in Anne Llewellyn Barstow, "Marguerite Porete: Heretical Mysticism and the Free Spirit," unpublished manuscript, p. 5.
89. Cited in Kraft in Wilson, ed., pp. 123, 124.
90. St. Teresa, *Life,* p. 194.
91. St. Teresa, *Life,* p. 193, and *Spiritual Testimonies,* p. 325.
92. St. Teresa, *Life,* p. 194.
93. Cited in Howard in Wilson, ed., p. 154.
94. Cited in Howard in Wilson, ed., p. 164.
95. Saint Teresa wrote an autobiography and devotional works: *The Way of Perfection, The Interior Castle, The Book of Foundations,* and *Meditations on the Songs.*
96. St. Teresa, *Life,* p. 247.

97. St. Teresa, *Life*, p. 44.
98. St. Teresa, *Spiritual Testimonies*, p. 352.

PART III: **3. Authority Outside the Institutional Church**

1. Given the patriarchal framework of Christianity, Mary's emergence as an important aspect of the Catholic faith has inspired all sorts of speculation, by historians, anthropologists, theologians, and psychologists. Historians and anthropologists have suggested that Mariolatry (Mary's "cult") replaced the worship of female deities common to all parts of the Mediterranean world. There is no question that her worship superseded that of many of the ancient goddesses and that their festivals, symbols, and images became hers. Psychologists see all of men's fantasies come together in one idealized woman: adoring mother, virginal bride. Theologians would emphasize that she at one and the same time satisfies man's desire for absolutes, and his need to give concepts concrete form. See for example Yrjö Hirn, *The Sacred Shrine: A Study of the Poetry and Art of the Catholic Church* (Boston: Beacon Press, n.d.), p. 478, and Rosemarie Thee Morewedge, ed., *The Role of Woman in the Middle Ages* (Albany: SUNY, at Binghamton, 1975), p. xiii.
2. See Eleanor Como McLaughlin, "Equality of Souls, Inequality of Sexes: Woman in Medieval Theology," in Rosemary Radford Ruether, ed., *Religion and Sexism: Images of Woman in the Jewish and Christian Traditions* (New York: Simon and Schuster, 1974), p. 220.
3. Cited in Wiegand, p. 33.
4. Cited in Wiegand, p. 33.
5. Christine de Pizan, *The Book of the City of Ladies*, translated by Earl Jeffrey Richards (New York: Persea Books, 1982), p. 218.
6. Cited in Marina Warner, *Alone of All Her Sex: The Myth and the Cult of the Virgin Mary* (New York: Pocket Books, 1978), p. 214.
7. Bede, *A History of the English Church and People*, translated by Leo Sherley-Price (Baltimore: Penguin Books, 1960), p. 235. The twelfth, thirteenth, and fourteenth centuries continued to produce saintly well-born women, honored in particular for their care of the poor and the sick, for example: Anne of Bohemia, Saint Margaret of Hungary, and Queen Isabel of Portugal.
8. Jørgensen, Vol. I, p. 156.
9. For a description of the order, see Barbara Obrist, "The Swedish Visionary: Saint Bridget," in Wilson, ed., p. 249.
10. Cited in Jørgensen, Vol. I, p. 199.
11. Cited in Jørgensen, Vol. II, p. 175.
12. Cited in Jørgensen, Vol. II, p. 175.
13. Jørgensen, Vol. II, pp. 259 and 249, and Obrist in Wilson, ed., pp. 139–41.
14. Cited in Jørgensen, Vol. II, p. 263.
15. Cited in Judith Hook, "St. Catherine of Siena," *History Today*, vol. 30 (July 1980), p. 32.
16. Cited in Piero Misciatelli, *The Mystics of Siena*, translated by M. Peters-Roberts (New York: D. Appleton and Company, 1930), p. 104.
17. Cited in Joseph Berrigan, "The Tuscan Visionary: Saint Catherine of Siena," in Wilson, ed., pp. 265, 266.

18. Roughly 350 of her letters to the prominent women and men of her day have survived. See for example Berrigan, in Wilson, ed., pp. 261, 245, 259.
19. Cited in Berrigan, in Wilson, ed., p. 258.
20. Margery Kempe, *The Book of Margery Kempe*, edited by W. Butler-Bowden (New York: The Devin-Adair Company, 1944), p. 25.
21. Kempe, p. 83; for examples of her visions of Jesus and Mary, see pp. 192, 279.
22. See Ronald Bainton, *Women of the Reformation from Spain to Scandinavia* (Minneapolis: Augsburg Publishing House, 1977), p. 31.
23. See Fritz Rörig, *The Medieval Town* (Berkeley: University of California Press, 1967), p. 116; Southern, *Church*, p. 325; Ernest McDonnell, *The Beguines and Beghards in Medieval Culture* (New York: Octagon Books, 1969), figures for Namur, pp. 69–70. In their zeal and their numbers the Beguines are part of what historians identify as *die Frauenfrage*, "women question": Why are the numbers of women out of all proportion to the numbers of men who responded in this period of religious enthusiasm? For speculation on this question, see Brenda Bolton, "Mulieres Sanctae," in Susan Mosher Stuard, ed., *Women in Medieval Society* (Philadelphia: The University of Pennsylvania Press, 1976).
24. McDonnell has compiled the names of some of the best-known *mulieres sanctae*, around whom beguinages were established. Most came from the more well-to-do families of the towns, especially in Flanders and the Rhineland. See McDonnell, pp. 98–99, note.
25. Eileen Power, trans., *The Goodman of Paris (Ménagier de Paris)* (London: George Routledge & Sons, Ltd., 1928), p. 210.
26. Even the origin of the name "Beguine" is believed to have been derogatory, either synonymous with beggar or taken from "Albigensian" (the French heresy suppressed in the first two decades of the fourteenth century). McDonnell gives another theory that links the name to Lambert le Begue, a preacher and reformer in Liège, who died in 1177.
27. Cited in Gwendolyn Bryant, "The French Heroic Beguine: Marguerite Porete," in Wilson, ed., p. 211. For a description of the unique qualities of her book, see Dronke, *Women Writers*, especially pp. 217–28.
28. See Dronke, *Women Writers*, pp. 223–24.
29. Cited in Bryant in Wilson, ed., p. 219; see also Dronke, *Women Writers*, pp. 220–21.
30. Bryant in Wilson, ed., p. 204. The book did not die with the burning and her death. After it was translated into Latin, then Italian and English, scribes forgot the heretic and the heresy. *The Mirror of Simple Souls* became an anonymous book of meditations that would influence later mystics, like the Englishman Richard Rolle and the German Thomas à Kempis. When it was first reprinted in the twentieth century, editors hypothesized a male author. The historians Anne Barstow and Robert Lerner suggest that had she written within the confines of a monastery like her contemporary Mechthild of Magdeburg, or had she had the support of a male advocate as Schwester Katri had Meister Eckert, she would have become a model of piety not of heresy. Barstow, "Porete," pp. 11–12; Robert E. Lerner, *The Heresy of the Free Spirit in the Later Middle Ages* (Berkeley: University of California Press, 1972), p. 1. A woman alone could not claim such access to God or preach of it to others.

31. Cited in Southern, *Church,* p. 330; see Bryant in Wilson, ed., for an outline of the Bull *Ad nostrum* and the list of eight errors of belief, p. 207.
32. Walter L. Wakefield, *Heresy, Crusade and Inquisition in Southern France, 1100–1250* (Berkeley: University of California Press, 1974), p. 71.
33. Sherrin Marshall Wyntjes, "Women in the Reformation Era," in Renate Bridenthal and Claudia Koonz, eds., *Becoming Visible: Women in European History* (Hopewell, N.J.: Houghton Mifflin Company, 1977), p. 169.

PART III: 4. Authority Given and Taken Away: The Protestant and Catholic Reformations

1. Cited in Roland Bainton, Women of the Reformation in Germany and Italy (Boston: Beacon Press, 1971), p. 69.
2. Cited in Patricia Higgins, "The Reactions of Women with Special Reference to Women Petitioners," in Brian Manning, ed., *Politics, Religion and the English Civil War* (London: Edward Arnold, 1973), p. 216.
3. Cited in Jane Dempsey Douglass, "Women and the Continental Reformation," in Ruether, ed., *Religion and Sexism,* p. 305.
4. For her life, see Bainton, *Germany and Italy,* pp. 125–42.
5. For a description of her final confrontation with Henry and her capitulation, see John Foxe, *Actes and Monuments* (Boston: AMS Press Inc., 1965), p. 166.
6. Bainton, *Spain to Scandinavia,* p. 212.
7. For her life, see Bainton, *Germany and Italy,* pp. 235–50.
8. For the other patrons of Bernardo Ochino, the general of this new order, see Bainton, *Germany and Italy,* pp. 171–232. One example is Vittoria Colonna (1490–1547), the friend of the artist Michelangelo.
9. See, for France, Nancy L. Roelker, "The Appeal of Calvinism to French Noblewomen in the Sixteenth Century," *The Journal of Interdisciplinary History,* Vol. II, no. 4 (Spring 1972), p. 398.
10. Cited in Roland Bainton, *Women of the Reformation in France and England* (Boston: Beacon Press, 1975), p. 197.
11. Carolly Erickson, *Bloody Mary* (Garden City, N.Y.: Doubleday & Company, Inc., 1978), p. 309.
12. Bainton, *France and England,* p. 205. It should be noted that Mary did not change two key features of Henry VIII's policies: She remained head of the English Church before the Pope, and she allowed the nobility to keep the lands confiscated from the Church.
13. Foxe, p. 209.
14. Erickson, *Mary,* p. 460.
15. Cited in Natalie Zemon Davis, *Society and Culture in Early Modern France* (Stanford, Calif.: Stanford University Press, 1975), p. 295.
16. Cited in Bainton, *France and England,* p. 60.
17. Cited in Bainton, *France and England,* p. 67.
18. Cited in Bainton, *France and England,* p. 67.
19. Historians have just begun to document the relative numbers of women and men involved in all of the various aspects of the Protestant Reformation. Except for

Nancy Roelker's study of French noblewomen and Claus Peter Clasen's study of the Anabaptists, *Anabaptism: A Social History 1525–1618* (Ithaca, N.Y.: Cornell University Press, 1972), the work on individual towns and isolated congregations reveals no discernible pattern. Women of every educational level, of all types of occupations joined; sometimes they were in the majority, sometimes the minority. See for example, Davis, *Society and Culture,* pp. 80–81, and Keith Thomas, "Women and the Civil War Sects," in Trevor Aston, ed., *Crisis in Europe 1560–1660* (New York: Basic Books, Inc., Publisher, 1965), pp. 336, 337, note; see also Keith L. Sprunger, "God's Powerful Army of the Weak: Anabaptist Women of the Radical Reformation," in Richard L. Greaves, ed., *Triumph over Science* (Westport, Conn.: Greenwood, 1985).

20. Cited in David Weigall, "Women Militants in the English Civil War," *History Today,* Vol. XXII, no. 6 (June 1972), p. 437.
21. Mabel Richmond Brailsford, *Quaker Women, 1650–1690* (London: Duckworth, 1915), p. 268.
22. Cited in Bainton, *Germany and Italy,* p. 106.
23. Cited in Bainton, *Germany and Italy,* p. 73.
24. Cited in Thomas, in Aston, ed., p. 338.
25. Cited in Brailsford, p. 153.
26. Cited in Isabel Ross, *Margaret Fell: Mother of Quakerism* (New York: Longmans, 1949), pp. 7, 13.
27. Cited in Ross, *Fell,* p. 15.
28. Cited in Ross, *Fell,* p. 15.
29. Cited in Ross, *Fell,* p. 49.
30. Cited in Ross, *Fell,* p. 58.
31. Cited in Ross, *Fell,* p. 155.
32. Cited in Ross, *Fell,* p. 199; she had previously suggested to him that the plague, the fire of 1666, and the Dutch wars were punishments for his persecutions.
33. Cited in Ross, *Fell,* p. 40.
34. Margaret Fell, *Womens Speaking* (Amherst: Mosher Book & Tract Company, 1980), p. 4.
35. Fell, p. 11.
36. Fell, p. 9.
37. Fell, p. 16.
38. Mary Teresa Neylan, trans., *The Writings of St. Angela Merici* (Rome: Ursulines of the Roman Union, 1969), pp. 30, 31–32.
39. Merici, p. 12.
40. Merici, pp. 15–16.
41. The foundresses of other such orders were also honored by the Church: for example, Montaigne's niece Saint Joan de Lestonnac (1556–1640), who founded the Sisters of Notre Dame of Bordeaux, and Saint Madeleine Sophie Barat (1779–1865), the founder of the Society of the Sacred Heart.
42. Edith Deen, *Great Women of the Christian Faith* (New York: Harper & Row, Publishers, 1959), p. 335.
43. Cited in Monseigneur Bougaud, *History of St. Vincent de Paul,* Vol. I, translated by Rev. Joseph Brady (New York: Longmans, Green, and Co., 1899), p. 310.

44. Lucy Ridgely Seymer, *A General History of Nursing* (London: Faber & Faber Limited, 1932), p. 59.
45. Cited in Bougaud, Vol. II, p. 115.
46. Cited in Bainton, *Germany and Italy,* p. 152.
47. Foxe, p. 396.
48. Foxe, p. 366.
49. Cited in Davis, *Society and Culture,* p. 79.
50. Cited in Bainton, *Germany and Italy,* p. 146.
51. Hans J. Hillerbrand, ed., *The Protestant Reformation* (New York: Harper & Row, Publishers, 1968), p. 150.
52. Cited in Davis, *Society and Culture,* p. 92.
53. Cited in Bainton, *Spain to Scandinavia,* p. 70.
54. Cited in Bainton, *France and England,* p. 188.
55. Hillerbrand, ed., p. 151.
56. Hillerbrand, ed., p. 147.
57. Hillerbrand, ed., p. 148.
58. Hillerbrand, ed., p. 151.
59. Cited in Phyllis Mack, "Women as Prophets During the English Civil War," *Feminist Studies,* vol. 8, no. 1 (Spring 1982), p. 30.
60. Cited in Mack, p. 22.
61. Cited in Mack, pp. 28–29.
62. Cited in Mack, p. 28.
63. Cited in Hannelore Sachs, *The Renaissance Woman,* translated by Marianne Hertzfeld (New York: McGraw-Hill Book Company, 1971), p. 48.
64. Cited in Bainton, *Germany and Italy,* p. 247.
65. Cited in Weigall, p. 436, and in Higgins in Manning, ed., p. 212.
66. Cited in Higgins in Mannings, ed., p. 213.
67. Women's groups continued to have limited responsibilities and to meet separately until 1907. Only in 1943 did a woman act as clerk of the Yearly Meeting, the governing body of the Quakers.
68. P. S. Lewis, *Later Medieval France: The Polity* (New York: St. Martins Press, 1968), p. 206.
69. See Enid McLeod, *The Order of the Rose: The Life and Ideas of Christine de Pizan* (Totowa, N.J.: Rowman and Littlefield, 1976), pp. 55–56.
70. See Power, *Nunneries,* pp. 674–75.
71. Cited in Power, *Nunneries,* p. 676.
72. See for example Rev. J. Waterworth, trans., *The Canons and Decrees of the Sacred and Ecumenical Council of Trent* (New York: Catholic Publication Society Company, 1848), p. 242.
73. Cited in Mary Daly, *The Church and the Second Sex* (New York: Harper & Row, Publishers, 1975), p. 104.
74. Cited in Daly, p. 104.
75. Cited in Daly, p. 106. Despite the efforts of the Church, the order did not die. A group of teachers obtained permission from Pope Benedict XIV in the mid-eighteenth century to follow a modified version of her rule. In the twentieth century, the

papacy allowed the teaching sisters to mention her name and to acknowledge her as their foundress.

76. Cited in Power, *Nunneries,* p. 676.
77. Cited in Bougaud, Vol. I, p. 285.
78. Cited in Bougaud, Vol. I, p. 284.
79. *The Autobiography of St. Thérèse of Lisieux* (Garden City, N.Y.: Doubleday, 1957), p. 20.
80. Most extreme of these examples is the young peasant girl Maria Goretti (1890–1902) canonized in 1950. She had joined no order, had received no holy visions, she was a simple Catholic girl who at the age of eleven was stabbed to death by her rapist.
81. Cited in Hirn, p. 311. On the reasoning of theologians, see for example Joan M. Ferrante, *Woman as Image in Medieval Literature from the Twelfth Century to Dante* (New York: Columbia University Press, 1975), p. 108.
82. For official Church policies, see Waterworth, *Council of Trent,* p. 47.

PART III: **5. Traditional Images Redrawn**

1. Cited in Vern L. Bullough, "Medieval Medical and Scientific Views of Women," in John Leyerle, ed., "Marriage in the Middle Ages," *Viator,* vol. 4 (1973), p. 487.
2. Cited in Ian Maclean, *The Renaissance Notion of Woman: A Study of the Fortunes of Scholasticism and Medical Science in European Intellectual Life* (New York: Cambridge University Press, 1980), pp. 9–10.
3. Cited in Thomas, in Aston, ed., p. 340.
4. Cited in Katharine M. Rogers, *The Troublesome Helpmate: A History of Misogyny in Literature* (Seattle: University of Washington Press, 1966), p. 146.
5. Cited in Davis, *Society and Culture,* p. 125, and in Bainton, *Spain to Scandinavia,* p. 73.
6. Cited in Atkinson, *Kempe,* p. 184.
7. Scott-Moncrieff, ed., *Letters,* p. 45.
8. Cited in Bynum, *Jesus,* p. 145.
9. Cited in Hugo Rahner, S.J., *Saint Ignatius Loyola* (New York: Herder and Herder Inc., 1960), p. 12.
10. The Eastern Church took a less negative view of marriage for its clergy. This was not, however, the issue that caused the separation of the Eastern and Western branches of Christianity. Differences in observance, ritual, doctrine, and jurisdiction existed from the eighth century. The acknowledged separation of 1054 came over the question of the authority of the Roman Pope. See Southern, *Church,* pp. 53–72.
11. Waterworth, *Council of Trent,* p. 270.
12. Cited in Douglass in Ruether, ed., *Religion and Sexism,* p. 296.
13. Douglass in Ruether, ed., *Religion and Sexism,* p. 298.
14. Cited in George H. Tavard, *Woman in Christian Tradition* (Notre Dame: University of Notre Dame Press, 1973), p. 173.
15. Cited in Douglass in Ruether, ed., *Religion and Sexism,* p. 296.

16. Cited in Douglass in Ruether, ed., *Religion and Sexism,* p. 299.
17. Cited in Jean-Louis Flandrin, "Contraception, Marriage and Sexual Relations in the Christian West," in Robert Forster and Orest Ranum, eds., *Biology of Man in History* (Baltimore: The Johns Hopkins University Press, 1975), pp. 35–36.
18. Cited in Christopher Hill, *The World Turned Upside Down: Radical Ideas During the English Revolution* (London: Temple Smith, 1972), p. 250.
19. Cited in Flandrin in Forster and Ranum, eds., *Biology,* p. 167.
20. Cited in Louis B. Wright, *Middle-Class Culture in Elizabethan England* (Chapel Hill, N.C.: University of North Carolina Press, 1935), p. 222.
21. Cited in John Noonan, *Contraception* (Cambridge, Mass.: Belknap Press, 1965), p. 313.
22. Cited in Bainton, *Germany and Italy,* p. 55.
23. Cited in Bainton, *Germany and Italy,* p. 154.
24. Cited in Deen, p. 96.
25. Cited in Bainton, *Germany and Italy,* p. 26.
26. Cited in Douglass in Ruether, ed., *Religion and Sexism,* p. 301.
27. Some Protestant sects allowed divorce, but in circumstances almost as limited as those of the Catholic faith. See Part V, "Women of the Walled Towns," for examples.
28. Cited in Bainton, *Germany and Italy,* p. 43.
29. Cited in Tavard, p. 174.
30. Cited in Margaret George, "From Goodwife to Mistress: The Transformation of the Female in Bourgeois Culture," *Science and Society,* Vol. XXXVII, no. 2 (Summer 1973), p. 166.
31. Cited in George Edwin Fussell and K. R. Fussell, *The English Countrywoman: A Farmhouse Social History 1500–1900* (London: A. Melrose, 1953), p. 48.
32. Cited in Flandrin in Forster and Ranum, eds., *Biology,* p. 218.
33. Cited in Doris Mary Stenton, *The English Woman in History* (New York: The Macmillan Company, 1957), p. 109.
34. Cited in Flandrin in Forster and Ranum, eds., *Biology,* p. 127.
35. Cited in Douglass in Ruether, ed., *Religion and Sexism,* p. 297.
36. Cited in John A. Phillips, *Eve: The History of an Idea* (New York: Harper & Row, Publishers, 1984), p. 105.
37. Cited in L. V. Schnucker, "The English Puritans and Pregnancy, Delivery and Breast-Feeding," *History of Childhood Quarterly,* Vol. I, no. 4 (Spring 1974), p. 651.
38. Cited in Noonan, p. 423, note.
39. Cited in Tavard, p. 173.
40. Cited in Wyntjes, in Bridenthal and Koonz, eds., p. 174.
41. Cited in George, p. 169.
42. Cited in Deen, p. 96.
43. Barbara Beuys, *Familienleben in Deutschland: Neue Bilder aus der deutschen Vergangenheit* (Reinbek bei Hamburg: Rowohlt, 1980), p. 224.
44. Cited in Deen, p. 325.
45. Cited in Bainton, *France and England,* p. 88.

PART III: **6. The Legacy of the Protestant Reformation**

1. Cited in Carole Levin, "Women in the Book of Martyrs as Models of Behavior in Tudor England," *International Journal of Women's Studies,* vol. 4, no. 2 (March/April 1981), p. 202.
2. Cited in E. M. Williams, "Women Preachers of the Civil War," *The Journal of Modern History,* Vol. I, no. 4 (December 1929), p. 568.
3. Cited in Higgins in Manning, ed., p. 212.
4. Cited in Higgins in Manning, ed., pp. 210, 211.
5. Cited in Joyce L. Irwin, "Anna Maria van Schurman: The Star of Utrecht," in J. R. Brink, ed., *Female Scholars: A Tradition of Learned Women Before 1800* (Montreal: Eden Press Women's Publications, 1980), p. 79.

PART IV. WOMEN OF THE CASTLES AND MANORS:
CUSTODIANS OF LAND AND LINEAGE

PART IV: **1. From Warrior's Wife to Noblewoman: The Ninth to the Seventeenth Centuries**

1. The use of literature as a source for a historical period is always problematic. In this section literary sources are referred to when they can be corroborated from more traditional historical sources such as legal or ecclesiastical documents. On the use of the literature of this period as a source for women's lives and the attitudes they suggest, see Penny Schine Gold, *The Lady and the Virgin: Image, Attitude, and Experience in Twelfth-Century France* (Chicago: University of Chicago Press, 1985), pp. xvii, 3–4, 12, 20–26, 37–42, 103–104, 148. By the end of the thirteenth century all of the major oral works had been given written form. Of the longer genres, the epic, saga, chanson de geste, and romance all had been part of the entertainment of the castles, stories sung to the chieftains and their thanes, to the feudal lords and their vassals. The following have been most frequently referred to in this chapter: Joan M. Ferrante, trans., *Guillaume d'Orange: Four Twelfth-Century Epics* (New York: Columbia University Press, 1974); Jessie Crosland, trans., *Raoul de Cambrai* (London: Chatto & Windus, 1926); C. K. Scott Moncrieff, trans., *The Song of Roland* (Ann Arbor: The University of Michigan Press, 1960); W. S. Merwin, trans., *The Poem of the Cid,* and Helen M. Mustard, trans., *The Nibelungenlied,* in *Medieval Epics* (New York: The Modern Library, 1963); Magnus Magnusson and Hermann Palsson, trans., *Njal's Saga* (New York: Penguin Books, 1977); Chrétien de Troyes, *Arthurian Romances,* translated by W. W. Comfort (New York: Dutton, 1978) including *Lancelot, Erec et Enide, Cligés;* Gottfried von Strassburg, *Tristan and Isolde with the Surviving Fragments of the Tristan of Thomas,* translated by A. T. Hatto (New York: Penguin Books, 1978). For a general description of the evolution of the different forms and their characteristics, see Geoffrey Brereton, *A Short History of French Literature* (Baltimore: Penguin Books, 1968), pp. 12ff.

PART IV: **2. Constants of the Noblewoman's Life**

1. *Raoul*, p. 150.
2. See *Njal's Saga*, p. 88; *Raoul*, p.126.
3. Cited in Pierre Riché, *Daily Life in the World of Charlemagne*, translated by Jo Ann McNamara (Philadelphia: University of Pennsylvania Press, 1978), p. 79.
4. On the origins of the concept of "nobility" by birth, the idea of "lineage," and the evolution of a clearly defined noble class from the ninth to the twelfth centuries, see especially Georges Duby, *The Chivalrous Society*, translated by Cynthia Postan (Berkeley: University of California Press, 1977), pp. 76–77, 98, 198, 173–75, 178–85, and Maurice Keen, *Chivalry* (New Haven: Yale University Press, 1984), pp. 24–26.
5. See Henry S. Bennett, *The Pastons and Their England* (Cambridge: Cambridge University Press, 1970), p. 119.
6. M. H. Keen, *The Laws of War in the Late Middle Ages* (Toronto: University of Toronto Press, 1965), gives an excellent description of this last kind of warfare. For a discussion of all the changes in fighting and the ways in which they are reflected in literature, see R. Howard Bloch, *Medieval French Literature and Law* (Berkeley: University of California Press, 1977).
7. Ferrante, ed., *Guillaume*, p. 267.
8. Edward Noble Stone, trans., *The Song of William* (Seattle: University of Washington Press, 1951), p. 51.
9. *Lancelot*, p. 302.
10. See Wolfram Von Eschenbach, *Parzival*, translated by Helen M. Mustard and Charles E. Passage (New York: Vintage Books, 1961), p. 125.
11. *Erec*, p. 7.
12. *Erec*, p. 68.
13. *Parzival*, p. 92.
14. *Aucassin and Nicolette* in Angel Flores, ed., *Medieval Age* (New York: Dell Publishing Co., Inc., 1963), p. 442.
15. *Tristan*, p. 146.
16. Eileen Power, *Medieval English Nunneries* (Cambridge: Cambridge University Press, 1922), pp. 429–30.
17. Joseph Dahmus, *Seven Medieval Queens: Vignettes of Seven Outstanding Women of the Middle Ages* (Garden City, N.Y.: Doubleday, 1972), p. 130.
18. Lloyd deMause, "The Evolution of Childhood," in Lloyd deMause, ed., *The History of Childhood* (New York: Harper & Row, Publishers, 1974), p. 33.
19. Geoffrey Brereton, ed. and trans., *Froissart Chronicles* (New York: Penguin Books Ltd., 1978), p. 61.
20. Dorothy Atkinson, "Society and Sexes in the Russian Past," in Dorothy Atkinson, Alexander Dallin and Gail Warshofsky Lapidus, eds., *Women in Russia* (Stanford, Calif.: Stanford University Press, 1977), p. 20.
21. See Dahmus, p. 78, or JoAnn McNamara and Suzanne Wemple, "The Power of Women Through the Family in Medieval Europe: 500–1100," in Mary Hartman and Lois W. Banner, eds., *Clio's Consciousness Raised: New Perspectives on the History of Women* (New York: Harper & Row, Publishers, 1974), p. 107.
22. Dahmus, p. 181.

23. See Dahmus, p. 287.
24. Cited in Dahmus, p. 120.
25. Cited in Dahmus, p. 137. For these queens, see in addition to Dahmus, Eleanor Shipley Duckett, *Medieval Portraits from East and West* (Ann Arbor: The University of Michigan Press, 1972), Charles E. Odegaard, "The Empress Engelberge," *Speculum: A Journal of Medieval Studies,* Vol. XXVI, no. 1 (January 1951), pp. 77–103, and the survey history by Pauline Stafford, *Queens, Concubines, and Dowagers: The King's Wife in the Early Middle Ages* (Athens, Ga.: The University of Georgia Press, 1983).
26. G. N. Garmonsway, trans., *The Anglo-Saxon Chronicle* (New York: E. P. Dutton & Co., Inc., 1962), p. 96.
27. Earlier English queens of Wessex and Mercia, like Ine and Cynethyth, had fought for their husbands, also minting their own coins, a symbol of their importance in the realm.
28. See Doris M. Stenton, *The English Woman in History* (New York: The Macmillan Company, 1957), and F. T. Wainwright, "Aethelfled, Lady of the Mercians," in *Scandinavian England: Collected Papers of F. T. Wainwright,* H. P. R. Finberg, ed. (Chichester, West Sussex: Phillimore and Co., Ltd., 1975), and the most recent work on these women, Christine Fell, *Women in Anglo-Saxon England and the Impact of 1066* (Bloomington, Ind.: Indiana University Press, 1984).
29. Cited in Keen, *War,* p. 89. See Stenton, and Marion F. Facinger, "A Study of Medieval Queenship: Capetian France 987–1237," *Studies in Medieval and Renaissance History,* vol. 5 (1968), pp. 3–48.
30. See Dahmus, pp. 276–327.
31. See Patricia Higgins, "The Reactions of Women, with Special Reference to Women Petitioners," in Brian Manning, ed., *Politics, Religion and the English Civil War* (London: Edward Arnold, 1973), and Maurice Ashley, *The Stuarts in Love* (New York: Macmillan, 1964).
32. Cited in Roland H. Bainton, *Women of the Reformation in France and England* (Boston: Beacon Press, 1975), p. 86.
33. For the women of the Fronde, see Dorothy Anne Liot Backer, *Precious Women: A Feminist Phenomenon in the Age of Louis XIV* (New York: Basic Books, Inc., Publishers, 1974), and G. P. Gooch, *Courts and Cabinets* (New York: Alfred A. Knopf, 1946).
34. See J. C. Russell, *Late Ancient and Medieval Population* (Philadelphia: The American Philosophical Society, 1958), p. 19; and David Herlihy, "The Medieval Marriage Market," *Medieval and Renaissance Studies,* Series No. 6, Dale B. J. Randall, ed. (Durham, N.C.: Duke University Press, 1976), p. 18.
35. Margaret Wade Labarge, *Saint Louis: Louis IX Most Christian King of France* (Boston: Little, Brown and Company, 1968), p. 56.
36. *Froissart,* p. 256.
37. *Froissart,* p. 257.
38. *Froissart,* p. 259.
39. See Georges Duby, *Medieval Marriage: Two Models from Twelfth-Century France,* translated by Elborg Forster (Baltimore: The Johns Hopkins University Press, 1978), pp. 92, 5.

40. Cited in Johannes Jørgensen, *Saint Bridget of Sweden,* Vol. I, translated by Ingeborg Lund (New York: Longmans Green and Co., 1954) p. 48.
41. Cited in John T. Noonan, "Power to Choose" in John Leyerle ed., "Marriage in the Middle Ages," *Viator,* Vol. 4 (1973), p. 247.
42. For the evolution of Church policy on marriage, see Jean Brissaud, *A History of French Private Law* (Boston: Little, Brown and Company, 1912), p. 99ff., and James A. Brundage, "Concubinage and Marriage in Medieval Canon Law," *Journal of Medieval History,* Vol. I (1975), pp. 1–17, and Henry Ansgar Kelly, *Love and Marriage in the Age of Chaucer* (Ithaca, N.Y.: Cornell University Press, 1975) pp. 164–78; see also Rev. J. Waterworth, trans., *The Canons and Decrees of the Sacred and Oecumenical Council of Trent* (New York: Catholic Publication Society Company, 1848) pp. 196–98.
43. Germanic groups continued to give a gift to her family. Such was Viking practice into the thirteenth century. In England the bride's parents received a gift compensating them for the expenses of her upbringing. See P. G. Foote and D. M. Wilson, *The Viking Achievement* (New York: Praeger Publishers, 1970) pp. 112–13; and Dorothy Whitelock, ed., *English Historical Documents* (New York: Oxford University Press, 1968), p. 431.
44. Cited in Lorraine Lancaster, "Kinship in Anglo-Saxon Society (Seventh Century to Early Eleventh)," in Sylvia L. Thrupp, ed., *Early Medieval Society* (New York: Appleton-Century Crofts, 1967) p. 27.
45. Dahmus, p. 113.
46. For a survey description of practice see Diane Owen Hughes, "From Brideprice to Dowry in Mediterranean Europe," in Marion A. Kaplan, ed., *The Marriage Bargain: Women and Dowries in European History* (New York: Harrington Park Press, 1985); for more specialized descriptions see Diane Owen Hughes, "Urban Growth and Family Structure in Medieval Genoa," *Past and Present,* vol. 66 (February, 1975) pp. 13–14, and Herlihy, "Medieval Marriage," p. 7.
47. See for example, *The Secular Spirit: Life and Art at the End of the Middle Ages* (New York: E. P. Dutton & Co., Inc., 1975) p. 254.
48. See Inga Dåhlsgard, *Women in Denmark: Yesterday and Today* (Copenhagan: Det Danski Selskab, 1980) p. 23; Atkinson in Atkinson et al., p. 30, note; J.-L. Flandrin, *Families in Former Times: Kinship, Household and Sexuality,* translated by Richard Southern (New York: Cambridge University Press, 1979) pp. 130–33.
49. Meg Bogin, *The Women Troubadours* (New York: W. W. Norton and Company, 1976) p. 145.
50. Cited in G. G. Coulton, *Medieval Panorama: The English Scene from Conquest to Reformation* (Cambridge: Cambridge University Press, 1939) p. 644.
51. *Froissart,* p. 253.
52. C. H. Talbot, ed. and trans., *The Life of Christina of Markyate: A Twelfth Century Recluse* (Oxford: Clarendon Press, 1959) p. 69.
53. Talbot, ed., *Christina,* p. 69. This kind of marriage proved even more significant for young men, who as early as the ninth and tenth centuries used it as a means to rise. See studies by Duby, *Medieval Marriage,* pp. 10–12, and Constance B. Bouchard, "The Origins of the French Nobility: A Reassessment," *American Historical Review,* vol. 86, no. 3 (June 1981), pp. 514–29.

54. See for example, Peter Laslett, *The World We Have Lost* (New York: Charles Scribner's Sons, 1965) pp. 48–49.
55. See Pauline Stafford, "Sons and Mothers: Family Politics in the Early Middle Ages," in Derek Baker, ed., *Medieval Women* (Oxford: Basil Blackwell, 1978), pp. 96–97.
56. *Anglo-Saxon Chronicle,* pp. 256–57.
57. See Elizabeth A. R. Brown, "Eleanor of Aquitaine: Parent, Queen and Duchess," in William W. Kibler, ed., *Eleanor of Aquitaine: Patron and Politician* (Austin, Tex.: University of Texas Press, 1976) pp. 17–18; Robert Fawtier, *Capetian Kings of France,* translated by Lionel Butler (London: Macmillan, 1965), p. 36; Felipe Fernández-Armesto, *Ferdinand and Isabella* (New York: Taplinger Publishing Company, 1975) pp. 119–23.
58. For examples of laws from the eleventh to the sixteenth century, see the following: for France, Carl Ludwig von Bar, *A History of Continental Criminal Law* (Boston: Little, Brown and Company, 1916), p. 70; for Carolingian Law, Carlo Calisse, *A History of Italian Law,* translated by Layton B. Register (New York: Augustus Kelley, 1969), p. 548; for England, Sir Frederick Pollack and Frederic William Maitland, *The History of English Law Before the Time of Edward I,* Vol. II (London: Cambridge University Press, 1923), pp. 490–91.
59. Waterworth, *Council of Trent,* p. 202.
60. See Heath Dillard, "Women in Reconquest Castile: The Fueros of Sepulveda and Cuenca," in Susan Mosher Stuard, ed., *Women in Medieval Society* (Philadelphia: University of Pennsylvania Press, 1976), p. 80, and Dillard's recent book, *Daughters of the Reconquest: Women in Castilian Town Society, 1100–1300* (New York: Cambridge University Press, 1985).
61. Katharine Simms, "Women in Norman Ireland," in Margaret MacCurtain and Donncha Ó'Corráin, eds., *Women in Irish Society: The Historical Dimension* (Dublin: Arlen House, 1978) p. 17.
62. Bennett, pp. 31–32.
63. See Pollack and Maitland, Vol. II, p. 390.
64. Atkinson in Atkinson et al., eds., p. 30, note.
65. Cited in Marc Bloch, *Feudal Society,* Vol. I, translated by L. A. Manyon (Chicago: University of Chicago Press, 1966) p. 227.
66. See Facinger, p. 14, and Duby, *Medieval Marriage,* pp. 75–79.
67. See Duby, *Medieval Marriage,* p. 87, and Joseph and Francis Gies, *Life in a Medieval Castle* (New York: Harper & Row, Publishers, 1974), pp. 57–60.
68. This ritual is the origin of the term "curfew," *couvre-feu.*
69. Hermann Kellenbenz, "Technology in the Age of the Scientific Revolution 1500–1700," translated by John Nowell, in Carlo M. Cipollo, ed., *The Fontana Economic History of Europe: The Sixteenth and Seventeenth Centuries* (Glasgow: William Collins Sons & Co., 1976) p. 250.
70. Alice Kemp-Welch, *Of Six Medieval Women* (London: Macmillan and Co., Ltd., 1913) p. 93.
71. Kemp-Welch, p. 97.
72. The books are in fact parchment rolls, called membranes, and cover seven months in 1265.
73. Gies, *Castle,* p. 105.

74. For statistics for the "upper gentry" see Miriam Slater, "The Weightiest Business: Marriage in an Upper-Gentry Family in 17th Century England," *Past and Present*, no. 72 (August 1976), p. 27.
75. Laslett, *World Lost*, p. 7.
76. See Isabel Ross, *Margaret Fell: Mother of Quakerism* (New York: Longmans, 1949) p. 12.
77. See Christine de Pizan, *The Treasure of the City of Ladies or the Book of the Three Virtues*, translated by Sarah Lawson (New York: Penguin Books Ltd., 1985), pp. 128–29.
78. Cited in James Bruce Ross and Mary Martin McLaughlin, eds., *The Portable Medieval Reader* (New York: The Viking Press, 1966) p. 128.
79. The descriptions of foods come from a variety of sources: Riché, *Daily Life*, pp. 172–77; Coulton, *Panorama*, p. 313; Dorothy Hartley, *Lost Country Life* (New York: Pantheon Books, 1979), p. 232; Margaret Labarge, *A Baronial Household in the 13th Century* (New York: Barnes and Noble, 1966).
80. Labarge, *Baronial Household*, pp. 90–93, 96, and Georges Duby, *Rural Economy and Country Life in the Medieval West*, translated by Cynthia Postan (Columbia, S.C.: University of South Carolina Press, 1968), p. 227; and G. G. Coulton, *The Medieval Village* (Cambridge: Cambridge University Press, 1926), p. 114.
81. See Gies, *Castle*, pp. 101, 103, and Labarge, *Baronial Household*, pp. 195, 196.
82. See Lucien Febvre, *Life in Renaissance France*, edited and translated by Marian Rothstein (Cambridge, Mass.: Harvard University Press, 1977), p. 96; Keen, *War*, p. 205; Ross, *Fell*, pp. 264–65.
83. Labarge, *Baronial Household*, p. 64.
84. Kemp-Welch, p. 99.
85. Kate Campbell Hurd-Mead, *A History of Women in Medicine: From the Earliest Times to the Beginning of the 19th Century* (Haddam, Conn.: The Haddam Press, 1938) p. 401.
86. Cited in R. W. Southern, *Western Society and the Church in the Middle Ages* (New York: Penguin Books, 1970) p. 263.
87. See David Herlihy, "Land, Family and Women in Continental Europe, 701–1200," in Stuard, ed., *Medieval Society*, p. 28; on average their donations represented about 11 or 12 percent of the gifts, see Table 3, p. 31.
88. Penny S. Gold, "Women and Family in the Alienation of Property in Anjou, 1000–1249: Problems in Method and Interpretation" Appendix, p. 6, unpublished manuscript, April 1979. See also Gold, *Lady*, pp. 134–40 for a more complete study of donations.
89. Penelope Johnson, "Agnes of Burgundy: A Notable Medieval Monastic Patron," paper delivered at The Berkshire Conference on the History of Women, held at Mount Holyoke College, South Hadley, Massachusetts, August 25, 1978.
90. Kemp-Welch, p. 102.
91. Cited in Stenton, p. 33.
92. Dahmus, p. 272.
93. Cited in G. E. Mingay, *The Gentry: The Rise and Fall of a Ruling Class* (New York: Longman Group, Ltd., 1976) p. 136.

94. Dorothy Gardiner, *English Girlhood at School: A Study of Women's Education Through Twelve Centuries* (London: Oxford University Press, 1929) pp. 88–89.
95. See Michael Jones and Malcolm Underwood, "Lady Margaret Beaufort," *History Today*, vol. 35 (August 1985), pp. 23–30.
96. Mary Martin McLaughlin, "Survivors and Surrogates: Children and Parents from the Ninth to the Thirteenth Centuries," in De Mause, ed., *History of Childhood*, p. 125.
97. Gardiner, p. 126.
98. Keith Thomas, *Religion and the Decline of Magic* (New York: Charles Scribner's Sons, 1971) p. 28.
99. R. V. Schnucker, "The English Puritans and Pregnancy, Delivery and Breastfeeding," *History of Childhood Quarterly*, Vol. I, no. 4 (Spring 1974), p. 643.
100. Charles Jackson, ed., *The Autobiography of Mrs. Alice Thornton of East Newton, Co. York*, Surtees Society, Vol. LXII (Edinburgh: 1873), p. 95.
101. Cited in Ross, *Fell*, p. 351.
102. Cited in Joseph E. Illick, "Child-Rearing in Seventeenth Century England and America," in De Mause, ed., *History of Childhood*, p. 305.
103. Jackson, ed., *Thornton*, p. 125.
104. See Ross, *Fell;* facing p. 343.

PART IV: **3. Power and Vulnerability**

1. For the evolution of "lineage" and family names, see in particular two articles by Georges Duby in *Medieval Marriage*, especially p. 10, and "Lineage, Nobility and Chivalry in the Region of Mâcon during the Twelfth Century" in Robert Forster and Orest Ranum, eds., *Family and Society: Selections from the Annales*, translated by Elborg Forster and Patricia M. Ranum (Baltimore: The Johns Hopkins University Press, 1976), pp. 19, 26. This change in naming and identifying the family meant the revival of the narrower lineal, and thus conjugal, definition of family that had prevailed in Greece and Rome.
2. In eleventh-century Spain it required twenty-five ounces of gold to buy two war horses. By the middle of the fifteenth century a member of a company of mounted warriors would expect to spend twenty-six months' wages for his mount. See Pierre Bonnassie, "A Family of the Barcelona Countryside and Its Economic Activities," in Thrupp, ed., *Early Medieval*, p. 121; Malcolm Vale, *War and Chivalry: Warfare and Aristocratic Culture in England, France and Burgundy at the End of the Middle Ages* (Athens, Ga.: The University of Georgia Press, 1981), p. 126.
3. See, for example, Mâconnais in southern Burgundy in Duby, *Chivalrous Society*, pp. 73–74.
4. A few regions did not alter their customs until after the thirteenth century. The Fuero Jusco of Spain in the thirteenth century still retained equal inheritance for daughters and sons, as did some parts of Germany. For Spain, see Julia O'Faolain and Lauro Martines, eds., *Not in God's Image: Women in History from the Greeks to the Victorians* (New York: Harper & Row, Publishers, 1973), p. 148; for Germany see Bloch, Vol. I, p. 204.
5. Cited in Duby, *Rural Economy*, p. 383.

6. Historians suggest that this was also a period of fewer surviving sons. Demographic pressures thus contributed to the phenomenon of female "lords." Noble families limited offspring; there is a naturally greater attrition of infant boys than girls; the centuries saw more or less continuous fighting, with estimates of losses in the Crusades, for example, of from 50,000 to 500,000 European men. See Bogin, pp. 34–35, and JoAnn McNamara and Suzanne F. Wemple, "Sanctity and Power: The Dual Pursuit of Medieval Women," in Renate Bridenthal and Claudia Koonz, eds., *Becoming Visible: Women in European History* (Hopewell, N.J.: Houghton Mifflin Company, 1977). For general descriptions of the establishment of the right of inheritance and the ideas of "lineage" and "patrimony" see the following: Bloch, Vol. I; Duby, *Rural Economy,* and Flandrin. For more specialized discussions, see Emmanuel Le Roy Ladurie, "Family Structures and Inheritance Customs in Sixteenth-Century France," and Joan Thirsk, "The European Debate on Customs of Inheritance, 1500–1700," in Jack Goody, Joan Thirsk and E. P. Thompson, eds., *Family and Inheritance: Rural Society in Western Europe 1200–1800* (New York: Cambridge University Press, 1978); for France, Duby, *Family and Society* and *Chivalrous Society;* Robert Hajdu, "Family and Feudal Ties in Poitou 1100–1300" in the *Journal of Interdisciplinary Studies,* Vol. VIII, no. 1 (Summer 1977), pp. 117–39; for England, Sidney Painter, "The Family and the Feudal System in 12th Century England," *Speculum,* vol. 35, no. 1 (January 1960), pp. 1–16; for the different patterns in Russia see Atkinson in Atkinson et al., eds., p. 11.
7. Although historians have found examples of women exercising feudal authority from as early as the tenth century, there is an increase in the numbers during the eleventh and twelfth centuries. See, for example, the study of southern France and of Spain in Herlihy in Stuard, ed., Table II, p. 27; for France see Duby, *Chivalrous Society,* p. 109; Stenton, pp. 5–6; and Lancaster, p. 34 for Anglo-Saxon and Viking examples in England. In Central Europe in the eleventh and twelfth centuries women gained lands and used them not so much for their families as to enhance the power and influence of the papacy. For example, Matilda, Countess of Tuscany (1046–1115), became heir to the family lands on the death of her older brother and sister. In the conflict between the Holy Roman Emperor Henry IV and Pope Gregory VII, she used her powers on the side of the Church. With Abbot Hugh of Cluny she oversaw the reconciliation between the Pope and the Emperor in 1077 at Canossa, one of her castles. The land she granted the Pope became the nucleus of the future Papal States. In Spain, Urraca (c. 1080–1126) reigned as queen of Leon and Castile, inheriting as her father's heir, later ruling for her son; see the biography, Kevin B. Reilly, *The Kingdom of Léon-Castilla Under Queen Urraca 1109–1126* (Princeton: Princeton University Press, 1982).
8. For their lives see John T. Appleby, *Henry II: The Vanquished King* (New York: Macmillan Co., 1962); Facinger, Marion Meade, *Eleanor of Aquitaine: A Biography* (New York: Hawthorn Books, Inc., 1977); Brown in Kibler, ed.
9. She had been named for her mother Aenor; she was "the other Aenor" or "Alia-Aenor."
10. On this reasoning of Eleanor's, see Facinger, p. 8.
11. On Louis' reluctance to consider divorce, see Meade, pp. 108–109; and Steven

Runciman, *A History of the Crusades* Vol. II (New York: Harper & Row, Publishers, 1967), p. 279.

12. See Meade, p. 144.
13. Meade, p. 187.
14. The children: William, born August 1153 (died 1156); Henry, February 1155; Matilda, June 1156; Richard, September 1157; Geoffrey, September 1158; Eleanor, September 1161; Joanna, October 1165; John, December 1166. See Brown in Kibler, ed., p. 16.
15. Meade, pp. 280–81.
16. Cited in Dahmus, p. 225.
17. Cited in Dahmus, p. 225.
18. 100,000 silver marks: Meade, p. 320.
19. From the eleventh to the thirteenth centuries, outside Europe in the Crusader Kingdoms of the eastern Mediterranean, the great magnates survived and flourished. There for almost 200 years they lived out a warrior's feudal dream: rich territories gained by right of conquest, held by right of inheritance, with their kings too far away to contest or curtail their power or their authority. In this complete warrior world, women fulfilled every role from the most powerful to the most vulnerable, from pawns to rulers in their own right. For an account of the Crusades and the kingdoms established see Steven Runciman's three-volume history, especially Volumes II and III. For more specialized studies of Agnes of Courtney and Melisende, see Bernard Hamilton, "Women in the Crusader States: the Queens of Jerusalem 1100–1190," in Baker, ed.
20. Alcuin cited in Riché, *Daily Life,* p. 204. For praise of Amalaswintha, daughter of King Theodoric the Ostrogoth, Judith, Queen of the Franks, Adehlheid, Empress to Otto I, Adela, daughter of the Norman William the Conqueror, see Joan M. Ferrante, "The Education of Women in the Middle Ages in Theory, Fact and Fantasy," in Patricia H. Labalme, ed., *Beyond Their Sex: Learned Women in the European Past* (New York: New York University Press, 1980).
21. Cited in Sybille Harksen, *Women in the Middle Ages,* translated by Marianne Herzfeld (New York: Abner Schram, 1975), p. 13.
22. For descriptions of Mahaut's library see Kemp-Welch, pp. 85–86; for her books and other women's (into the fourteenth century), see Susan Groag Bell, "Medieval Women Book Owners: Arbiters of Lay Piety and Ambassadors of Culture," *Signs,* vol. 7, no. 4 (Summer 1982), pp. 747–60.
23. There is no clear evidence of Eleanor's role in this cultural change. Brown suggests that Henry II's court had more significance than hers as a center. See Brown in Kibler, ed., p. 19.
24. Brown in Kibler, ed., introduction, p. 6.
25. See Chrétien de Troyes, p. 270.
26. For a description of gowns such as these see *Tristan,* p. 185.
27. *Lancelot,* p. 291.
28. Cited in Gardiner, p. 54, note.
29. Gardiner, p. 62.
30. *Secular Spirit,* p. 81. The ladies of the sixteenth and seventeenth century created

the full range of stitches preserved in the embroidery known as crewel (wool or linen) and worked on all varieties of finer cloth for dresses, wall hangings, upholstery, and bedclothes.

31. Bogin, p. 13.
32. See Peter Dronke, "The Provençal Trobairitz Castelloza" in Katharina M. Wilson, ed., *Medieval Women Writers* (Athens, Ga.: The University of Georgia Press, 1984); see also Matilda Tomaryn Bruckner, "Na Castelloza, Trobairitz, and Troubadour Lyric," *Romance Notes*, Vol. XXV, no. 3 (Spring 1985), pp. 239–53. This volume also has a very good bibliography on the subject of the trobairitz and other women poets of these centuries.
33. Bogin, p. 81.
34. Bogin, p. 131.
35. Bogin, p. 103.
36. Bogin, p. 131.
37. Cited in Peter Dronke, *The Medieval Lyric* (New York: Cambridge University Press, 1977), p. 105.
38. Cited in Dronke, *Medieval Lyric,* p. 105.
39. Cited in Dronke, *Medieval Lyric,* p. 106; the literal translation of the passage is "in a (or the) husband's place" and could also mean, "as a husband to me." The authors are grateful to Joan Ferrante for bringing this to their attention.
40. Bogin, p. 85; see also Keen, *Chivalry,* p. 21.
41. Bogin, p. 155.
42. Bogin, p. 111.
43. Cited in Dronke, *Medieval Lyric,* p. 106.
44. Bogin, p. 107.
45. Bogin, p. 85.
46. See Flores, ed., *Medieval Age,* p. 185.
47. Bogin, p. 95.
48. Bogin, p. 95.
49. Marie de France, *Milun,* p. 164.
50. Robert Hanning and Joan Ferrante, eds. and trans., *The Lais of Marie de France* (New York: E. P. Dutton, 1978), p. 29; the Ferrante and Hanning translation and edition of her lais gives biographical information, as does Ferrante's article "The French Courtly Poet, Marie de France," in Wilson, ed.
51. *The Purgatory of St. Patrick, Aesop's Fables.*
52. Marie de France, *Milun,* p. 16.
53. Marie de France, *Guigemar,* p. 30.
54. Marie de France, *Milun,* p. 170.
55. See *Lancelot,* p. 330; *Tristan,* pp. 276–277.
56. Marie de France, *Lanval,* p. 108.
57. Marie de France, *Lanval,* p. 121.
58. Marie de France, *Lanval,* p. 120
59. Marie de France, *Lanval,* p. 108.
60. Marie de France, *Lanval,* p. 112.
61. Marie de France, *Eliduc,* p. 204.
62. Marie de France, *Yonec,* pp. 141, 142, 143.

63. Marie de France, *Lanval,* p. 108; see also *Yonec;* and *Guigemar,* p. 45.
64. Marie de France, *Milun,* p. 176.
65. Marie de France, *Equitan,* p. 60.
66. Marie de France, *Laüstic,* p. 157.
67. Marie de France, *Laüstic,* p. 158.
68. Marie de France, *Eliduc,* p. 226.
69. Marie de France, *Laüstic,* pp. 155, 158.
70. Marie de France, *Chaitivel,* p. 182.
71. Magnusson and Palsson, *Njal,* p. 73.
72. Ferrante, ed., *Guillaume,* p. 230.
73. *Parzival,* p. 43.
74. *Tristan,* p. 89. There is no suggestion of effeminacy as this description comes just after Tristan has slain a dragon.
75. *Tristan,* p. 110.
76. *Lancelot,* p. 287.
77. On the evolution of the concept of chivalry and its connection to the creation of a separate nobility based on lineage and a code of behavior, see Keen, *Chivalry,* Introduction, Chapter VIII, and Conclusion, especially pp. 2, 42, 143, 145, 160–61.
78. *Froissart,* p. 67.
79. The Order of the Garter in England, c. 1344; of the Star in France in 1351; of St. Michael in 1469; of the Golden Fleece in 1429; and the Order of the Crescent of the Dukes of Burgundy in 1448. See Duby, *Chivalrous Society,* for a description of the relationship between knighthood, nobility, and heredity, pp. 95–96, 158–70, and Keen, *Chivalry,* pp. 26, 145–46, 153–60.
80. See Keen, *Chivalry,* for descriptions of the evolution of the ethical aspects of chivalry, and for summaries of the principal guides, pp. 2, 4–13. See also Vale, pp. 25–26.
81. Cited in C. T. Allmand, ed., *Society at War: The Experience of England and France During the Hundred Years War* (New York: Barnes and Noble, 1973), pp. 26–27.
82. Cited in Painter, p. 145.
83. Painter, p. 146.
84. Cited in Johan Huizinga, *The Waning of the Middle Ages* (Garden City, N.Y.: Doubleday & Company, Inc., 1954), p. 75. From the late twelfth to the fifteenth century, men wrote guides and handbooks of behavior, the most popular being Ramon Lull's *Le Libre del Orde de Cauayleria* (written in the thirteenth century, used into the fifteenth); the English *Book of St. Albans* gave levels of gentility and the requirements of each.
85. See Keen, *War,* pp. 245–47.
86. *Le Jouvençel,* cited in Vale, p. 30.
87. Fernández, p. 98.
88. Fernández, p. 99; for the analysis of the gradual shift, see Keen, *War,* pp. 245–47; see also Keen, *Chivalry,* pp. 238–47. Vale gives an example from Burgundy, pp. 147–54.
89. Cited in Régine Pernoud, *The Retrial of Joan of Arc,* translated by J.M. Cohen (New York: Harcourt Brace and Company, 1955), p. 139.
90. See Duby, *Rural Economy,* pp. 260–61 and Georges Duby, *The Early Growth of*

the European Economy: Warriors and Peasants from the Seventh to the Twelfth Century, translated by Howard B. Clarke (Ithaca, N.Y.: Cornell University Press, 1974), p. 221.

91. See Marc Bloch, *French Rural History: An Essay on Its Basic Characteristics,* translated by Janet Sondheimer (Berkeley: University of California Press, 1966) p. 120, and Henry Kamen, *The Iron Century: Social Change in Europe 1550–1660* (New York: Praeger Publishers, 1971), p. 73.
92. For the most complete description of the changes in the design of houses in these centuries, see Margaret Wood, *The English Medieval House* (London: Phoenix House, 1965).
93. Kemp-Welch, p. 107.
94. Gladys S. Thomson, *Life in a Noble Household, 1671–1700* (Ann Arbor: University of Michigan Press, 1959), p. 383.
95. Harksen, p. 16.
96. Labarge, *Household,* p. 35.
97. Joan Evans, ed., *The Flowering of the Middle Ages* (New York: McGraw-Hill Book Company, 1966), p. 172.
98. Horace Dewey and Ann Kleimole, eds. and trans., *Russian Private Law XIV–XVII Centuries: An Anthology of Documents* (Ann Arbor: The University of Michigan, 1973), pp. 222–23.
99. See Duby, *Family and Society,* pp. 31–32.
100. Duby, *Family and Society,* pp. 31–32.
101. Bonnassie in Thrupp, ed., p. 121.
102. Labarge, *Saint Louis,* p. 235.
103. Margaret Fell inherited £3,000 and the manor of Marsh Grange from her father as her marriage portion. Ross, *Fell,* p. 5. See also Mingay, pp. 11–12, for examples of sixteenth- and seventeenth-century bequests to daughters by English gentry families.
104. J. P. Cooper, "Patterns of Inheritance and Settlement by Great Landowners from the Fifteenth to the Eighteenth Centuries," in Goody et al., eds., *Family and Inheritance,* p. 258.
105. This development emerges in the studies done by Diane Owen Hughes. See "Brideprice" in Kaplin, ed. Many historians have contributed to the analysis of these shifting patterns of European inheritance and their effects on women's access to land. For summary descriptions, see Jack Goody and J. P. Cooper in Goody et al., eds., *Family and Inheritance,* and the two articles by Jo Ann McNamara and Suzanne Wemple in Hartman and Banner, eds., and Bridenthal and Koonz, eds. Monographs have been written for regions of France, Italy, Spain, Germany, England, and Russia. Aside from those already cited in the text see the following: for France, Duby, *Rural Economy,* Ralph Giesey, "Rules of Inheritance and Strategies of Mobility in Pre-Revolutionary France," *American Historical Review,* vol. 82, no. 2 (April 1977), pp. 271–89; for Italy, David Herlihy, "Life Expectancies for Women in Medieval Society," in Rosemarie Thee Morewedge, ed., *The Role of Woman in the Middle Ages* (Albany, N.Y.: SUNY at Binghamton, 1975); for Spain, Dillard, "Reconquest," in Stuard, ed.; for Germany, K. Leyser, "The German Aristocracy from the 9th to the early 12th Century: A Historical and Cultural Sketch," *Past and Present,* no. 41 (December 1968), pp. 25–53; for Russia, Atkinson in Atkinson et al., eds.

106. *Secular Spirit*, p. 254.
107. Dewey and Kleimole, eds., p. 220.
108. Margaret Spufford, "Peasant Inheritance Customs and Land Distribution in Cambridgeshire from the Sixteenth to the Eighteenth Centuries," in Goody et al., eds., *Family and Inheritance*, p. 57.
109. Alan Macfarlane, *The Origins of English Individualism: The Family, Property and Social Transition* (New York: Cambridge University Press, 1979), p. 134.
110. Labarge, *Household*, p. 13.
111. Cooper in Goody et al., eds., *Family and Inheritance*, p. 256.
112. Flandrin, p. 14.
113. Isabella and Ferdinand were unlucky with their children: Juan the son died at nineteen in 1497 and only one daughter's male child survived, Carlos, or Charles, born to Juana in 1500.
114. See Fernández, pp. 14, 83.
115. For the life of Isabella, see the joint biography by Fernández.
116. Cited in Stenton, p. 20; see also Whitelock, ed., *Documents*, pp. 408, 429.
117. Slater, p. 53.
118. Stenton, p. 38–39.
119. Duby, *Rural Economy*, p. 234.
120. Cited in Natalie Zemon Davis, "Ghosts, Kin, and Progeny: Some Features of Family Life in Early Modern France," in Alice S. Rossi, Jerome Kagan, and Tamara K. Haveren, eds., *The Family* (New York: W. W. Norton & Company Inc., 1978), p. 108.
121. Cited in Stenton, p. 38.
122. Stenton, p. 36.
123. Cited in Stenton, p. 51.
124. See Sue Sheridan Walker, "Widow and Ward: The Feudal Law of Child Custody in Medieval England," in Stuard, ed.
125. Gies, *Castle*, p. 77.
126. Walker, in Stuard, ed., p. 160.
127. Cited in Duckett, *Portraits*, p. 215.
128. Cited in Duckett, *Portraits*, p. 216.
129. Jackson, ed., *Thornton*, p. 234.
130. Much has been written describing the variations in provisions for widows from the tenth to the seventeenth centuries. See in particular the following: for the Vikings, L. M. Larson, *The Earliest Norwegian Laws* (New York: Columbia University Press, 1935); for Spain, Lucy A. Sponslar, *Women in the Medieval Spanish Epic and Lyric Traditions* (Lexington, Ky.: The University Press of Kentucky, 1975); for Ireland, Simms in MacCurtain and Ó'Corráin, eds.; for Russia, George Vernadsky, trans., *Medieval Russian Laws* (New York: W. W. Norton & Company, Inc., 1969); for England, Mingay; and for France, Brissaud.
131. Cited in Walker in Stuard, ed., p. 171, note.
132. Walker in Stuard, ed., p. 169, note.
133. See Pearl Hogrefe, "Legal Rights of Tudor Women and their Circumvention by Men and Women," *The Sixteenth Century Journal*, Vol. III, no. 1 (April 1972), pp. 104–105.

134. Cited in Slater, p. 51.
135. Leyser, p. 51.
136. Cited in Labarge, *Saint Louis,* p. 151.
137. Cited in Jones and Underwood, p. 26.
138. Cited in Dahmus, p. 257.
139. Cited in Dahmus, p. 260.
140. Cited in Dahmus, p. 260.

PART IV: 4. The New Flowering of Ancient Traditions

1. See Gold, *Lady,* for discussion of this "ambivalence," in law, religion, and literature, pp. 145–52. See Dronke, *Medieval Lyric,* and Joan Ferrante and George Economou, eds., *In Pursuit of Perfection* (Port Washington, N.Y.: Kennikat Press, 1975) for discussions of these aspects of lyrics and romances.
2. Cited from Riché, *Daily Life,* p. 98; Ferrante, ed., *Guillaume,* p. 149; *Raoul,* p. 115; *Flamenca* cited in G. G. Coulton, *Life in the Middle Ages,* Vol. III (New York: The Macmillan Co., 1930), p. 35.
3. *Nibelungenlied,* p. 256.
4. *Lancelot,* p. 329.
5. *Tristan,* p. 172.
6. *Tristan,* p. 280.
7. Cited in Joan M. Ferrante, *Woman as Image in Medieval Literature from the Twelfth Century to Dante* (New York: Columbia University Press, 1975), p. 68; this book explores this aspect of the lyrics and the romances.
8. *Tristan,* p. 195; see also p. 230.
9. In Provençal, the language of the troubadours of southern France, the word sometimes used to describe the lady was *res*—in Latin, literally "thing." See Bogin, p. 55 and p. 55, note.
10. *Lancelot,* p. 278. See also Keen, *Chivalry,* on this effect of love, pp. 116–17.
11. R. H. Bloch, pp. 144–45. For examples of the "pain" of love in the lyrics see for example: Frederick Golden, "The Array of Perspectives in Early Courtly Love Lyric," in Ferrante and Economou, eds., pp. 73–80.
12. Cited in Dronke, *Medieval Lyric,* p. 134.
13. *Tristan,* p. 294.
14. *Tristan,* p. 227.
15. For a Danish version, see Olivia E. Coolidge, *Legends of the North* (Boston: Houghton Mifflin Company, 1951), pp. 146–47.
16. Ferrante, ed., *Guillaume,* p. 151.
17. *Erec,* p. 33.
18. *Lancelot,* p. 279.
19. *Tristan,* p. 277.
20. *Nibelungenlied,* p. 300.
21. *Tristan,* p. 277.
22. *Tristan,* p. 280.
23. Flores, ed., p. 152.
24. Cited in Claire Richter Sherman, "Taking a Second Look: Observations on the Iconography of a French Queen, Jeanne de Bourbon (1338–1378)," in Norma

Broude and Mary D. Garrard, eds., *Feminism and Art History: Questioning the Litany* (New York: Harper & Row, Publishers, 1982), p. 112. See Facinger, on the diminution of the French queen's power, pp. 18–19, 20.

25. *Froissart,* p. 118.
26. *Froissart,* p. 109.
27. *Froissart,* p. 109.
28. A study of the use by noblewomen of seals from 1150–1350, the symbol of the landholder's authority, reveals the same changes: From affixing their seal to their own documents in the twelfth century to using their seal in conjunction with a male member of the family by the thirteenth century. The emblem on the seal continued to identify the woman with her husband's or her father's lineage, but the image of the woman on the device changed: from a queen or noblewoman standing as if in authority to a "lady" with her hawk, her body posed gracefully at ease. Brigitte Bedoz Rezak "Women, Seals and Power in Medieval France 1150–1350," paper delivered at Fordham University, New York City, March 16, 1985.
29. See Calisse, p. 519.
30. Cited in Ian Maclean, *The Renaissance Notion of Woman: A Study of the Fortunes of Scholasticism and Medical Science in European and Intellectual Life* (New York: Cambridge University Press, 1980), p. 76.
31. Cited in Hogrefe, "Legal Rights," p. 98.
32. See Atkinson in Atkinson et al., eds.
33. Cited in Coulton, *Panorama,* p. 617.
34. See Stenton, pp. 30–31.
35. See McNamara and Wemple in Stuard, ed., pp. 103–104, and Riché, *Early Life,* pp. 54–56, for discussion of the evolution of these policies.
36. Marriage had not been considered a sacrament before. Augustine had disallowed it because it led to sexual intercourse. See Waterworth, *Council of Trent,* for provisions of Council of Trent canons, pp. 194–95.
37. For details of these and similar incidents see McNamara and Wemple in Stuard, ed., Duby, *Medieval Marriage,* and Angela M. Lucas, *Women in the Middle Ages: Religion, Marriage and Letters* (Brighton, England: The Harvester Press Ltd., 1983).
38. The control of marriage had always been an issue of hegemony fought over by religious and secular rulers. Jack Goody in his book, *The Development of the Family and Marriage in Europe* (New York: Cambridge University Press, 1983), suggests that the Church's policies on marriage, adultery, and divorce had been intended to create more heirless families and thus more opportunities for the Church to acquire property, on which all power ultimately rested. See especially pp. 32–47 for statement of this argument.
39. Protestant sects would make other compromises. For Catholic policy, see especially Duby, *Medieval Marriage,* Chapter 2; the policy of 1215 is summarized in McNamara and Wemple in Stuard, ed., p. 112.
40. See Eleanor Como McLaughlin, "Equality of Souls, Inequality of Sexes: Woman in Medieval Theology," in Rosemary Radford Ruether, ed., *Religion and Sexism: Images of Woman in the Jewish and Christian Traditions* (New York: Simon and Schuster, 1974), pp. 227–28; Duby, *Medieval Marriage,* pp. 41–42; Maclean, *Renaissance Notion,* p. 15.

41. *Froissart,* p. 243.
42. See for example the laws of thirteenth-century Languedoc in Leah Lydia Otis, *Prostitution in Medieval Society* (Chicago: University of Chicago Press, 1985).
43. Dillard, "Reconquest," in Stuard, ed., p. 85.
44. See Waterworth, *Council of Trent,* p. 203.
45. See Atkinson in Atkinson et al., eds., p. 21.
46. *Froissart,* p. 309.
47. *Froissart,* p. 310.
48. *Froissart,* p. 312.
49. In this instance the husband was victorious and the wife vindicated, but it is this sort of reasoning and the fears of such dishonor that led suspicious husbands of the fourteenth, fifteenth, and sixteenth centuries to speak of chastity belts for their wives. The idea of "belt" is a misnomer, as the first illustration from a military manuscript of 1405 shows a leather-covered metal contrivance much like what men wear for support and protection of their genitals. A few examples of the apparatus still exist in the museums of Europe. There is, however, no evidence of their use. See Harvey Graham, *Eternal Eve: The History of Gynecology and Obstetrics* (Garden City, N.Y.: Doubleday & Company Inc., 1951), pp. 122–23, and Harold Speert, *Iconographia Gyniatrica: A Pictorial History of Gynecology and Obstetrics* (Philadelphia: F. A. Davis, 1973), pp. 452–53.
50. For stories of adulterous wives and their fates, see especially John F. Benton, "Clio and Venus, An Historical View of Courtly Love" in F. X. Newman, ed., *The Meaning of Courtly Love* (Albany, N.Y.: SUNY Press, 1968).
51. See Michael M. Sheehan, "The Influence of Canon Law on the Property Rights of Married Women in England," *Medieval Studies,* Vol. XXV (1963), pp. 110–16.
52. Backer, p. 245.
53. Cited in Slater, p. 44.
54. See Fawtier, p. 53.
55. See Stenton, p. 67.
56. Cited in Coulton, *Life,* Vol. III, p. 16.
57. *Njal,* p. 267.
58. *Tristan,* p. 66.
59. *Tristan,* p. 278.
60. *Erec,* p. 32.
61. *Erec,* p. 64.
62. Ferrante, ed., *Guillaume,* p. 278.
63. *Raoul,* p. 32.
64. *Nibelungenlied,* p. 281.
65. See *Tristan,* pp. 148, 147.
66. For a clear and complete survey of this didactic literature, see Diane Bornstein, *The Lady in the Tower: Medieval Courtesy Literature for Women* (Hamden, Conn.: Archon Books, 1983).
67. Cited in Bornstein, p. 49.
68. Cited in Rosemary Barton Tobin, "Vincent of Beauvais on the Education of Women," *Journal of the History of Ideas,* vol. 35 (July–September 1974), p. 486, note.
69. See Bornstein, pp. 59–60.
70. See Coulton, *Panorama,* p. 617; Coulton, *Life,* Vol. III, p. 7.

71. Thomas Wright, ed., *The Book of the Knight of La Tour-Landry* (London: Kegan Paul, Trench, Trubner & Co., Ltd., 1906), p. 79; note that there is a 1973 reprint.
72. *La Tour-Landry,* p. 10.
73. Cited in Gies, *Castle,* p. 79.
74. *La Tour-Landry,* p. 117.
75. *La Tour-Landry,* p. 128.
76. *La Tour-Landry,* p. 55.
77. *La Tour-Landry,* p. 58.
78. Bell, "Book Owners," pp. 756–57.
79. Cited in Bennett, p. 42.
80. Cited in Gardiner, p. 119.
81. Cited in David Hunt, *Parents and Children in History: The Psychology of Family Life in Early Modern France* (New York: Harper & Row, Publishers, 1972), p. 73.
82. Cited in Slater, p. 34.
83. For discussion of this seeming contradiction see Sylvia Huot, "Seduction and Sublimation: Christine de Pizan, Jean de Meun, and Dante," *Romance Notes,* Vol. XXV, no. 3 (Spring 1985), pp. 361–73; for the full text see Pizan, *Three Virtues.* Summaries are in Ruth Kelso, *Doctrine for the Lady of the Renaissance* (Urbana, Ill.: University of Illinois Press, 1956), pp. 236–46, and in Bornstein, pp. 68–69, 85–87.
84. Pizan, *Three Virtues,* p. 161.
85. Pizan, *Three Virtues,* p. 68.
86. See Pizan, *Three Virtues,* pp. 86–105.
87. See Pizan, *Three Virtues,* pp. 45–48, 55, 58.
88. Pizan, *Three Virtues,* pp. 55, 56.
89. Anne de Beaujeu (c. 1441–1522), the late fifteenth-century regent for her brother Charles VIII of France, had two copies of Christine de Pizan's book (Bornstein, p. 71). When she came to write a similar treatise for her daughter Suzanne, de Beaujeu repeated the same attitudes and presented the same image of the ideal wife. See summary in Kelso, pp. 212, 227, 246–47 and Bornstein.
90. Lucy (Apsley) Hutchinson, *Memoirs of the Life of Colonel Hutchinson,* edited by Rev. Julius Hutchinson (London: John C. Nimmo, 1985), p. 24.
91. *Hutchinson,* p. 25.
92. *Hutchinson,* p. 35.
93. Cited in Stenton, p. 166.
94. Cited in Stenton, p. 166.
95. Pizan, *Three Virtues,* pp. 62–64.
96. Pizan, *Three Virtues,* p. 64.

PART V. WOMEN OF THE WALLED TOWNS: PROVIDERS AND PARTNERS

PART V: 1. The Townswoman's Daily Life: The Twelfth to the Seventeenth Centuries

1. Fernand Braudel, *The Structures of Everyday Life: Civilization & Capitalism 15th–18th Century,* Vol. I, translated by Siân Reynolds (New York: Harper & Row, Publishers, 1981), p. 51.

2. Carlo M. Cipolla, *Before the Industrial Revolution: European Society and Economy 1000–1700* (New York: W. W. Norton & Company Inc., 1976), p. 283. For population figures, in addition to the works already cited, see Fernand Braudel, *Capitalism and Material Life* translated by Miriam Kochan (New York: Harper & Row, Publishers, 1973), pp. 414–17, 431; Henry Kamen, *The Iron Century: Social Change in Europe 1550–1660* (New York: Praeger Publishers, 1971), p. 19; J. C. Russell "Population in Europe 500–1500," in Carlo M. Cipolla, ed., *The Fontana Economic History of Europe: The Middle Ages* (Glasgow: William Collins Sons and Co. Ltd., 1975), pp. 34–35; Fritz Rörig, *The Medieval Town* (Berkeley: University of California Press, 1967), pp. 112–13. See these texts also for descriptions of urban development in these centuries: Braudel, *Structures,* Vol. I, Chapter III.
3. Pope Pius II, cited in James Bruce Ross and Mary Martin McLaughlin, eds., *The Portable Renaissance Reader* (New York: Penguin Books, 1980), p. 211.
4. Pietro Aretino cited in Ross and McLaughlin, eds., *Renaissance Reader,* p. 241.
5. See Braudel, *Capitalism,* p. 435.
6. Braudel, *Capitalism,* pp. 375–76 and Fernand Braudel, *The Mediterranean and the Mediterranean World in the Age of Philip II,* Vol. I, translated by Siân Reynolds (New York: Harper & Row, Publishers, 1973), p. 427, for relative populations; see Roger Mols, "Population in Europe 1500–1700," in Carlo M. Cipolla, ed., *The Fontana Economic History of Europe: The Sixteenth and Seventeenth Centuries* (Glasgow: William Collins Sons & Co., 1976), pp. 43–44, for population densities.
7. For comparisons of north Italian towns and those in other parts of Europe, see David Herlihy, "The Tuscan Town in the Quattrocento: A Demographic Profile," Paul Maurice Clogan, ed., *Medievalia et Humanistica,* n.s. 1 (1970), pp. 84–85.
8. See David Herlihy, "Population, Plague and Social Change in Rural Pistoia 1201–1430," *The Economic History Review,* 2nd series, Vol. XVIII, no. 2 (August 1965), p. 231.
9. See Richard Goldthwaite, "The Florentine Palace as Domestic Architecture," *American Historical Review,* vol. 77, no. 4 (October 1972), p. 997.
10. Geoffrey Brereton, ed. and trans., *Froissart Chronicles* (New York: Penguin Books Ltd., 1978), p. 239.
11. Marcelin Defourneaux, *Daily Life in Spain in the Golden Age,* translated by Newton Branch (London: George Allen and Unwin Ltd., 1970), p. 96.
12. Sylvia L. Thrupp, *The Merchant Class of Medieval London (1300–1500)* (Ann Arbor: University of Michigan Press, 1977 ed.), p. 41; see also Ross and McLaughlin, eds., *Renaissance Reader,* p. 188.
13. See for descriptions of houses: Braudel, *Structures,* Vol. I, pp. 266–82; for London, Thrupp, *Merchant,* p. 130, 131, note; for Amsterdam and Paris, Braudel, *Capitalism,* pp. 201–203, 218; for France, *The Secular Spirit: Life and Art at the End of the Middle Ages* (New York: E. P. Dutton & Co., Inc., 1975), pp. 16, 17; for Nuremberg, see Gerald Strauss, *Nuremberg in the Sixteenth Century: City Politics and Life Between Middle Ages and Modern Times* (New York: John Wiley & Sons Inc., 1966), p. 25.
14. See Goldthwaite for a description of the evolution of the "private house," pp. 982–89 and his book: *The Building of Renaissance Florence: An Economic and Social History* (Baltimore: The Johns Hopkins University Press, 1985), pp. 13–26;

for Datini, see Iris Origo, *The Merchant of Prato: Francesco di Marco Datini* (London: Jonathan Cape, 1957), p. 136.

15. See Barbara Beuys, *Familienleben in Deutschland: Neue Bilder aus der deutschen Vergangenheit* (Reinbek bei Hamburg: Rowohlt, 1980), pp. 140–41; *Secular Spirit*, pp. 15, 19; Defourneaux, pp. 61–62.
16. Kamen, p. 52.
17. For south German towns, see Merry E. Wiesner, *Working Women in Renaissance Germany* (New Brunswick, N.J.: Rutgers University Press, 1986), p. 191; Christiane Klapisch, "Household and Family in Tuscany in 1427," in Peter Laslett, ed., *Household and Family in Past Time* (New York: Cambridge University Press, 1974), p. 273. In Pistoia in 1427 female-headed households made up 21.2 percent of the total households; see Herlihy, "Tuscan Town," p. 101.
18. D.V. Glass, "Notes on the Demography of London at the End of the Seventeenth Century," *Daedalus: Historical Population Studies,* vol. 97, no. 2 (Spring 1968), pp. 584, 586.
19. See Mols in Cipolla, ed., *Sixteenth and Seventeenth Centuries,* for the traditional description of the overall pattern; for specific ratios for towns in Germany, Switzerland, France, and Spain see J. C. Russell, *Late Ancient and Medieval Population* (Philadelphia: Transactions of the American Philosophical Society, 1958), p. 16; David Herlihy, "Life Expectancies for Women in Medieval Society" in Rosemarie Thee Morewedge, ed., *The Role of Women in the Middle Ages* (Albany, N.Y.: SUNY at Binghamton, 1975), pp. 12–13, 22; Glass, "Demography," pp. 584, 586; Benjamin Z. Kedar, "The Genoese Notaries of 1382: The Anatomy of an Urban Occupational Group," in Harry A. Miskimin, David Herlihy, A. L. Udovitch, eds., *The Medieval City* (New Haven: Yale University Press, 1977), p. 9; Wiesner, *Working Women,* p. 5.
20. Doris Mary Stenton, *The English Woman in History* (New York: The Macmillan Company, 1957 ed.), p. 85.
21. Martha C. Howell, "Working Women in Early Modern Europe: A Cross-Cultural Approach," paper delivered at The Berkshire Conference on the History of Women, held at Vassar College, Poughkeepsie, N.Y., June 1981.
22. Eileen Power, *Medieval Women,* edited by M. M. Postan (New York: Cambridge University Press, 1975) p. 109.
23. See Wiesner, *Working Women,* pp. 83–84.
24. Emmanuel LeRoy Ladurie, *Carnival in Romans* (New York: George Braziller, Inc., 1979), p. 4.
25. Sarah C. Maza, *Servants and Masters in Eighteenth-Century France: The Uses of Loyalty* (Princeton, N.J.: Princeton University Press, 1983), p. 61, note.
26. By the eighteenth century the relationship would have changed to the less familial one of wage contract between servant and employer. See Maza, for description of this change, p. 14, and Cissie Fairchilds, *Domestic Enemies: Servants & Their Masters in Old Regime France* (Baltimore: The Johns Hopkins University Press, 1984), pp. 17–18.
27. Eileen Power, trans., *The Goodman of Paris (Le Ménagier de Paris)* (London: George Routledge & Sons, Ltd., 1928), pp. 16–17.
28. Origo, *Merchant,* p. 206, note; Herlihy, "Tuscan Town," p. 109, note. Slavery

reemerged as a practice of Mediterranean Europe in this period. See Part II, "Women of the Fields," for a description of the end of slave labor in rural areas.

29. Origo, *Merchant,* p. 254.
30. Cited in Gene Brucker, ed., *The Society of Renaissance Florence* (New York: Harper & Row, Inc., 1971), p. 223.
31. Cited in Origo, *Merchant,* p. 209.
32. On the sexual vulnerability of servant women, see for example, David Nicholas, *The Domestic Life of a Medieval City: Women, Children and the Family in Fourteenth Century Ghent* (Lincoln, Neb.: University of Nebraska Press, 1985), pp. 67–68; Fairchilds, *Domestic Enemies,* pp. 86–88, 165. From the fourteenth century on, young maids in this predicament were a staple of popular tales. See especially the Italian versions in Giovanni Boccaccio, *The Decameron,* translated by G. H. McWilliam (New York: Penguin Books, 1975), fourth story on the tenth day.
33. In fourteenth-century Ghent, the child could inherit from the mother and her kin. See Nicholas, pp. 154–55.
34. See Susan Mosher Stuard, "Women in Charter and Statute Law: Medieval Ragusa," in Susan Mosher Stuard, ed., *Women in Medieval Society* (Philadelphia: University of Pennsylvania Press, 1976) and Julius Kirshner and Anthony Molho, "The Dowry Fund and the Marriage Market in Early Quattrocento Florence," *The Journal of Modern History,* vol. 50, no. 3 (September 1978), p. 429; Yvonne Maguire, *The Women of the Medici* (London: George Routledge & Sons, Ltd., 1927), pp. 45, 62; Jacob Bronowski and Bruce Mazlish, *The Western Intellectual Tradition* (New York: Harper & Row, Publishers, 1962), p. 5.
35. See Walter Minchinton, "Patterns and Structure of Demand 1500–1750," in Cipolla, ed., *Sixteenth and Seventeenth Centuries,* p. 97.
36. Gene Brucker, ed., *Two Memoirs of Renaissance Florence: the Diaries of Buonaccorso Pitti & Gregorio Dati,* translated by Julia Martines (New York: Harper & Row, Publishers, 1967), pp. 26–27.
37. Mary Elizabeth Perry, " 'Lost Women' in Early Modern Seville: the Politics of Prostitution," *Feminist Studies,* vol. 4, no. 1 (February 1978), p. 200.
38. Kamen, p. 69.
39. See James Bruce Ross and Mary Martin McLaughlin, *Medieval Women,* unpublished manuscript, pp. 38–39; Natalie Zemon Davis, *Society and Culture in Early Modern France* (Stanford: Stanford University Press, 1975), p. 291, note; for South German towns see Wiesner, *Working Women,* p. 93.
40. *Pitti and Dati,* pp. 26–27.
41. Maryanne Kowaleski, "Exeter, Bristol and Dartmouth," paper delivered at The Berkshire Conference on the History of Women, held at Smith College, Northampton, Massachusetts, June 1984.
42. See Strauss, pp. 7–8.
43. Cissie C. Fairchilds, *Poverty and Charity in Aix-en-Provence 1640–1789* (Baltimore: The Johns Hopkins University Press, 1976), p. 75.
44. Ross and McLaughlin, *Medieval Women,* p. 39.
45. Davis, *Society and Culture,* p. 291, note.
46. Historians call manufacturing organized in this way bye industries.
47. Historians are just beginning to collect information on the extent of this informal

retailing. In Leyden in the sixteenth century 15 percent of the women who worked for pay were involved in some kind of retailing. For Leyden, see Martha Howell, "Working Women," unpublished manuscript, and her book *Women, Production and Patriarchy in Late Medieval Cities* (Chicago: The University of Chicago Press, 1986), pp. 87–94; for English towns, Pearl Hogrefe, *Tudor Women: Commoners and Queens* (Ames, Iowa: The Iowa State University Press, 1975), p. 44; for Paris, see Ross and McLaughlin, *Medieval Women,* p. 12, and p. 27 for Brussels; for Florence, Judith Brown, "Working Women in Early Modern Europe," unpublished manuscript, pp. 3–4; German towns, in Hannelore Sachs, *The Renaissance Woman* (New York: McGraw-Hill Book Company, 1971), pp. 42–43, and Wiesner, *Working Women,* pp. 111–42; a good summary for English and French towns is in Louise A. Tilly and Joan W. Scott, *Women, Work and Family* (New York: Holt, Rinehart and Winston, 1978), p. 49.

48. Ross and McLaughlin, *Medieval Women,* p. 195.
49. See Barton C. Hacker, "Women and Military Institutions in Early Modern Europe: A Reconnaissance," *Signs,* vol. 6, no. 4 (Summer 1981), pp. 643–71.
50. Cited in Brucker, ed., *Renaissance Florence,* pp. 191–92.
51. Brucker, ed., *Renaissance Florence,* pp. 190–91.
52. G. R. Quaife, *Wanton Wenches and Wayward Wives: Peasants and Illicit Sex in Early Seventeenth Century England* (New Brunswick, N.J.: Rutgers University Press, 1979, p. 150.
53. For Florence see Brucker, ed., *Renaissance Florence,* pp. 196–98, 200–201; for Venice, Fernando Henriques, *Prostitution and Society: A Survey, Primitive, Classical and Oriental,* Vol. II (New York: The Citadel Press, 1965), p. 89, and Brian Pullan, *Rich and Poor in Renaissance Venice* (Cambridge, Mass.: Harvard University Press, 1971), p. 382; for Seville, Perry and Ruth Pike, *Aristocrats and Traders: Sevillian Society in the Sixteenth Century* (Ithaca, N.Y.: Cornell University Press, 1972), pp. 201, 208.
54. See Lydia Leah Otis, *Prostitution in Medieval Society* (Chicago: University of Chicago Press, 1985), p. 64.
55. For information on Rome see Georgina Masson, *Courtesans of the Italian Renaissance* (New York: St. Martin's Press, 1976), pp. 25–26.
56. See Henriques, p. 50; Otis, p. 54.
57. See Henriques, p. 78; Vern and Bonnie Bullough, *An Illustrated Social History of Prostitution* (New York: Crown Publishers, Inc., 1978), p. 136; Otis, pp. 20–37, 54n, 201, 203, 210.
58. See Andrew McCall, *The Medieval Underworld* (London: Hamish Hamilton, 1979), p. 191, note, for streets in fourteenth-century London and sixteenth-century Paris, indicating areas where prostitutes practiced their trade, like Slut's Hole and Rue Gattecon (Scratchcunt Street).
59. *Pitti and Dati,* p. 190.
60. See, for example, Henriques, pp. 47–52, 83–85; Ladurie, *Carnival,* p. 224; Sachs, pp. 52–53; Strauss, p. 212; Otis, pp. 30–31, 55–59, 79–83, 94–98; for regulations in other German towns see Wiesner, *Working Women,* pp. 98–104.
61. See Bullough, *Prostitution,* p. 142, and Otis, pp. 42–43.
62. Cited in Sachs, p. 53.

63. Cited in Perry, p. 206.
64. Otis, pp. 84–88.
65. Henriques, pp. 52–53, Defourneaux, p. 224; see also Perry, pp. 207–10, and Pike, pp. 204–205.
66. Cited in Julia O'Faolain and Lauro Martines, eds., *Not in God's Image: Women in History from the Greeks to the Victorians* (New York: Harper & Row, Publishers, 1973), p. 294.
67. Masson, pp. 129, 137. The Venetian house was attached to the hospital and between 1553 and 1620 had between 220 and 400 members, Pullan, p. 378. See also Mary Martin McLaughlin, "Survivors and Surrogates: Children and Parents from the Ninth to the Thirteenth Centuries," in Lloyd de Mause, ed., *The History of Childhood* (New York: Harper & Row, Publishers, 1974), p. 159, note. James A. Brundage, "Prostitution in the Medieval Canon Law," *Signs,* vol. 1, no. 4 (Summer 1976), p. 842; McCall, pp. 189–90; for French foundations see Otis, pp. 72–76.
68. Hugo Rahner, S.J., *Saint Ignatius Loyola* (New York: Herder and Herder Inc., 1960), p. 80.
69. For information on Venice's refuges, see Pullan, pp. 386–90; on Rome, see Rahner, pp. 13–20, Masson, pp. 132–33; for Florence, see Brucker, ed., *Renaissance Florence,* pp. 211–12; and Ruth P. Liebowitz, "Voices from Convents: Nuns and Repentant Prostitutes in Late Renaissance Italy," paper delivered at The Berkshire Conference on the History of Women, held at Mount Holyoke College, South Hadley, Massachusetts, August 1978; for Aix-en-Provence, see Fairchilds, *Charity,* p. 19; for Seville, see Perry, p. 198.
70. Hogrefe, *Tudor Women,* p. 90.
71. Carlo Calisse, *A History of Italian Law,* translated by Layton B. Register (New York: Augustus Kelley, 1969), p. 373.
72. Cited in Stenton, p. 65. In south German towns, women rarely attacked people when they stole. See Wiesner, *Working Women,* pp. 108–109.
73. Cited in Ross and McLaughlin, eds., *Renaissance Reader,* p. 543.
74. Origo, *Merchant,* p. 68.
75. Natalie Zemon Davis, "Women in the Crafts in Sixteenth-Century Lyon," *Feminist Studies,* vol. 8, no. 1 (Spring 1982); p. 52; Davis, *Society and Culture,* p. 332, note.
76. Maza, p. 82.
77. See Maza, p. 64.
78. For Italy, see Klapisch in Laslett, ed., *Past Time,* p. 277; for Germany, Russell, *Medieval Population,* pp. 53, 55; for France, Ladurie, *Carnival* p. 3; for England, Peter Laslett, *The World We Have Lost* (New York: Charles Scribner's Sons, 1965), p. 1. For a discussion of the recent literature on family size and the persistence of the "nuclear" family pattern, see Nicholas, pp. 7–12.
79. See Braudel, *Capitalism,* pp. 201–202.
80. G. G. Coulton, *Medieval Panorama: The English Scene from Conquest to Reformation* (Cambridge: Cambridge University Press, 1939), p. 310.
81. The information on women's participation in guilds on this and subsequent pages comes from studies of individual town records. See in particular Ross and McLaughlin, *Medieval Women,* for Germany and France; for Germany, Howell, *Women,*

Rudolf Hübner, *A History of Germanic Private Law* (Boston: Little, Brown and Company, 1918), Nicholas, and Rörig; for France, Davis, "Crafts," Frances and Joseph Gies, *Women in the Middle Ages* (New York: Thomas Y. Crowell Company, 1978); for England, Hogrefe, *Tudor Women*, and Alice Clark, *Working Life of Women in the Seventeenth Century* (New York: Harcourt, Brace & Howe, 1920).

82. See Rörig, p. 115.
83. See Ross and McLaughlin, *Medieval Women*, pp. 24–26.
84. See Miriam Usher Chrisman, *Lay Culture, Learned Culture: Books and Social Change in Strasbourg 1480–1599* (New Haven: Yale University Press, 1983), pp. 22–23; Clark, p. 167.
85. See for example, Barbara B. Diefendorf, "Widowhood and Remarriage in Sixteenth-Century Paris," paper delivered at The Berkshire Conference on the History of Women, held at Vassar College, Poughkeepsie, New York, June 1981.
86. See Wiesner, *Working Women*, Chapter 4; Sachs, p. 42.
87. See for example, Ross and McLaughlin, *Medieval Women*, pp. 32, 27; Davis, "Crafts," pp. 62–64; Nicholas, pp. 94–95, 99, 102, and Steven Ozment, *When Fathers Ruled: Family Life in Reformation Europe* (Cambridge, Mass.: Harvard University Press, 1983), p. 13.
88. Clark, pp. 201–208, 215–19.
89. Joan Evans, *Life in Medieval France* (New York: Phaidon Publishers, Inc., 1969), p. 48.
90. Ross and McLaughlin, *Medieval Women*, p. 24.
91. See Thrupp, *Merchant*, p. 172.
92. Tilly and Scott, p. 35.
93. Marian K. Dale, "The London Silkwomen of the Fifteenth Century," *Economic History Review*, Vol. IV (1932–1934), pp. 326–27.
94. Ross and McLaughlin, *Medieval Women*, p. 19.
95. Maria Fowler, "Women as Secular Musicians in Medieval France," unpublished manuscript, p. 10.
96. Ross and McLaughlin, *Medieval Women*, pp. 24–26; for Cologne see Howell, *Women*, pp. 95–97, 124–29; for Frankfurt-am-Main, see Beuys, pp. 147, 150; for Frankfurt, Ernest McDonnell, *The Beguines and Beghards in Medieval Culture* (New York: Octagon Books, 1969), p. 85.
97. Ross and McLaughlin, *Medieval Women*, pp. 7–10.
98. See Ross and McLaughlin, *Medieval Women*, p. 13.
99. For Ragusa, Stuard in Stuard, ed., *Medieval Society*, p. 200; for Nuremberg, Strauss, p. 79; for London, Thrupp, *Merchant*, p. 51.
100. Thrupp, *Merchant*, p. 139.
101. For information on the houses, living arrangements, and possessions see the following: for Germany, Beuys, p. 141; for Italy, Goldthwaite, "Palace," p. 1004; for England, Thrupp, *Merchant*, pp. 132, 138–41; and Braudel, *Capitalism*, pp. 213–25, and *Secular Spirit*, pp. 21–22.
102. Origo, *Merchant*, pp. 260–63, 270–71.
103. Cited in Maguire, pp. 64, 65.
104. Origo, *Merchant*, pp. 281–88.

105. *Ménagier,* pp. 36, 310.
106. See *Secular Spirit,* pp. 30, 32; Janet Shirley, trans., *A Parisian Journal* (Oxford: Clarendon Press, 1968), p. 37.
107. See *Ménagier,* p. 21; Origo, *Merchant,* pp. 289–90; for England, Dorothy Hartley, *Lost Country Life* (New York: Pantheon Books, 1979), p. 101; for Nuremberg, Walter French, *Medieval Civilization as Illustrated by the Fastnachtspiele of Hans Sachs* (Baltimore: Johns Hopkins Press, 1925), p. 55.
108. Origo, *Merchant,* p. 241.
109. Strauss, p. 93.
110. Cited in Clark, p. 37; for other wives see Nicholas, pp. 44, 81–82; Stanley Chojnacki, "Patrician Women in Early Renaissance Venice," *Studies in the Renaissance,* Vol. XXI (1974), p. 198; Maguire; Janet Ross, ed. and trans., *Letters of the Early Medici as told in their Correspondence* (Boston: The Gorham Press, 1911), pp. 10, 48, 63; Diefendorf, p. 14; Beuys; Hogrefe, *Tudor Women,* p. 67.

PART V: **2. Dangers and Remedies**

1. Shirley, ed., *Journal,* pp. 332–33.
2. Joseph and Frances Gies, *Life in a Medieval City* (New York: Thomas Y. Crowell & Company, 1973), p. 192.
3. Braudel, *Mediterranean,* Vol. I, p. 328.
4. Shirley, ed., *Journal,* p. 322.
5. Herlihy, "Tuscan Town," p. 85.
6. Henry S. Lucas, "The Great European Famine of 1315, 1316, and 1317," *Speculum,* Vol. V (1930), pp. 353–54.
7. Shirley, ed., *Journal,* p. 140.
8. *Pitti and Dati,* p. 42.
9. Cited in Roland Bainton, *Women of the Reformation in Germany and Italy* (Boston: Beacon Press, 1971), p. 263.
10. See C. V. Wedgewood, *The Thirty Years War* (Garden City, N.Y.: Doubleday & Company, Inc., 1961), pp. 493–96.
11. Wedgewood, p. 400.
12. Davis, *Society and Culture,* p. 301, note.
13. Nuremberg had fifty public privies. See Strauss, p. 192.
14. Muriel Joy Hughes, *Women Healers in Medieval Life and Literature* (New York: King's Crown Press, 1943), p. 29.
15. See Braudel, *Capitalism,* p. 240.
16. Kamen, pp. 25, 29.
17. See Philip Ziegler, *The Black Death* (New York: Harper & Row, Publishers, 1969), p. 84; Braudel, *Mediterranean,* Vol. I, p. 333; Mols in Cipolla, ed., *Sixteenth and Seventeenth Centuries,* p. 75; Ladurie, *Carnival,* p. 3; Braudel, *Capitalism,* pp. 50, 46, 48.
18. *Decameron,* pp. 54–56.
19. Shirley, ed., *Journal,* p. 131.
20. Cited in Kamen, p. 30.
21. Cited in Maguire, p. 38.

22. Marvin Lowenthal, trans., *The Memoirs of Glückel of Hameln* (New York: Schocken Books, 1977), p. 51.
23. See Herlihy in Morewedge, ed., pp. 14, 22; for south German towns, see Wiesner, *Working Women,* p. 5.
24. Christiane Klapisch-Zuber has suggested that the practice of leaving girls with a wet nurse longer than boys made their survival less sure. See her book *Women, Family and Ritual in Renaissance Italy* (Chicago: University of Chicago Press, 1985), pp. 102, 105–106.
25. See *Pitti and Dati,* pp. 112, 127, 128.
26. See Klapisch, *Family and Ritual,* p. 158; Thrupp, *Merchant,* p. 231, and J-L. Flandrin, *Families in Former Times: Kinship, Household and Sexuality,* translated by Richard Southern (New York: Cambridge University Press, 1979), p. 29.
27. In contrast, women from artisan or laborers' families married later (in their twenties) and brought fewer pregnancies to term.
28. Saint Teresa of Avila, *The Book of Her Life,* in *The Collected Works,* Vol. I translated by Kieran Kavanaugh and Otilio Rodriguez (Washington, D.C.: ICS Publications, 1976), p. 1.
29. Origo, *Merchant,* p. 230.
30. *Pitti and Dati,* p. 132.
31. Cited in Natalie Zemon Davis, "Ghosts, Kin, and Progeny: Some Features of Family Life in Early Modern France," in Alice S. Rossi, Jerome Kagan, Tamara K. Hareven, eds., *The Family* (New York: W. W. Norton & Company, Inc., 1978), p. 99.
32. Cited in Origo, *Merchant,* p. 303.
33. For York, see Ursula M. Cowgill, "The People of York 1538–1815," *Scientific American,* vol. 222, no. 1 (January 1970), pp. 104, 106, 110; see *Pitti and Dati,* p. 22.
34. Rates rise and fall for Genoa in the twelfth century, for Venice from the sixteenth to the eighteenth centuries, for London from the fourteenth to the sixteenth centuries, in French towns in the seventeenth century, and in Geneva in the sixteenth and seventeenth centuries. Only the fact of higher mortality among the poorer members of the towns and communes holds consistently. See for example, J. C. Russell, *Medieval Population,* p. 21; Diane Owen Hughes, "Urban Growth and Family Structure in Medieval Genoa," *Past and Present,* vol. 66 (February 1975), p. 23, note; Elizabeth Wirth Marvick, "Nature Versus Nurture: Patterns and Trends in Seventeenth-Century French Child-Rearing" in de Mause, ed., p. 283; David Herlihy, *Medieval and Renaissance Pistoia* (New Haven: Yale University Press, 1967), p. 98; Herlihy, *Pistoia* p. 98; Thrupp, *Merchant,* pp. 206, note, 231; Richard W. and Dorothy C. Wertz, *Lying-In: A History of Childbirth in America* (New York: Schocken Books, 1977), p. 20; Mols in Cipolla, ed., *Sixteenth and Seventeenth Centuries,* p. 69.
35. Cited in Origo, *Merchant,* p. 309.
36. *Glückel,* p. 144.
37. For a description of the evolution of this kind of drama, see John Gassner, ed., *Medieval and Tudor Drama* (New York: Bantam Books, 1968), p. 44; and Alfred

W. Pollard, ed., *Miracle Plays, Moralities and Interludes* (Oxford: Clarendon Press, 1895), pp. xxxi–xxxv.

38. Granger Ryan and Helmut Ripperger, trans., *The Golden Legend of Jacobus de Voragine,* p. 221.
39. To create this composite Mary meant combining many parts of the Gospels: Matthew 26:6–13; 27:56 and 61, 28:1; Luke 7:37–50, 24:10; John 2:1–10; 11:1–45, 12:1–8, 19:25, 20:1; Mark 15:40, and 47, 16:1.
40. Already at the end of the sixth century Pope Gregory the Great acknowledged her holiness and had combined three of the women of the Gospels as the Magdalen—Mary's sister, the anointing sinner, and the witness of the resurrection.
41. In addition to the *Golden Legend* on the stories and worship of Mary Magdalen, see for example; Helen Meredith Garth, "Saint Mary Magdalene in Medieval Literature," *The Johns Hopkins University Studies in Historical and Political Science,* Series LXVII, no. 3 (1950), pp. 12–15, 29, 33–37, 89; A. C. Cawley, ed., *Everyman and Medieval Miracle Plays* (New York: E. P. Dutton & Co. Inc., 1959), pp. 82, 136; and Perry, p. 205.
42. Kamen, p. 12.
43. See Defourneaux.
44. Edward Noble Stone, trans., "Adam," in Barrett H. Clark, ed., *World Drama,* Vol. I (New York: D. Appleton and Co., 1933), p. 312.
45. *Golden Legend,* p. 205.
46. The stories came from the many apocryphal works such as the Protevangel of James and the pseudo-Matthew omitted from authorized versions of the Bible.
47. In the eighteenth century Benedict XIII made the feast day, "Lady of Sorrows," official and "Stabat Mater" commemorating Mary's sadness part of the liturgy.
48. In the twentieth century this maternal image of Mary was emphasized even at Vatican II with the additional title granted to her in 1964, "Mother of the Church."
49. Gregory of Tours made the first collection at the end of the sixth century. The tales multiplied in the fifteenth century, Voragine's *Golden Legend* and Johannes Herolt's *Miracles of the Blessed Virgin* becoming the collections most widely disseminated and repeated.
50. Marina Warner, *Alone of All Her Sex: The Myth and the Cult of the Virgin Mary* (New York: Pocket Books, 1978), p. 115. See also on the meaning of her worship. Penny Schine Gold, *The Lady and the Virgin: Image, Attitude and Experience in Twelfth-Century France* (Chicago: University of Chicago Press, 1985), pp. 68–75.
51. Cited in Ross and McLaughlin, eds., *Renaissance Reader,* pp. 228–29. Despite the popularity of this image, the Assumption became dogma only in 1950; only in 1954 did "Queen of Heaven" become one of the Virgin's official titles.
52. For descriptions of the evolution of the popular worship and official veneration of the Virgin and her many faces, see Part III, "Women of the Churches," and in addition to the works cited in the text, the following: Yrjö Hirn, *The Sacred Shrine: A Study of the Poetry and Art of the Catholic Church* (Boston: Beacon Press, 1932); Geoffrey Ashe, *The Virgin* (London: Routledge & Kegan Paul, 1976); R. W. Southern, *The Making of the Middle Ages* (New Haven: Yale University Press, 1963); Emile Mâle, *The Gothic Image: Religious Art in France of the Thirteenth Century,* translated by Dora Nussey (New York: Harper & Row, Publishers, 1958); Ian

Maclean, *Woman Triumphant: Feminism in French Literature 1610–1652* (Oxford: Clarendon Press, 1977).

PART V: **3. The World of Commercial Capitalism: The Thirteenth to the Seventeenth Centuries**

1. See Braudel, *Mediterranean,* Vol. I, p. 427, and Braudel, *Capitalism,* pp. 375–76. Northern Italy was the exception. By the end of the thirteenth century and the beginning of the fourteenth century, 26.3 percent of the region's population lived in ten of its towns; see Herlihy, "Tuscan Town," pp. 84–85. Demographers estimate that 10–15 percent of Germany's people lived a town life in 1340, 10 percent of England's population in 1377, Beuys, p. 125; Herlihy, "Tuscan Town," p. 84. Even in the 1700s in France, 16 percent of the people had left the land and moved to the urban areas. In addition, before 1500, 90–95 percent of the towns had only 2,000 inhabitants or fewer.
2. See Georges Duby, *The Early Growth of the European Economy: Warriors and Peasants from the Seventh to the Twelfth Century,* translated by Howard B. Clarke (Ithaca, N.Y.: Cornell University Press, 1974), p. 263; as Coulton remarked: "The fact is, that the capitalist system is far older than it has often been represented," *Panorama,* p. 652.
3. Martha C. Howell, "Citizenship and Gender: The Problem of Women's Political Status in Late Medieval Cities of Northern Europe," New York: Fordham University, March 16, 1985, unpublished. Historians have suggested that the rise of capitalism caused this and the other adverse developments. Howell and Wiesner disagree. They identify the cause of the diminution of rights, status, and opportunity not with capitalism as it is broadly defined, but with the specific shift from household production to large-scale enterprise that was characteristic of early capitalist industry. See Howell, *Women,* pp. 30–36, and Wiesner, *Working Women,* p. 3.
4. See Howell, *Women,* pp. 36, 43.
5. *Ménagier,* p. 88.
6. *Decameron,* p. 135.
7. Robert Hellman and Richard F. O'Gorman, eds., *Fabliaux: Ribald Tales from Old France* (New York: Thomas Y. Crowell Company, 1965), p. 151.
8. *Glückel,* p. 2.
9. See concluding sections in Howell, *Women,* and Wiesner, *Working Women,* on the significance of the "ideology of patriarchy."
10. On marriage ages see Mols in Cipolla, ed., *Sixteenth and Seventeenth Centuries,* p. 72; Cowgill; Thrupp, *Merchant,* p. 193; Organization of American Historians, Materials on Women, "The European Marriage Pattern," unpublished manuscript.
11. See Diane Owen Hughes, "Kinsmen and Neighbors in Medieval Genoa," in Miskimin et al., eds., p. 26; René Girard, "Marriage in Avignon in the Second Half of the Fifteenth Century," *Speculum,* Vol. XXVIII, no. 1 (January 1953), passim, pp. 485–98.
12. Owen Hughes in Miskimin et al., eds., p. 23.
13. Girard, pp. 488, 486.
14. Tilly and Scott, p. 43; for Paris examples, see also Diefendorf, p. 32, note.

15. For information about marriage customs and statutes in Ragusa/Dubrovnik see Stuard in Stuard, ed.
16. *Pitti and Dati,* p. 46.
17. A total of nineteen interrelationships; see Lawrence Stone, *The Family, Sex and Marriage* (New York: Harper & Row, Publishers, 1977), p. 131.
18. Tilly and Scott, p. 41.
19. The practice began as early as the twelfth century in Pisa. By the thirteenth century *exclusio propter dotem*—the exclusion of a daughter from inheritance because of the gift of her dowry—had become common throughout Italy. See Eleanor S. Riemer, "Women, Dowries, and Capital Investment in Thirteenth-Century Siena," in Marion A. Kaplan, ed., *The Marriage Bargain: Women and Dowries in European History* (New York: Harrington Park Press, 1985), pp. 62, 65, and Diane Owen Hughes, "From Brideprice to Dowry in Mediterranean Europe," in Kaplan, ed., pp. 32–35. Fourteenth-century Florentine statutes specifically excluded women from all inheritance except their dowry. Susan Mosher Stuard comments, "Law, Society and Women in Medieval and Renaissance Italy," New York, Convention of the American Historical Association, December 28, 1985.
20. Cited in Patricia Labalme, ed., *Beyond Their Sex: Learned Women of the European Past* (New York: New York University Press, 1980), p. 3. In addition to the works cited in the text, see Owen Hughes, in Miskimin et al., eds., pp. 16–19; J. P. Cooper, "Patterns of inheritance and settlement by great landowners from the fifteenth to the eighteenth centuries," in Jack Goody, Joan Thirsk and E. P. Thompson, eds., *Family and Inheritance: Rural Society in Western Europe 1200–1800* (New York: Cambridge University Press, 1978), pp. 280–82.
21. See *Pitti and Dati,* pp. 114, 123, 134; see also Riemer in Kaplan, ed., p. 64, and Susan Mosher Stuard, "Dowry Increase and Increments in Wealth in Medieval Rugusa (Dubrovnik)," *Journal of Economic History,* vol. 16, no. 4 (December 1981), pp. 795–812.
22. See Alison Hanham, ed., *The Cely Letters 1472–1488* (New York: Oxford University Press, 1975), p. xv.
23. Cited in Maurice Ashley, *The Stuarts in Love* (New York: The Macmillan Company, 1964), p. 26.
24. See Girard, p. 486.
25. *Glückel,* p. 96.
26. See *Pitti and Dati,* p. 17.
27. See Kirshner, and Molhs, especially pp. 409, 435–38; Klapisch-Zuber, *Family and Ritual,* pp. 213–24.
28. See Pullan, pp. 164–65, 184–85.
29. For Spain, see Heath Dillard, "Women in Reconquest Castile: The Fueros of Sepulveda and Cuenca," in Stuard, ed., pp. 77–78; for Ragusa, see Stuard in Stuard, ed., pp. 200–04; for Prato, see Origo, *Merchant,* p. 203; for Venice see Chojnacki, "Patrician Women," p. 194, and Stanley Chojnacki, "Dowries and Kinsmen in Early Renaissance Venice," *Journal of Interdisciplinary History,* vol. 4 (Spring 1975), p. 572; for Siena, see Riemer in Kaplan, p. 66.
30. See Ozment, pp. 28–30; for Genoa, see Owen Hughes in Miskimin et al., eds., *Medieval City,* pp. 18–19; for Florence, Lauro Martines, "A Way of Looking at

Women in Renaissance Florence," *The Journal of Medieval and Renaissance Studies,* vol. 4, no. 1 (Spring 1974), p. 20, note; French towns, David Hunt, *Parents and Children in History: The Psychology of Family Life in Early Modern France* (New York: Harper & Row, Publishers, 1972), pp. 60–61, and Davis in Rossi et al., eds., p. 107; Geneva, William E. Monter, "Women in Calvinist Geneva (1550–1800)," *Signs,* vol. 6, no. 2 (Winter 1980), p. 194.

31. For marriage ages, see Klapisch in Laslett, ed., *Household and Family,* p. 272, and Klapisch-Zuber, *Family and Ritual,* pp. 110–11 for the Florentine example; for other towns see the following: Riemer in Kaplan, ed., p. 69; Herlihy, "Tuscan Town," and David Herlihy, "The Medieval Marriage Market," in Dale B. J. Randall, ed., *Medieval and Renaissance Studies Series,* no. 6 (Durham, N.C.: Duke University Press, 1976); Chojnacki "Patrician Women"; Thrupp, *Merchant;* for Protestants see J. Hajnal, "European Marriage Patterns in Perspective," in D. V. Glass and D. E. Eversley, eds., *Population in History* (London: E. Arnold, 1965), p. 114.
32. Maguire, pp. 70–72.
33. Sachs, p. 22; see also Origo, *Merchant,* pp. 265–66; Herlihy, *Pistoia,* for descriptions of more modest weddings.
34. *Medici Letters,* p. 108.
35. *Medici Letters,* p. 110.
36. *Medici Letters,* p. 133.
37. See Maguire, pp. 139–41.
38. See for Italy, Calisse, p. 582; for France, Jean Brissaud, *A History of French Private Law* (Boston: Little, Brown and Company, 1912), pp. 171–72, and Diefendorf, p. 5; for Germany, Hübner, pp. 632–38, 643, and Beuys, p. 144.
39. See for Germany and Switzerland, Ozment, pp. 92–93, 83–85, Davis, *Society and Culture,* p. 90, note, and Jane Dempsey Douglass, "Women and the Continental Reformation" in Rosemary Radford Ruether, ed., *Religion and Sexism: Images of Woman in the Jewish and Christian Traditions* (New York: Simon and Schuster, 1974), pp. 303–04; for England, Stone, *Family,* pp. 37–38.
40. See Girard, pp. 493–97 for Avignon examples; for Italy, see Owen Hughes in Kaplan, ed., p. 36, and Chojnacki, "Patrician Women," p. 188.
41. See Kamen, pp. 389, 410.
42. See Pullan, p. 390; Strauss, p. 197, and Fairchilds, *Charity,* p. 19, for other examples of dowry and trousseau contributions.
43. Harold Speert, *Iconographia Gyniatrica: A Pictorial History of Gynecology and Obstetrics* (Philadelphia: F. A. Davis, 1973), p. 497.
44. Cited in Ross and McLaughlin, eds., *Renaissance Reader,* p. 351.
45. This shift in attitude—from sympathy with the poor and charity as a means of doing good, to fear and charity as a way to insure order—is discussed in Kamen, pp. 405–406; in Braudel, *Capitalism,* p. 40, and extensively documented in Fairchilds, *Charity,* see especially ix–x.
46. Cited in Pullan, p. 299.
47. See Davis, "Crafts," p. 51.
48. See Fairchilds, *Charity,* pp. 29–34 for Aix-en-Provence; Pullan, for Venice, pp. 239–40; Davis, *Society and Culture,* pp. 39–51, for Lyons.
49. *Pitti and Dati,* p. 233.

50. When the English town of Norwich did a census of the indigent poor in 1570, many of the 832 women were destitute because they had been abandoned by their husbands. See Stone, *Family,* pp. 38–39, and Kamen, p. 389. See also Fairchilds, *Charity,* p. 73, on the other categories of poor.
51. Cited in Richard C. Trexler, "The Foundlings of Florence 1395–1455," *History of Childhood Quarterly,* Vol. I, no. 2 (Fall 1973), p. 269.
52. Cited in Trexler, "Foundlings," p. 275.
53. Eighteenth- and nineteenth-century accounts from these institutions indicate that relegation to the foundling home meant leaving the baby and some money to pay the wet nurse, the woman in the country who contracted with the town authorities to feed and care for the infants. In fact, most of these babies died, and no one was surprised. See Olwen H. Hufton, *The Poor of Eighteenth-Century France 1750–1789* (Oxford: Clarendon Press, 1974), pp. 326–32, and William L. Langer, "Infanticide: A Historical Survey," *History of Childhood Quarterly* (Winter 1974), pp. 353–65.
54. For Nuremberg, see Merry E. Wiesner, "Early Modern Midwifery: A Case Study," *International Journal of Women's Studies,* vol. 6, no. 1 (January/February 1983), p. 39; for English towns see M. J. Tucker, "The Child as Beginning and End: Fifteenth and Sixteenth Century English Childhood," de Mause, ed., p. 244, and Barbara A. Kellum, "Infanticide in England in the Later Middle Ages," *History of Childhood Quarterly,* Vol. I, no. 3 (Winter 1974), p. 371.
55. Cited in Quaife, p. 52.
56. For descriptions of young women's choices and circumstances see Quaife, pp. 60–63; David Levine and Keith Wrightson, "The Social Context of Illegitimacy in Early Modern England," unpublished manuscript; Fairchilds, "Sexual Attitudes," pp. 628–29 and 635; Langer gives figures for Nuremberg, see p. 356.
57. See George C. Homans, *English Villagers of the Thirteenth Century* (New York: W. W. Norton & Company, 1975), p. 172; E. P. Thompson, "The Grid of Inheritance: A Comment," in Goody et al., eds., p. 351, note; Carl Ludwig von Bar, *A History of Continental Criminal Law* (Boston: Little, Brown and Company, 1916), pp. 166–67.
58. See R. H. Helmholtz, "Infanticide in the Province of Canterbury during the Fifteenth Century," *History of Childhood Quarterly,* Vol. II, no. 3 (Winter 1975), pp. 382–83; Kellum, p. 370; and John T. McNeill and Helena M. Gamer, *Medieval Handbooks of Penance,* Records of Civilization Sources and Studies No. 29 (New York: Columbia University Press, 1938), p. 96, for an example of the penances assigned. For the severe attitude see Inga Dahlsgård, *Women in Denmark Yesterday and Today* (Copenhagen: Det Danske Selskab, 1980), pp. 24–25; McLaughlin in de Mause, ed., p. 158, note; *Pitti and Dati,* p. 147.
59. This custom is evident among the Germanic peoples of the eighth century, and continuing in Scotland and Norway, and into the 1930s in Yugoslavia. Surveys of English parish records from 1550 to 1820 list 1,855 brides for whom the baptismal records of their first-born children suggest that almost 50 percent of them were pregnant at the time of their marriage. See P. E. H. Hair, "Bridal Pregnancy in Rural England in Earlier Centuries," *Population Studies,* Vol. XX, no. 2 (November 1966), pp. 235–36.
60. See Fairchilds, "Sexual Attitudes," for these conclusions.

61. Cited in Origo, *Merchant,* p. 93.
62. See for Nuremberg, Langer, p. 356, and Wiesner, "Midwifery," p. 38; English towns, R. V. Schnucker, "The English Puritans and Pregnancy, Delivery and Breast-Feeding," *History of Childhood Quarterly,* Vol. I, no. 4 (Spring 1974), p. 654, note; for France and Switzerland, Monter, "Geneva," pp. 196–97.
63. Cited in Quaife, p. 103.
64. Much of the information that follows comes from the round-table discussion, "Working Women in Early Modern Europe: A Cross-Cultural Approach," with Merry Wiesner Wood on Nuremberg, Natalie Zemon Davis on Lyons, Judith Brown on Florence, Martha Howell on Dutch and Flemish cities, and Nancy Adamson on London, The Berkshire Conference on the History of Women, held at Vassar College, Poughkeepsie, New York, June 1981. In their books Howell and Wiesner see a relative increase in women's guild membership in the early centuries as long as production was tied to the household. See for example, Howell, *Women,* pp. 43, 87, 152–55, 161–62, and Wiesner, *Working Women,* pp. 3, 33.
65. The major entries are cited in Ross and McLaughlin, *Medieval Women,* p. 20.
66. Clark, p. 155.
67. Davis, "Crafts," p. 68. For the pattern of gradual exclusion in fourteenth- and fifteenth-century Cologne see Wiesner, *Working Women,* pp. 133, 137, 157, 185, and Howell, *Women,* pp. 115–16; for Ghent see Nicholas, pp. 98–101.
68. Clark, p. 167.
69. For a discussion of the phenomenon in the sixteenth and seventeenth centuries, see the following: for Geneva, Monter, "Geneva," p. 200; for France, Tilly and Scott, p. 49; for Germany, Jean H. Quataert, "The Shaping of Women's Work in Manufacturing Guilds, Households, and the State in Central Europe 1648–1870," *The American Historical Review,* vol. 90, no. 5 (December 1985), pp. 1126–33.
70. For descriptions of this practice in the thirteenth and fourteenth centuries see Ross and McLaughlin, *Medieval Women,* pp. 6–10.
71. See Ross and McLaughlin, *Medieval Women,* pp. 30–33, and Gies and Gies, *Women.*
72. For descriptions of his firm, see Ross and McLaughlin, *Medieval Women,* pp. 35–36, and Florence Edler de Roover, "Andrea Banchi, Florentine Silk Manufacturer and Merchant in the Fifteenth Century," in William M. Bowsky, ed., *Studies in Medieval and Renaissance History,* Vol. III (Lincoln, Neb.: University of Nebraska Press, 1966), pp. 241–43, 260–61.
73. See Roover, p. 248 and note.
74. Roover, p. 247.
75. Figures for 1663 dramatically show the shifts in employment from women to men. Though only 38 percent of the workers in wool, men represented 96 percent of the weavers in one section of Florence and 65 percent in another. They represented 84 percent of the 14,034 of the workers in silk, 78 percent of the master weavers, 59 percent of the weaver apprentices. Judith C. Brown and Jordan Goodman, "Women and Industry in Florence," *Journal of Economic History,* Vol. XL, no. 1 (March 1980), pp. 79–80.
76. For Leyden see Howell, *Women,* pp. 71–72, 88–90; see also Wiesner, *Working Women,* pp. 174–85.

77. For descriptions of these developments see the following: for Italian towns, Ross and McLaughlin, *Medieval Women,* pp. 34–35, Roover, pp. 251–53; for English towns, Dale, pp. 331–32, 331, note, Kowaleski, pp. 11–12, Clark, p. 158, Power, *Medieval Women,* p. 60; for German and Dutch towns, Howell, "Working Women."
78. See for example, Davis, *Society and Culture,* p. 291, note; Kamen, pp. 361–62, 384.
79. Cited in E. A. McArthur, "Women Petitioners and the Long Parliament," *English Historical Review,* no. XCIII (January 1909), p. 701. See also David Weigall, "Women Militants in the English Civil War," *History Today,* Vol. XXII, no. 6 (June 1972), p. 435.
80. Cited in McArthur, pp. 202–203.
81. Cited in Patricia Higgins, "The Reactions of Women, with Special Reference to Women Petitioners," in Brian Manning, ed., *Politics, Religion and the English Civil War* (London: Edward Arnold, 1973), p. 201, see also p. 203; McArthur, p. 707.
82. Cited in Higgins, p. 203.
83. Cited in McArthur, p. 708.
84. Germaine Greer, *The Obstacle Race: The Fortunes of Women Painters and Their Work* (New York: Farrar, Straus & Giroux, 1979), p. 160.
85. Cited in Ann Sutherland Harris and Linda Nochlin, *Women Artists 1550–1950* (New York: Alfred A. Knopf, 1978), p. 25.
86. On the unorthodoxy of the painting, the artist's identification with the subject, and the details of the trial of her accused rapist, see Mary D. Garrard, "Artemisia and Susanna," in Norma Broude and Mary D. Garrard, eds., *Feminism and Art History: Questioning the Litany* (New York: Harper & Row, Publishers, 1982), p. 165.
87. The husband disappears from the historical records; there are references to one daughter Prudentia and perhaps another, listed in the Roman census of 1624 as Palmira.
88. Cited in Harris and Nochlin, p. 119.
89. Cited in Harris and Nochlin, p. 120.
90. Longhi, Tintoretto, and Van Dyck all had daughters and trained them as painters. Others married and began artistic dynasties. Mayken Verhulst trained her grandson, the artist Jan Breughel. See Greer for the list of factors that deterred women artists before the nineteenth century.
91. Cited in Harris and Nochlin, p. 25.
92. Cited in Harris and Nochlin, p. 31.
93. Cited in Harris and Nochlin, p. 107.
94. Harris and Nochlin, p. 159.
95. Most of the names of women medical practitioners listed in this section come from the monograph by M. J. Hughes, pp. 139–47; her classifications have not been followed, however.
96. See Kate Campbell Hurd-Mead, *A History of Women in Medicine from the Earliest Times to the Beginning of the Nineteenth Century* (Haddam, Conn.: The Haddam Press, 1938), p. 521.
97. M. J. Hughes, p. 83.
98. Cited in M. J. Hughes, p. 83.
99. Mead, p. 404.
100. For the description of the trial and her remarks see James Bruce Ross and Mary

Martin McLaughlin, eds., *The Portable Medieval Reader* (New York: The Viking Press, 1966), pp. 636–39.

101. Cited in M. J. Hughes, pp. 85–86, note.
102. Cited in Clark, p. 260. For other English examples see p. 260; for Church edicts see Mary Chamberlain, *Old Wives' Tales: Their History, Remedies and Spells* (London: Virago Press Limited, 1981), p. 43; for France see Davis, "Crafts," p. 69.
103. See Speert, pp. 102–16.
104. For fees, see Wiesner, "Midwifery," pp. 27–28, 29, 36, and her book *Working Women,* pp. 55–64; see R. L. Petrelli, "The Regulation of French Midwifery during the *Ancien Regime," Journal of the History of Medicine,* vol. 26 (1971), p. 281 and passim; see also Thomas Benedek, "The Changing Relationship Between Midwives and Physicians During the Renaissance," *Bulletin and History of Medicine,* vol. 51, no. 4 (Winter 1977), pp. 550–64.
105. Speert, p. 71.
106. Mead, p. 395.
107. Scholars have argued over female authorship. Given the knowledge of women's ailments and of childbirth—areas usually closed to thirteenth-century male practitioners, even master physicians—an experienced female practitioner must have been involved in the writing even if she was not the sole author. Trotula explains that it was this need for women to deal with women's ailments, given the sex's modesty and reluctance to speak with a male physician, that brought her to the study of medicine. See for example, John F. Benton, "Trotula, Women's Problems, and the Professionalization of Medicine in the Middle Ages," *Bulletin of the History of Medicine,* vol. 59 (1985), pp. 30–53.
108. Elizabeth Mason-Hohl, trans., *The Diseases of Women by Trotula of Salerno* (Los Angeles: The Ward Ritchie Press, 1940), pp. 1, 2.
109. She is usually known by her maiden name "Bourgeois."
110. Mead, p. 420.
111. Mead, fn. p. 429.
112. Bourgeois had lost favor at court when one of her patients, the Duchess of Orléans, died of puerperal fever, probably because the surgeon had not removed all of the placenta. She wrote her text to prove her expertise.
113. See *Trotula,* p. 23.
114. Cited in Carolyn Merchant, *The Death of Nature: Women, Ecology and the Scientific Revolution* (New York: Harper & Row, Publishers, 1980), p. 153.
115. See James Hitchcock, ed., "A Sixteenth Century Midwife's License," *Bulletin of the History of Medicine,* vol. 41 (January/February 1967), p. 76; see also Thomas R. Forbes, "Midwifery and Witchcraft," *Journal of the History of Medicine,* vol. 17 (April 1962), p. 280, and Chamberlain, *Remedies,* p. 54, for sixteenth- and seventeenth-century restrictions and prosecutions; see Wiesner, *Working Women,* p. 69, for examples from Germany.
116. Cited in Clark, p. 282.
117. Called the widow's portion, the *donatio propter nuptias,* the *sponsalitium.*
118. Cited in Davis, "Crafts," p. 54.
119. Russian customs into the seventeenth century remain the exception, giving the responsibility for the widow to the son. The charters for Novgorod and Pskov in the

late fifteenth century enjoined sons to act and care for widowed mothers. In Pskov a son could be disinherited if he did not provide for both parents when they were old. See George Vernadsky, trans., *Medieval Russian Laws* (New York: W. W. Norton & Company, Inc., 1969), p. 86. For other information on widows, Owen Hughes in Kaplan, ed., gives a complete analysis of the evolution of the widow's portion, tracing it to the *Morgengabe* of Germanic and Celtic marriages, pp. 29–30; for Italian examples see Owen Hughes, "Urban Growth," pp. 15, 24–25; Herlihy, "Marriage," p. 8; Maguire, pp. 132–33, Calisse, pp. 579, 625; for France, Girard, p. 487, Cooper in Goody et al., eds., p. 225, John H. Mundy, "The Influence of Women in the Medieval Economy—Comments," paper delivered at The Berkshire Conference on the History of Women, held at Mount Holyoke College, South Hadley, Massachusetts, August 24, 1978; Diefendorf, pp. 6, 8–9, 10–12; for England, Cooper in Goody et al., eds., p. 296, Thrupp, *Merchant,* p. 106, and Kowaleski, p. 2; for Ghent in Belgium see Nicholas, pp. 28, 119.

120. See Eleanor S. Riemer, "Women, Dowries, and Capital Investment in Thirteenth-Century Siena," in Kaplan, ed.
121. For Italy, see *Pitti and Dati,* p. 50, Chojnacki, "Patrician Women," p. 199; for Yugoslavia (Ragusa/Dubrovnik) Stuard in Stuard, ed., pp. 200–203, Riemer in Kaplan, ed., pp. 64–65, 75; for France, Diefendorf, pp. 18–25, Davis, *Society and Culture,* p. 69, Brissaud, p. 157.
122. London wills studied for the years 1271–1330 show wives to have been 61.4 percent of the survivors. The 1427 *catasto* (census for tax purposes) for Florence and its environs identifies women as heads of 21.1 percent of the households. In Florence the percentage increased as the population aged, from 11.2 percent ages 33–37, to 25.2 percent ages 43–47, to over 50 percent over the age of fifty. Parisian tax rolls for 1527 show 12 percent of the households headed by females. In sixteenth-century Seville the percentage rose to 20 percent in the 1534 census. See the following: Harry A. Miskimin, "The Legacies of London 1259–1330," in Miskimin et al., eds.; Klapisch-Zuber, *Women,* pp. 120–21; Herlihy, *Pistoia,* p. 117; Herlihy, "Tuscan Town," p. 97; Herlihy in Martines, ed., p. 149; Diefendorf, p. 3, note; Pike, p. 7.
123. See Clark, pp. 32–34.
124. Sachs, p. 42.
125. Monter, "Geneva," p. 201.
126. See Owen Hughes in Miskimin et al., eds., p. 142; Harksen, p. 11, Kowaleski, p. 5ff, Hogrefe, *Tudor Women,* p. 37, Dale, pp. 327–28, Clark, pp. 30–31.
127. Beuys, p. 147.
128. Hogrefe, *Tudor Women,* pp. 40–41.
129. Ann Crabb, "Motherhood and Power in Renaissance Florence: Alessandra Macinghi Strozzi and her Sons, 1441–1471," Fordham University, New York City, March 15, 1985.
130. Cited in Martines, p. 25.
131. Cited in Martines, p. 25.
132. See also Richard A. Goldthwaite, *Private Wealth in Renaissance Florence: Four Florentine Families* (Princeton, N.J.: Princeton University Press, 1969), especially pp. 45–59.
133. Ross, ed., *Medici Letters,* p. 245.

134. *Glückel,* p. 34.
135. *Glückel,* p. 255.
136. *Glückel,* p. 264.

PART V: **4. The Invisible and Visible Bonds of Misogyny**

1. "Adam," in B. H. Clark, ed., Vol. I, p. 313. This is typically the message of the plays showing the Fall. For example, in another cycle, the York Coopers play of the fifteenth and sixteenth centuries, Adam realizes his sin and his nakedness right away. He quickly accuses Eve of "this bad bargain," and is sorry that he believed her stories. Cawley, p. 23.
2. "Adam," in B. H. Clark, ed., Vol. I, p. 311.
3. See Bede Jarrett, O.P., *Social Theories of the Middle Ages* (Westminster, Md.: The Newman Bookshop, 1942), p. 81, and Thomas Frederick Crane, ed., *The Exempla or Illustrative Stories from the Sermones Vulgares of Jacques de Vitry* (New York: Burt Franklin, 1971), p. 235; even more important because of its wide circulation in these centuries as a source of anecdotes and parables used by priests and friars was the *Gesta Romanorum* (c. 1340). It gave the same picture of the female sex.
4. Cited in E. T. Healy, *Women According to Saint Bonaventure* (New York: Georgian Press, 1955), p. 46.
5. The first versions of the fabliaux are from the thirteenth and fourteenth centuries. In their crudest form, they appeared in the towns all across Europe. They portray women as a group, rarely as individuals, and almost always use them to teach a negative lesson about the female sex. A number of articles have been written on the audience for the fabliaux and their misogyny. See the bibliography by Roberta L. Krueger and E. Jane Burns for "Women in Medieval French Literature," in *Romance Notes,* Vol. XXV, no. 3 (Spring 1985), pp. 375–90 and especially Norris J. Lacy, "Fabliau Women," in the same volume, pp. 318–27, and Sarah Melhado White, "Sexual Language and Human Conflict in Old French Fabliaux," *Comparative Studies in Society and History,* vol. 24, no. 2 (1982), pp. 185–210. For historians' discussion of them as a source for townsmen's attitudes toward women, see for example Howell, *Women,* pp. 181–83.
6. *Ménagier,* p. 50.
7. See Judith C. Brown, *Immodest Acts: The Life of a Lesbian Nun in Renaissance Italy* (New York: Oxford University Press, 1986), pp. 133–34.
8. See examples for Nuremberg in Strauss, p. 113.
9. M. Margaret Newett, "The Sumptuary Laws of Venice in the Fourteenth and Fifteenth Centuries" in T. F. Tout and James Tait, eds., *Historical Essays by Members of Owens College, Manchester* (London: Longmans Green and Co., 1902), p. 247.
10. Cited in Brucker, ed., *Renaissance Florence,* p. 181.
11. For additional examples see the following: for Italy, Klapisch-Zuber, *Women,* pp. 242–43, Owen Hughes in Kaplan, ed., p. 42, Origo, *Merchant,* pp. 260–69; for Germany, Sybille Harksen, *Women in the Middle Ages,* translated by Marianne Herzfeld (New York: Abner Schram, 1975), p. 22, Sachs, p. 33, Strauss, pp. 113–14, Philippe Erlanger, *The Age of Courts and Kings: Manners and Morals 1558–1715* (New York: Harper & Row, Publishers, 1967), p. 257; for England, Thrupp, *Mer-*

chant, pp. 146–48; for France, R. Turner Wilcox, *The Mode in Costume* (New York: Charles Scribner's Sons, 1944), p. 96; for Spain, Defourneaux, pp. 56–58.

12. Cited in Hübner, p. 67.
13. See Katharine M. Rogers, *The Troublesome Helpmate: A History of Misogyny in Literature* (Seattle: University of Washington Press, 1966), p. 70.
14. Cited in Origo, *Merchant,* p. 259; for other examples, see Henriques, p. 45, Sachs, p. 52.
15. *Decameron,* p. 201.
16. *Decameron,* p. 318.
17. See "The Miller and the Two Clerics," in Hellman and O'Gorman, eds., p. 57. The image of insatiability took a particularly loathsome form when men turned their humor to old women. "Unchastity" is portrayed in relief on the twelfth-century Charterhouse near Avignon as a naked hag having intercourse with a goat. See Henry Kraus, "Eve and Mary: Conflicting Images of Medieval Women," in Broude and Garrard, eds., p. 79. The popular storytellers of the fabliaux and the clerical author of the *Fifteen Joys of Marriage* mocked the old widow, no less insatiable but now lusting in a decrepit body. The old woman always rejoices in her husband's death, now ready for "the delight and pleasure of young flesh." Elisabeth Abbott, trans., *The Fifteen Joys of Marriage* (London: The Orion Press, 1959), p. 192. See also Hellman and O'Gorman, eds., pp. 148, 152, 156.
18. The term "cuckold" came from the cuckoo, who lays eggs in other birds' nests and they, unsuspecting, raise them.
19. See Hellman and O'Gorman, eds., pp. 171–77.
20. For examples of laws see for Italy, Calisse, pp. 438, 571; for France, Girard, p. 495, Brissaud, pp. 138–40; for England, Stone, *Family,* pp. 22, 145; for Denmark, Dahlsgård, p. 42; for Switzerland, Monter, "Geneva," p. 192.
21. In the years of famine 1430–1439, 66.3 percent of the abandoned babies at Florence's San Gallo were female. Trexler, "Foundlings," p. 268. The mortality rate for girls was higher than that for boys. See Richard C. Trexler, "Infanticide in Florence: New Sources and First Results," *History of Childhood Quarterly,* Vol. I, no. 7 (Summer 1973), pp. 101–102.
22. See Herlihy in Martines, ed., p. 146; see also the figures for Pistoia, Herlihy, "Tuscan Town," p. 101.
23. Kirshner and Mallio, "Dowry Fund," p. 420.
24. The attacks continued into the next year, when he was accused, convicted, and executed. Brucker, ed., *Renaissance Florence,* pp. 152–53.
25. Cited in Ross and McLaughlin, *Medieval Women,* p. 13.
26. Margery Kempe, *The Book of Margery Kempe,* edited by W. Butler-Bowden (New York: The Devin-Adair Company, 1944), p. 49.
27. See cases from Brucker, ed., *Renaissance Florence.*
28. Brucker, ed., *Renaissance Florence,* p. 121.
29. See Clark, p. 191.
30. Cited in Ozment, p. 3.
31. "About the Men Who Came to Heaven," Brewer ed., in *Medieval Comic Tales,* translated by Peter Rickard, Alan Deyermond, Peter King, David Blamires, Michael

Lapidge, and Derek Brewer (Cambridge, Eng.: D. S. Brewer Ltd., 1972), pp. 53–54.

32. Abbott, *Fifteen Joys,* pp. 43, 137.
33. Hans Sachs, *Seven Shrovetide Plays,* translated by E. V. Ouless (London: H. F. W. Deane & Sons, The Year Book Press Ltd., 1930), p. 33.
34. Cited in Davis, *Society and Culture,* p. 116.
35. See for example, Crane, ed., *Vitry,* pp. 222–23, 231; William Woods, *England in the Age of Chaucer* (New York: Stein and Day Publishers, 1976), pp. 97–98.
36. *Decameron,* pp. 715–17.
37. *Ménagier,* p. 150.
38. *Ménagier,* p. 145.
39. *Ménagier,* p. 52.
40. *Ménagier,* p. 168; see White, "Sexual Language," for an earlier version, p. 201.
41. Brucker, ed., *Renaissance Florence,* pp. 38–39.
42. Cited in Eugene F. Policelli, "Medieval Women: A Preacher's Point of View," *International Journal of Women's Studies,* Vol. I, no. 3 (May/June, 1978), p. 287; see also Margaret Leah King, "Caldiera and the Barbaros on Marriage and the Family: Humanist Reflections of Venetian Realities," *The Journal of Medieval and Renaissance Studies,* vol. 6, no. 1 (Spring 1976), pp. 32–34, for summary of Barbaro's secular view in his treatise *De re uxoria.*
43. Cited in Origo, *Merchant,* p. 251.
44. Hogrefe, *Tudor Women,* p. 147.
45. Cited in James Bruce Ross, "The Middle-Class Child in Urban Italy, Fourteenth to Early Sixteenth Century," in de Mause, ed., p. 204.
46. Cited in Policelli, p. 286.
47. *Ménagier,* pp. 40, 147.
48. "Adam," in B. H. Clark, ed., Vol. I, p. 305.
49. "Adam," B. H. Clark, ed., Vol. I, p. 305.
50. Cited in Origo, *Merchant,* p. 194.
51. *Ménagier,* p. 108.
52. This image and this theory of political hierarchy became a commonplace of seventeenth-century political writing.
53. *Decameron,* p. 721.
54. Sachs, *Shrovetide,* p. 53.
55. See Policelli, pp. 286–87.
56. *Ménagier,* p. 188.
57. *Ménagier,* p. 189.
58. See Origo, *Merchant,* p. 385.
59. *Ménagier,* pp. 171–72.
60. *Ménagier,* p. 173.
61. *Ménagier,* pp. 171–72.
62. *Glückel,* p. 142.
63. On the establishment of schools for girls, see the summary in Phyllis Stock, *Better Than Rubies: A History of Women's Education* (New York: G. P. Putnam's Sons, 1978), Chapter II; for Florence, see Jacques LeGoff, "The Town as an Agent of

Civilisation 1200–1500," in Cipolla, ed., *The Middle Ages,* p. 85; for London, see Thrupp, *Merchant,* pp. 156, 171, and for Paris, see Harksen, p. 17; for Geneva, Douglas in Ruether, ed., *Religion and Sexism,* p. 304.

64. *Ménagier,* p. 43.
65. Titled "Patient Griselda" in English and French versions. On the popularity of the tale, see Klapisch-Zuber, *Women,* p. 228, note.

Bibliography

This bibliography is selective and is designed to give the reader the names of the most useful and most accessible secondary works. It is arranged as follows:

1. Works about women useful to many parts.
2. Works about women (part by part) essential to the analysis and narrative.
3. General works (part by part), including information about women not found elsewhere or giving necessary context to their lives.

References to primary sources and to specialized articles and monographs may be found in the Notes.

Works About Women Useful to Many Sections

Abrams, Lynn, and Elizabeth Harvey, eds. *Gender Relations in German History: Power, Agency and Experience from the Sixteenth to the Twentieth Century*. Durham, N.C.: Duke University Press, 1997.

Akkerman, Tjitske, and Siep Stuurman, eds. *Perspectives on Feminist Political Thought in European History: From the Middle Ages to the Present*. New York: Routledge, 1998.

Alec, Margaret. *Hypatia's Heritage: A History of Women in Science From Antiquity Through the Nineteenth Century*. Boston: Beacon Press, 1986.

Ariès, P., and Georges Duby, eds. *A History of Private Life*. 5 vols. Translated by A. Goldhammer. Cambridge, Mass.: Harvard University Press, 1987–1991.

Atkinson, Dorothy, Alexander Dallin, and Gail Warshofsky Lapidus, eds. *Women in Russia*. Stanford, Calif.: Stanford University Press, 1977.

Baranski, Zygmunt G., and Shirley W. Vinall, eds. *Women and Italy: Essays on Gender, Culture and History*. New York: St. Martin's Press, 1991.

Baskin, Judith R., ed. *Jewish Women in Historical Perspective*. Detroit: Wayne State University Press, 1991.

Becker-Cantarino, Barbara. "Feminist Consciousness and Wicked Witches: Recent Studies on Women in Early Modern Europe," *Signs*, vol. 20, no. 1 (Autumn 1994), pp. 152–75.

Benjamin, Marina. *A Question of Identity: Women, Science and Literature*. [women and the history of science] New Brunswick, N.J.: Rutgers University Press, 1993.

Bremer, Jan, and Lourens van den Bosch, eds. *Between Poverty and the Pyre: Moments in the History of Widowhood*. New York: Routledge, 1995.

Bridenthal, Renate, Susan Mosher Stuard, and Merry Wiesner, eds. *Becoming Visible: Women in European History*. 3rd ed. Boston: Houghton Mifflin, 1998.

Brink, Jean R., Allison P. Coudert, and Maryanne C. Horowitz, eds. *The Politics of Gender in Early Modern Europe*. Kirksville, MO: Sixteenth Century Journal Publishers, 1989.

Broude, Norma, and Mary D. Garrard. *The Expanding Discourse: Feminism and Art History*. New York: Icon Editions, 1992.

Carpenter, Jennifer, and Sally-Beth MacLean, eds. *Power of the Weak: Studies on Medieval Women*. Urbana: University of Illinois Press, 1995.

Clements, Barbara Evans, Barbara Alpern Engel, and Christine D. Worobec, eds. *Russia's Women: Accommodation, Resistance, Transformation*. Berkeley: University of California Press, 1991.

Dahlsgård, Inga. *Women in Denmark Yesterday and Today*. Copenhagen: Det Danske Selskab, 1980.

Duby, Georges, and Michelle Perrot, eds. *A History of Women in the West*, 5 vols. Cambridge, Mass.: Harvard University Press, 1994.

Duckett, Eleanor Shipley. *Medieval Portraits from East and West*. Ann Arbor: University of Michigan Press, 1972.

Erler, Mary, and Maryanne Kowaleski, eds. *Women and Power in the Middle Ages*. Athens: The University of Georgia Press, 1988.

Fildes, Valerie. *Wet Nursing: A History From Antiquity to the Present*. New York: Basil Blackwell, 1988.

Gibson, Wendy. *Women in Seventeenth-Century France*. Basingstoke: Macmillan, 1989.

Gold, Barbara K., Paul Allen Miller, and Charles Platter, eds. *Sex and Gender in Medieval and Renaissance Texts: The Latin Tradition*. Albany: State University of New York Press, 1997.

Goscilo, Helena, and Beth Holmgren, eds. *Russia, Women, Culture*. Bloomington: University of Indiana Press, 1996.

Gottlieb, Beatrice. *The Family in the Western World from The Black Death to the Industrial Age*. New York: Oxford University Press, 1993.

Green, Monica. "Female Sexuality in the Medieval West," *Trends in History*, vol. 4 (1990), pp. 127–58.

Hufton, Olwen. *The Prospect Before Her: A History of Women in Western Europe, 1500–1800*. New York: Alfred A. Knopf, 1996.

Journal of Women's History. "Special Issue on Irish Women," vol. 6/7, no. 4/1 (Winter/Spring 1995).

Kertzer, David I., and Richard P. Saller, eds. *The Family in Italy from Antiquity to the Present*. New Haven: Yale University Press, 1991.

Labarge, Margaret Wade. *A Small Sound of the Trumpet: Women in Medieval Life*. Boston: Beacon Press, 1986.

Laqueur, Thomas. *Making Sex: Body and Gender From the Greeks to Freud*. Cambridge, Mass.: Harvard University Press, 1990.

Laurence, Anne. *Women in England, 1500–1760: A Social History*. New York: St. Martin's Press, 1994.

Levin, Carole, and Jeanie Watson, eds. *Ambiguous Realities: Women in the Middle Ages and Renaissance*. Detroit: Wayne State University Press, 1987.

Leyser, Henrietta. *Medieval Women: A Social History of Women in England, 450–1500*. New York: St. Martin's Press, 1995.

MacCurtain, Margaret, and Mary O'Dowd, eds. *Women in Early Modern Ireland*. New York: Columbia University Press, 1991.

Marshall, Kimberly, ed. *Rediscovering the Muses: Women's Musical Traditions*. Boston: Northeastern University Press, 1993.

Marshall, Rosalind K. *Virgins and Viragos. A History of Women in Scotland from 1080–1980*. Chicago: University of Chicago Press, 1983.

Maynes, Mary Jo, ed. *Gender, Kinship, Power: A Comparative and Interdisciplinary History*. New York: Routledge, 1996.

McNamara, Jo Ann. *Sisters in Arms: Catholic Nuns Through Two Millennia*. Cambridge, Mass.: Harvard University Press, 1996.

Micale, Mark S. *Approaching Hysteria: Disease and Its Interpretations*. Princeton, N.J.: Princeton University Press, 1995.

Murray, Jacqueline, and Konrad Eisenbichler, eds. *Desire and Discipline: Sex and Sexuality in the Premodern West*. Buffalo, N.Y.: University of Toronto Press, 1996.

Offen, Karen, Ruth Roach Pierson, and Jane Rendall, eds. *Writing Women's History: International Perspectives*. Bloomington: Indiana University Press, 1991.

Parsons, John Carmi, and Bonnie Wheeler, eds. *Medieval Mothering*. New York: Garland, 1996.

Partner, Nancy F., ed. *Studying Medieval Women: Sex, Gender, Feminism*. Cambridge, Mass.: Medieval Academy of America, 1993.

Pendle, Karin, ed. *Women and Music: A History*. Bloomington: Indiana University Press, 1991.

Pushkareva, Natalia. *Women in Russian History: From the Tenth to the Twentieth Century*. Translated and edited by Eve Levin. New York: M. E. Sharpe, 1997.

Rosenthal, Joel, ed. *Medieval Women and the Sources of Medieval History*. Athens: University of Georgia Press, 1990.

Ruether, Rosemary, and Eleanor McLaughlin, eds. *Women of Spirit*. New York: Simon and Schuster, 1979.

Ruether, Rosemary Radford, ed. *Religion and Sexism: Images of Woman in the Jewish and Christian Traditions*. New York: Simon and Schuster, 1974.

Scott, Joan Wallach. *Gender and the Politics of History*. New York: Columbia University Press, 1988.

Shahar, Shulamith. *Childhood in the Middle Ages*. London: Routledge, 1990.

Shapiro, Ann-Louise, ed. *Feminists Revision History*. New Brunswick, N.J.: Rutgers University Press, 1994.

Sheils, W. J., and Diana Wood, eds. *Women in the Church*. Cambridge, Mass.: Basil Blackwell, 1990.

Smart, Carol, ed. *Regulating Womanhood: Historical Essays on Marriage, Motherhood and Sexuality*. London: Routledge, 1992.

Stafford, Pauline. *Queens, Concubines and Dowagers: The King's Wife in the Early Middle Ages*. Athens: The University of Georgia Press, 1983.

Stanley, Autumn. *Mothers and Daughters of Invention: Notes Towards a Revised History of Technology*. Metuchen, N.J.: Scarecrow Press, 1993.

Stuard, Susan Mosher, ed. *Women in Medieval Society*. Philadelphia: University of Pennsylvania Press, 1976.

Stuard, Susan Mosher, ed. *Women in Medieval History and Historiography*. Philadelphia: University of Pennsylvania Press, 1987.

Van Deursen, A. T. *Plain Lives in a Golden Age: Popular Culture, Religion and Society in Seventeenth-Century Holland*. Translated by Maarten Ultee. New York: Cambridge University Press, 1991.

Waithe, Mary Ellen, ed. *Medieval, Renaissance, and Enlightenment Women Philosophers AD 500–1600*. Boston: Klumer, 1989.

Walker, Sue Sheridan, ed. *Wife and Widow in Medieval England*. Ann Arbor: University of Michigan Press, 1994.

Warner, Marina. *Monuments and Maidens: The Allegory of the Female Form*. New York: Atheneum, 1985.

Warnock, Mary, ed. *Women Philosophers*. London: J. M. Dent, 1996.

Wiesner, Merry E. *Women and Gender in Early Modern Europe*. New York: Cambridge University Press, 1993.

Wilson, Katharina, ed. *Medieval Women Writers*. Athens: The University of Georgia Press, 1984.

Wunder, Heide. *He is the Sun, She is the Moon: Women in Early Modern Germany*. Translated by Thomas Dunlap. Cambridge, Mass.: Harvard University Press, 1998.

The following journals have useful articles and reviews: *Feminist Studies*, *Gender and History*, *Journal of Women's History*, *Signs: Journal of Women in Culture and Society*, *Women's History Review*.

The following web sites are useful resources:

Diotima—Materials for the Study of Women and Gender in the Ancient World.
http://www.uky.edu/ArtsScience/Classics/Gender.html

Dutch Women's History (Vereniging voor Vrouwengeschiedenis, Amsterdam)
http://www.let.ruu.nl/hist/info/VVG

GABRIEL—Gateway to Europe's National Libraries.
http://portico.bl.uk/gabriel/

International Federation for Research in Women's History—Access to the newsletter of the international organization of women historians' group
http://www.arts.unimelb.edu.au/Dept/History/ifrwh

Italian Women's History (Societa italiana delle storiche)
http://www.idg.fi.cnr.it/wwwdonna/storiche.htm

Medieval Feminist Index—Indexes over 300 journals and essay collections, 1994–97, for the period 450–1500 C.E.
http://www.haverford.edu/library/reference/mschaues/mfi/

ViVa—A bibliography of articles in women's history from 60 journals from 1995 to the present.
http://www.iisg.hl/~ womhist

Women and Gender in Early Modern Europe—A web site and access to H-Frauen-L, an internet discussion forum

http://www.h-net.msu.edu./~frauen-l

Women Writers Project at Brown University—Texts of women's literary works from England, Scotland, Ireland, Wales, and the colonies before 1830.

http://www.wwp.brown.edu

PART I TRADITIONS INHERITED

Works About Women

Agonito, Rosemary, ed. *History of Ideas on Women*. New York: Capricorn, 1977.

Archer, Leonie J., Susan Fischler, and Maria Wyke, eds. *Women in Ancient Societies*. New York: Routledge, 1994.

Arjava, Antti. *Women and Law in Late Antiquity*. New York: Oxford University Press, 1996.

Barber, Elizabeth Wayland. *Women's Work: The First 20,000 Years: Women, Cloth, and Society in Early Times*. New York: W. W. Norton and Company, 1994.

Barrett, Anthony A. *Agrippina: Sex, Power, and Politics in the Early Empire*. New Haven: Yale University Press, 1996.

Bauman, Richard A. *Women and Politics in Ancient Rome*. London: Routledge, 1992.

Beard, Mary, and John Henderson. "With this body I thee worship: Sacred prostitution in antiquity," *Gender and History*, vol. 9, no. 3 (1997), pp. 480–503.

Bitel, Lisa M. *Land of Women: Tales of Sex and Gender from Early Ireland*. Ithaca, N.Y.: Cornell University Press, 1996.

Blok, Josine, and Peter Mason, eds. *Sexual Asymmetry: Studies in Ancient Society*. Amsterdam: J. C. Gieben, 1987.

Blundell, Sue. *Women in Ancient Greece*. Cambridge, Mass.: Harvard University Press, 1995.

Blundell, Sue, and Margaret Winsom, eds. *The Sacred and the Feminine in Ancient Greece*. New York: Routledge, 1998.

Brownmiller, Susan. *Against Our Will: Men, Women and Rape*. New York: Simon and Schuster, 1975.

Cameron, Averil, and Amélie Kuhrt, eds. *Images of Women in Antiquity*. Detroit: Wayne State University Press, 1983.

Cantarella, Eva. *Pandora's Daughters: The Role and Status of Women in Greek and Roman Antiquity*. Baltimore: Johns Hopkins University Press, 1987.

Carroll, Berenice A., ed. *Liberating Women's History: Theoretical and Critical Essays*. Urbana: University of Illinois Press, 1976.

Chodorow, Nancy. *The Reproduction of Mothering: Psychoanalysis and the Sociology of Gender*. Berkeley: University Of California Press, 1978.

Clark, Elizabeth A. "Patrons, Not Priests: Gender and Power in Late Ancient Christianity," *Gender and History*, vol. 2, no. 3 (Autumn 1990), pp. 253–73.

Clark, Gillian. *Women in the Ancient World*. New York: Oxford University Press, 1989.

———. *Women in Late Antiquity: Pagan and Christian Lifestyles*. New York: Oxford University Press, 1993.

Cloke, Gillian. *This Female Man of God: Women and Spiritual Power in the Patristic Age, AD 350–450*. New York: Routledge, 1995.

Coon, Lynda L. *Sacred Fictions: Holy Women and Hagiography in Late Antiquity*. Philadelphia: University of Pennsylvania Press, 1997.

Cooper, Kate. *The Virgin and the Bride: Idealized Womanhood in Late Antiquity*. Cambridge, Mass.: Harvard University Press, 1996.

Dahlberg, Frances, ed. *Woman the Gatherer*. New Haven: Yale University Press, 1981.

D'Avino, Michele. *The Women of Pompeii*. Translated by Monica Hope Jones and Luigi Nusco. Napoli: Loffredo, 1967.

Demand, Nancy H. *Birth, Death, and Motherhood in Classical Greece*. Baltimore: Johns Hopkins University Press, 1994.

Diaz-Andreu, Margarita, and Marie Louise Stig, eds. *Excavating Women: A History of Women in European Archaeology*. New York: Routledge, 1998.

Dixon, Susanne. *The Roman Mother*. Norman: University of Oklahoma Press, 1988.

Duckett, Eleanor Shipley. *Women and Their Letters in the Early Middle Ages*. Ann Arbor: University of Michigan Press, 1965.

Dzielska, Maria. *Hypatia of Alexandria*. Translated by F. Yra. Cambridge, Mass.: Harvard University Press, 1995.

Eckenstein, Lina. *Women Under Monasticism*. New York: Russell and Russell, 1963 [1896].

Elm, Susanna. *"Virgins of God": The Making of Asceticism in Late Antiquity*. New York: Oxford University Press, 1994.

Evans, John K. *War, Women and Children in Ancient Rome*. London: Routledge, 1991.

Fantham, Elaine, Helene Peet Foley, Natalie Boymel Kampen, Sarah B. Pomeroy, and H. A. Shapiro. *Women in the Classical World: Image and Text*. New York: Oxford University Press, 1994.

Gero, Joan M., and Margaret W. Conkey, eds. *Engendering Archaeology: Women and Prehistory*. Oxford: Basil Blackwell, 1991.

Gimbutas, Maria. *The Civilization of the Goddess: The World of Old Europe*. San Francisco: HarperCollins, 1991.

Grant, Michael. *Cleopatra*. New York: Simon and Schuster, 1972.

Greene, Ellen, ed. *Re-Reading Sappho: Reception and Transmission*. Berkeley: University of California Press, 1996.

Hager, Lori D. *Women in Human Evolution*. London: Routledge, 1997.

Hallett, Judith P. "Sappho and Her Social Context: Sense and Sensuality," *Signs*, vol. 4, no. 3 (Spring 1979), pp. 447–64.

Hawley, R., and B. Levick, eds. *Women in Antiquity: New Assessments*. New York: Routledge, 1995.

Henry, Madeleine Mary. *Prisoner of History: Aspasia of Miletus and her Biographical Tradition*. New York: Oxford University Press, 1995.

Henry, Sondra, and Emily Taitz. *Written out of History: Our Jewish Foremothers*, 2nd ed., rev. Fresh Meadows, N.Y.: Biblio Press, 1983.

Hollis, Stephanie. *Anglo-Saxon Women and the Church: Sharing a Common Fate*. Rochester, N.Y.: Boydell Press, 1992.

Holum, Kenneth G. *Theodosian Empresses: Women and Imperial Dominion in Late Antiquity*. Berkeley: University of California Press, 1989.

Horney, Karen. *Feminine Psychology*. New York: Norton, 1973.

Hrdy, Sarah Blaffer. *The Woman That Never Evolved*. Cambridge, Mass.: Harvard University Press, 1981.

Hughes-Hallett, Lucy. *Cleopatra: Histories, Dreams and Distortions*. New York: Harper & Row, 1990.

Jochens, Jenny. *Women in Old Norse Society*. Ithaca, N.Y.: Cornell University Press, 1995.

Just, Roger. *Women in Athenian Law and Life*. New York: Routledge, 1989.

Kampen, Natalie. "Hellenistic Artists: Female," *Archeologia Classica*, vol. XXVII, no. 1 (1975), pp. 9–17.

Kaplan, Marion A., ed. *The Marriage Bargain: Women and Dowries in European History*. New York: Haworth Press, 1985.

Kessler, Evelyn S. *Women: An Anthropological View*. New York: Holt, Rinehart and Winston, 1976.

Kleinbaum, Abby Wettan. *The War Against the Amazons*. New York: McGraw-Hill, 1983.

Larson, Jennifer. *Greek Heroine Cults*. Madison: University of Wisconsin Press, 1995.

Leacock, Eleanor Burke. *Myths of Male Dominance: Collected Articles on Women Cross-Culturally*. New York: Monthly Review Press, 1981.

Lefkowitz, Mary R., and Maureen B. Fant, eds. *Women's Life in Greece and Rome: A Source Book in Translation*. 2nd ed. Baltimore: Johns Hopkins University Press, 1995.

Loewe, Raphael. *The Position of Women in Judaism*. London: Society for the Propagation of Christian Knowledge and the Hillel Foundation, 1966.

Loraux, Nicole. *The Experiences of Tiresias: The Feminine and the Greek Man*. Princeton, N.J.: Princeton University Press, 1995.

Lyons, Deborah. *Gender and Immortality: Heroines in Ancient Greek Myth and Cult*. Princeton, N.J.: Princeton University Press, 1997.

MacDonald, Margaret Y. *Early Christian Women and Pagan Opinion: The Power of the Hysterical Woman*. New York: Cambridge University Press, 1996.

Macurdy, Grace Harriet. *Hellenistic Queens: A Study of Woman-Power in Macedonia, Seleucid Syria, and Ptolemaic Egypt*. Baltimore: Johns Hopkins University Press, 1932.

Markale, Jean. *Women of the Celts*. Translated by A. Mygind, C. Hauch, and P. Henry. London: Gordon Cremonesi, 1975.

Martin, M. Kay, and Barbara Voorhies. *Female of the Species*. New York: Columbia University Press, 1975.

Massey, Michael. *Women in Ancient Greece and Rome*. New York: Cambridge University Press, 1988.

McAuslan, Ian, and Peter Walcot, eds. *Women in Antiquity*. Oxford: Oxford University Press, 1996.

McNamara, Jo Ann. "Wives and Widows in Early Christian Thought," *International Journal of Women's Studies*, vol. 2, no. 6 (November/December 1979), pp. 575–92.

———. *A New Song: Celibate Women in the First Three Centuries of Christianity*. New York: Haworth Press, 1983.

McNamara, Jo Ann, and John E. Halborg with E. Gordon Whatley. *Sainted Women of the Dark Ages*. Durham, N.C.: Duke University Press, 1992.

Morris, Joan. *The Lady Was a Bishop*. New York: Macmillan, 1973.

Neusner, Jacob. "Mishnah on Women: Thematic or Systemic Description," *Marxist Perspectives*, vol. 3, no. 1 (Spring 1980), pp. 78–98.

Okin, Susan Moller. *Women in Western Political Thought*. Princeton, N.J.: Princeton University Press, 1979.

Oost, Stewart Irvin. *Galla Placidia Augusta: A Biographical Essay*. Chicago: University of Chicago Press, 1968.

Otwell, John H. *And Sarah Laughed: The Status of Women in the Old Testament*. Philadelphia: Westminster Press, 1977.

Peradotto, John, and J. P. Sullivan, eds. *Women in the Ancient World: The Arethusa Papers*. Albany, N.Y.: State University of New York Press, 1984.

Phillips, John A. *Eve. The History of an Idea*. New York: Harper & Row, 1984.

Pomeroy, Sarah B. *Goddesses, Wives, Whores and Slaves: Women in Classical Antiquity*. New York: Schocken, 1975.

———. *Women in Hellenistic Egypt from Alexander to Cleopatra*. New York: Schocken, 1984.

———. *Women's History and Ancient History*. Chapel Hill: University of North Carolina Press, 1991.

Rabinowitz, Nancy, and Amy Richlin, eds. *Feminist Theory and the Classics*. New York: Routledge, 1993.

Reeder, Ellen D. *Pandora: Women in Classical Greece*. Princeton, N.J.: Princeton University Press, 1995.

Reiter, Rayna R., ed. *Toward an Anthropology of Women*. New York: Monthly Review Press, 1975.

Rosaldo, Michelle Zimbalist, and Louise Lamphere, eds. *Women, Culture, and Society*. Stanford, Calif.: Stanford University Press, 1974.

Ross, Margaret Clunies. "Concubinage in Anglo-Saxon England," *Past and Present*, vol. 108 (August 1985), pp. 3–34.

Salisbury, Joyce E. *Perpetua's Passion: The Death and Memory of a Young Roman Woman*. New York: Routledge, 1997.

Sanday, Peggy Reeves. *Female Power and Male Dominance: On the Origins of Sexual Inequality*. Cambridge: Cambridge University Press, 1981.

Scholer, David M., ed. *Women in Early Christianity*. New York: Garland, 1993.

Sealey, Raphael. *Women and Law in Classical Greece*. Chapel Hill: University of North Carolina Press, 1990.

Taafe, Lauren K. *Aristophanes and Women*. London, New York: Routledge, 1993.

Talalay, Lauren E. "A Feminist Boomerang: The Great Goddess of Greek Prehistory," *Gender and History*, vol. 6, no. 2 (August 1994), pp. 165–83.

Tavard, George H. *Women in the Christian Tradition*. Notre Dame, Ind.: University of Notre Dame Press, 1973.

Treggiari, Susan. *Roman Marriage*. New York: Oxford University Press, 1991.

Webster, Graham. *Boudica: The British Revolt Against Rome 60 A.D.* Totowa, N.J.: Rowman and Littlefield, 1978.
Wemple, Suzanne Fonay. *Women in Frankish Society: Marriage and the Cloister 500 to 900.* Philadelphia: University of Pennsylvania Press, 1981.
Witt, R. E. *Isis in the Graeco-Roman World.* Ithaca, N.Y.: Cornell University Press, 1971.

General Works

This list is by no means comprehensive; it includes the general works most helpful for this section.

Attwater, Donald. *Dictionary of Saints.* Baltimore: Penguin, 1966.
Bateson, Mary. *Origin and Early History of Double Monasteries.* Royal Historical Society Transactions, New Series 13. London, 1899.
Brown, Peter. *The Body and Society: Men, Women, and Sexual Renunciation in Early Christianity.* New York: Columbia University Press, 1988.
Bynum, Caroline Walker. *Resurrection of the Body in Western Christianity, 200–1336.* New York: Columbia University Press, 1995.
Dinnerstein, Dorothy. *The Mermaid and the Minotaur: Sexual Arrangements and Human Malaise.* New York: Harper & Row, 1976.
Douglas, Mary. *Purity and Danger: An Analysis of Concepts of Pollution and Taboo.* Harmondsworth, U.K.: Penguin, 1966.
Duckett, Eleanor Shipley. *Medieval Portraits from East and West.* Ann Arbor: University of Michigan Press, 1972.
Fiorenza, Elizabeth S. *In Memory of Her: A Feminist Theological Reconstruction of Christian Origins.* New York: Crossroad, 1983.
Foote, P. G., and D. M. Wilson. *The Viking Achievement.* New York: Praeger, 1970.
Gimbutas, Marija. *The Goddesses and Gods of Old Europe: 6500–3500 B.C. Myths and Cult Images*, rev. ed. London: Thames and Hudson, 1982.
Halperin, David M., John J. Winkler, and Froma I. Zeitlin, eds. *Before Sexuality: The Construction of Erotic Experience in the Ancient Greek World.* Princeton, N.J.: Princeton University Press, 1990.
Lancaster, Jane Beckman. *Primate Behavior and the Emergence of Human Culture.* New York: Holt, Rinehart and Winston, 1975.
Lee, Patrick C., and Robert Sussman Stewart, eds. *Sex Differences: Cultural and Developmental Dimensions.* New York: Urizen, 1976.
Maccoby, Eleanor Emmons, and Carol Nagy Jacklin. *The Psychology of Sex Differences*, 2 vols. Stanford, Calif.: Stanford University Press, 1974.
Money, John, and Anke A. Ehrhardt. *Man and Woman, Boy and Girl: The Differentiation and Dimorphism of Gender Identity from Conception to Maturity.* Baltimore: Johns Hopkins University Press, 1972.
Pagels, Elaine. *The Gnostic Gospels.* New York: Random House, 1979.
———. *Adam, Eve, and the Serpent.* New York: Random House, 1988.
Rawson, Philip. *Primitive Erotic Art.* New York: G. P. Putnam, 1973.
Riché, Pierre. *Daily Life in the World of Charlemagne.* Translated by Jo Ann McNamara. Philadelphia: University of Pennsylvania Press, 1978.

Richlin, Amy, ed. *Pornography and Representation in Greece and Rome*. New York: Oxford University Press, 1992.
Stephens, William N. "A Cross-Cultural Study of Menstrual Taboos," *Genetic Psychology Monographs*, vol. 64 (1961), pp. 385–416.
Tanner, Nancy. *On Becoming Human*. Cambridge: Cambridge University Press, 1981.
Veith, Ilza. *Hysteria: The History of a Disease*. Chicago: University of Chicago Press, 1965.

PART II WOMEN OF THE FIELDS

Works About Women

Allauzen, Marie. *La Paysanne française aujourd'hui*. Paris: Editions Gonthier, 1967.
Ankarloo, Bengt, and Gustav Henningsen, eds. *Early Modern European Witchcraft: Centres and Peripheries*. Oxford: Clarendon Press, 1993.
Barry, Jonathan, Marianne Hester, and Gareth Roberts, eds. *Witchcraft in Early Modern Europe: Studies in Culture and Belief*. New York: Cambridge University Press, 1996.
Barstow, Anne Llewellyn. *Witchcraze: A New History of the European Witch Hunts*. New York: Pandora, 1994.
Behringer, Wolfgang. *Witchcraft Persecutions in Bavaria: Popular Magic, Religious Zealotry and Reason of State in Early Modern Europe*. Translated by J. C. Grayson and David Lederer. New York: Cambridge University Press, 1997.
Bennett, Judith M. *Women in the Medieval English Countryside: Gender and Household in Brigstock before the Plague*. New York: Oxford University Press, 1987.
———. *Ale, Beer, and Brewsters in England: Women's Work in a Changing World, 1300–1600*. New York: Oxford University Press, 1996.
Cashmere, John. "Sisters together: Women without men in seventeenth-century French village culture," *Journal of Family History*, vol. 21 (1996), pp. 44–62.
Chamberlain, Mary. *Fenwomen: a Portrait of Women in an English Village*. London: Virago Press Limited, 1977.
———. *Old Wives' Tales: Their History, Remedies and Spells*. London: Virago Press Limited, 1981.
Clements, Barbara Evans, Barbara Alpern Engel, and Christine D. Worobec, eds. *Russia's Women: Accommodation, Resistance, Transformation*. Berkeley: University of California Press, 1991.
Cornelisen, Ann. *Women of the Shadows*. Boston: Little, Brown, 1976.
Davies, Owen. "Urbanization and the decline of witchcraft: An examination of London," *Journal of Social History*, vol. 30, no. 3 (1997), pp. 597–618.
Engel, Barbara Alpern. *Between the Fields and the City: Women, Work, and Family in Russia, 1861–1914*. New York: Cambridge University Press, 1994.
Fairchilds, Cissie. "Female Sexual Attitudes and the Rise of Illegitimacy: A Case Study," *Journal of Interdisciplinary History*, vol. VIII, no. 4 (Spring 1978), pp. 627–67.
Farnsworth, Beatrice, and Lynne Viola, eds. *Russian Peasant Women*. New York: Oxford University Press, 1992.

Favret-Saada, Jeanne. *Deadly Words: Witchcraft in the Bocage*. Translated by Catherine Cullen. New York: Cambridge University Press, 1980.

Gies, Frances. *Joan of Arc: The Legend and the Reality*. New York: Harper & Row, 1981.

Ginzburg, Carlo. *Ecstasies: Deciphering the Witches' Sabbath*. Translated by Raymond Rosenthal. New York: Pantheon Books, 1991.

Hanawalt, Barbara. *The Ties That Bound: Peasant Families in Medieval England*. New York: Oxford University Press, 1986.

Hardy, Sheila. *Diary of a Suffolk Farmer's Wife, 1854–69: A Woman of Her Time*. London: Macmillan, 1992.

Henningsen, Gustav. *The Witches' Advocate: Basque Witchcraft and the Spanish Inquisition 1609–1640*. Reno: University of Nevada Press, 1980.

Hill, Bridget. *Women, Work, and Sexual Politics in Eighteenth-Century England*. New York: Basil Blackwell, 1989.

———. "The Marriage Age of Women and the Demographers," *History Workshop Journal*, vol. 28 (1989), pp. 129–47.

Hudson, Pat, and W. R. Lee, eds. *Women's Work and the Family Economy in Historical Perspective*. Manchester, U.K.: Manchester University Press, 1990.

Jonas, Raymond A. *Industry and Politics in Rural France: Peasants of the Isère, 1870–1914*. Ithaca, N.Y.: Cornell University Press, 1994.

Kieckhefer, Richard. *European Witch Trials: Their Foundation in Popular and Learned Culture, 1300–1500*. Berkeley: University of California Press, 1976.

Klaits, Joseph. *Servants of Satan: The Age of the Witch Hunt*. Bloomington: Indiana University Press, 1985.

Kors, Alan C., and Edward Peters, eds. *Witchcraft in Europe 1100–1700: A Documentary History*. Philadelphia: University of Pennsylvania Press, 1972.

Larner, Christina. *Enemies of God: The Witchhunt in Scotland*. Baltimore: Johns Hopkins University Press, 1981.

Liu, Tessie P. *The Weaver's Knot: The Contradictions of Class Struggle and Family Solidarity in Western France, 1750–1914*. Ithaca, N.Y.: Cornell University Press, 1994.

Martin, Ruth. *Witchcraft and the Inquisition in Venice, 1550–1650*. Oxford: Basil Blackwell, 1989.

Midelfort, H. C. Erik. *Witch Hunting in Southwestern Germany 1562–1684: The Social and Intellectual Foundations*. Stanford, Calif.: Stanford University Press, 1972.

Pernoud, Régine. *The Retrial of Joan of Arc*. Translated by J. M. Cohen. New York: Harcourt, Brace and Company, 1955.

———. *Joan of Arc, by Herself and Her Witnesses*. Translated by Edward Hyams. London: Macdonald, 1964.

Quaife, G. R. *Wanton Wenches and Wayward Wives: Peasants and Illicit Sex in Early Seventeenth Century England*. New Brunswick, N.J.: Rutgers University Press, 1979.

Roper, Lyndal. *Oedipus and the Devil: Witchcraft, Sexuality and Religion in Early Modern Europe*. New York: Routledge, 1994.

Sayer, Karen. *Women of the Fields: Representations of Rural Women in the Nineteenth Century*. New York: St. Martin's Press, 1995.

Sharpe, J. A. "Witchcraft and Women in Seventeenth-Century England: Some Northern Evidence," *Continuity and Change*, vol. 6, no. 2 (1991), pp. 179–99.

Soman, Alfred. "The Parlement of Paris and the Great Witch Hunt 1565–1640," *Sixteenth Century Journal*, vol. IX, no. 2 (1978), pp. 31–44.

Tilly, Louise A., and Joan W. Scott. *Women, Work and Family*. New York: Holt, Rinehart and Winston, 1978.

Valenze, Deborah. *The First Industrial Woman*. New York: Oxford University Press, 1994.

Warner, Marina. "The Absent Mother: Women Against Women in Old Wives' Tales," *History Today*, vol. 41 (April 1991), pp. 22–28.

Wheeler, Bonnie, and Charles T. Wood, eds. *Fresh Verdicts on Joan of Arc*. New York: Garland, 1996.

Whitney, Elspeth. "The witch 'she'/the historian 'he': Gender and the historiography of the European witch-hunts," *Journal of Women's History*, vol. 7, no. 3 (1995/96), pp. 77–101.

General Works

Amussen, Susan Dwyer. *An Ordered Society: Gender and Class in Early Modern England*. New York: Basil Blackwell, 1988.

Arensberg, Conrad. *The Irish Countryman*. Garden City, N.Y.: The Natural History Press, 1968.

Beresford, Maurice, and John G. Hurst, eds. *Deserted Medieval Villages Studies*. New York: St. Martin's Press, 1971.

Bloch, Marc. *Feudal Society*. 2 vols. Translated by L. A. Manyon. Chicago: University of Chicago Press, 1966.

Bonfield, L., R. M. Smith, and Keith Wrightson, eds. *The World We Have Earned: Histories of Population and Social Structure*. New York: Oxford University Press, 1986.

Bouton, Cynthia A. *The Flour War: Gender, Class, and the Community in Late Ancien Regime French Society*. University Park: Pennsylvania State University Press, 1993.

Braudel, Fernand. *Capitalism and Material Life 1400–1800*. Translated by Miriam Kochan. New York: Harper & Row, 1973.

———. *The Structures of Everyday Life: Civilization & Capitalism 15th–18th Century*. 3 vols. Translated by Siân Reynolds. New York: Harper & Row, 1981.

Christian, William A., Jr. *Person and God in a Spanish Valley*. New York: Seminar Press, 1972.

Clark, Stuart. *Thinking with Demons: The Idea of Witchcraft in Early Modern Europe*. New York: Oxford University Press, 1997.

Davis, Natalie Zemon. *Society and Culture in Early Modern France*. Stanford, Calif.: Stanford University Press, 1975.

Duby, Georges. *Rural Economy and Country Life in the Medieval West*. Translated by Cynthia Postan. Columbia: University of South Carolina Press, 1968.

———. *The Early Growth of the European Economy: Warriors and Peasants from the Seventh to the Twelfth Century*. Translated by Howard B. Clarke. Ithaca, N.Y.: Cornell University Press, 1974.

Erickson, Carolly. *The Medieval Vision*. New York: Oxford University Press, 1976.
Flandrin, J.-L. *Families in Former Times: Kinship, Household and Sexuality*. Translated by Richard Southern. New York: Cambridge University Press, 1979.
Freeman, Margaret B. *Herbs for the Mediaeval Household for Cooking, Healing and Divers Uses*. New York: The Metropolitan Museum of Art, 1979.
Friedl, Ernestine. *Vasilika: A Village in Modern Greece*. New York: Holt, Rinehart and Winston, 1962.
Goody, Jack, Joan Thirsk, and E. P. Thompson, eds. *Family and Inheritance: Rural Society in Western Europe 1200–1800*. New York: Cambridge University Press, 1978.
Goubert, Pierre. "The French Peasantry of the Seventeenth Century: A Regional Sample," in Trevor Aston, ed. *Crisis in Europe 1560–1660*. New York: Basic Books, 1965.
Gutmann, Myron P. *War and Rural Life in the Early Modern Low Countries*. Princeton, N.J.: Princeton University Press, 1980.
Heaton, Eliza Putnam. *By-Paths in Sicily*. New York: E. P. Dutton, 1920.
Hélias, Pierre-Jakez. *The Horse of Pride: Life in a Breton Village*. Translated by June Guicharnaud. New Haven: Yale University Press, 1978.
Hufton, Olwen H. *The Poor of Eighteenth Century France 1750–1789*. Oxford: Clarendon Press, 1974.
Kaser, Karl. "The Balkan Joint Family Household: Seeking its Origins," *Continuity and Change*, vol. 9, no. 1 (May 1994), pp. 45–68.
Laslett, Peter, ed., with Richard Wall. *Household and Family in Past Time*. New York: Cambridge University Press, 1974.
Levin, Eve. *Sex and Society in the World of the Orthodox Slavs, 900–1700*. Ithaca, N.Y.: Cornell University Press, 1989.
Levine, David. *Family Formation in an Age of Nascent Capitalism*. Toronto: Ontario Institute for Studies in Education, 1977.
McArdle, Frank. *Altopascio: A Study in Tuscan Rural Society 1587–1784*. New York: Cambridge University Press, 1978.
McNeill, John T., and Helena M. Gamer, eds. *Handbooks of Penance*. Records of Civilization: Sources and Studies No. 29. New York: Columbia University Press, 1938.
Mitterauer, Michael, and Reinhard Sieder. *The European Family: Patriarchy to Partnership from the Middle Ages to the Present*. Translated by Karla Oosterveen and Manfred Horzinger. Oxford: Basil Blackwell, 1982.
Reiter, Rayna R. "Men and Women in the South of France: Public and Private Domains," in Rayna R. Reiter, ed. *Toward an Anthropology of Women*. New York: Monthly Review Press, 1975.
Riché, Pierre. *Daily Life in the World of Charlemagne*. Translated by Jo Ann McNamara. Philadelphia: University of Pennsylvania Press, 1978.
Riddle, John M. *Contraception and Abortion from the Ancient World to the Renaissance*. Cambridge, Mass.: Harvard University Press, 1992.
Segalen, Martine. *Love and Power in the Peasant Family: Rural France in the Nineteenth Century*. Translated by Sarah Matthews. Chicago: University of Chicago Press, 1983.

Sommestad, Lena. "Rethinking Gender and Work: Rural Women in the Western World," *Gender and History*, vol. 7, no. 1 (April 1995), pp. 100–05.
St. Erlich, Vera. *Family in Transition: A Study of 300 Yugoslav Villages*. Princeton, N.J.: Princeton University Press, 1966.
Thomas, Keith. *Religion and the Decline of Magic*. New York: Charles Scribner's Sons, 1971.
Ziegler, Philip. *The Black Death*. New York: Harper & Row, 1969.

PART III WOMEN OF THE CHURCHES

Works About Women

Ahlgren, Gillian T. W. *Teresa of Avila and the Politics of Sanctity*. Ithaca, N.Y.: Cornell University Press, 1996.
Arenal, Electra, and Stacey Schlau, eds. *Untold Sisters: Hispanic Nuns in Their Own Words*. Albuquerque: University of New Mexico Press, 1989.
Atkinson, Clarissa W. *Mystic and Pilgrim: The Book and the World of Margery Kempe*. Ithaca, N.Y.: Cornell University Press, 1983.
Bainton, Ronald H. *Women of the Reformation in Germany and Italy*. Boston: Beacon Press, 1971.
———. *Women of the Reformation in France and England*. Boston: Beacon Press, 1975.
———. *Women of the Reformation from Spain to Scandinavia*. Minneapolis: Augsburg Publishing House, 1977.
Baker, Derek, ed. *Medieval Women*. Oxford: Basil Blackwell, 1978.
Baldauf-Berdes, Jane J. *Women Musicians of Venice: Musical Foundations, 1525–1855*. New York: Oxford University Press, 1993.
Beilin, Elaine V., ed. *The Examinations of Anne Askew*. New York: Oxford University Press, 1996.
Berman, Constance H., Charles W. Connell, and Judith Rice Rothschild, eds. *The Worlds of Medieval Women: Creativity, Influence, and Imagination*. Morgantown: West Virginia University Press, 1985.
Bilinkoff, Jodi. *The Avila of St. Teresa: Religious Reform in a Sixteenth-Century City*. Ithaca, N.Y.: Cornell University Press, 1989.
Bornstein, Daniel, and Roberto Ruscoin, eds. *Women and Religion in Medieval and Renaissance Italy*. Translated by Margery J. Schneider. Chicago: University of Chicago Press, 1996.
Bowers, Jane, and Judith Tick, eds. *Women Making Music: The Western Art Tradition 1150–1950*. Urbana: University of Illinois Press, 1986.
Brailsford, Mabel Richmond. *Quaker Women 1650–1690*. London: Duckworth, 1915.
Brooten, Bernadette J. *Love between Women: Early Christian Responses to Female Homoeroticism*. Chicago: University of Chicago Press, 1996.
Brown, Judith C. *Immodest Acts: The Life of a Lesbian Nun in Renaissance Italy*. New York: Oxford University Press, 1986.
Bynum, Caroline Walker. *Holy Feast and Holy Fast: The Religious Significance of Food to Medieval Women*. Berkeley: University of California Press, 1987.

Clark, Anne L. *Elisabeth of Schönau: A Twelfth-Century Visionary*. Philadelphia: University of Pennsylvania Press, 1992.

Coon, Lynda L., Katherine J. Haldane, and Elisabeth W. Sommer, eds. *That Gentle Strength: Historical Perspectives on Women in Christianity*. Charlottesville: University Press of Virginia, 1990.

Daly, Mary. *The Church and the Second Sex*. New York: Harper & Row, 1975.

Davis, Natalie Zemon. "City Women and Religious Change in Sixteenth Century France," in Dorothy McGuigan, ed., *A Sampler of Women's Studies*. Ann Arbor: University of Michigan Center for Continuing Education of Women, 1973.

Douglass, Jane Dempsey. *Women, Freedom, and Calvin*. Philadelphia: Westminster Press, 1985.

Dronke, Peter. *Women Writers of the Middle Ages: A Critical Study of Texts from Perpetua to Marguerite Porete*. New York: Cambridge University Press, 1985.

Eckenstein, Lina. *Woman Under Monasticism*. New York: Russell & Russell, 1963 [1896].

Elkins, Sharon K. *Holy Women of Twelfth-Century England*. Chapel Hill: University of North Carolina Press, 1988.

Erickson, Carolly. *Bloody Mary*. Garden City, N.Y.: Doubleday, 1978.

Fife, Robert Herndon. *Hroswitha of Gandersheim*. New York: Columbia University Press, 1947.

Finnegan, Sr. Mary Jeremy, O.P. *The Women of Helfta: Scholars and Mystics*. Athens: University of Georgia Press, 1991.

Flanagan, Sabina. *Hildegard of Bingen, 1098–1179: A Visionary Life*. New York: Routledge, 1989.

Freeman, Elizabeth. "The Public and Private Functions of Heloise's Letters," *Journal of Medieval History*, vol. 23, no. 1 (1997), pp. 15–28.

Gilchrist, Roberta. *Gender and Material Culture: The Archeology of Religious Women*. New York: Routledge, 1994.

Gold, Penny Schine. *The Lady and the Virgin: Image, Attitude, and Experience in Twelfth Century France*. Chicago: University of Chicago Press, 1985.

Greaves, Richard L., ed. *Triumph Over Silence: Women in Protestant History*. Westport, Conn.: Greenwood Press, 1985.

Harrison, Wes. "The Role of Women in Anabaptist Thought and Practice: The Hutterite experience of the sixteenth and seventeenth centuries," *Sixteenth Century Journal*, vol. 23 (1992), pp. 49–70.

Hildegard of Bingen. *The Letters of Hildegard of Bingen*. vol. I. Edited and translated by Joseph L. Baird and Radd K. Ehrman. New York: Oxford University Press, 1994.

Hirn, Yrjö. *The Sacred Shrine: A Study of the Poetry and Art of the Catholic Church*. Boston: Beacon Press, n.d.

Holsinger, Bruce Wood. "The Flesh of the Voice: Embodiment and the Homoerotics of Devotion in the Music of Hildegard of Bingen (1098–1179)," *Signs*, vol. 19, no. 1 (Autumn 1993), pp. 92–125.

Irwin, Joyce, ed. *Women in Radical Protestantism*. New York: E. Mellen, 1979.

Jansen, Sharon L. *Dangerous Talk and Strange Behavior: Women and Popular Resistance to the Reforms of Henry VIII*. New York: St. Martin's Press, 1996.

Johnson, Penelope D. *Equal in Monastic Profession: Religious Women in Medieval France*. Chicago: University of Chicago Press, 1991.

Jørgensen, Johannes. *Saint Bridget of Sweden*. Translated by Ingeborg Lund. New York: Longmans, Green & Co., 1954.

Kendrick, Robert L. *Celestial Sirens: Nuns and Their Music in Early Modern Milan*. New York: Oxford University Press, 1996.

Kunze, Bonnelyn Young. *Margaret Fell and the Rise of Quakerism*. Stanford, Calif.: Stanford University Press, 1994.

Levin, Carole. "Women in *The Book of Martyrs* as Models of Behavior in Tudor England," *International Journal of Women's Studies*, vol. IV, no. 2 (March/April 1981), pp. 196–207.

Linehan, Peter. *The Ladies of Zamora*. University Park: Pennsylvania State University Press, 1997.

Lochrie, Karma. *Margery Kempe and Translations of the Flesh*. Philadelphia: University of Pennsylvania Press, 1991.

Mack, Phyllis. *Visionary Women: Ecstatic Prophecy in Seventeenth-Century England*. Berkeley: University of California Press, 1992.

Maclean, Ian. *The Renaissance Notion of Woman: A study of the fortunes of scholasticism and medical science in European intellectual life*. New York: Cambridge University Press, 1980.

Magray, Mary Peckham. *The Transforming Power of the Nuns: Women, Religion and Cultural Change in Ireland, 1750–1900*. New York: Oxford University Press, 1998.

Marshall, Sherrin, ed. *Women in Reformation and Counter-Reformation Europe: Public and Private Worlds*. Bloomington: Indiana University Press, 1989.

Matter, E. Ann. "Discourses of Desire: Sexuality and Christian Women's Visionary Narratives," *Journal of Homosexuality*, vol. 18 (1990), pp. 119–31.

Matter, E. Ann, and John Coakley, eds. *Creative Women in Medieval and Early Modern Italy*. Philadelphia: University of Pennsylvania Press, 1995.

McDonnell, Ernest. *The Beguines and Beghards in Medieval Culture*. New York: Octagon Books, 1969.

McNamara, Jo Ann, and John E. Halborg, eds., with E. Gordon Whatley. *Sainted Women of the Dark Ages*. Durham, N.C.: Duke University Press, 1992.

Monson, Craig A. *The Crannied Wall: Women, Religion, and the Arts in Early Modern Europe*. Ann Arbor: University of Michigan Press, 1992.

———. *Disembodied Voices: Music and Culture in an Early Modern Italian Convent*. Berkeley: University of California Press, 1995.

Morris, Joan. *The Lady Was a Bishop*. New York: Macmillan, 1973.

Newman, Barbara. *Sister of Wisdom: St. Hildegard's Theology of the Feminine*. Berkeley: University of California Press, 1987.

d'Oin, Marguerite. *The Writings of Margaret of Oingt*. Translated by Renate Blumenfeld-Kosinski. Newburyport, Mass.: Focus Information Group, 1990.

Petroff, Elizabeth Alvilda, ed. *Medieval Women's Visionary Literature*. Oxford: Oxford University Press, 1986.

———, ed. *Body and Soul: Essays on Medieval Women and Mysticism*. New York: Oxford University Press, 1994.

Power, Eileen. *Medieval English Nunneries*. Cambridge: Cambridge University Press, 1922.
Rapley, Elizabeth. *The Dévotes: Women and the Church in Seventeenth Century France*. Montreal: McGill-Queen's University Press, 1990.
Roelker, Nancy L. "The Appeal of Calvinism to French Noblewomen in the Sixteenth Century," *The Journal of Interdisciplinary History*, vol. II, no. 4 (Spring 1972), pp. 391–418.
Roper, Lyndal. *The Holy Household: Women and Morals in Reformation Augsburg*. Oxford: Clarendon Press, 1991.
Ross, Isabel. *Margaret Fell: Mother of Quakerism*. New York: Longmans, Green & Co., 1949.
Saint-Saëns, Alain, ed. *Religion, Body, and Gender in Early Modern Spain*. San Francisco: Mellon Research University Press, 1991.
Schulenburg, Jane Tibbetts. *Forgetful of Their Sex: Female Sanctity and Society, ca. 500–1100*. Chicago: University of Chicago Press, 1998.
Snyder, C. Arnold, and Linda A. Huebert Hecht, eds. *Profiles of Anabaptist Women: Sixteenth-Century Reforming Pioneers*. Waterloo, Ont.: Wilfrid Laurier University Press, 1996.
Surtz, Ronald E. *Writing Women in Late Medieval and Early Modern Spain: The Mothers of Saint Teresa of Avila*. Philadelphia: University of Pennsylvania Press, 1995.
Tavard, George H. *Woman in Christian Tradition*. Notre Dame, Ind.: University of Notre Dame Press, 1973.
Thomas, Keith. "Women and the Civil War Sects," in Trevor Aston, ed., *Crisis in Europe 1560–1660*. New York: Basic Books, 1965.
van der Meer, S. J. Haye. *Women Priests in the Catholic Church? A Theological-Historical Investigation*. Translated by Arlene and Leonard Swidler. Philadelphia: Temple University Press, 1973.
Venarde, Bruce L. *Women's Monasticism and Medieval Society: Nunneries in France and England, 890–1215*. Ithaca, N.Y.: Cornell University Press, 1997.
Warner, Marina. *Alone of All Her Sex: The Myth and the Cult of the Virgin Mary*. New York: Pocket Books, 1978.
Wemple, Suzanne Fonay. *Women in Frankish Society: Marriage and the Cloister 500–900*. Philadelphia: University of Pennsylvania Press, 1981.
Wood, Diana, ed. *Martyrs and Martyrologies: Papers Read at the 1992 Summer Meeting and the 1993 Winter Meeting of the Ecclesiastical History Society*. Cambridge, Mass.: Blackwell Publishers, 1993.
Wood, Jeryldene. *Women, Art, and Spirituality: The Poor Clares of Early Modern Italy*. New York: Cambridge University Press, 1996.

General Works

Bartlett, Anne Clark. *Male Authors, Female Readers: Representation and Subjectivity in Middle English Devotional Literature*. Ithaca, N.Y.: Cornell University Press, 1995.
Bitel, Lisa M. *Isle of the Saints: Monastic Settlement and Christian Community in Early Ireland*. Ithaca, N.Y.: Cornell University Press, 1990.

Bougaud, Monseigneur. *History of St. Vincent de Paul*, 2 vols. Translated by Rev. Joseph Brady. New York: Longmans, Green & Co., 1899.

Bynum, Caroline Walker. *Jesus as Mother: Studies in the Spirituality of the High Middle Ages*. Berkeley: University of California Press, 1982.

———. *Fragmentation and Redemption: Essays on Gender and the Human Body in Medieval Religon*. New York: Zone Books, 1991.

Carlson, Eric Josef. *Marriage and the English Reformation*. Boston: Blackwell, 1994.

Clasen, Claus Peter. *Anabaptism: A Social History 1525–1618*. Ithaca, N.Y.: Cornell University Press, 1972.

Coakley, John. "Friars as Confidants of Holy Women in Medieval Dominican Hagiography," in Renate Blumenfeld-Kosinski and Timea Szell, eds. *Images of Sainthood in Medieval Europe*. Ithaca, N.Y.: Cornell University Press, 1991.

Davis, Natalie Zemon. *Society and Culture in Early Modern France*. Stanford, Calif.: Stanford University Press, 1975.

Dunn, E. Catherine. "Popular Devotion in the Vernacular Drama of Medieval England," in Paul Maurice Clogan, ed., *Medievalia et Humanistica: Studies in Medieval and Renaissance Culture*, New Series, no. 4 (1973).

Elliott, Dyan. *Spiritual Marriage: Sexual Abstinence in Medieval Wedlock*. Princeton, N.J.: Princeton University Press, 1993.

Erickson, Carolly. *The Medieval Vision*. New York: Oxford University Press, 1976.

Hampton, Joel F. *Recording Marriage and Society in Reformation Germany*. New York: Cambridge University Press, 1995.

Jedin, Hubert. *A History of the Council of Trent*, 2 vols. Translated by Ernest Graf. St. Louis: B. Herder Book Co., 1961.

Kieckhefer, Richard. *Unquiet Souls: Fourteenth-Century Saints and Their Religious Milieu*. Chicago: University of Chicago Press, 1984.

Kingdon, Robert M. *Adultery and Divorce in Calvin's Geneva*. Cambridge, Mass.: Harvard University Press, 1995.

Kleinberg, Aviad M. *Prophets in Their Own Country: Living Saints and the Making of Sainthood in the Later Middle Ages*. Chicago: University of Chicago Press, 1992.

Lerner, Robert E. *The Heresy of the Free Spirit in the Later Middle Ages*. Berkeley: University of California Press, 1972.

Matter, E. Ann. *The Voice of My Beloved: The Song of Songs in Western Medieval Christianity*. Philadelphia: University of Pennsylvania Press, 1990.

McGinn, Bernard, ed. *Meister Eckhart and the Beguine Mystics: Hadewijch of Brabant, Mechthild of Magdeburg, and Marguerite Porete*. New York: Continuum, 1994.

McLaughlin, Eleanor. "The Heresy of the Free Spirit and Late Medieval Mysticism," in Paul Maurice Clogan, ed., *Medievalia et Humanistica*, New Series, no. 4 (1973).

McSheffrey, Shannon. *Gender and Heresy: Women and Men in Lollard Communities, 1420–1530*. Philadelphia: University of Pennsylvania Press, 1995.

Moxey, Keith. *Peasants, Warriors, and Wives: Popular Imagery in the Reformation*. Chicago: University of Chicago Press, 1989.

Nalle, Sara T. *God in La Mancha: Religious Reform and the People of Cuenca, 1500–1650*. Baltimore: Johns Hopkins University Press, 1992.

Rahner, Hugo S. J. *Saint Ignatius Loyola*. New York: Herder and Herder Inc., 1960.

Schnucker, R. V. "The English Puritans and Pregnancy, Delivery and Breast-Feeding," *History of Childhood Quarterly*, vol. I, no. 4 (Spring 1974), pp. 637–58.
———. "Elizabethan Birth Control and Puritan Attitudes," *Journal of Interdisciplinary History*, vol. V, no. 4 (Spring 1975), pp. 655–67.
Weinstein, Donald, and Rudolph Bell. *Saints and Society*. Chicago: University of Chicago Press, 1982.
Wood, Charles T. "The Doctor's Dilemma: Sin, Salvation, and the Menstrual Cycle in Medieval Thought," *Speculum*, vol. 56 (1981), pp. 710–27.

PART IV WOMEN OF THE CASTLES AND MANORS

Works About Women

Backer, Dorothy Anne Liot. *Precious Women: A Feminist Phenomenon in the Age of Louis XIV*. New York: Basic Books, 1974.
Baker, Derek, ed. *Medieval Women*. Oxford: Basil Blackwell, 1978.
Bandel, Betty. "The English Chronicler's Attitude Toward Women," *Journal of the History of Ideas*, vol. 16 (1955), pp. 113–18.
Bell, Susan Groag. "Medieval Women Book Owners: Arbiters of Lay Piety and Ambassadors of Culture," *Signs*, vol. 7, no. 4 (Summer 1982), pp. 742–68.
Berman, Constance H., Charles W. Connell, and Judith Rice Rothschild. *The Worlds of Medieval Women: Creativity, Influence, Imagination*. Morgantown: West Virginia University Press, 1985.
Bogin, Meg. *The Women Troubadours*. New York, W. W. Norton and Company, 1976.
Bornstein, Diane. *The Lady in the Tower: Medieval Courtesy Literature for Women*. Hamden, Conn.: Archon Books, 1983.
Bruckner, Matilda Tomaryn. "Na Castelloza, Trobairitz, and Troubadour Lyric," *Romance Notes*, vol. XXV, no. 3 (Spring 1985), pp. 239–53. This entire issue is devoted to courtly literature, and it also has an excellent bibliography.
Bruckner, Matilda Tomaryn, Laurie Shepard, and Sarah White, eds. and translators. *Songs of the Women Troubadours*. New York: Garland Publishing, 1995.
Brundage, James A. "Concubinage and Marriage in Medieval Canon Law," *Journal of Medieval History*, vol. 1 (1975), pp. 1–17.
Chibnall, Marjorie. *The Empress Matilda: Queen Consort, Queen Mother and Lady of the English*. Oxford: Basil Blackwell, 1991.
Collis, Louise. *Memoirs of a Medieval Woman: The Life and Times of Margery Kempe*. New York: Thomas Y. Crowell Company, 1964.
Dhuoda. *Handbook for William: A Carolingian Woman's Counsel for Her Son*. Translated by Carol Neel. Lincoln: University of Nebraska Press, 1991.
Dillard, Heath. *Daughters of the Reconquest: Women in Castilian Town Society, 1102–1300*. New York: Cambridge University Press, 1989.
Duby, Georges. *Women of the Twelfth Century*. 2 vols. Translated by Jean Birrell. Chicago: University of Chicago Press, 1997.
Duggan, Anne J., ed. *Queens and Queenship in Medieval Europe*. Rochester, N.Y.: Boydell Press, 1997.

Erickson, Amy Louise. *Women and Property in Early Modern England.* New York: Routledge, 1993.

Facinger, Marion F. "A Study of Medieval Queenship: Capetian France 987–1237," *Studies in Medieval and Renaissance History*, vol. 5 (1968), pp. 3–48.

Fell, Christine E. *Women in Anglo-Saxon England*, and Clark, Cecily, and Elizabeth Williams. *The Impact of 1066*. Bloomington: Indiana University Press, 1984. [Published together]

Ferrante, Joan. *Woman as Image in Medieval Literature from the Twelfth Century to Dante*. New York: Columbia University Press, 1975.

———. *To the Glory of Her Sex: Women's Roles in the Composition of Medieval Texts*. Bloomington: Indiana University Press, 1997.

Frakes, Jerold C. *Brides and Doom: Gender, Property, and Power in Medieval German Women's Epic*. Philadelphia: University of Pennsylvania Press, 1994.

Gold, Penny Schine. *The Lady and the Virgin: Image, Attitude, and Experience in Twelfth Century France*. Chicago: University of Chicago Press, 1985.

Jesch, Judith. *Women in the Viking Age*. Woodbridge, N.J.: Boydell Press, 1991.

Jones, Ann Rosalind. *The Currency of Eros: Women's Love Lyric in Europe, 1540–1620*. Bloomington: Indiana University Press, 1990.

Jones, Michael, and Malcolm Underwood. "Lady Margaret Beaufort," *History Today*, vol. 35 (August 1985), pp. 23–30.

Jørgensen, Johannes. *Saint Bridget of Sweden*. Translated by Ingeborg Lund. New York: Longmans, Green & Co., 1954.

Kibler, William W., ed. *Eleanor of Aquitaine: Patron and Politician*. Austin: University of Texas Press, 1976.

Krueger, Roberta L. *Women Readers and the Ideology of Gender in Old French Verse Romance*. New York: Cambridge University Press, 1993.

Liss, Peggy K. *Isabel the Queen: Life and Times*. New York: Oxford University Press, 1992.

McCash, June Hall, ed. *The Cultural Patronage of Medieval Women*. Athens: University of Georgia Press, 1996.

Mickel, Emanuel J., Jr. *Marie de France*. New York: Twayne Publishers, 1974.

Mirrer, Louise. *Women, Jews and Muslims in the Texts of Reconquest Castile*. Ann Arbor: University of Michigan Press, 1996.

Morewedge, Rosemarie Thee, ed. *The Role of Woman in the Middle Ages*. Albany, N.Y.: State University of New York Press, 1972.

Nicholson, Helen. "Women on the Third Crusade," *Journal of Medieval History*, vol. 23, no. 4 (1997), pp. 335–49.

Owen, Douglas David Roy. *Eleanor of Aquitaine: Queen and Legend*. Oxford: Blackwell, 1993.

Parker, Roszika. *The Subversive Stitch: Embroidery and the Making of the Feminine*. London: Women's Press, 1986.

Parsons, John C., ed. *Medieval Queenship*. New York: St. Martin's Press, 1994.

———. *Eleanor of Castile: Queen and Society in Thirteenth-Century England*. New York: St. Martin's Press, 1997.

Pouncy, Carolyn Johnston, ed. and translator. *The Domostroi: Rules for Russian House-*

holds in the Time of Ivan the Terrible. Ithaca, N.Y.: Cornell University Press, 1994.

Sherman, Claire Richter. "The Queen in Charles V's 'Coronation Book': Jeanne de Bourbon and the 'Ordo Ad Reginam Benedicendam,'" *Viator: Medieval and Renaissance Studies*, vol. VIII (1977), pp. 254–97.

Solterer, Helen. "Figures of Female Militancy in Medieval France," *Signs*, vol. 16, no. 3 (Spring 1991), pp. 522–49.

Sponslar, Lucy A. *Women in the Medieval Spanish Epic and Lyric Traditions*. Lexington: The University Press of Kentucky, 1975.

Stafford, Pauline. *Queen Emma and Queen Edith: Queenship and Women's Power in Eleventh-Century England*. Malden, Mass.: Blackwell, 1997.

Walker, Sue Sheridan, ed. *Wife and Widow in Medieval England*. Ann Arbor: University of Michigan Press, 1993.

General Works

Bennett, H. S. *The Pastons and Their England: Studies in an Age of Transition*. New York: Cambridge University Press, 1975.

Bloch, Marc. *Feudal Society*, 2 vols. Translated by L. A. Manyon. Chicago: The University of Chicago Press, 1966.

Bloch, R. Howard. *Medieval French Literature and Law*. Berkeley: University of California Press, 1977.

———. *Medieval Misogyny and the Invention of Western Romantic Love*. Chicago: University of Chicago Press, 1991.

Bouchard, Constance B. "The Origins of the French Nobility: A Reassessment," *American Historical Review*, vol. 86, no. 3 (June 1981), pp. 501–32.

Boucher, François. *A History of Costume in the West*. Translated by John Ross. London: Thames and Hudson, 1967.

Boyer, Marjorie Nice. "A Day's Journey in Medieval France," *Speculum*, vol. XXVI, no. 4 (October 1951), pp. 597–608.

Brissaud, Jean. *A History of French Private Law*. Boston: Little, Brown and Company, 1912.

Calisse, Carlo. *A History of Italian Law*. Translated by Layton B. Register. New York: Augustus Kelley, 1969.

DeMause, Lloyd, ed. *The History of Childhood*. New York: Harper & Row, 1974.

Dronke, Peter. *The Medieval Lyric*. New York: Cambridge University Press, 1977.

Duby, Georges. *The Chivalrous Society*. Translated by Cynthia Postan. Berkeley: University of California Press, 1977.

———. *Medieval Marriage: Two Models from Twelfth-Century France*. Translated by Elborg Forster. Baltimore: Johns Hopkins University Press, 1978.

———. *Love and Marriage in the Middle Ages*. Translated by Jane Dunnett. Chicago: University of Chicago Press, 1994.

Goody, Jack, Joan Thirsk, and E. P. Thompson, eds. *Family and Inheritance: Rural Society in Western Europe 1200–1800*. New York: Cambridge University Press, 1978.

Gravdal, Kathryn. *Ravishing Maidens: Writing Rape in Medieval French Literature and Law*. Philadelphia: University of Pennsylvania Press, 1991.

Hajdu, Robert. "Family and Feudal Ties in Poitou 1100–1300," *Journal of Interdisciplinary Studies*, vol. VIII, no. 1 (Summer 1977), pp. 117–39.

Hanning, Robert W. *The Individual in Twelfth-Century Romance*. New Haven: Yale University Press, 1977.

Keen, Maurice. *Chivalry*. New Haven: Yale University Press, 1984.

Keyser, K. "The German Aristocracy from the ninth to the early twelfth century: A Historical and Cultural Sketch," *Past and Present*, no. 41 (December 1968), pp. 25–53.

Labarge, Margaret Wade. *A Baronial Household in the Thirteenth Century*. New York: Barnes and Noble, 1966.

Mingay, G. E. *The Gentry: The Rise and Fall of a Ruling Class*. New York: Longman Group Limited, 1976.

Morris, Katherine. *Sorceress or Witch?: The Image of Gender in Medieval Iceland and Northern Europe*. Lanham, Md.: University Press of America, 1991.

Mundy, John Hine. *Men and Women at Toulouse in the Age of the Cathars*. Toronto: Pontifical Institute of Mediaeval Studies, 1990.

Newman, F. X., ed. *The Meaning of Courtly Love*. Albany, N.Y.: State University of New York Press, 1968.

Riché, Pierre. *Daily Life in the World of Charlemagne*. Translated by Jo Ann McNamara. Philadelphia: University of Pennsylvania Press, 1978.

Rosenthal, Joel T. *Patriarchy and Families of Privilege in Fifteenth-Century England*. Philadelphia: University of Pennsylvania Press, 1991.

Rossi, Alice S., Jerome Kagan, and Tamara K. Hareven, eds. *The Family*. New York: W. W. Norton & Company, 1978.

Runciman, Steven. *A History of the Crusades*, 3 vols. New York: Harper & Row, 1967.

Sheehan, Michael M. "The Influence of Canon Law on the Property Rights of Medieval Women in England," *Mediaeval Studies*, vol. XXV (1963), pp. 109–24.

———. *Marriage, Family and Law in Medieval Europe: Collected Studies*. Edited by James K. Farge. Toronto/Buffalo: University of Toronto Press, 1996.

Solterer, Helen. *The Master and Minerva: Disputing Women in French Medieval Culture*. Berkeley: University of California Press, 1995.

Spring, Eileen. *Law, Land and Family: Aristocratic Inheritance in England, 1300 to 1800*. Chapel Hill: University of North Carolina Press, 1993.

Thomson, Gladys Scott. *Life in a Noble Household 1641–1700*. Ann Arbor: University of Michigan Press, 1959.

Vale, Malcolm. *War and Chivalry: Warfare and Aristocratic Culture in England, France and Burgundy at the End of the Middle Ages*. Athens: The University of Georgia Press, 1981.

Wood, Margaret. *The English Medieval House*. London: Phoenix House, 1965.

PART V WOMEN OF THE TOWNS

Works About Women

Adelman, Howard. "Rabbis and Reality: Public Activities of Jewish Women in Italy during the Renaissance and Catholic Restoration," *Jewish History*, vol. 5 (1992).

Ashe, Geoffrey. *The Virgin*. London: Routledge & Kegan Paul, 1976.
Ashley, Kathleen, and Pamela Schiengorn, eds. *Interpreting Cultural Symbols: St. Anne in Late Medieval Society*. Athens: University of Georgia Press, 1990.
Atkinson, Clarissa W. *The Oldest Vocation: Christian Motherhood in the Middle Ages*. Ithaca, N.Y.: Cornell University Press, 1991.
Bainton, Roland. *Women of the Reformation in Germany and Italy*. Boston: Beacon Press, 1971.
Barron, Caroline M., and Anne F. Sutton, eds. *Medieval London Widows, 1300–1500*. Rio Grande, Ohio: Hambledon Press, 1994.
Blackbourn, David. *Marpingen: Apparitions of the Virgin Mary in Nineteenth-Century Germany*. New York: Alfred A. Knopf, 1994.
Broude, Norma, and Mary D. Garrard, eds. *Feminism and Art History: Questioning the Litany*. New York: Harper & Row, 1982.
Brown, Judith C., and Jordan Goodman. "Women and Industry in Florence," *Journal of Economic History*, vol. XL, no. 1 (March 1980), pp. 73–80.
Brundage, James A. "Prostitution in the Medieval Canon Law," *Signs*, vol. I, no. 4 (Summer 1976), pp. 825–45.
Cahn, Susan. *Industry of Devotion: The Transformation of Women's Work in England, 1500–1660*. New York: Columbia University Press, 1987.
Cameron, Keith. *Louise Labé: Renaissance Poet and Feminist*. New York: St. Martin's Press, 1990.
Cereta, Laura. *Collected Letters of a Renaissance Feminist*. Transcribed, translated, and edited by Diana Robin. Chicago: University of Chicago Press, 1997.
Clark, Alice. *Working Life of Women in the Seventeenth Century*. New York: Routledge, 1992.
Cohen, Sherrill. *The Evolution of Women's Asylums Since 1500: From Refuges for Ex-Prostitutes to Shelters for Battered Women*. New York: Oxford University Press, 1992.
Cohen, Thomas V., and Elizabeth S. Cohen. *Words and Deeds in Renaissance Rome: Trials before the Papal Magistrates*. Toronto: University of Toronto Press, 1993.
Cohn, Samuel K., Jr. *Women in the Streets: Essays on Sex and Power in Renaissance Italy*. Baltimore: Johns Hopkins University Press, 1996.
Dale, Marian K. "The London Silkwomen of the Fifteenth Century," *Economic History Review*, vol. IV (1932–1934), pp. 324–35.
Davis, Natalie Zemon. "Women in the Crafts in Sixteenth-Century Lyon," *Feminist Studies*, vol. VIII, no. 1 (Spring 1982), pp. 47–80.
———. *Women on the Margins: Three Seventeenth-Century Lives*. Cambridge, Mass.: Harvard University Press, 1995.
Edwards, Robert R., and Vickie Ziegler, eds. *Matrons and Marginal Women in Medieval Society*. Rochester, N.Y.: Boydell & Brewer, 1995.
Farmer, Sharon. "Persuasive Voices: Clerical Images of Medieval Wives." *Speculum*, vol. 61 (1986), pp. 521–26.
———. "Down and Out and Female in Thirteenth-Century Paris," *American Historical Review*, vol. 103, no. 2 (June 1988), pp. 344–72.

Fiorenza, Elisabeth Schussler, and M. Shawn Copeland, eds. *Violence Against Women*. Maryknoll, N.Y.: Orbis Books, 1994.

Folger Collective on Early Women Critics, ed. *Women Critics 1660–1820: An Anthology*. Bloomington: Indiana University Press, 1995.

Furst, Lilian R., ed. *Women Physicians and Healers: Climbing a Long Hill*. Lexington: University Press of Kentucky, 1997.

Garrard, Mary D. *Artemisia Gentileschi*. Princeton, N.J.: Princeton University Press, 1989.

Gelbart, Nina Rattner. *The King's Midwife: A History and Mystery of Madame du Coudray*. Berkeley: University of California Press, 1998.

George, Margaret. *Women in the First Capitalist Society: Experiences in Seventeenth-Century England*. Urbana: University of Illinois Press, 1988.

Goodell, William. *A Sketch of the Life and Writings of Louyse Bourgeois*. Philadelphia: Collins, Printer, 1876.

Gowing, Laura. *Domestic Dangers: Women, Words, and Sex in Early Modern London*. New York: Clarendon Press, 1996.

———. "Secret Births and Infanticide in Seventeenth-Century England," *Past & Present*, vol. 156 (August 1997), pp. 87–115.

Green, Monica H. "Documenting Medieval Women's Medical Practice," in Luis Garcia Ballester, et.al., eds. *Practical Medicine from Salerno to the Black Death*. New York: Cambridge University Press, 1994.

Greer, Germaine. *The Obstacle Race: The Fortunes of Women Painters and Their Work*. New York: Farrar, Straus & Giroux, 1979.

Grossinger, Christa. *Picturing Women in Late Medieval Art*. Manchester, U.K.: Manchester University Press, 1997.

Hacker, Barton C. "Women and Military Institutions in Early Modern Europe: A Reconnaissance," *Signs*, vol. V1, no. 4 (Summer 1981), pp. 643–71.

Hafter, Daryl M., ed. *European Women and Preindustrial Craft*. Bloomington: Indiana University Press, 1995.

Hanawalt, Barbara A., ed. *Women and Work in Preindustrial Europe*. Bloomington: Indiana University Press, 1986.

Harris, Ann Sutherland, and Linda Nochlin. *Women Artists 1550–1950*. New York: Alfred A. Knopf, 1978.

Haskins, Susan Mary. *Mary Magdalen: Myth and Metaphor*. New York: Harcourt, Brace, 1994.

Healy, E. T. *Women According to Saint Bonaventure*. New York: Georgian Press, 1955.

Herlihy, David. *Opera Muliebria: Women and Work in Medieval Europe*. New York: McGraw-Hill, 1990.

Hirn, Yrjö. *The Sacred Shrine: A Study of the Poetry and Art of the Catholic Church*. Boston: Beacon Press, n.d.

Hotchkiss, Valerie R. *Clothes Make the Man: Female Cross Dressing in Medieval Europe*. New York: Garland, 1996.

Howell, Martha C. *Women, Production, and Patriarchy in Late Medieval Cities*. Chicago: University of Chicago Press, 1986.

———. *The Marriage Exchange: Property, Social Place, and Gender in Cities of the Low Countries, 1300–1550*. Chicago: University of Chicago Press, 1998.

Jacobs, Fredrika Herman. *Defining the Renaissance Virtuosa: Women Artists and the Language of Art History and Criticism*. New York: Cambridge University Press, 1997.

Jameson, Anne Brownell, ed. *Legends of the Madonna*. New York: Houghton, Mifflin, 1896.

Jansen, Katherine L. "Mary Magdalen and the Mendicants: The preaching of penance in the late Middle Ages," *Journal of Medieval History*, vol. 21 (1995), pp. 1–25.

Jordan, Constance. *Renaissance Feminism: Literary Texts and Political Models*. Ithaca, N.Y.: Cornell University Press, 1990.

Jordan, William C. *Women and Credit in Pre-Industrial and Developing Societies*. Philadelphia: University of Pennsylvania Press, 1993.

Karras, Ruth Mazo. *Common Women: Prostitution and Sexuality in Medieval England*. New York: Oxford University Press, 1996.

Kermode, Jennifer, and Garthine Walker, eds. *Women, Crime and the Courts in Early Modern England*. Chapel Hill: University of North Carolina Press, 1994.

King, Margaret L., and Albert Rabil, Jr., eds. *Her Immaculate Hand: Selected Works by and about the Women Humanists of Quattrocento Italy*. Binghamton, N.Y.: Center for Medieval and Early Renaissance Studies, State University of New York at Binghamton, 1981.

King, Margaret L. *Women of the Renaissance*. Chicago: University of Chicago Press, 1991.

Kittell, Ellen E. "Guardianship over Women in Medieval Flanders: A Reappraisal," *Journal of Social History*, vol. 31, no. 4 (Summer 1998), pp. 897–930.

Klapisch-Zuber, Christiane. *Women, Family, and Ritual in Renaissance Italy*. Translated by Lydia Cochrane. Chicago: University of Chicago Press, 1985.

Kloek, Els, Nicole Teeuwen, and Marijke Huisman, eds. *Women of the Golden Age: An International Debate on Women in Seventeenth-Century Holland, England and Italy*. Hilversum: Verloren, 1994.

Kuehn, Thomas. "Understanding Gender Inequality in Renaissance Florence: Personhood and Gifts of Maternal Inheritance by Women," *Journal of Women's History*, vol. 8, no. 2 (1996), pp. 58–80.

Labalme, Patricia H., ed. *Beyond Their Sex: Learned Women of the European Past*. New York: New York University Press, 1980.

Lacy, Norris J. "Fabliau Women," *Romance Notes*, vol. XXV, no. 3 (Spring 1985), pp. 318–27. This volume also contains a useful bibliography on this subject.

Lawner, Lynne. *Lives of the Courtesans: Portraits of the Renaissance*. New York: Rizzoli International Publishers, 1981.

Marland, Hilary, ed. *The Art of Midwifery. Early Modern Midwives in Europe*. New York: Routledge, 1993.

Migiel, Marilyn, and Juliana Schiesari, eds. *Refiguring Women: Perspectives on Gender and the Italian Renaissance*. Ithaca, N.Y.: Cornell University Press, 1991.

Monter, E. William. "Women in Calvinist Geneva (1550–1800)," *Signs*, vol. VI, no. 2 (Winter 1980), pp. 189–208.

Nicholas, David. *The Domestic Life of a Medieval City: Women, Children, and the Family in Fourteenth Century Ghent*. Lincoln: University of Nebraska Press, 1985.

Otis, Leah Lydia. *Prostitution in Medieval Society*. Chicago: University of Chicago Press, 1985.

Peters, Christine. "Single Women in Early Modern England: Attitudes and Expectations," *Continuity and Change*, vol. 12, no. 3 (1997), pp. 325–45.

Plowden, Alison. *Women All On Fire: The Women of the English Civil War*. Gloucestershire, U.K.: Sutton Publishers, 1998.

Quataert, Jean H. "The Shaping of Women's Work in Manufacturing: Guilds, Households, and the State in Central Europe, 1648–1870," *American Historical Review*, vol. 90, no. 5 (December 1985), pp. 1122–48.

Rosenthal, Margaret F. *The Honest Courtesan: Veronica Franco, Citizen and Writer in Sixteenth-Century Venice*. Chicago: University of Chicago Press, 1992.

Rossiaud, Jacques. *Medieval Prostitution*. Translated by Lydia G. Cochrane. New York: Basil Blackwell, 1988.

Rubik, Margarete. *Early Women Dramatists, 1550–1800*. New York: St. Martin's Press, 1998.

Rublack, Ulinka. "Pregnancy, Childbirth and the Female Body in Early Modern Germany," *Past & Present*, no. 150 (February 1996), pp. 84–110.

Russell, Diane H., with Bernadine Barnes. *Eva/Ava: Women in Renaissance and Baroque Prints*. New York: The Feminist Press, 1990.

Smith, Hilda L. *Reason's Disciples: Seventeenth-Century English Feminists*. Urbana: University of Illinois Press, 1982.

Stanley, Jo, ed. *Bold in Her Breeches: Women Pirates Across the Ages*. San Francisco: Pandora, 1995.

Stock, Phyllis. *Better Than Rubies: A History of Women's Education*. New York: G. P. Putnam's Sons, 1978.

Strozzi, Alessandra. *Selected Letters of Alessandra Strozzi*. Translated by Heather Gregory. Berkeley: University of California Press, 1997.

Todd, Janet, and Elizabeth Spearing, eds. *Counterfeit Ladies: The Life and Death of Mal Cutpurse; The Case of Mary Carleton*. London: William Pickering, 1994.

Warner, Marina. *Alone of All Her Sex: The Myth and the Cult of the Virgin Mary*. New York: Pocket Books, 1978.

Welu, James A., ed. *Judith Leyster: A Dutch Master and Her World*. New Haven: Yale University Press, 1993.

Wiesner, Merry E. *Working Women in Renaissance Germany*. New Brunswick, N.J.: Rutgers University Press, 1986.

Wilson, Katharina M. *Women Writers of the Renaissance and Reformation*. Athens: University of Georgia Press, 1986.

Wiltenburg, Joy. *Disorderly Women and Female Power in the Street Literature of Early Modern England and Germany*. Charlottesville: University Press of Virginia, 1992.

General Works

Ashley, Maurice. *The Stuarts in Love with Some Reflections on Love and Marriage in the Sixteenth and Seventeenth Centuries*. New York: Macmillan, 1964.

Bar, Carl Ludwig von. *A History of Continental Criminal Law*. Boston: Little, Brown and Company, 1916.

Bates, Alfred, ed. *The [German] Drama: Its History, Literature and Influence on Civilization*, vol. X. New York: Smart and Stanley, 1903.

Braudel, Fernand. *Capitalism and Material Life 1400–1800*. Translation by Miriam Kochan. New York: Harper & Row, 1973.

———. *The Structures of Everyday Life: Civilization and Capitalism 15th–18th Century*, 3 Vols. Translation by Siân Reynolds. New York: Harper & Row, 1981.

Brissaud, Jean. *A History of French Private Law*. Boston: Little, Brown and Company, 1912.

Cadden, Joan. *Meanings of Sex Difference in the Middle Ages: Medicine, Science, and Culture*. New York: Cambridge University Press, 1993.

Calisse, Carlo. *A History of Italian Law*. Translated by Layton B. Register. New York: Augustus Kelley, 1969.

Chrisman, Miriam Usher. *Lay Culture, Learned Culture: Books and Social Change in Strasbourg 1480–1599*. New Haven: Yale University Press, 1982.

Continuity and Change. Special Issue, "Charity and the Poor in Medieval and Renaissance Europe," vol. 3, no. 1 (May 1988), pp. 153–314.

Cowgill, Ursula M. "The People of York 1538–1815," *Scientific American*, vol. 222, no. 1 (January 1970), pp. 104–12.

Davis, Natalie Zemon. *Society and Culture in Early Modern France*. Stanford, Calif.: Stanford University Press, 1975.

Defourneaux, Marcelin. *Daily Life in Spain in the Golden Age*. Translated by Newton Branch. London: Allen and Unwin Ltd., 1970.

De Mause, Lloyd. *The History of Childhood*. New York: Harper & Row, 1974.

Fairchilds, Cissie. *Poverty and Charity in Aix-en-Provence 1640–1789*. Baltimore: Johns Hopkins University Press, 1976.

———. *Domestic Enemies: Servants and Their Masters in Old Regime France*. Baltimore: Johns Hopkins University Press, 1984.

Ferguson, Margaret W., Maureen Quilligan, and Nancy Vickers, eds. *Rewriting the Renaissance: The Discourses of Sexual Difference in Early Modern Europe*. Chicago: University of Chicago Press, 1986.

Frank, Grace. *The Medieval French Drama*. Oxford: Oxford University Press, 1954.

French, Walter. *Mediaeval Civilization as Illustrated by the Fastnachtspiele of Hans Sachs*. Baltimore: Johns Hopkins University Press, 1925.

Goldberg, Jonathan, ed. *Queering the Renaissance*. Durham, N.C.: Duke University Press, 1994.

Goldthwaite, Richard. *Private Wealth in Renaissance Florence: Four Florentine Families*. Princeton, N.J.: Princeton University Press, 1969.

———. "The Florentine Palace as Domestic Architecture," *American Historical Review*, vol. 77, no. 4 (October 1972), pp. 997–1012.

Goody, Jack, Joan Thirsk, and E. P. Thompson, eds. *Family and Inheritance: Rural Society in Western Europe 1200–1800*. New York: Cambridge University Press, 1978.

Gregg, Joan Young. *Devils, Women, and Jews: Reflections of the Other in Medieval Sermon Stories*. Albany: State University of New York Press, 1997.

Hanawalt, Barbara A. *Growing Up in Medieval London*. New York: Oxford University Press, 1993.

———. *"Of Good and Ill Repute": Gender and Social Control in Medieval England*. New York: Oxford University Press, 1998.

Henderson, Katherine Usher, and Barbara F. McManus. *Half Humankind: Contexts and Texts of the Controversy About Women in England, 1540–1640*. Urbana: University of Illinois Press, 1985.

Herlihy, David. *Medieval and Renaissance Pistoia*. New Haven: Yale University Press, 1967.

———. "The Tuscan Town in the Quattrocento, A Demographic Profile," in Paul Maurice Clogan, ed. *Medievalia et Humanistica*, New Series, vol. I (1970).

———. "The Medieval Marriage Market," in Dale B. J. Russell, ed. *Medieval and Renaissance Studies*. Durham, N.C.: Duke University Press, 1976.

Hübner, Rudolf. *A History of Germanic Private Law*. Boston: Little, Brown and Company, 1918.

Hufton, Olwen H. *The Poor of Eighteenth-Century France 1750–1789*. Oxford: Clarendon Press, 1974.

Hughes, Diane Owen. "Urban Growth and Family Structure in Medieval Genoa," *Past & Present*, vol. 66 (February 1975), pp. 3–28.

Hunt, Margaret R. *The Middling Sort: Commerce, Gender and the Family in England, 1680–1780*. Berkeley: University of California Press, 1996.

Kingdon, Robert M. *Adultery and Divorce in Calvin's Geneva*. Cambridge, Mass.: Harvard University Press, 1995.

LeGoff, Jacques. *The Birth of Purgatory*. Translated by Arthur Goldhammer. Chicago: University of Chicago Press, 1986.

Molho, Anthony. *Marriage Alliance in Late Medieval Florence*. Cambridge, Mass.: Harvard University Press, 1994.

Newett, M. Margaret. "The Sumptuary Laws of Venice in the Fourteenth and Fifteenth Centuries," in T. F. Tout and James Tait, eds. *Historical Essays by Members of Owens College Manchester*. London: Longmans, Green & Co., 1902.

Ogilvie, Sheilagh, and Markus Cerman. *European Proto-Industrialization*. Cambridge: Cambridge University Press, 1996.

Origo, Iris. *The Merchant of Prato: Francesco di Marco Datini*. London: Jonathan Cape, 1957.

Pardaithé-Galabrun, Annik. *The Birth of Intimacy: Private and Domestic Life in Early Modern Paris*. Translated by Jocelyn Phelps. Philadelphia: University of Pennsylvania Press, 1991.

Perry, Mary Elizabeth. *Gender and Disorder in Early Modern Seville*. Princeton, N.J.: Princeton University Press, 1990.

Pollard, Alfred W. *English Miracle Plays, Moralities and Interludes*. Oxford: Clarendon Press, 1895.

Pullan, Brian. *Rich and Poor in Renaissance Venice*. Cambridge, Mass.: Harvard University Press, 1971.

Quayle, Eric. *Old Cook Books: An Illustrated History*. New York: E. P. Dutton, 1978.

Redon, Odile, Françoise Sabban, and Silvano Serventi. *The Medieval Kitchen: Recipes*

From France and Italy. Translated by Edward Schneider. Chicago: University of Chicago Press, 1998.

Romano, Dennis. *Housecraft and Statecraft: Domestic Service in Renaissance Venice, 1400–1600*. Baltimore: Johns Hopkins University Press, 1996.

Rossi, Alice S., Jerome Kagan, and Tamara K. Hareven, eds. *The Family*. New York: W. W. Norton, 1978.

The Secular Spirit: Life and Art at the End of the Middle Ages. New York: E. P. Dutton, 1975.

Smith, Hilda. "Gynecology and Ideology in Seventeenth-Century England," in Berenice A Carroll, ed. *Liberating Women's History: Theoretical and Critical Essays*. Urbana: University of Illinois Press, 1976.

Stuard, Susan Mosher. "Dowry Increase and Increments in Wealth in Medieval Ragusa (Dubrovnik)," *Journal of Economic History*, vol. 16, no. 4 (December 1981), pp. 795–812.

Taylor, Larissa. *Soldiers of Christ: Preaching in Late Medieval and Reformation France*. New York: Oxford University Press, 1992.

Thrupp, Sylvia L. *The Merchant Class of Medieval London [1300–1500]*. Ann Arbor: University of Michigan Press, 1977.

Trexler, Richard C. "Infanticide in Florence: New Sources and First Results," *History of Childhood Quarterly*, vol. I, no. 1 (Summer 1973), pp. 98–116.

———. "The Foundlings of Florence 1395–1455," *History of Childhood Quarterly*, vol. I, no. 2 (Fall 1973), pp. 259–84.

Turner, James Grantham. *Sexuality and Gender in Early Modern Europe: Institutions, Texts, Images*. New York: Cambridge University Press, 1993.

Van Deursen, Arie Theodorus. *Plain Lives in a Golden Age: Popular Culture, Religion, and Society in Seventeenth-Century Holland*. Translated by Maarten Ultee. New York: Cambridge University Press, 1991.

Walker, Garthine. "Rereading Rape and Sexual Violence in Early Modern England," *Gender and History*, vol. 10, no. 1 (April 1998), pp. 1–25.

Watt, Jeffrey R. *The Making of Modern Marriage: Matrimonial Control and the Rise of Sentiment in Neuchâtel, 1550–1800*. Ithaca, N.Y.: Cornell University Press, 1993.

Weissler, Chava. "The Traditional Piety of Ashkenazic Women," in Arthur Greene, ed. *Jewish Spirituality II*. New York: Crossroad, 1987.

White, Sarah Melhado. "Sexual Language and Human Conflict in Old French Fabliaux," *Comparative Studies in Society and History*, vol. 24, no. 2 (1982), pp. 185–210.

INDEX

Illustration Credits

PART I

1. Photograph by E. Dania.
2. The Metropolitan Museum of Art, Rogers Fund, 1964. (64.11.7)
3. The Metropolitan Museum of Art, Fletcher Fund, 1931. (31.11.10)
4. Hirmer Verlag München (56.1087). Rome, Museo dela terre.
5. National Museum of Denmark. Photograph by Niels Elswing.
6. Scala/Art Resource, NY
7. Fitzwilliam Museum, Cambridge, England.
8. Reproduced by Courtesy of the Trustees of the British Museum (BM Cat. Sculpture 1117).
9. Autun-Saone et Loire. Haut Relief. Provenant de la Cathédrale "Eve." © 1987 ARS, N.Y./Arch. Phot. Paris/S.P.A.D.E.M.
10. Musée Municipal, Semur-en-Auxois, France.

PART II

1. Giraudon/Art Resource, NY
2. *Tacuinum sanitatis*, fol. 62. Bildarchiv der Österreichischen Nationalbibliothek, Vienna, Austria.
3. National Museum of Denmark. Photograph by Niels Elswing. Research by Frantz Wendt.
4. Magyar Fotómüveszek Szövetsége, Budapest, Hungary. Photograph by Ernó Vadas.
5. Print: Eric Johnson for American Heritage Publishing Company.
6. "La Vielle" by Françoise Duparc. Musée des Beaux Arts, Marseille, France.
7. Kungliga Biblioteket, Stockholm, Sweden. *History Today* Archives.
8. American Russian Institute, Inc., San Francisco, California.
9. Photo from *Women of the Shadows* by Ann Cornelisen. Copyright © 1976 by Ann Cornelisen. Reprinted by permission of the Wallace Literary Agency, Inc.

PART III

1. Tracing made from the original manuscript of *Hortus deliciarum*. The work itself was destroyed in the Franco-Prussian War of 1870.
2. Hildegard von Bingen, Wisse die Wege-Scivias, Otto Müller Verlag, Salzburg, Austria.
3. Burgundy Breviary, British Library (Harley MS 2897, F. 340b).
4. Bildarchiv Foto Marburg.
5. From *Livre de la vie active*, 1425. C.M.T. Assistance Publique, Paris, France.
6. From *Het Bloedig Tooneel* by Tilleman van Bracht, part II, Amsterdam, The Netherlands, 1685. Beinecke Rare Book and Manuscript Library, Yale University, New Haven, Connecticut.
7. Portrait from the Convent of the Carmelites in Seville, Spain. Photograph by MAS, Barcelona, Spain.
8. © Rijksmuseum-Stichting, Amsterdam, The Netherlands.
9. National Museum of Denmark. Photograph by Niels Elswing.

PART IV

1. From *De nobilitatibus, sapientiis, et prudentiis regum* by Walter de Milemere, The Bodleian Library. Reproduced by kind permission of the Governing Body of Christ Church, Oxford, England.
2. The Pierpont Morgan Library, New York, Ms. M. 917, p. 65.
3. "Codex Manesse," Grosse Heidelberger Liederhandschrift. Co. Pal. Germ. (848 B1.397v). Universitätsbibliothek, Heidelberg, Federal Republic of Germany.
4. Bildarchiv Foto Marburg.
5. The Pierpont Morgan Library, New York, Ms. M. 691, f. 131v.
6. Musée de Cluny, France. Photograph by Cliché des Musées Nationaux, Paris, France.
7. Giraudon/Art Resource, NY
8. Bibliothéque Nationale, Paris, France (MS. 5073, Fol. 117v).

PART V

1. "The Birth of the Virgin" by the Master of the Life of Mary, Cologne, Germany (1463–1480). Alte Pinothek, Munich, Federal Republic of Germany.
2. *Beggar Woman* by Jacques Callot. R. L. Baumfeld Collection. Photograph © Board of Trustees, National Gallery of Art, Washington.
3. *Tacuinum sanitatis*, Fol. 105v. Bildarchiv der Österreichischen Nationalbibliothek, Vienna, Austria.
4. © Rijksmuseum-Stichting, Amsterdam, Netherlands.
5. Skulptuengalerie, Staatliche Museen, Preussischer Kulturbesitz, Berlin, Federal Republic of Germany. Photograph by Jörg P. Anders.
6. Marvin Lowenthal, trans., *The Memoirs of Glückel of Hameln* (New York: Schocken Books, 1977). Reprinted with permission of the publishers.

7. *Tacuinum sanitatis*, Fol. 82v. Bildarchiv der Österreichischen Nationalbibliothek, Vienna, Austria.
8. Fol. 174. Bildarchiv der Österreichischen Nationalbibliothek, Vienna, Austria.
9. *La Pittura*. Hampton Court. Copyright Reserved to Her Majesty Queen Elizabeth II.

CPSIA information can be obtained
at www.ICGtesting.com
Printed in the USA
BVOW03s1156040117
472502BV00001B/3/P